1986 YEARBOOK OF SCIENCE AND THE FUTURE

1986 YEARBOOK OF SCIENCE AND THE FUTURE

Encyclopædia Britannica, Inc.
Chicago

Auckland Geneva London Manila Paris Rome Seoul Sydney Tokyo Toronto

Library of Congress Catalog Card Number: 69-12349
International Standard Book Number: 0-85229-430-1
International Standard Serial Number: 0096-3291

The University of Chicago
The Yearbook of Science and the Future
is published with the editorial advice of the faculties of
the University of Chicago.

1986
YEARBOOK OF SCIENCE AND THE FUTURE

Director of Yearbooks	Bruce L. Felknor
Editor	David Calhoun
Associate Editor	Charles Cegielski
Editorial Staff	Daphne Daume, Karen Justin, Arthur Latham
Art Director	Cynthia Peterson
Senior Picture Editor	Kathy Nakamura
Picture Editor	Kathryn Creech
Layout Artist	Dale Horn
Illustrators	Anne H. Becker, John L. Draves
Art Production	Richard A. Roiniotis
Art Staff	Daniel M. Delgado, Patricia A. Henle, Paul Rios, Lillian Simcox
Director, Yearbooks Copy Production	J. Thomas Beatty
Manager, Copy Department	Anita Wolff
Senior Copy Editor	Julian Ronning
Copy Staff	Robert Curley, Ellen Finkelstein, Anthony L. Green, Anne B. Hedblom, Patrick Joyce, Lawrence D. Kowalski, Elizabeth Laskey, Carol Smith
Manager, Copy Control	Mary C. Srodon
Copy Control Staff	Marilyn Barton, Mayme R. Cussen
Manager, Composition and Page Makeup	Melvin Stagner
Composition Staff	Griselda Chaidez, Duangnetra Debhavalya, Morna Freund, Van Jones, John Krom, Jr., Thomas Mulligan, Gwen Rosenberg, Tammy Tsou
Page Makeup Staff	Marsha Check, *supervisor* Michael Born, Jr., Philip Rehmer, Michael Schramm, William Soltis, Danette Wetterer, Beverly Woods
Director, Editorial Computer Services	Michael J. Brandhorst
Computer Services Staff	Steven Bosco, Joanna Jacobsen, Carole Paulson, Arnell Reed, Vincent Star
Manager, Index Department	Frances E. Latham
Assistant Manager	Rosa E. Casas
Index Staff	Christopher G. Boucek, Steven Monti, Martin L. White
Librarian	Terry Miller
Associate Librarian	Shantha Uddin
Assistant Librarian	David W. Foster
Secretarial Staff	Dorothy Hagen, Lavonne Nannenga

Editorial Administration

Editor-in-Chief
Philip W. Goetz

Executive Editor
Margaret Sutton

Director of Budgets and Controller
Verne Pore

Director of Editorial Production
Karen M. Barch

CONTENTS

Feature Articles

DIVING FOR THE ANCIENT PAST

by Robert L. Hohlfelder

A last frontier for archaeologists, the ocean floor, is yielding treasures that date back to the civilizations of ancient Greece.

In the spring of 1900 two small Greek fishing boats struggled through heavy seas on their homeward voyage from Tunisia to Syme, an island in the Aegean Sea. These vessels, under the command of Capt. Demetrios Kondos, had been searching for sponges off the African coast. Bad weather had plagued the entire operation. Forced to leave their fishing grounds before both hulls were filled with sponges, the two boats had sailed for home during a storm and could not escape the high waves and violent north wind. Finally, Kondos spotted the desolate island of Antikythera off the coast of the Peloponnesus in Greece and made a run for shelter. He dropped anchor along a deserted section of the southern coast of this Greek island to wait out the worst of the storm.

As a man of the sea Captain Kondos was always the opportunist. So much of his life as a sponge fisherman depended on luck and chance. Now, obliged by the elements to stand at anchor along an unfamiliar coast, he made the best of a bad situation and sent one of his hard-helmet divers below to explore the bottom as soon as conditions permitted. The sea was always unpredictable. Even in such an unlikely spot a rich bed of sponges might be found.

As the diver, Elias Stadiatis, descended into the depths, he quickly realized that the water at this location was too deep for sponge collecting. He touched the bottom at about 55 meters (180 feet), well beyond the safe limits at which he could work. In 1900 no one knew why exposure to such depths produced debilitating and often fatal sickness, but divers like Stadiatis who lived with those dangers daily were aware they existed. His time on the bottom that day would have to be very short.

As Stadiatis's eyes adjusted to the dim light of the bottom coming through the portal of his helmet, he must have suddenly thought he was not alone! All around him were shadows and shapes in various postures and positions. Arms and heads were vaguely visible on the ocean floor. He grabbed an arm sticking from the bottom and yanked it free. With his trophy held firmly in his grip, he excitedly jerked on his safety line, signaling his companion on the boat above to pull him to the surface. Although he could not possibly have realized the implications of his find, when he reached the surface still clutching his bronze prize he had opened a new avenue of scientific exploration. His recovery would spark the beginnings of underwater archaeology. With that one find, made inadvertently by a Greek sponge diver, diving for the ancient past had begun.

Pioneering efforts

Elias Stadiatis's act of snatching an ancient artifact from the ocean floor and rushing it to the surface hardly qualifies as archaeology. Nor was his recovery even the first of its kind. For hundreds of years fishermen throughout the Mediterranean from time to time would pull precious art objects or amphorae, the ceramic barrels or shipping crates of antiquity, from the sea. Free divers and later hard-helmet commercial divers like the men who worked for Captain Kondos added to this inventory. Yet today, with the hindsight of more than eight decades, we know that it was Stadiatis's discovery and the subsequent recovery of scores of precious

ROBERT L. HOHLFELDER, *Professor of History at the University of Colorado, has been a member of underwater archaeological research teams in the Mediterranean Sea.*

artifacts from what was a Roman treasure ship filled with plunder from Greece that captured the imagination of the world and alerted everyone to the potential of the seas as a storehouse of history. The Mediterranean and, indeed, all the seas of the world had been revealed as an unexplored repository of mankind's past. Scientists wondered what secrets awaited discovery on the ocean floor.

Eighty-five years later we have some of the answers to this question. Since Antikythera, equally spectacular chance discoveries have been made in unexpected locations and circumstances. Some have even occurred in the last few years in very shallow water. In 1972 Stefano Mariottini was on a holiday on the Adriatic Sea near the town of Riace in Calabria, Italy. While spearfishing in 7½ meters (25 feet) of water, he chanced to see a bronze arm sticking up from the ocean floor. This accidental sighting led to the recovery of two exquisite bronze statues now called the Riace bronzes. They rival any statues from antiquity discovered to date.

Off the coast of Israel near 'Athlit, Yehoshua Ramon noticed a piece of bronze jutting from the sand during a routine underwater survey conducted in 1981. He immediately contacted his professor at the Center for Maritime Studies, Elisha Linder, who in turn organized a systematic search and recovery effort centering on Ramon's find. The visible bronze protrusion turned out to be a small section of an entire ancient warship's battle ram. This unique artifact, the first such instrument of war ever discovered, is now the centerpiece of the National Maritime Museum of Israel in Haifa. The subject of intense scholarly interest, it promises to reveal much about ancient metallurgy, technology, and ship construction.

Chance discoveries of this sort are obviously still possible today. As commercial and recreational use of the sea increases throughout the world, opportunities for such finds also increase. But as exciting as such recoveries have been in the past or may be in the future, this type of random salvage of artifacts and works of art does not qualify as archaeology. The archaeologist must have the opportunity to study and record objects in the context in which they are discovered and in association with other cultural remains. When observed in this way, an artifact can reveal more about itself and the society that produced it. Although this type of recording was not done at Antikythera, that project suggested that careful supervision of divers by archaeologists might yield results comparable to land excavations.

In the decades following Antikythera professional archaeologists on board ships attempted on several occasions to control and coordinate the activities of divers working below. The results were not completely satisfactory. The only diving equipment available early in this century was the hard-helmet suit. It was cumbersome to use and restricted the mobility of the diver, who was linked to the surface by his air line and burdened by the heavy weight of his suit. Visibility through a single portal in the helmet was also too limited. Since this equipment was difficult and often dangerous to use, archaeologists were content to supervise from the surface. In such circumstances, no matter how carefully divers were trained

Greek sponge divers like those at the top right made the first discoveries of treasures from the ancient past on the seafloor. The Riace bronzes (above and right) were found in the Adriatic Sea in 1972. The two statues, each complete and more than six feet tall, are believed to have been made in Greece during the 5th century BC.

Ancient Greek bronze statue of Zeus or Poseidon (above), considered by many the finest bronze statue in existence, was discovered by Greek Navy divers in the Aegean Sea in 1928. A bronze battle ram (right) from an ancient warship was found half buried in sand off the coast of Israel in 1981 and pulled up to surface with the aid of divers (above right).

or briefed, the archaeologist in charge never had the opportunity to visit personally the site being explored. Obviously, any system of research that prohibits the investigator from seeing or controlling his experiment directly cannot be totally satisfactory.

In spite of the limitations of such methods of exploration, there were some achievements during the 1920s and 1930s. Greek Navy divers recovered two spectacular bronze statues from the sea off the northern coast of the island of Euboea at Cape Artemisium. One of the statues, the Zeus or Poseidon of Artemisium recovered in the 1928 expedition, is generally considered the finest bronze statue in existence, although the Riace bronzes are serious rivals for that honor. Unfortunately, the treasure ship that had been transporting this statue and, presumably, numerous other masterpieces of Greek sculpture was not found. Before a systematic search for it could begin, tragedy struck. One of the divers died as a result of working in the 35 meters (120 feet) of water. Further exploration was immediately curtailed. As authorities noted at that time, there was no urgency for recovering whatever other treasures might be off Artemisium. The sea had protected this cache for two thousand years and would continue to do so until some later date when recovery could be undertaken more easily and safely. Surprisingly, as of 1985 the Greek Archaeological Service had not yet decided to renew this investigation.

Farther to the east in the Mediterranean another pioneer effort in underwater archaeology began in the 1930s. At Tyre in Lebanon Antoine Poidebard began undersea investigations of the submerged portions of that city. As with so many other ancient cities built on the coast, sections of Tyre had been inundated over the centuries by relative sea-level changes usually resulting from land subsidence. Employing aerial reconnaissance to identify points of archaeological interest, Poidebard directed both hard-helmet and free divers to investigate the underwater ruins. His

Divers explore a shipwreck in the Mediterranean Sea. Each is using a self-contained underwater breathing apparatus, or scuba. Developed by Jacques-Yves Cousteau and Émile Gagnan in 1943, the scuba allowed divers much greater freedom in their work underwater.

Courtesy, The Cousteau Society, Inc., 930 West 21st Street, Norfolk, Va. 23517, a membership-supported environmental organization divers also experimented with underwater photography as a research and recording tool. Poidebard's work at Tyre opened another distinct branch of underwater archaeology. Not only were shipwrecks and their cargoes available to the archaeologist, but even drowned sections of ancient cities could be explored systematically.

The scuba breakthrough

Although these two projects and others undertaken before the late 1950s achieved some success, they were clearly pioneer efforts severely limited by the available technology. The development of self-contained underwater-breathing apparatus, or scuba as it is commonly known, by Jacques-Yves Cousteau and Émile Gagnan in 1943 dramatically changed man's ability to work beneath the sea. For the first time it was possible to work underwater without reliance on the surface for a source of air. With this equipment a diver also enjoyed almost total freedom of movement. This development must rank as the single most important breakthrough to date in man's efforts to explore and exploit the sea. For the underwater archaeologist and other oceanographic researchers as well, a new chapter of discovery had opened.

Within just a few years of scuba's development amateur divers in the Mediterranean, particularly off the coast of France, had begun to discover and report the cultural riches of the sea. Regrettably, in their enthusiasm and ignorance they would often bring artifacts indiscriminately to the surface with no consideration for their archaeological context. Such activity was little more than looting. These sport divers, and far too many of their counterparts today throughout the world, failed to realize how much information could be irrevocably lost by the random removal of ancient objects from the seafloor. When scientists do not know the context of an artifact's discovery, its scientific value is significantly diminished. In some instances the damage can not be undone. For example, very often the only visible trace of an ancient shipwreck will be an amphora or fragments of those containers lying on the ocean floor. The removal of those cultural artifacts thus obliterates forever any record of the shipwreck site.

In spite of some zealous excesses in the early years of scuba exploration the public became aware of the potential of diving for the ancient past. Scholars, too, recognized the possibilities of translating the procedures of land archaeology to a marine environment. The moment was right for the marriage of scuba and archaeology.

The first major effort at this integration was attempted by the discoverer of scuba, Jacques Cousteau. He was instrumental in organizing and executing the first scientific shipwreck excavation to employ scuba and other state-of-the-art equipment such as underwater television. Following leads supplied by a local lobster fisherman, Cousteau and his associates discovered an ancient shipwreck site in 1952 near Marseille, France, lying in depths of about 40 meters (140 feet) in the Mediterranean Sea off the island of Grand Congloué. His team was guided by a noted French scholar, Fernand Benoit. Over the next few years this expedition discovered about 10,000 amphorae and fragments from this site. Many of these artifacts bore the name of Marcus Sestius, the Roman entrepreneur whose ship had foundered there. For most of the decade these investigations and discoveries captured the imagination of the European public and fostered the cause of underwater archaeology.

By today's professional standards this excavation would not be acceptable. One must remember, however, that at Grand Congloué procedures and techniques for underwater archaeology were being formulated as the work proceeded. Mistakes were made. No plan of the site was completed, and Benoit and his colleagues did not determine if more than one wreck had been discovered. But technical shortcomings aside, this expedition demonstrated the practical use of scuba for underwater archaeology and should be recognized for what it was—another important step in the maturation of a scientific discipline.

Engineer and inventor Edwin Link (below) designed his second "Sea Diver" research vessel (bottom) to undertake underwater explorations off the coast of Israel at the site of Caesarea, a city built by Herod the Great several years before the birth of Jesus.

At about the same time across the Atlantic Edwin Link had turned his considerable energies and talents to the new field of oceanographic research with scuba. This distinguished inventor and engineer began in the 1950s to search for shipwreck sites in the Caribbean with Mendel Peterson of the Smithsonian Institution. In 1956 Link began excavations at the partially submerged city of Port Royal in Jamaica, which had been inundated in 1692 following a violent earthquake. Working from his research vessel, "Sea Diver," he achieved some important discoveries.

But Link's real interest in underwater archaeology was in the Mediterranean. Responding to an invitation from the Israeli government, he constructed a new "Sea Diver" research vessel, which was designed with underwater archaeology in mind. In May 1960 he sailed to Israel to begin underwater explorations at Caesarea (often referred to as Caesarea Palaestinae or Caesarea Maritima). This city, located on the coast of Israel midway between Tel Aviv and Haifa, had been built as an interna-

Photos, courtesy, Marilyn C. Link, Harbor Branch Foundation, Inc.

tional port by Herod the Great shortly before the birth of Christ and was the largest and most important Mediterranean port still readily accessible to archaeologists. Unlike so many other major ports of antiquity still in service today, Caesarea saw no significant use after 1291, when it was abandoned by the Crusaders and destroyed by the Mamluks.

Link's Caesarea expedition extended Poidebard's work at Tyre in the 1930s and later at Sidon, also in Lebanon, in the 1940s. Both men were interested in the exploration of submerged coastal settlements. Link's divers, however, used scuba. Unfortunately, neither of the archaeologists in charge, Charles Fritsch or Immanuel Ben-Dor, were trained in the use of this equipment. Once again, as at Tyre and Sidon, the directing scientists could not personally visit the site or supervise the excavations. Nevertheless, during the few days of good weather when sea conditions permitted work, Link's team demonstrated the desirability and practicability of scuba for underwater investigations of the partially sunken cities of the Mediterranean and, perhaps as significant, fixed the exact location of Herod's harbor and announced to the world that incredible ruins were still visible.

At the same time that Link was diving at Caesarea, to the north off the coast of Turkey all the various pioneer efforts at underwater archaeology were coming together at last. At Cape Gelidonya in 1960 archaeologists trained in scuba were for the first time conducting the actual explorations and excavations themselves. The credit for this first project that can truly be called underwater archaeology belongs to Peter Throckmorton, a U.S. archaeologist with long experience in underwater research, and George F. Bass, then a graduate student at the University of Pennsylvania and in 1985 a professor of anthropology at Texas A&M University. Responding to information relayed to him by Throckmorton concerning the location of an early shipwreck, Bass formed a team of diving archaeologists to explore the suspected site located off Cape Gelidonya.

At Cape Gelidonya on the southern coast of Turkey archaeologists in 1960 performed the actual explorations and excavations themselves for the first time. One of the boats in the expedition (top) is attached by a line to another research vessel. Above, George Bass, director of the expedition, and his wife apply plastic preservatives to bits of wood from a shipwreck found at the site and dated about 1200 BC.

Photos, Peter Throckmorton

The results of this operation, which lasted for several years, were spectacular in all ways. The earliest shipwreck uncovered to that time, dated about 1200 BC, was found along with much of its cargo. Bass developed techniques of excavation and exploration that are still followed. Equally important, with this excavation he established the tradition of diving archaeologists. No longer would middlemen be permitted in underwater archaeology. Like their colleagues on land, archaeologists who worked in the sea would personally direct operations.

Bass's excavation at Cape Gelidonya spawned a rich legacy of other underwater shipwreck explorations in Turkey. In 1984 he began another early Bronze Age shipwreck excavation near the town of Kas on the coast of Turkey. He also explored numerous Turkish wreck sites and trained a cadre of Turkish and U.S. divers to assist him. Many of his former students and current associates are themselves leading figures in the field of underwater archaeology and have conducted explorations throughout the world. Bass created the major research institute in the world for the study of ancient shipping and shipwreck archaeology, the Institute of Nautical Archaeology, currently located at Texas A&M. This center stands at the forefront of underwater archaeology.

This short survey of the development of underwater archaeology has been selective and has consciously, although regrettably, omitted the contributions of many other pioneers such as Nicholas Flemming, Honor Frost, Gerhard Kapitän, and many more individuals who nurtured this new field of research. Any attempt to cover recent work from Cape Gelidonya to the present would not be possible in this essay. During the last 25 years expeditions around the world have demonstrated fully the potential of the oceans to add to our understanding of the past. In the Mediterranean, still the center for such research, progress has been impressive. Michael Katzev, one of Bass's students and associates, discovered, raised, and restored a 4th-century-BC merchant ship off Kyrenia in Cyprus. Joseph Shaw of the University of Toronto led an important underwater exploration at Kenchreai, Greece, in the 1960s. Several important shipwrecks were discovered off the coast of France and excavated with the highest possible standards. None was more impressive than the wreck of Madrague de Giens, published by André Tchernia and Patrice Pomey. Along the Italian coast Anna McCann and John Oleson of the University of Victoria explored several submerged Etruscan and Roman port sites. Elisha Linder and Avner Raban of the University of Haifa began underwater archaeology in Israel in the 1960s and have enjoyed many successes since then. It is a happy circumstance that it is no longer easy to list every important underwater archaeological project in the Mediterranean. The volume of research recently conducted or currently in progress is the most meaningful barometer of the vitality of this new discipline.

Caesarea Ancient Harbour Excavation Project

A good way to illustrate the current status of diving for the ancient past is to discuss one major excavation, the underwater investigations currently in progress at Caesarea, the Israeli site Edwin Link first examined in 1960. The present exploration is being undertaken by the Caesarea Ancient Harbour Excavation Project (CAHEP), a tri-national consortium of the Universities of Haifa, Victoria (in Canada), Colorado, and Maryland.

In many ways Caesarea is unique. There has been no significant habitation there for the past eight centuries. Much of the land area of the ancient city and practically all of its massive harbor complex are readily available to the archaeologist. Also a great deal about the life of the city is known from a variety of literary sources from all periods of history. The most important one for CAHEP is the Jewish historian Flavius Josephus. Writing about 80 years after Caesarea's completion in approximately 10 BC, he offers a detailed and vivid description of the main harbor's design and construction.

According to Josephus engineers in the service of King Herod had to overcome formidable obstacles to construct Sebastos, as Caesarea's harbor facility was called in antiquity. For the first time in history they fashioned an artificial harbor from a straight, sandy coastline bereft of natural anchorages or protected moorings. Before Caesarea artificial harbors had been carved into the coast. Never before had one been built out into the sea. To accomplish this Herculean task, the engineers lowered massive blocks of stone 15 by 5.5 by 2.7 meters (50 by 18 by 9 feet) into the open sea to build up a huge breakwater some 60 meters (200 feet) wide. When this enclosing arm for the harbor breached the surface, it was bisected down its center axis by a seawall surmounted with towers. The outer side, the one exposed to the brutal winter storms that can batter the Israeli coast, was left unfinished. The interior face was finished and featured docking facilities and a warehouse built into the wall and protected by it. The mouth of the harbor was adorned with a cluster of three colossal statues standing on platforms on either side of the entrance. When Sebastos was completed, Josephus claimed that it rivaled Athens's harbor, Piraeus, in size.

Before Link's expedition Josephus's account seemed unbelievable. Could such an immense, sophisticated harbor have been constructed 2,000 years ago given the available technology? This question, along with Link's report of extensive submerged ruins, provided CAHEP's starting point. Beginning in 1980 a research team representing many disciplines worked with volunteer divers from nine countries in conducting land and underwater explorations along Caesarea's coastline. In 1984 some 120 staff and volunteer divers conducted underwater excavations at various points near the harbor entrance and along the southern face of the northern breakwater. In addition, land probes were conducted in an inner harbor area that is now completely silted in and partially covered with tourist shops.

In some respects CAHEP's task was simplified because the location of the harbor was established by Link and the submerged ruins are well defined and in water not deeper than about 10 meters (35 feet). No elaborate oceanographic search or survey equipment was needed to discover zones of archaeological interest. Also because divers are working in such shallow water most of the time, problems with the exotic diseases of diving, such as the "bends," are minimal. Long dive shifts are possible, and elaborate communications systems between the ocean floor and the surface are not needed. The shallow depth, however, does mean that fieldwork can be adversely influenced by surface conditions of the sea, a problem archaeologists working in deep water do not face. Storms and heavy seas can reduce visibility and create unsatisfactory and often unsafe conditions. Several days each season are lost to the weather.

CAHEP employed survey and excavation equipment first developed by George Bass at Cape Gelidonya in 1960 but with various innovations, modifications, and improvements. The basic excavation tools were the air lift and dredge. Both pieces of equipment use air or water introduced under pressure into a tube to create a suction that enables divers to suck away sand, mud, gravel, small stones, and other bottom debris so that the area can be excavated. Both devices are usually compared to a vacuum cleaner in the way they operate to clear a trench. Probing instruments such as sonar penetrators and water or air jets provided some idea of features of the ancient harbor below the present ocean floor. Air bags, balloons made of heavy plastic, were carried to the bottom in a deflated condition and then filled with air from a diver's regulator. They are capable of lifting heavy objects from the ocean floor and transporting them to the surface.

Each summer CAHEP constructed a small diving platform or barge of plywood sheets mounted on a wooden frame affixed to oil drums. The barge was towed into a central location for most of the archaeological fieldwork planned for that season. Once in place it remained there for the entire season of exploration, secured by four anchors extending from each corner. It served as the control center for all underwater operations, the place from which divers entered and left the water. Radio communi-

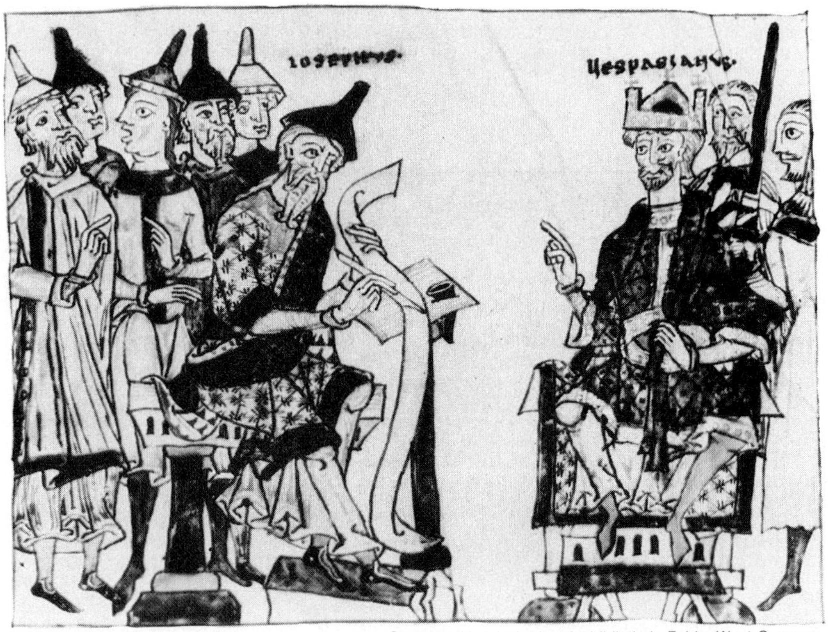

Jewish scholar and historian Flavius Josephus, shown before the Roman emperor Vespasian in this 14th-century miniature, wrote in about 70 AD a detailed description of the design and construction of the main harbor at Caesarea. This information has proved valuable to present-day explorers of the site.

Courtesy, Hessische Landesbibliothek, Fulda, West Germany

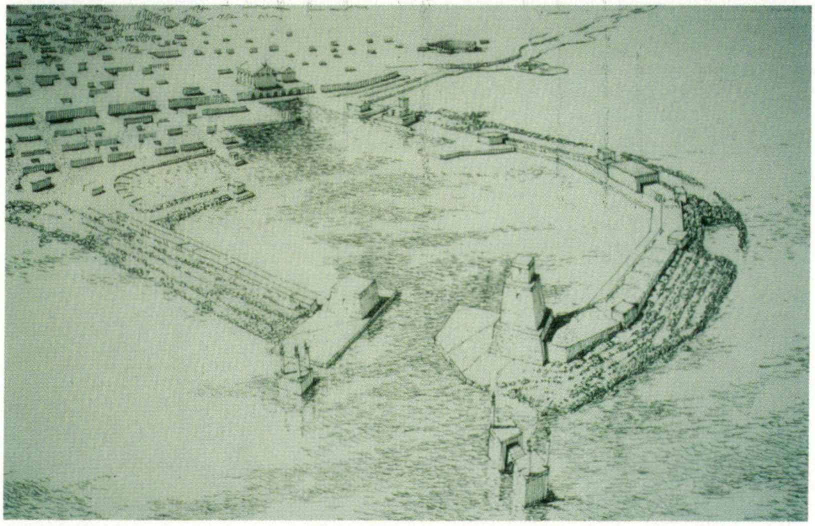

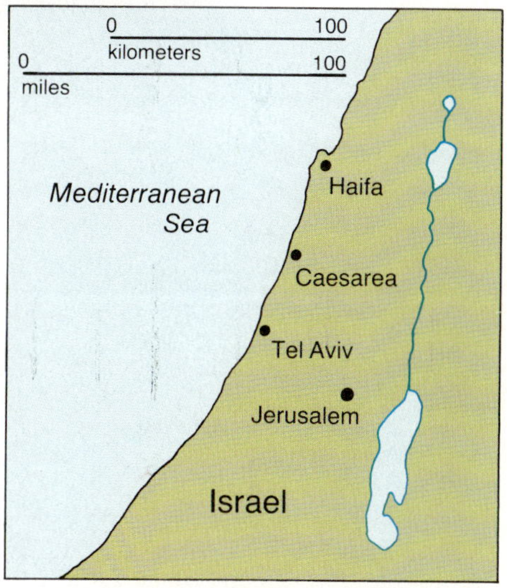

cation with the shore was provided. Various safety and emergency items were on board should a crisis arise. A compressor for filling tanks along with repair facilities and a storage locker for all personal diving gear were located on shore.

Using this equipment for the past five summers, CAHEP's staff of hundreds of volunteers and professional personnel logged thousands of hours of bottom time in an effort to unravel Caesarea's underwater secrets. These investigations added considerable information to our understanding of the history of the city and of ancient Palestine, the region that it served, as well as to our knowledge of ancient maritime life in general. CAHEP also uncovered data about the engineering technology employed in the construction of Caesarea's harbors and about the complex design of those facilities. The complexity of the designs was impressive and largely unexpected.

The harbor of ancient Caesarea is seen in the foreground of this aerial view looking northeast along the coast of Israel (top). Above, a map shows the location of Caesarea in relation to present-day cities of Israel, and at the left is a drawing of a reconstruction of the harbor's outer basin.

(Top) Robert L. Hohlfelder; (bottom, left) drawing by architects from the University of Maryland

(Top) Robert L. Hohlfelder; (bottom, left) Mark Little; (bottom, right) Harry Wadsworth

Diving platform for the Caesarea expedition is moored in the outer basin (top). A diver (above left) fills an air bag with air supplied from the surface through hoses. (Above right) A diver uses jets of air from his tank to probe the seafloor.

Photos, Mark Little

Divers use a metal detector on the hull of a Roman merchant ship (top left) discovered on the seafloor north of the harbor at Caesarea. In order to measure and sketch the remains of the shipwreck (bottom left and right), divers prepare a grid to be placed over it (top right).

Josephus did not exaggerate his account of the harbor built by Herod. Based on the discoveries by CAHEP, if anything, he understated this engineering triumph. Blocks even larger than his dimensions of 50 by 18 by 9 feet were found. Many of them were made of concrete, a fact Josephus did not mention. He may not have known or recognized the differences between this new building material and cut stones, or he may simply have been indifferent to such matters. Some of these concrete blocks ignored by Josephus were actually poured in the sea and allowed to harden in place underwater. Wooden construction frames built on shore were floated into position and then weighted until they sank to the ocean floor. The liquid concrete was then poured through tubes into the frames until the water was displaced. When the concrete had hardened, the new block could become the foundation for a second one.

Although concrete that could set underwater had been developed in Italy more than 100 years before Caesarea, it had not been used before anywhere in the Roman world as extensively. Yet, without this new building material, Caesarea's main harbor could not have been completed on such a grand scale or with such innovative design features. In this respect Herod's enterprise was a daring experiment.

Much of the raw building material, such as the wood for the concrete construction frames, came through the Mediterranean by ship from the west. It is difficult today to imagine shipping such common items such long distances when local substitutes are readily available. But in antiquity, even during Rome's dominance when a superb road network was in operation, transport by sea was easier, cheaper, and more reliable. CAHEP's discovery that the wooden construction frames, or at least the lumber to build them, came from Europe, confirms and expands another assertion made by Josephus. He claimed that all construction materials for Caesarea had come from elsewhere. Most scholars had assumed that he was referring to the marble facing for the city's grand buildings, but it has now become evident that more than marble was imported.

Concrete was used in the construction of both the enclosing arms of Herod's harbor. On the southern breakwater blocks were placed at or above the waterline along the outer, unfinished face at precisely the point where wave attack from winter storms would have been most damaging. This placement, of course, was not accidental.

This breakwater exhibits other structural innovations not mentioned by Josephus. The whole structure rests on an apron or foundation of rubble wider than the breakwater itself. This rubble layer increased the durability of the main structure. With such an apron in place the undercutting action of waves and currents was greatly reduced.

Herod's engineers further protected the southern breakwater from wave attack by constructing a parallel rubble wall some 13 to 18 meters (45 to 60 feet) seaward of the main one. This secondary or subsidiary breakwater was constructed in unconnected sections and was intended to absorb most of the force of incoming waves. When the sea did roll over this secondary wall and crashed into the outer face of the main breakwater, it did far less damage and produced far less spray to wash

(Top) Robert L. Hohlfelder; (bottom) Mark Little

Sluice channel (top) is one of several that were cut through the southern breakwater of the outer harbor at Caesarea. This was done to allow sand-free water from the crests of waves to enter the harbor. This water helped to create an outflowing current that prevented silt and sand from entering and blocking the harbor's mouth. Warehouse (bottom), seen from within, was built on the inner side of the breakwater.

over the seawall. The subsidiary breakwater, the unfinished outer face of the main one, and the seawall together probably assured the safety of cargo stored in the warehouses built on the inner side of the breakwater.

Another interesting design element of this harbor was a system integrated into the breakwater itself that was intended to reduce silting in the enclosed basin. A series of sluice channels was cut through the breakwater to allow sand-free water from the crests of waves to enter the harbor area. This water helped to create a scouring outflowing current that blocked silt and sand from entering the harbor's mouth. The huge quantity of artifacts found underwater near and in the entrance channel testifies to the success of this experiment. The outflowing current carried many cultural remains lost or dropped into the basin toward the mouth and the open sea beyond. When the current met the sea at the harbor entrance, the artifacts were finally deposited on the ocean floor. There, they swirled around with the current until they were buried or lodged in the rocks of the breakwaters.

CAHEP's investigations discovered three other harbors not described by Josephus. The main one, which he does describe in detail, was connected by a channel to an inner harbor that had been carved from the sand dunes. Although it had fallen out of use by the Byzantine era (about AD 500), it was functioning in Herod's day and during most of the Roman occupation of Caesarea. The bay south of the outer basin of Herod's har-

Ceramic jug (far right), figurine (right), and lamp (below) are among the thousands of artifacts uncovered by the explorations at Caesarea.

Photos, Mark Little

Mark Little

The Dive Shop at Caesarea is a scene of activity in the morning. Archaeologists believe that Caesarea will be followed by many other major discoveries in oceans and seas throughout the world.

bor also served as a fair-weather anchorage. To the north of the northern breakwater a fourth harbor existed as well. This one, facing north, had been built late in the 4th century BC or early in the next one. It had been abandoned early in the first century BC but was revived by Herod. These four facilities, all in operation during the reign of Herod, made Caesarea one of the largest ports, if not the largest, in the Mediterranean at the time of the birth of Christ.

Future prospects

CAHEP's explorations to date have confirmed the essence of Josephus's account and added unique information to our understanding of ancient technology. In addition, the excavations on land and underwater have uncovered thousands of cultural remains—pottery of all types and ages, statuary, instruments of daily living, lamps, ships' stores, various metal objects, and several ancient shipwrecks to be investigated in the future. Yet, the search for history in Herod's great city has only really started. Twenty-five years of underwater investigations of various sorts have only scratched the surface. There is much more to learn about this important capital and its relationship to the sea, particularly from the submerged portions of the port.

In this regard Caesarea exemplifies the status of underwater archaeology throughout the Mediterranean. During the past quarter century salvage has become science. Archaeologists have moved from the random plucking of artifacts from the seafloor to the systematic search and recovery of the physical record of antiquity. Diving for the ancient past is now firmly established as a scholarly discipline. But as impressive as the first decades have been, the future promises greater things. The oceans are slowly yielding their secrets, and this unique harvest will most certainly continue.

THE COSMOS ACCORDING TO IRAS

by Michael G. Hauser

A spectacular new view of the universe provided by the Infrared Astronomical Satellite is changing our understanding of the solar system, nearby stars, the Milky Way, and distant galaxies.

__MICHAEL G. HAUSER__ is Head of the Infrared Astrophysics Branch, Laboratory for Extraterrestrial Physics, NASA-Goddard Space Flight Center, Greenbelt, Maryland, and a member of the Joint IRAS Science Working Group. The views expressed in this article are the author's own and are not official views of the National Aeronautics and Space Administration.

(Overleaf) Jet Propulsion Laboratory; (inset) illustration by Seth Shostak, Kapteyn Laboratorium, The Netherlands

Fascination with the universe beyond the Earth—what it contains and what it implies about our world and our own existence—is an ancient and continuing human preoccupation. In recent decades new observing techniques have greatly expanded our knowledge and understanding of the cosmos. A major step in this exploration was taken during 1983 with the flight of the Infrared Astronomical Satellite (IRAS), the first spacecraft designed for astronomical observations in the infrared portion of the electromagnetic spectrum; that is, at wavelengths longer than visible light but shorter than microwaves. Conceived, developed, and operated cooperatively by the space agencies of the United States (National Aeronautics and Space Administration), The Netherlands (Netherlands Agency for Aerospace Programs), and the United Kingdom (Science and Engineering Research Council), IRAS was a scientific and technological adventure of the highest order.

Advantages of infrared astronomy

Astronomers study the sky at infrared wavelengths for a number of reasons: because such observations reveal the location, character, and physical conditions of cool material in the cosmos, material typically found around or between stars in galaxies; because long-wavelength radiation readily penetrates regions obscured at shorter, visible wavelengths by the ubiquitous cosmic dust; and because much of the energy emitted as short-wavelength radiation by cosmic sources is absorbed by nearby dust and reradiated at infrared wavelengths. Study of basic phenomena occurring in dusty regions, such as the formation and death of stars, or of the energetics of systems as large as the galaxies themselves demands infrared observations.

Major elements of the IRAS instrument are identified in the schematic cutaway view on the facing page. A doughnut-shaped tank, suspended within an outer shell, contained liquid helium at 2 K. Solar panels attached to the shell powered the satellite, and a sunshade at the open end prevented direct sunlight from interfering with the measurements or boiling away the helium. The primary and secondary mirrors of the telescope focused infrared and visible radiation from the sky onto an array of detectors mounted in a focal-plane assembly (center right). Each of the rectangular apertures in the assembly allowed light to reach one of 62 infrared radiation detectors, while a filter behind each aperture passed only radiation of the desired wavelength. The eight visible-light detectors were located behind long narrow slits on each side of the array. (Bottom) The assembled IRAS instrument, minus its exterior insulating covers and solar panels, is suspended for testing prior to launch. The assembly was 3.6 meters (12 feet) long and 2.2 meters (7 feet) in diameter and weighed 1,020 kilograms (2,240 pounds) at launch.

Although infrared radiation was first discovered in sunlight by the astronomer Sir William Herschel in 1800, technological limitations prevented its substantial use in astronomical research for many years. Starting in the 1950s infrared observations were vigorously conducted from mountaintop observatories and high-altitude aircraft, balloons, and rockets. Astronomers learned much, but their knowledge of the infrared sky gained by those techniques was roughly equivalent to a map of the Earth's surface that depicted only the highest mountaintops: neither the less prominent peaks, nor the broad sweep of the continents, nor the marvels of the surrounding oceans were yet perceptible. Comprehensive infrared views of the sky on all scales, the equivalent of what man could do with his eyes and earthbound telescopes at visible wavelengths, were not possible prior to IRAS.

The primary aim of the IRAS mission was an unbiased survey of the entire sky in four broad wavelength ranges centered at 12, 25, 60, and 100 micrometers (one micrometer is a millionth of a meter, or four hundred-thousandths of an inch). The survey was designed to be sufficiently sensitive to detect numerous celestial sources beyond the Milky Way Galaxy, as well as within it and within our own solar system, assuring sampling of previously known and entirely new phenomena. It also included low-resolution infrared spectroscopy in the 8–22-micrometer

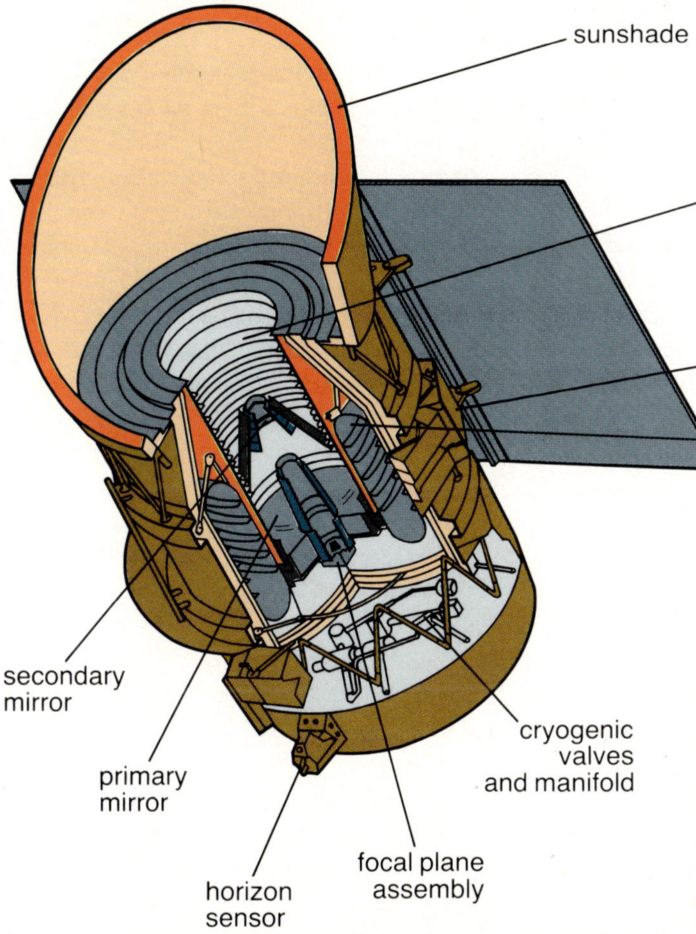
sunshade
baffle
experiment electronics
superfluid helium tank
secondary mirror
primary mirror
cryogenic valves and manifold
horizon sensor
focal plane assembly

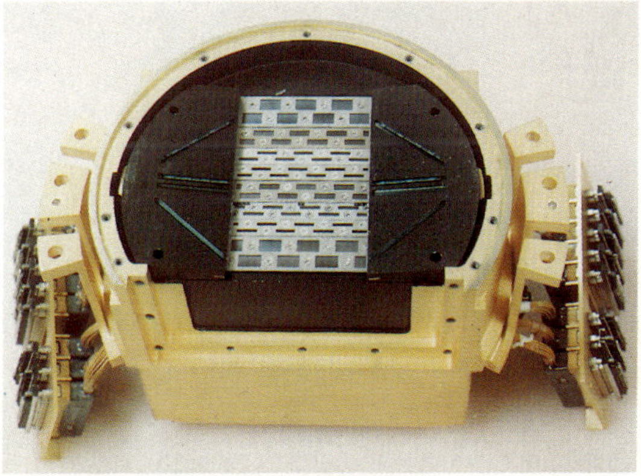

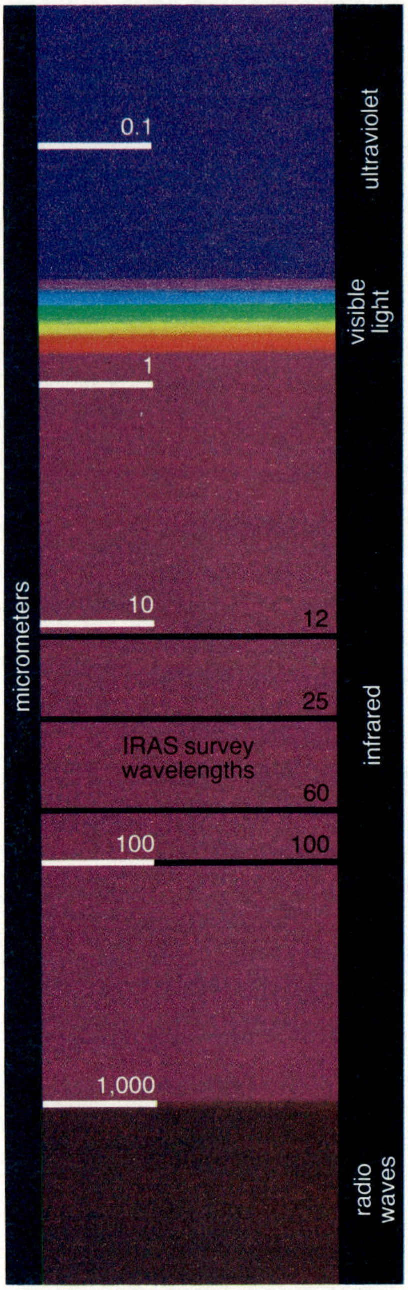

IRAS surveyed the sky in four wavelength ranges in the infrared portion of the electromagnetic spectrum. Although the bands are identified by their central wavelengths of 12, 25, 60, and 100 micrometers, the detectors in fact had rather broad sensitivities: 8–15, 17–30, 40–80, 80–120 micrometers, respectively.

range to aid in characterizing the brighter sources. For the first time the total brightness of the infrared sky on both large and small angular scales would be recorded. In addition, detailed observations of selected interesting sources would occupy a significant fraction of the mission.

All matter emits thermal, or heat, radiation, the amount and relative brightness at different wavelengths depending upon the temperature and the optical properties of the material. In general, the warmer the material, the shorter the wavelength at which it is brightest. For example, objects at room temperature, near 300 degrees kelvin (K; 80° F), glow brightest at about 10 micrometers, whereas material at 30 K (−405° F), near absolute zero, glow brightest at 100 micrometers. The wavelengths of the IRAS survey were hence particularly well-suited for measuring thermal emission from material ranging in temperature from tens of degrees kelvin, typical of the matter in space remote from stars, to hundreds of degrees kelvin, typical of solar-system material or clouds around young or old stars.

The IRAS spacecraft

The opacity and brightness of the Earth's atmosphere prevent sensitive surveys from the ground at wavelengths longer than a few micrometers. Scientists have long recognized the need for a space instrument to accomplish the kinds of objectives for which IRAS was designed, but formidable technological challenges existed even after space missions became possible. The biggest obstacle was cooling: the spacecraft's instrument for gathering and detecting infrared radiation had to operate at a temperature of a few degrees kelvin in order to eliminate interfering thermal emission from the instrument itself and allow best use of its highly sensitive infrared-radiation detectors. The solution to the cooling problem was a superfluid-helium Dewar, a vacuum-insulated storage vessel from which liquid helium slowly evaporated during the mission. Filled initially with some 500 liters (about 130 gallons) of helium, the Dewar kept the instrument at a stable 2 K (−455° F). So effective was the design of the Dewar that, although it was always in direct sunlight and earthshine, the net power reaching the helium was less than that radiated by the end of a person's thumb.

IRAS's astronomical instrument included an all-mirror Cassegrain telescope having a primary mirror 57 centimeters (22.5 inches) in diameter. The entire telescope—both its mirrors and supporting structure—was made of beryllium, a metal offering an attractive combination of lightness and strength. Three sensor systems shared the telescope's focal plane: a survey array consisting of 62 infrared and 8 visible-light detectors, a low-resolution spectrometer operating in the 8–22-micrometer range, and a photometer for high-angular-resolution measurements of selected sources at 50 and 90 micrometers. The visible-light sensors detected known bright stars to provide accurate pointing information. The infrared detectors in the survey array were arranged to map out a path ½-degree wide over the sky at each of the four wavelengths as the satellite moved in its orbit around the Earth. The radiation-sensitive elements were

JPL

Flaring rocket motors light the evening sky of January 25, 1983, as IRAS *begins its pioneering mission atop a Delta launch vehicle.*

photoconductors, specially doped pieces of silicon (for the 12- and 25-micrometer sensors) and germanium (for the 60- and 100-micrometer sensors), that produce a small electric current when infrared radiation is absorbed. So sensitive was this combination of telescope and detectors that thermal emission from a single asteroid ten kilometers (six miles) in diameter at twice the Earth–Sun distance, about 300 million kilometers (190 million miles), was readily detectable.

The mission

The long-awaited era of space infrared astronomy began at Vandenburg Air Force Base near Lompoc, California, where on January 25, 1983, IRAS lifted rapidly through a low layer of clouds atop a Delta rocket. The mission was virtually flawless from the start. The satellite reached its desired orbit—a circular, nearly polar path 900 kilometers (560 miles) in altitude—with an accuracy of 10 meters (33 feet). After two weeks of testing, the sky survey began on February 9 with all systems working well. From that day onward IRAS collected high-quality scientific data in both survey and special observing modes until the last drops of liquid helium boiled away on November 22. The electrical signals from the detectors were stored on a tape recorder aboard the spacecraft and transmitted to the ground twice a day during passes over a ground station at the Rutherford Appleton Laboratory in Chilton, England. Astronomers in Chilton made a preliminary examination of the data, while more extensive processing was carried out at the Jet Propulsion Laboratory in Pasadena, California.

A major consideration in the planning of IRAS was to achieve a highly reliable, as well as complete and sensitive, survey of the sky. Such reliability requires repeated measurements. The IRAS survey actually consisted of carefully designed, successive observations of each position in the sky over time scales ranging from seconds to months in order to identify and

Whole-sky maps of diffuse infrared emission (facing page, top) and discrete source counts (bottom) were compiled from IRAS survey data. Blue is 12 micrometers; green, 60; and red, 100. Yellow appears where both 60- and 100-micrometer emissions are bright (top) or sources are numerous (bottom). The plane of the Galaxy is the central horizontal feature. Galactic center is at the center of the images, and galactic north is up. Diffuse emission from interstellar dust in the plane of the galaxy dominates the top map. Several nearby star-forming clouds, including those of Orion (south of the plane at the extreme right) and Ophiuchus (just north of center), are prominent. Dust in our solar system produces the emission near the ecliptic plane, the light-blue, S-shaped feature passing through the image center. The Large Magellanic Cloud is the isolated yellow feature below and to the right of center. In the bottom map blue dots indicate mostly cool, giant stars within the Galaxy. Green and red dots close to the galactic plane show largely cool interstellar clouds or dust-embedded stars within our Galaxy, while green dots well away from the plane show external galaxies.

characterize moving and time-variable sources as well as fixed sources of infrared emission. IRAS scanned 98% of the sky on each of at least two separate orbits, over 96% on four or more, and 72% on at least six orbits. Only 2% was not scanned at all.

Appearance of the infrared sky

Visible images created from IRAS data reveal the sky to have a striking appearance at infrared wavelengths. The dominant large-scale features in these images are two bright intersecting bands, one associated with the ecliptic plane, the plane of the Earth's orbit around the Sun, and the other with the plane of the disk of our Galaxy. The glow surrounding the ecliptic plane arises from emission by interplanetary dust in the solar system. Since the Sun warms this dust to several hundred degrees kelvin, the emission is most prominent at wavelengths of 12 and 25 micrometers. The infrared emission from the interplanetary dust is distributed over the sky in much the same way as the visible zodiacal light, which is sunlight scattered by the same dust.

The emission from the plane of the Milky Way, a spectacularly bright narrow band around the sky, arises from interstellar dust. Since most of this dust is much cooler than that in the solar system, its emission is brightest at the longer wavelengths, especially 100 micrometers. In the infrared the Galaxy looks markedly different than its image in visible light. For example, the same dust that emits infrared radiation from the direction of the center of the Milky Way screens out visible light from stars at the center by a factor of a trillion. Hence, to the human eye

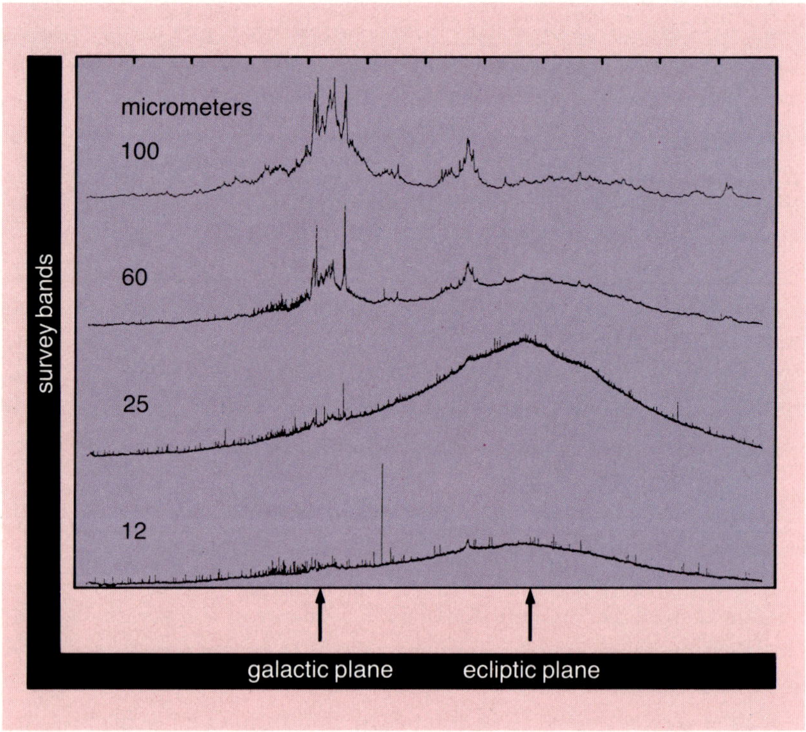

The traces at right display the brightness measured by single infrared detectors at each of the four IRAS survey wavelengths as the telescope scanned the sky on half of an orbit. Narrow spikes are signals from stars or other compact regions of emission. Emission from interplanetary dust produces the broad peak at the ecliptic plane, while interstellar dust in the Milky Way is responsible for the broad features at the galactic plane. Computers combined such data from all detectors to find sources and produce images, or maps, of the sky.

Adapted from information obtained from the Jet Propulsion Laboratory

Photos, JPL

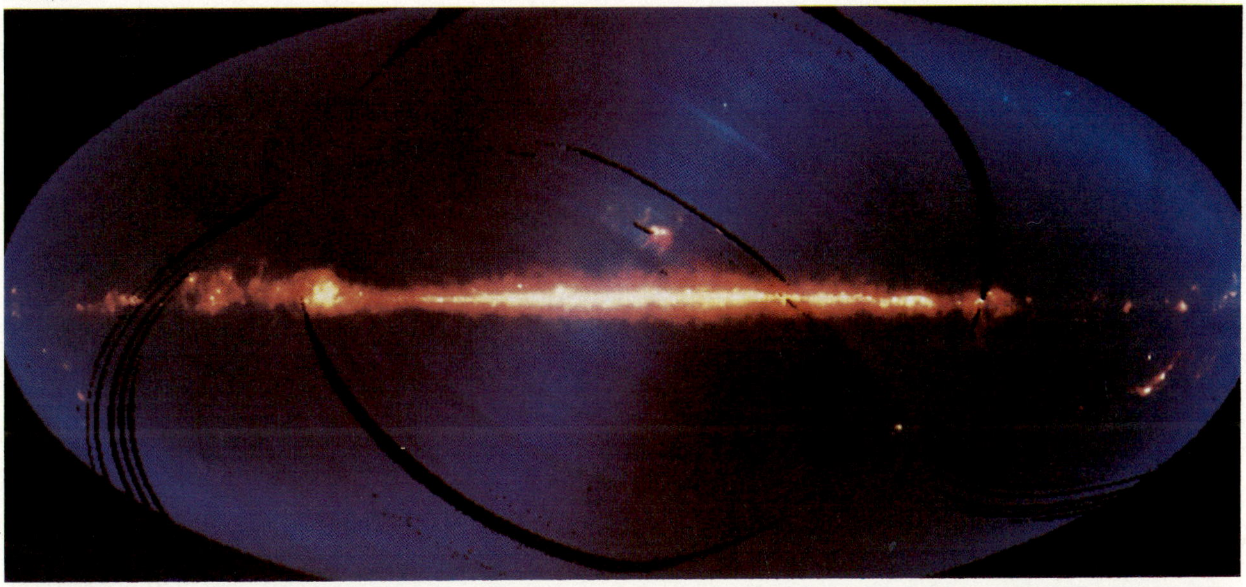

the galactic center appears dark, whereas at 100 micrometers it is the brightest region in the sky.

A widespread and striking feature of the infrared sky, particularly at 100 micrometers, is one that IRAS scientists dubbed "infrared cirrus": large complex patches and filaments of emission reminiscent of cloud patterns in Earth's atmosphere. Much of this emission is clearly correlated with known interstellar clouds of atomic and molecular gas and evidently arises from dust in these clouds. It is also possible that some originates from cold material in the outer part of the solar system.

In addition to these extended features, the infrared sky contains many "point sources," or starlike objects, just as the sky does in visible light. The character of the sources, however, changes dramatically with

Photos, JPL

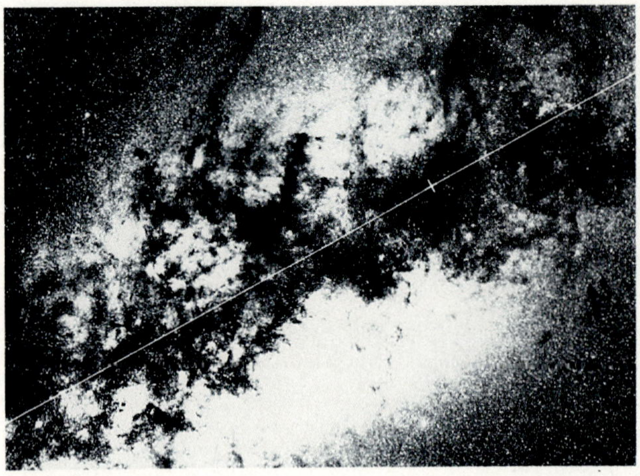

Views toward the center of the Milky Way in visible light (above) and in the infrared (right) are strikingly different. In the visible image nearby stars (white) predominate, whereas more distant stars are obscured by dust. The galactic plane lies along the white line, and the galactic center is marked with a cross. The infrared image—in which red indicates cool interstellar dust, yellow and green warmer material, and blue the warm dust in our solar system—generally reveals galactic material far from the Sun. The center of the Milky Way, 30,000 light-years from the Sun, lies in the central yellow region. The large yellow-green patches strung out close to the plane are diffuse nebulae within a few thousand light-years of the Sun containing dust and gas ionized by luminous young stars, while many of the small yellow sources are more distant examples of similar regions.

increasing wavelength. At 12 and 25 micrometers, giant stars having relatively cool envelopes—at 3,000–4,000 K, compared with 6,000 K for the Sun—dominate the point sources. These stars are within our Galaxy, that is, within about 60,000 light-years of the Sun, and appear strongly concentrated toward the plane of the Milky Way. Asteroids and comets are also prominent at these wavelengths but are distinguishable from stars in the IRAS data by their change in apparent position between successive sightings. At 60 and 100 micrometers, however, galaxies external to the Milky Way dominate the sky. These objects are typically at distances of tens of millions to hundreds of millions of light-years and are quite uniformly distributed over the sky. Dust-enshrouded stars within the Milky Way, both old and young objects, are also present in large numbers at the longer wavelengths. Overall, the discrete sources detected by the Infrared Astronomical Satellite have been compiled into catalogs containing more than a quarter of a million listings.

The solar system

The first thorough infrared look at the universe gave astronomers many surprises and triggered wide-ranging astrophysical investigations. Some of the earliest and most widely reported discoveries came from our own solar system, among them the observation of seven new comets. These finds were serendipitous: the comets turned up in a search of IRAS data for rare fast-moving asteroids, or Apollo objects, whose orbits cross that of the Earth. The comets were observable because they shed much more dust than comets had been thought to do, making them unexpectedly bright in the infrared. More new comets are likely to be found when the data are carefully analyzed for slowly moving objects.

The first comet found, named IRAS-Araki-Alcock after its independent codiscoverers, who included two amateur astronomers, was a splendid beginning. This object passed within a mere 4.7 million kilometers (three million miles) of the Earth, closer than any known comet in the past two centuries, permitting detailed examination both with IRAS and other instruments. IRAS-Araki-Alcock showed a surprisingly extensive dust tail in the infrared, a characteristic that IRAS observed repeatedly in new and previously known comets. In the case of the well-known comet Tempel 2, IRAS detected a tail stretching some 16° across the sky, although no tail appears in visible light. Increased knowledge of the production rate and character of cometary dust will enhance our understanding of the makeup of comets and of the origin of the interplanetary dust cloud that envelops the inner planets.

One of the most intriguing objects found in the search for fast-moving objects was first identified as an asteroid on the basis of its lack of a visible coma, the nebulous haze of dust and gas surrounding a comet's head. Astronomers soon realized that this object, named 1983 TB, shared the orbit of the well-known Geminid meteor stream and is likely the parent body for this stream. Since meteors are generally understood to originate from cometary debris, 1983 TB is apparently the head of a "dead" comet; that is, a comet whose frozen gases have completely evaporated. This is the first clear example of such an object, suggesting that some other Earth-crossing asteroids have a similar origin.

Asteroids themselves have become objects of increased interest over the past decade, partly because they are relatively primitive bodies carrying clues to the formation and evolution of the solar system and partly because their collisions with other bodies in the solar system influence the development of those bodies. In addition, asteroids represent potentially large sources of raw materials for future human endeavors. It is estimated that 30,000 asteroids will be identifiable by their movement between sightings in IRAS data, providing a major new resource for systematic investigation of the minor planets of our solar system.

The IRAS spacecraft also presented astronomers with the opportunity of finding previously unknown major planets, beyond the orbit of Pluto. To date, however, examination of its data has failed to reveal any clear-cut examples. IRAS was uniquely suited for detecting such objects well beyond the known planets: distant cold objects radiate strongly in the infrared due to their own internal energy sources but reflect very little sunlight and hence are extremely faint in visible light. For example, IRAS could sight a planet like Jupiter at a distance of 15,000 astronomical units (AU; one AU is the mean Earth–Sun distance), about 3,000 times Jupiter's actual distance from the Sun and much farther than any present or planned optical telescope could detect it. Unfortunately, distance measurements with IRAS data alone (using parallax, the change in apparent position of a celestial object due to the Earth's orbital motion) are limited to about 3,000 AU, requiring measurements with other instruments to place more distant objects. Nevertheless, IRAS's ability to see cold objects over a vast region of space in the vicinity of the Sun assures a vigorous search in the near future for unknown neighbors in the solar system.

Another major surprise in the solar system was the discovery of three faint bands of infrared radiation from interplanetary dust at the distance of the asteroid belt. These bands, which extend all around the sky, appear as small increases in brightness at locations about 10° above and below the ecliptic plane as well as near the plane itself. Such an effect could be produced by concentrations of dust particles in solar orbits

JPL

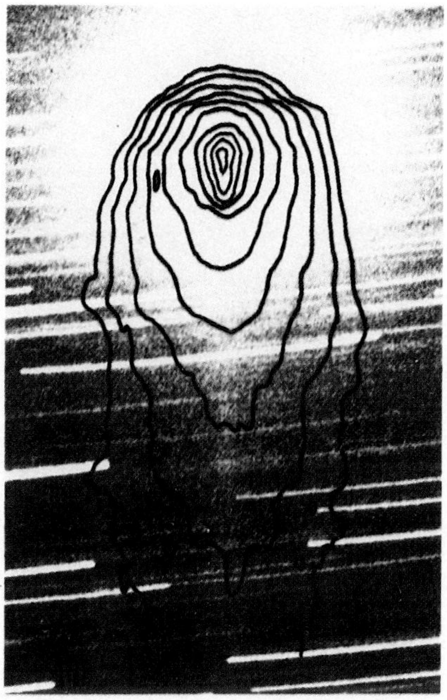

Comet IRAS-Araki-Alcock is shown in a visible-light photograph with contours of brightness at 60 micrometers superimposed. The Sun is in the direction of the top of the figure, and the comet is moving toward the right. The peak brightness of the infrared emission coincides with visible coma, and the contour lines follow the extended dust tail in the direction opposite the Sun.

tilted about 10° from the ecliptic plane and by others in the plane. The dust glows brightest at wavelengths that correspond to a temperature of 165–200 K (−160° to −100° F). This temperature matches that of dust at about 2.5 AU from the Sun, the position of the main asteroid belt. The location strongly indicates an asteroidal origin for the material, and some astronomers have suggested specific known groups of asteroids as likely sources. This development is particularly intriguing because the prevailing view is that interplanetary dust is material shed by comets.

The Vega phenomenon

Perhaps the most remarkable and intriguing discovery of the Infrared Astronomical Satellite to date has been the astonishingly high 60- and 100-micrometer brightness of Vega, 26 light-years distant and among the brightest stars in the night sky. The find is remarkable because Vega has been carefully studied and used by astronomers as a brightness standard at shorter wavelengths for years and intriguing because it implies that Vega is surrounded by ancient orbiting particles likely to be at least of pebble size, a few millimeters in diameter. Vega is thus the first star other than the Sun known to be orbited by solid objects of appreciable size that presumably accreted from material in the cloud from which the star formed. Since Vega is more massive and hence more short-lived than the Sun, the Vega system may be in a condition similar to that of the solar system long ago. The IRAS observations cannot establish whether there are solid objects comparable in size to comets, asteroids, or planets. Nevertheless, this discovery is a dramatic advance in the quest to understand our origins and possible uniqueness in the universe.

The findings about Vega exemplify the interplay between serendipity and detective work in science. Vega was one of several stars selected as calibration standards for IRAS. Once into its mission in space, however, IRAS found the star to be 30% brighter than expected at 25 micrometers, increasing to about 16 times brighter than expected at 100 micrometers. The excess long-wavelength emission implied the presence of material at

The faint bands of infrared emission from interplanetary dust unexpectedly discovered by IRAS appear in the specially enhanced image below as three horizontal blue lines centered on the ecliptic plane. The radiation arises from dust at a temperature of 165–200 K in the main asteroid belt between the orbits of Mars and Jupiter. Sources in the Milky Way show as white and yellow regions near the galactic plane.

JPL

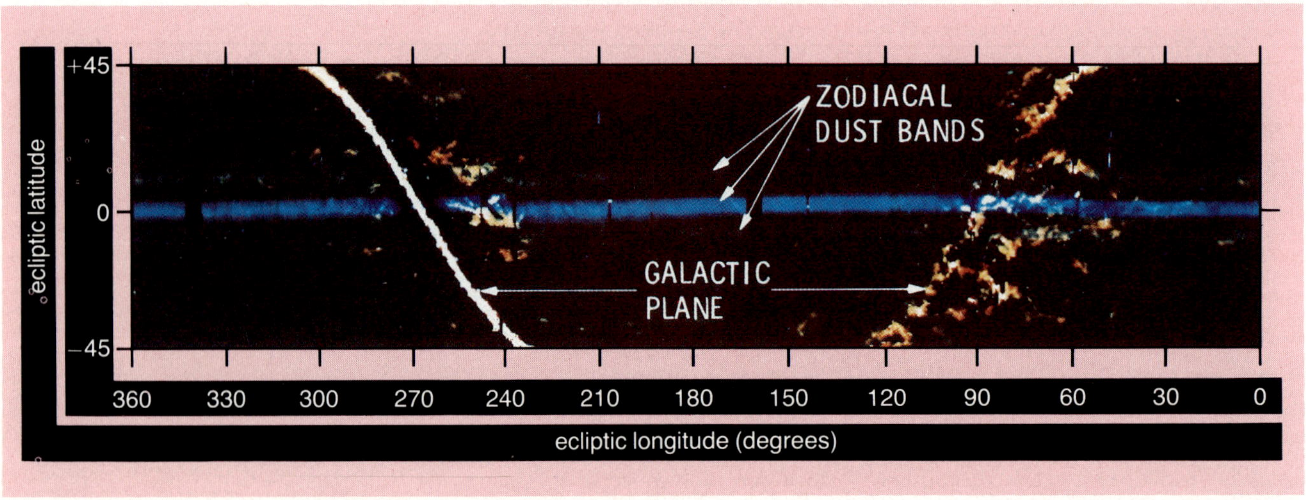

Illustration by Seth Shostak, Kapteyn Laboratorium, The Netherlands

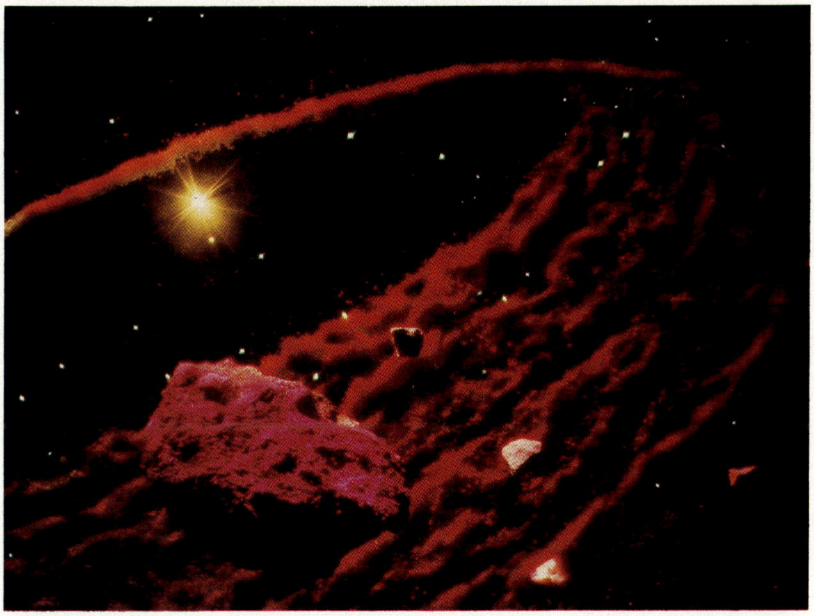

Artist's interpretation of a protoplanetary system around Vega was inspired by IRAS*'s observation of excessive 60- and 100-micrometer emission from the star. While the find implied that Vega is surrounded by orbiting particles of at least pebble size, it could not establish whether there are also solid objects comparable in size to asteroids or planets.*

a temperature of about 85 K, much cooler than the gaseous atmosphere of the star itself. Many special slow scans over Vega established that the emission was coming from the same direction at all wavelengths, making a chance superposition of Vega and a more distant source of long-wavelength radiation highly unlikely. The emission was perceptibly spread out compared with that from other stars, implying material at a distance of about 85 AU from Vega. The expected temperature of particles at 85 AU from Vega is 85 K (−305° F), in perfect agreement with the observed temperature. Two considerations lead to the inference that the particles are old and at least a millimeter in size: Vega is not shedding gas that could be condensing continuously into new small particles, and particles smaller than a millimeter either would have been blown away by Vega's radiation long ago or would have been slowed in their orbits by radiation and sent spiraling into the star. These observations imply a substantial amount of material orbiting Vega. If IRAS were to view our solar system from afar, the 100-micrometer emission from the interplanetary dust and all of the planets combined would be less than 1% of that from the Sun alone, *i.e.*, quite unnoticeable, compared with the 16-fold excess found in the Vega system.

This profoundly exciting discovery stimulated a rapid search of the data for similar cases. Fomalhaut, one of the brightest stars in the southern sky, was found to present an infrared picture very much like that of Vega, with a slightly larger ring of particles at a temperature of about 55 K (−360° F). Ground-based astronomers obtained striking optical evidence confirming the presence of a disk of material around the star Beta Pictoris, which also had been found by IRAS to have a long-wavelength excess. Perhaps even more remarkably, IRAS observations for the several hundred stars nearest the Sun show that about 20% have excess long-wavelength emission similar to that of Vega, Fomalhaut, and

Beta Pictoris. Although the evidence is less complete for these stars, taken together it strongly suggests that the condensation of solid particles of appreciable size commonly accompanies star formation.

Star birth in the Milky Way

Stars form within parts of clouds of interstellar gas and dust that collapse due to mutual gravitational attraction. Because of the dusty surroundings and low temperatures, visible-light observations reveal little about these regions. Astronomers would like to know what conditions cause clouds to fragment and collapse, what governs the distribution of masses of stars that do form, and what evolutionary stages occur as a protostar develops into a hydrogen-burning star, powered by nuclear fusion.

Prior to IRAS, studies of star birth dealt largely with stars much more massive and luminous than the Sun simply because they were easy to detect. One well-studied example is the Becklin-Neugebauer (BN) object located in the Great Nebula in the constellation of Orion. The BN object, now known to be a young massive star, is a powerful infrared source, about 100,000 times as luminous as the Sun, still buried in its natal cloud of gas and dust at a distance of only about 1,500 light-years from our solar system. This cloud was known from radio-astronomy maps to cover most of the sky occupied by the Orion constellation. The spectacular IRAS image of the same large region reveals widespread emission from cold

A disk of material orbiting the star Beta Pictoris is clearly seen nearly edge-on in the image at right, which was made in visible light by ground-based astronomers in 1984. The different colors in the image indicate different brightness levels. Beta Pictoris, a star 50 light-years from the Earth, had been found by IRAS to have an even larger excess of long-wavelength infrared radiation relative to its short-wavelength brightness than Vega. Observing from the Las Campanas Observatory in Chile, the astronomers used an occulting shield (visible as a dark circular object supported by crosshairs) to block light from the star and thus allow the much fainter light from the disk to be detected. The brightness distribution of the disk suggests that it may contain not only the small grains inferred from the IRAS data but possibly larger objects as well.

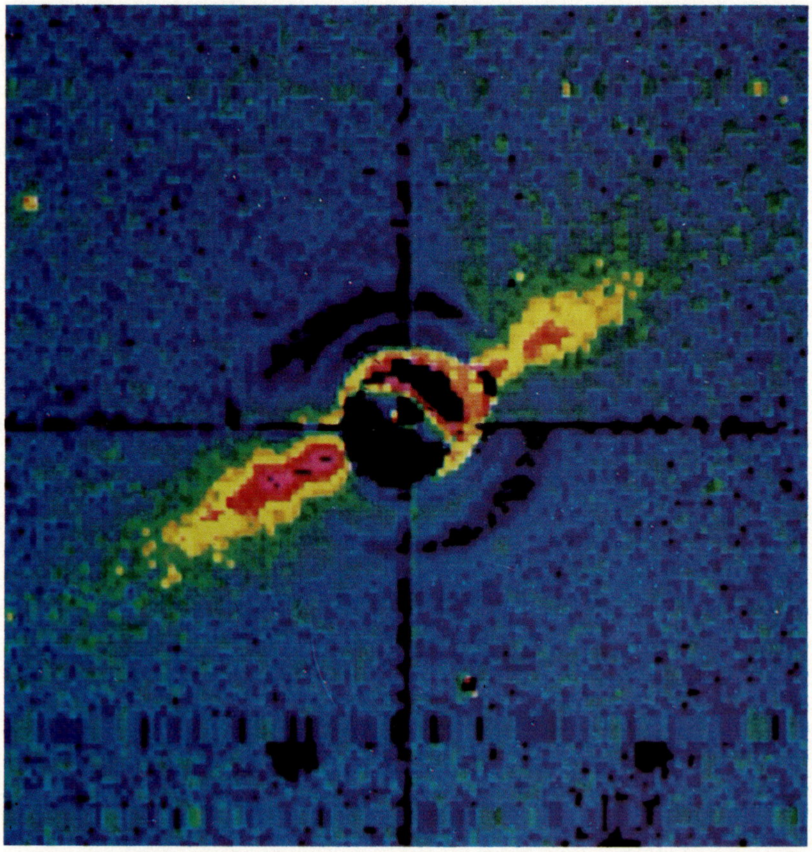

JPL

JPL

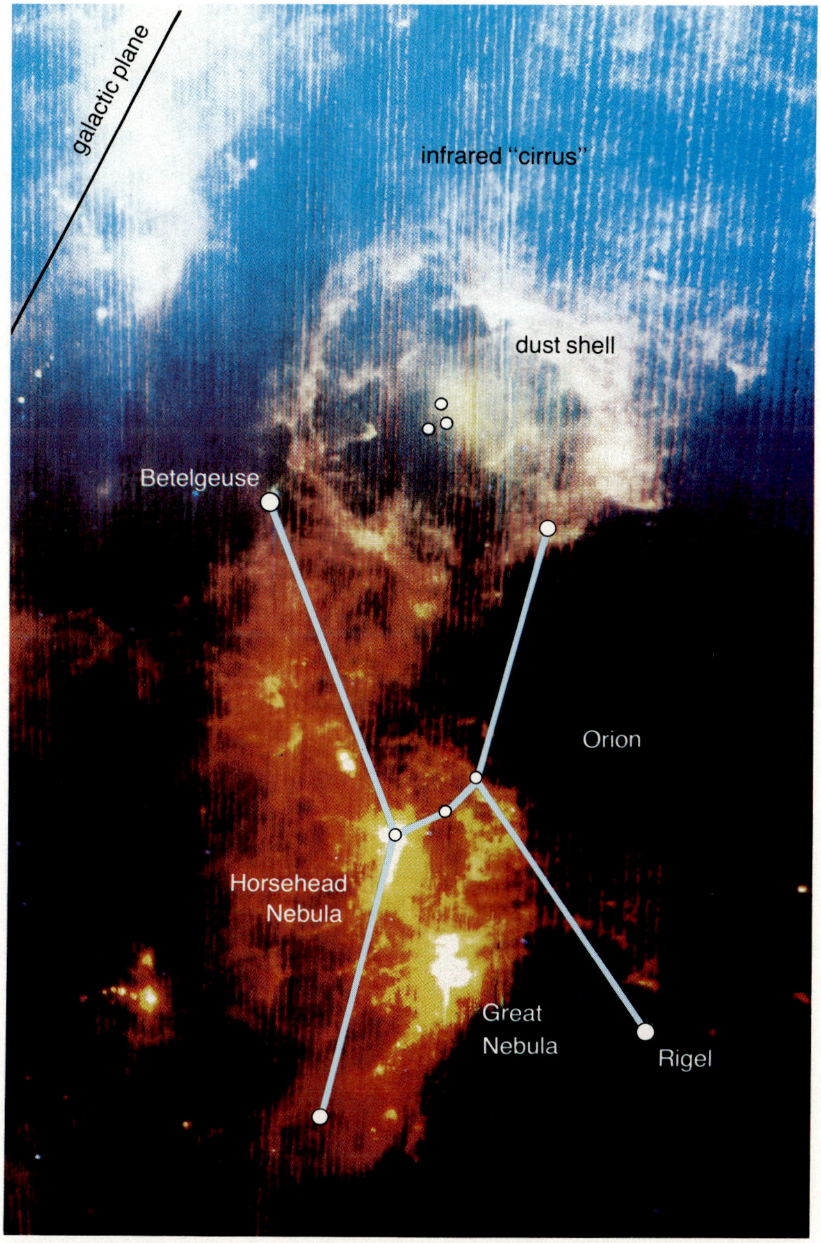

The enormous sky area outlined by the constellation Orion contains a giant, relatively close interstellar cloud complex of gas and dust that encompasses many star-forming regions. The IRAS *three-wavelength composite image at left identifies the cold dust of the cloud (red, less than 30 K) and warmer regions (yellow and white) containing luminous young stars and ionized hydrogen gas, such as the Great Nebula. Luminous young stars in the northern (upper) part of the region have blown away cloud material, producing a spectacular shell of emitting dust. In the northern part of the image the blue region represents strong emission from relatively warm dust in our solar system, while the patchy white "cirrus" is most likely cold interstellar material.*

dust highlighted by areas of especially high brightness due to heating by massive young stars either embedded in the cloud or nearby. Furthermore, the cloud contains a gigantic shell of dust blown away by luminous young stars. Detailed comparison of the Orion cloud at infrared and radio wavelengths will add greatly to our understanding of the activity and energetics in this nearby star-forming region. IRAS detected sources like the BN object throughout most of the disk of our Galaxy, providing a rich sample for systematic study of massive star formation.

In the case of low-mass stars, that is, stars comparable in mass to the Sun, one of the exciting facts emerging from IRAS is that many such stars are forming in seemingly quiescent dark clouds of interstellar material.

JPL

Three-wavelength IRAS image of a large sky area in the constellations of Taurus and Perseus reveals vigorous star-formation activity. Cool dust appears red (bright at 100 micrometers), with warmer material green (60 micrometers), blue (12 micrometers), or yellow-white (some combination of the three wavelengths). Many protostars or young low-mass stars appear as small points. The prominent central feature is the Pleiades, a visibly bright cluster of young stars still associated with a remnant of their natal cloud. The large yellow-white regions at the top are clouds of warm dust and hydrogen gas ionized by luminous young stars; the uppermost region is the California Nebula.

These clouds were previously known both from optical photographs, because their dust blocks visible light from more distant stars, and from measurements of radio emission from molecules contained in them. In many cases the location of strongest molecular emission coincides with an infrared source emitting energy at about the same rate as the Sun, much as expected for a protostar or newborn star of low mass. The colder infrared sources are probably concentrations of star material still contracting and warming toward conditions where nuclear reactions can begin. Others sources show a range of temperatures and a luminosity several times that of the Sun and are probably young stars still embedded in their natal cloud. Still others are relatively warm but of lower infrared luminosity than the dust-embedded sources and are often associated with a visible light source; these may be young low-mass stars that have already moved out of the dense core of the cloud in which they formed. Much more analysis is needed to create definitive categories of sources and to identify these categories with successive stages of stellar development. Nevertheless, the rich harvest of information from IRAS promises a detailed early evolutionary history of such stars as the Sun.

Beyond the Milky Way

The universe beyond our Galaxy, previously only sketchily familiar at infrared wavelengths, has been rendered in broad detail by IRAS. Nearby galaxies—those between 150,000 and several million light-years from the Milky Way, which include the Large and Small Magellanic clouds, our familiar giant spiral companion in Andromeda, and others in the Local Group of galaxies—have been fully mapped. When combined with optical- and radio-wavelength data, these maps are fertile ground for studying galactic-scale energetics, the distribution of material in the interstellar medium, and conditions for active star formation. Although astronomers also explore these same topics in the Milky Way, an external vantage point often simplifies the interpretation, and investigation of different kinds of systems gives valuable new perspectives.

At IRAS's detection limits (two billion light-years and beyond) are the quasars, the mysterious compact objects whose luminosities exceed those of whole galaxies by factors of hundreds. The IRAS survey catalog contains relatively few quasars, perhaps a few dozen, but because of the great interest in these objects, many previously known quasars were given special nonsurvey scrutiny. The survey data for a few quasars have been examined for clues as to why some are bright at both optical and radio wavelengths whereas others of equal optical brightness have little or no radio emission. An excess of 100-micrometer emission in the "radio-quiet" quasars suggests that these objects may differ from the "radio-loud" ones in being located within spiral galaxies, which in some as yet poorly understood way may be quenching their radio emission.

The Infrared Astronomical Satellite detected several tens of thousands of galaxies lying between the Local Group of galaxies and the quasars. It could not map individual galaxies at such distances, but it could measure their total energy emission at infrared wavelengths, crucial information for understanding galaxy energetics. Prior to IRAS it was known that certain galaxies, such as the irregular galaxy M82, radiate most of their energy at wavelengths of 50–100 micrometers. In the case of M82,

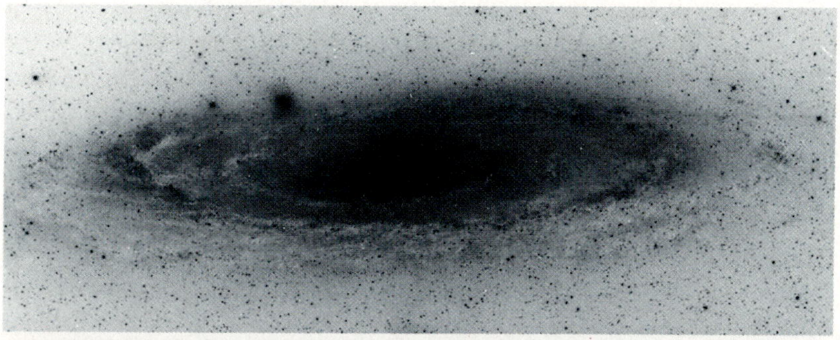

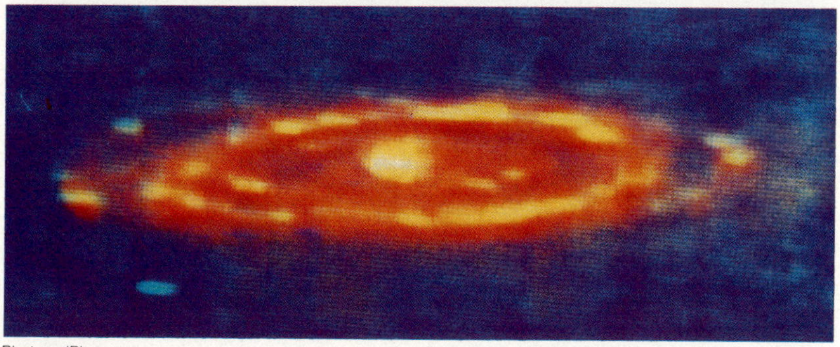

Photos, JPL

Two images of the spiral galaxy M31 in Andromeda, about two million light-years from the Sun, are compared: a visible-light photograph (top) and an IRAS composite image (bottom). In the optical photo, a negative, regions of bright starlight are black, while dust-obscured regions are light. In the IRAS image the coldest material is red, warmer material green or yellow, and the warmest material blue. The yellow ring, which coincides closely with a prominent dust lane in the optical photo, is a region of active star formation. The infrared radiation emanating from the nucleus may be coming from dust shed by old stars.

In visible light the irregular galaxy M82, located some ten million light-years from the Milky Way, shows a highly disturbed structure. Its dark patches are due to absorption by interstellar dust. Vigorous star formation, responsible for the high infrared brightness in this galaxy, is suspected to be the cause in many of the infrared-bright galaxies found by IRAS.

which is about ten times as luminous in the infrared as it is at shorter wavelengths, the explanation is that the luminosity is dominated by large numbers of massive young stars surrounded by the interstellar clouds from which they formed. Dust in these clouds absorbs most of the stars' visible and ultraviolet light, which warms the dust enough to radiate strongly in the far infrared. What causes such vigorous, large-scale star formation, or "starburst" activity, in some galaxies and just how common is this phenomenon are two major questions that IRAS's observations are ideally suited to address.

One fact that emerged clearly from early studies of the data is that some galaxies are more luminous at infrared wavelengths than in visible light by factors of 100 and more; *i.e.*, M82 is by no means an extreme example. By way of comparison, elliptical galaxies have so little infrared emission that very few were even seen by IRAS; the Andromeda galaxy (M31), an infrared-faint spiral, emits only a few percent of its total luminosity in the infrared; and the Milky Way, a fairly typical spiral, emits about half of its energy in the infrared. The most extreme examples of galaxies dominated by their infrared brightness may not yet be known: a galaxy faint to IRAS and having a visible luminosity a hundredth of its infrared luminosity or less would be too dim optically to have ever been recorded in previous galaxy catalogs or photographic sky-survey plates. Follow-up observations using large ground- or space-based telescopes are needed to determine whether IRAS sources not identifiable with known stars or galaxies can be associated with faint galaxies. This process has been carried out for a small sample of nine unidentified IRAS sources, and optically faint galaxies have been found at the positions of eight of them (the ninth was found to be faint "cirrus" within the Milky Way).

The galaxies in which infrared luminosity exceeds that in the visible are comparatively few: only about 1% of all galaxies emit more than twice as much in the infrared as in the visible. How can such galaxies be explained? Some may be undergoing intense bursts of star formation similar to that in M82. A clue to what might trigger starbursts arises when one examines the environs of the bright infrared galaxies: about one out of four appear to be colliding or merging galaxies, a ratio far higher than is found in the general population of galaxies. This characteristic suggests that strong gravitational interactions of galaxies can trigger bursts of star formation on a grand scale.

Some infrared-bright galaxies may have active centers like Seyfert galaxies, in which a compact primary energy source—perhaps in some cases a black hole—is embedded in dust. A possible example of this, although the interpretation is not certain, is the remarkably luminous galaxy Arp 220, which IRAS found to be not only 80 times brighter in the infrared than in the visible but also about 100 times more luminous than normal galaxies overall. The large sample of isolated galaxies and of galaxies in groups and clusters measured by IRAS promises some answers to the fascinating problems of galaxy energetics and the influence of surroundings on activity within galaxies.

G. E. Danielson, C. J. Lonsdale, and B. T. Soifer

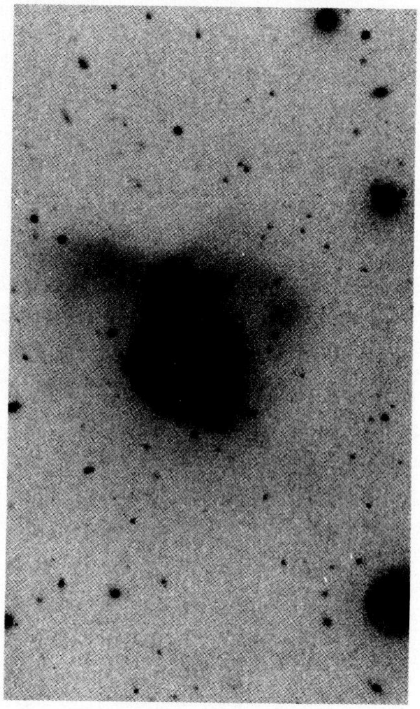

Arp 220, one of the most remarkably luminous galaxies detected by IRAS, is shown above in a visible-light negative image obtained with the Palomar Observatory five-meter telescope. Its disturbed central region and wispy tails suggest that it is a system of colliding galaxies.

Continuing the adventure

The IRAS survey of the infrared universe has been completed, and its magnificent images and catalog of the skies have been carefully prepared for serious study. Astronomers have already uncovered a few of the surprises lying hidden in cool cosmic matter ranging in distance from light-minutes to billions of light-years away. But the full intellectual fruits of this pioneering venture will be gathered only after years of probing: asking the right questions, finding the right way to wrest the answers from the inscrutable numbers, pursuing the critical follow-up observations, and relating these facts to the general body of astrophysical knowledge. The space agencies of both the United States and Europe are planning orbiting infrared observatories for the 1990s that can carry forward the exploration begun by IRAS. Rarely have measurements based on a single technique so suddenly illuminated such a diverse sample of the universe. It seems likely that the biggest surprises and insights are yet to come.

FOR ADDITIONAL READING

David A. Allen, *Infrared: The New Astronomy,* a Halsted Press Book (John Wiley and Sons, 1975).

Harm J. Habing and Gerry Neugebauer, "The Infrared Sky," *Scientific American* (November 1984, pp. 48–57).

G. Neugebauer and Robert B. Leighton, "The Infrared Sky," *Scientific American* (August 1968, pp. 50–65).

G. Neugebauer *et al.*, "Early Results from the Infrared Astronomical Satellite," *Science* (April 6, 1984, pp. 14–21).

Leif J. Robinson, "The Frigid World of IRAS—I," *Sky & Telescope* (January 1984, pp. 4–8).

Ronald A. Schorn, "The Frigid World of IRAS—II," *Sky & Telescope* (February 1984, pp. 119–124).

PERIODIC COMET SHOWERS: THE MAINSPRING OF EVOLUTION?

by Piet Hut

Mass extinctions of animals and plants have taken place at various times during the Earth's history. Comet showers triggered by a companion star to the Sun might be the cause.

During the last ten years it has become more and more apparent that life on Earth has not evolved gradually; instead, occasional worldwide catastrophes have played a large role in the process of evolution. The previously accepted view of a gradual evolutionary process in which plant and animal species develop very slowly, only to make way after many millions of years for new species, is now seen as only one aspect of the development of life on our planet. Equally important, according to the newer theory, are relatively sudden changes that took place during periods of time of at most a few million years—very little time on a paleontological scale. During these periods of rapid development changes that took place within species were much more drastic than the accumulated gradual changes that had occurred over the previous tens of millions of years.

A look at past geologic eras supports this theory. The last 570 million years are divided into three main eras: the Paleozoic, the Mesozoic, and the Cenozoic. Each of these is further divided into a number of shorter periods of about 50 million to 100 million years. These divisions are based on the findings of various fossils in rock layers of different ages. Every boundary between two periods is matched to the disappearance of large numbers of species from the fossil record and the subsequent appearance of many new ones. This is especially marked at the boundary lines between eras; in both cases, Paleozoic-Mesozoic and Mesozoic-Cenozoic, more than half of all the plant and animal species for which there is fossil evidence died out within only a few million years. For a long time the explanation of these catastrophic boundary periods has been one of the great riddles of paleontology.

The limited scope of this article precludes any detailed discussion of the explanations that have been proposed for these extinction events. Through the years many were devised, none of which was clearly more convincing than any of the others. The question to be resolved was often phrased in reference to the largest of the recent extinction events: "What killed the dinosaurs?"

PIET HUT *is a long-term member of the Institute for Advanced Study at Princeton, New Jersey.*

(Overleaf) Illustration by Ron Villani

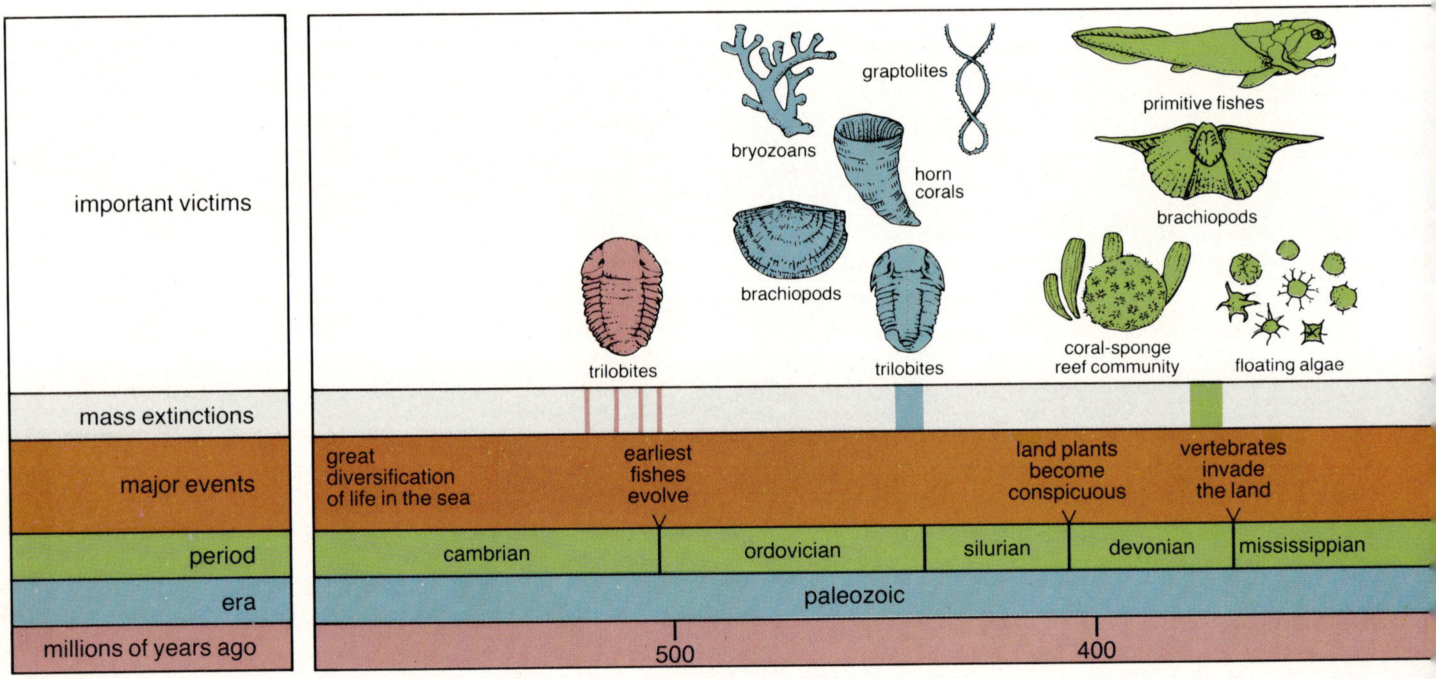

An answer to that question recently came from an unexpected quarter: research in nuclear physics. At the University of California at Berkeley an interdisciplinary team devised a method to date the various geologic periods more closely than had previously been possible. This team consisted of Luis Alvarez, a Nobel laureate in physics at Lawrence Berkeley Laboratory; his son, Walter Alvarez, professor of geology at the University of California at Berkeley; and Frank Asaro and Helen Michel, nuclear chemists at Lawrence Berkeley Laboratory. Their dating method involved measuring the amount of iridium present in rock. Iridium is a very rare element on the Earth's surface, because most of it sank to the core of the planet as the Earth cooled. Iridium is also rare in meteorites but not as rare as it is on the Earth's surface, since meteorites are composed of material very similar to the original material of Earth. The Alvarezes, Asaro, and Michel planned to measure the iridium present in particular strata and to determine from those measurements the number of micrometeorites that had rained down over millions of years. From this determination they hoped to infer how long it had taken for the strata to form, under the assumption that the rain of meteorites had remained constant throughout the history of the Earth.

Contrary to their expectations, the Alvarez team discovered an abnormally large amount of iridium in a very thin clay layer that marked the boundary between the Mesozoic and the Cenozoic eras; this layer closely coincides with the geologic time during which the dinosaurs died out. This amount of iridium, although still small in absolute value, was larger than normal amounts by a factor of about 100; it constituted about one-millionth of a percent of the total matter in the clay layer. The Alvarez team concluded that the massive extinctions at that time, 65 million years ago, were caused by the impact of an object from space with a diameter of a few miles. Such an asteroid or comet would contain enough iridium to spread a layer of it all around the Earth after the impact. A tremendous amount of energy must have been released by the impact of an object traveling at a speed of about 160,000 kilometers

Mass extinctions (colored lines and bars) have affected a large variety of marine plants and animals during the past 570 million years, ranging from single-celled algae and plankton to huge sea reptiles and whales. The best-known extinction took place at the end of the Cretaceous Period, about 65 million years ago. Most marine species were eliminated as were the dinosaurs on land. In some cases plant and animal groups recovered after crises and new species evolved, but other groups (bold type) vanished completely from the world's oceans.

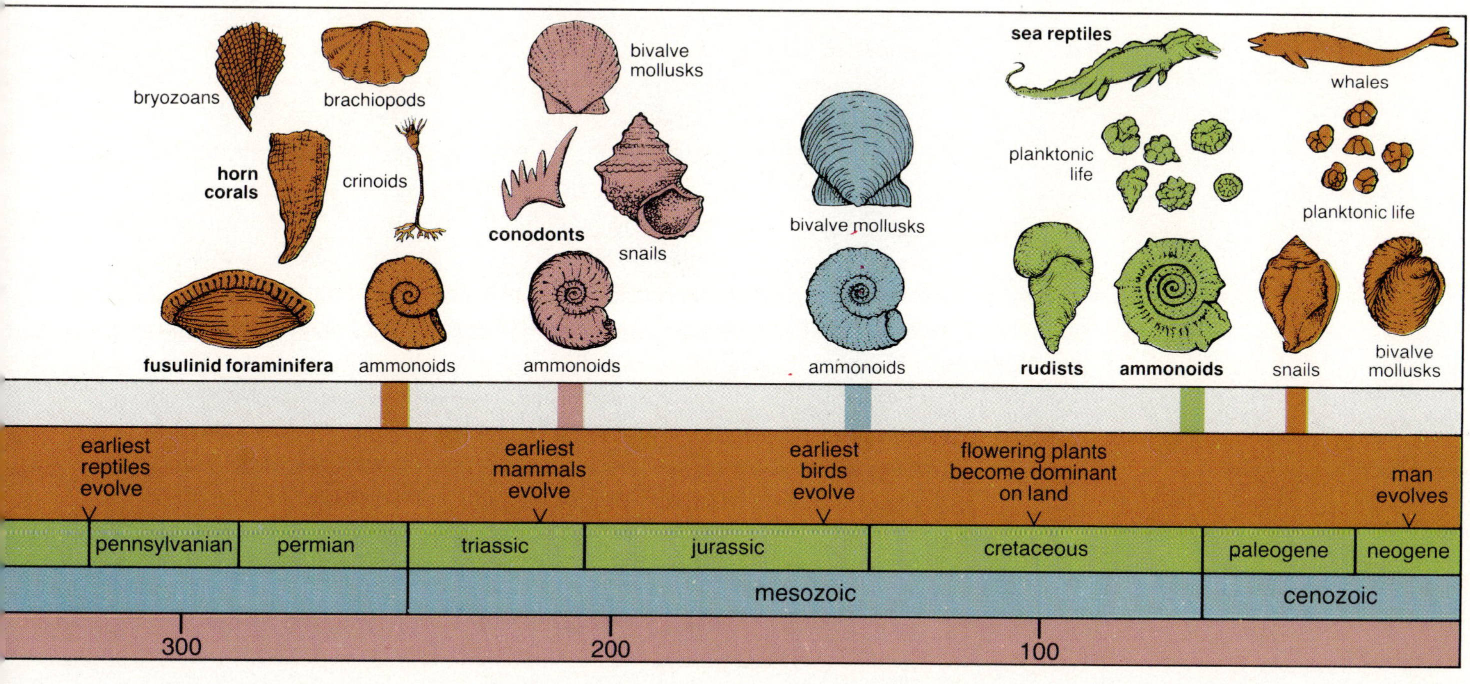

Lawrence Berkeley Laboratory

Walter Alvarez (second from right) holds a sample of clay from the layer that marks the boundary between the Mesozoic and Cenozoic eras, about 65 million years ago. Clay in this layer contains about 100 times more iridium than would normally be expected, leading to the conclusion that an extraterrestrial object rich in iridium struck the Earth at that time and caused the mass extinctions of dinosaurs and other animals. Alvarez is surrounded by other members of his research team (from left) Helen Michel, Frank Asaro, and Luis Alvarez.

(100,000 miles) per hour. It has been estimated that this energy must have been equal to 500 million megatons, an amount of explosive energy 10,000 times greater than that of all the nuclear weapons stockpiled by the United States and the Soviet Union combined. It is, therefore, not surprising that such an explosion could have deposited a worldwide layer of clay, nearly an inch thick, with an abnormally high iridium content. As it turned out, the predictions of the Alvarez team, made in 1979, were confirmed; within a few years in many places, on land and under the sea, similar layers also located at the Mesozoic-Cenozoic border were found to contain an unusually high percentage of iridium.

Impact of an asteroid or comet

What are the consequences for life on Earth of the impact of an interplanetary rock several miles in diameter? This question is unfortunately relevant to today's political situation, with tens of thousands of nuclear warheads in stockpiles ready to unleash a catastrophe not so different from that which might have destroyed the dinosaurs.

The consequences of the impact on Earth of a rock the size of a small mountain cannot be accurately predicted. One thing is certain, however; the Earth's atmosphere would be drastically altered, and it would require months at least for it to return to equilibrium. The circulation patterns of warm and cold currents in the air and the oceans would be disrupted; and no one can predict with any certainty how those disruptions would develop and what changes in patterns would occur before equilibrium was finally reestablished.

Calculating the effects of this impact on life on the Earth is even more complicated. At present, there is much controversy over the question of how and to what extent human activities affect the environment, involving such issues as acid rain and worldwide deforestation. It is difficult enough to make objective and comparable measurements of long-term developments in plant and animal populations; it is practically impossible to pinpoint which convoluted chains of ecological reactions have their origins in human activities.

Since it is so difficult to measure the consequences of human activity on the ecology of the planet, it is hardly surprising that scientists have no idea, not even a qualitative one, of the sudden and more dramatic effects of the impact of an extraterrestrial object. Many theories have been put forward, but their underlying assumptions vary so much that no idea can be formed of even the effect on world temperature. As an example of the uncertainties, even the direction of the change in temperature is a matter of controversy. Some theories predict a "greenhouse" effect. According to these, assuming that the comet or asteroid fell into the sea, so much steam would be added to the atmosphere that this damper atmosphere would absorb more infrared rays from the Sun and thus hinder the Earth's cooling. This would cause surface temperatures to climb, as the Earth's heat would be trapped under the air "blanket." Others predict that large amounts of dust and debris would be thrown up into the atmosphere by the impact; they maintain that even if the object landed in the sea, the force of the impact would blow part of the seafloor up into the air. This layer of dust would let in less sunlight than a clearer atmosphere, reflecting the Sun's heat away from the Earth with the result that surface temperature would drop sharply.

There can be no doubt that a collision between the Earth and a comet or asteroid would have serious consequences for life on Earth. After the discovery of a worldwide layer of iridium, deposited at roughly the same time as the extinction of the dinosaurs and therefore indicating that such a collision might have taken place, researchers naturally wondered whether there might be more such layers. The answer came in 1981, when the Alvarez team reported the discovery of a second iridium layer, in a rock stratum 34 million years old and thus in the later part of the Eocene Epoch (early Cenozoic Era)—also coinciding with a period of mass extinctions. In the spring of 1984 the discovery of yet a third iridium layer was announced. Chinese researchers found a high concentration of iridium in a rock layer at the Paleozoic-Mesozoic boundary, the time of one of the most severe mass extinctions (the trilobites, a group of marine arthropods, were one of the many types of animals that died out at that time).

These findings support the idea of a connection between impacts and extinctions. More support has come from a closer study of the first iridium layer discovered by the Alvarez team. This layer was found to contain, along with iridium, abnormally high percentages of gold and platinum, elements that also point to the impact of an asteroid or comet.

Periodic impacts?

In October 1983 two paleontologists at the University of Chicago, David Raup and John Sepkoski, announced a new discovery, one that put a different light on the question of mass extinctions. Sepkoski had worked for

Illustration by Don Davis; courtesy, Space Art International

One or more objects from space may have collided with the Earth at the end of the Cretaceous Period. Such impacts would have generated a huge dust cloud that prevented sunlight from reaching the Earth's surface, thereby suppressing photosynthesis and causing the death by starvation of the dinosaurs and many other forms of life.

years on an extremely detailed catalogue that included information about the extinctions of some 3,500 families of marine animals over geologic time. This extensive material clearly indicated several mass extinctions of those animals over the last 250 million years. The extinctions happened relatively quickly, during periods of at most a few million years and possibly much quicker.

The big surprise came when Raup and Sepkoski scrutinized this catalogue more closely, throwing out of their statistical analysis those families for which evidence was less solid. The result of this weeding-out process was that most of Sepkoski's work was put aside, leaving only 567 families for which the data were considered secure. The history of those families, now much better defined than in comparable studies, included the time of their origins, the rise and fall of the number of fossils from those families in the subsequent geologic periods, and the time at which each family died out. Raup and Sepkoski performed their statistical analysis on this select catalogue of families and found 12 different periods (each of which lasted at most a few million years) at which an abnormally high number of species died out. Most interestingly, these periods occurred at regular intervals, approximately once every 26 million years.

At first glance (and, indeed, at second glance as well), this result was puzzling. A few years earlier, in 1977, Alfred Fischer and Michael Arthur of Princeton University had made a similar suggestion of a periodicity in mass extinctions with a 32 million-year period. But without a quantitative analysis their ideas attracted few supporters. Now, however, the much stronger evidence for periodic extinctions could not be sidestepped, even though the connection between impacts and extinctions (not known to Fischer and Arthur in 1977) made such periodicity even more unlikely. If a time of mass extinctions was caused by the impact of an object from space, how could such extinctions be periodic? Both the object and the Earth circled the solar system for billions of years, and the distance between objects in the solar system is enormously greater than the diameters of the planets. Collisions may occasionally occur between comets or asteroids and planets such as the Earth, if the smaller objects cross the orbital paths of the larger ones. These collisions are the result of such complicated interactions among orbiting objects that, practically speaking, they cannot be predicted over time scales of several million years or longer. For this reason almost all scientists had considered it obvious that the impacts that had caused extinctions on Earth were random and unpredictable disasters.

A companion star?

A possible explanation of the puzzling 26 million-year periodicity of mass extinctions caused by impacts was put forward in December 1983, shortly after Raup and Sepkoski's findings were made available. Two groups of astrophysicists independently proposed that the Sun might have a companion star with such a wide orbit that it would take 26 million years to complete a full revolution around the Sun. This hypothesis is partially based on the fact that most stars in the universe are double stars; if our Sun is a single star, it is an exception. One version of the companion star theory was put forward by Daniel Whitmire of the University of Southwestern Louisiana and Albert Jackson of the Computer Sciences Corp. in Houston, Texas. The other version was proposed by Marc Davis and Richard Muller of the University of California at Berkeley along with Piet Hut of the Institute for Advanced Study in Princeton, New Jersey.

How could a double star be a sort of alarm clock for periodic mass extinctions? The central idea in this theory is that comets and not asteroids are responsible for the impacts that cause the extinctions. Most comets have very wide orbits that reach far beyond the planetary system; the farthest point in those orbits can be up to 1,000 times the distance

D. J. Roddy, U.S. Geological Survey

Barringer (or Meteor) Crater in Arizona, 180 meters (600 feet) deep and 1.2 kilometers (0.75 mile) wide, provides evidence that objects from space have collided with the Earth. Scientists estimate that a small asteroid about 45 meters (150 feet) in diameter created the hole some 25,000 years ago.

from the Sun to the outermost planet, Pluto. A typical observed comet normally passes through the planetary system once every million years or so. Most comets, however, never come close to the planetary system and slowly traverse the outer regions of the solar system on their wide orbits. Perturbed by passing stars, the cometary orbits will drift, and every now and then an orbit will intersect with the inner planetary system, bringing a "new" observable comet on its first passage close to the Sun. In fact, based on the appearances of new comets on extremely elongated orbits, the astronomer Jan Oort at Leiden Observatory in The Netherlands in 1950 proposed the existence of a large "cloud" of comets beyond the realm of the planets but still within the sphere of the Sun's influence. The existence of the Oort comet cloud, estimated to contain over 100,000,000,000 comets (and possibly 10 or 100 times more), is now generally accepted.

If there were a lightweight "second Sun" orbiting around our Sun, so far away that it might appear too faint even to be seen with binoculars, how would its orbit develop, and how would it affect the comets? These questions can be studied in turn.

A companion star that revolves around the Sun in a wide elliptical orbit once every 26 million years has been proposed as the explanation for the 26-million-year periodicity of mass extinctions on the Earth. According to the theory the companion passing near the Sun would cause the orbits of nearby comets to be shifted so that they would come much closer to the planets. Some of these comets would collide with the Earth, generating the huge dust clouds that cause the extinctions.

Illustration by Ron Villani

For the solar companion star to have an orbital period as long as 26 million years, its orbit would have to be very wide indeed, extending a large part of the way toward the nearest stars. (One light-year equals 9,460,500,000,000 kilometers or 5,878,000,000,000 miles). At present, the companion star is thought to be at a distance from the Sun of slightly more than two light-years, and it will approach the Sun to within less than half a light-year around 15,000,000 AD—the time at which the next extinction is expected. An orbit that takes the companion star light-years away from the Sun is not as stable as the much smaller planetary orbits. Many stars will pass close by such a companion, and once in about a million years a random passing star might even pass through its orbit around the Sun. This would cause the orbit of the companion to change slowly in an erratic fashion and, therefore, in the course of a billion years the companion would have changed its course considerably. Fortunately, though, even on such a slowly tumbling and jittering orbit a companion star has a negligible chance of entering the solar system. Such an intrusion would disturb the orbits of all the planets and would thereby induce significant temperature variations on the surfaces of the planets.

Because a companion star with a wide orbit is unlikely to penetrate the planetary system, it would not influence the orbits of the asteroids, which are located inside that system. Comets, on the other hand, orbiting well outside the planetary system, would be affected by the companion when it passed close to the Sun. It is relatively simple to estimate how many comets would be disturbed by such a passage. Such an estimate shows that the frequency of comets entering the planetary system during the companion's passage might easily be 100 times larger than normal. Presently, only a handful of new comets is detected each year, of which only one every few years is easily visible to the naked eye. During the passage of the companion near the Sun conditions would be rather different, since the perturbing influence of the companion would have increased the comet flux significantly during a period of about a million years. The above numbers would rise to a few new comets every day and a spectacular new comet about once a week. If there were observers during such a passage of the companion star, on a clear night such persons would be able to see several comets scattered across the sky without having to use a telescope.

Such a spectacle of several brilliant comet tails filling part of the sky might have poetic beauty, but it would not be harmless. Each comet that came close enough to the Sun to cross the Earth's orbit would have a tiny chance of colliding with the Earth itself. This chance would be no greater than one in a billion, but during the million years needed for the companion to pass by the Sun more than a billion comets would come close enough to the planetary system to threaten the Earth. Thus during that million years there would be a clear possibility of several collisions.

The theory that the Sun has a companion star that is responsible for a periodic shower of comets is not the only explanation proposed for periodic mass extinctions. Alternate theories have been put forward by Michael Rampino and Richard Stothers of the Goddard Institute for

Illustration by Ron Villani

Another explanation of mass extinctions involves the motion of the Sun perpendicular to the plane of the Milky Way Galaxy. Approximately once every 30 million years the Sun passes through the mid-plane of the Galaxy, a periodicity that closely matches that determined for the extinctions. At the mid-plane the Sun moves close to dense gas clouds. These clouds, some scientists believe, trigger comet showers on the Earth and other planets.

Space Studies in New York City, and also by Richard Schwartz and Philip James of the University of Missouri. Both of these theories explain the periodic impacts and mass extinctions in terms of the Sun's passage through the galactic plane. Roughly once every 60 million years the Sun completes an up-and-down movement inside the outer regions of the Milky Way Galaxy. This is a small perturbation in its main orbit of revolution around the galactic center, a revolution that takes more than 200 million years to complete. In the course of the complete oscillation perpendicular to the galactic disk, the Sun passes through the mid-plane of the Galaxy twice, once every 30 million years. This periodicity seems to match approximately that found by Raup and Sepkoski.

Rampino and Stothers explained the periodic comet showers as consequences of the solar system's passage close to the dense gas clouds that are most likely to reside close to the galactic mid-plane. The difficulty with this explanation is the fact that the Sun at present is passing through this mid-plane, but the next catastrophe is not expected for another 15 million years. The fact that the passage of the Sun through the galactic plane is thus out of phase with the periodic catastrophes was enough for Davis, Hut, and Muller to discard this explanation in favor of the theory of a companion to the Sun. Schwartz and James also realized this difficulty but proposed a different argument: each time the Sun is farthest from the galactic plane, the solar system might be more exposed to X-rays by lack of galactic shielding. They did not give quantitative estimates for this effect, however, and it is by no means clear that there exist X-rays strong enough to make a difference for life on Earth (nor did they give an explanation for the iridium anomalies).

At present the theory of a companion star has passed the tests put to it, but it does lack one major piece of evidence—the companion has not yet been found. In this sense the companion star theory is the most falsifiable of those so far proposed. If the companion is not found after several years of intensive search and if this search is complete enough to preclude the existence of a solar companion heavy enough to cause sizable comet showers, then the theory will lose credibility. On the other hand, if the companion is found, the theory will be vindicated.

Further developments

The theory of periodic comet showers was generally received with skepticism when it was proposed in December 1983. However, this is a normal reaction in science to new speculations; the first thing to do with a new idea is to try to shoot it down, and only when no serious flaws can be found can a theory be taken seriously. To date, the theory has stood up well against all attacks. And, what is more, within a month of the proposal, partial confirmation was obtained from additional geologic evidence. Walter Alvarez and Richard Muller decided to use the available literature to investigate the age distribution of impact craters on the surface of the Earth. They found indications for a periodicity of 28 million years in the ages of previously dated craters, a time span well within the range of uncertainty of the species extinction periodicity discovered by Raup and Sepkoski. The crater ages were also in phase with the mass extinctions, further evidence that a greater chance of impacts and accomanying mass extinction can be expected around 15,000,000 AD.

One month before the publication of all the relevant articles on the companion star theory in the April 19–25, 1984, issue of the British journal *Nature,* a conference was held in Berkeley on the subject of periodic comet showers. Organized by Luis Alvarez, Frank Asaro, Helen Michel, and David Raup, the meeting was a change from the usual scientific conferences, which tend to deal with highly specialized subdivisions of already specialized fields of research. Present at this conference were paleontologists, geophysicists, astronomers, nuclear physicists, chemists, and others. Afterward, several groups began the search for the companion star, and theorists began working to refine the details of how the theory works. These calculations are quite complex, involving the use of much computer time to determine the exact interactions between the Sun, the companion, comets, and planets.

The last word is not yet in on periodic extinctions and their causes. Whichever explanation proves to be correct, there can be no doubt that the consequences of these extinctions on life on Earth, and on its evolution, are far-reaching. This was underscored by Raup and Sepkoski in the closing lines of their original article, published in February 1984:

> The implications of periodicity for evolutionary biology are profound. The most obvious is that the evolutionary system is not "alone" in the sense that it is partially dependent upon external influences more profound than the local and regional environmental changes normally considered. Much has been written about the "bottlenecking" effect of mass extinction. With kill rates for species estimated to have been as high as 77% and 96% for the largest extinctions, the biosphere is forced through narrow bottlenecks and the recovery from these events is usually accompanied by fundamental changes in biotic composition. Without these perturbations, the general course of macroevolution could have been very different.

THE INFLATIONARY UNIVERSE

by Marek Demianski

Advances in modern cosmology and high-energy physics suggest that in its earliest instants, just a fraction of a second after the big bang, the universe passed through a critically important phase of enormous expansion.

Modern study of the universe began with the discovery that the Milky Way is only one of very many galaxies. About 1915 Vesto Slipher, working at the Lowell Observatory in Flagstaff, Arizona, noticed that the light received from galaxies is shifted to varying degrees toward the red end of the spectrum. He interpreted this displacement as due to the motion of distant galaxies away from the Milky Way; moreover, the amount of displacement was seen to be directly proportional to the velocity of recession. In acoustics a similar shift in the observed frequency of sound, called the Doppler shift, had been known for some time. Shortly thereafter Edwin Hubble and M. L. Humason at Mount Wilson Observatory in California combined Slipher's information with their own meticulous distance measurements for dozens of galaxies to show that distant galaxies are receding with velocities proportional to their distance. Hubble also noticed that this expansion is isotropic—*i.e.,* that it is going on in the same way no matter which direction one looks—and that over a sufficiently large scale galaxies are distributed uniformly. The discovery that the universe as a whole is expanding ranks among the most important discoveries of the 20th century.

The idea of an evolving universe prompted the physicist and cosmologist George Gamow in the 1940s to ask a profound question; namely, how were the different chemical elements created in the course of cosmic evolution? To explain the present abundance of elements, Gamow reformulated the big bang model in its modern version, postulating that the early universe was a very dense, very hot, and expanding fireball. Although Gamow failed to explain the observed abundance of elements with this model, he made an important prediction. If the early universe was indeed very hot, then even now there should be observable relics of the primeval fireball in the form of microwave radiation filling all space—a slight warmth lingering in a once very hot oven.

In 1965 exactly such a radiation was unexpectedly discovered by Arno Penzias and Robert Wilson at Bell Laboratories. When its spectrum was measured, it was found to have a thermal character corresponding to radiation at a temperature of three degrees kelvin (three degrees above absolute zero). Subsequent measurements of this relic radiation has provided important cosmological information. Its temperature taken in different parts of the sky is almost exactly the same, indicating that the universe at the moment when the radiation was emitted was extremely uniform.

The expansion of the universe and the existence of the microwave background radiation are the most important observational foundations of the standard big bang model. In order to develop this model and study the history of the universe, one needs additional ingredients: the general theory of relativity, thermodynamics, and elementary particle physics. General relativity provides dynamical equations that relate the mean energy density (the amount of energy in a unit volume of space) of the universe to its global structure; under very general conditions it "predicts" that the universe evolved from a singular state characterized by infinite density, infinite temperature, and zero volume. This singular

__MAREK DEMIANSKI__ is Professor of Physics at the Institute for Theoretical Physics, University of Warsaw. He recently served as Visiting Professor at the Center for Relativity, University of Texas, Austin.

(Overleaf) Illustration by Ron Villani

state occurred about 10 billion–18 billion years ago. Thermodynamics and elementary particle physics are needed to relate the temperature of the universe to the mean energy density and mean pressure.

Revelations of the standard big bang model

Using these various tools and observations, the standard big bang model attempts to reconstruct the history of the universe starting from the singular state. At first the universe expanded very fast, but at high densities and temperatures processes among elementary particles also occurred very fast, and matter was in a state of thermal equilibrium, everywhere absorbing and emitting radiation at the same uniform rate. It is not clear if matter just immediately after the big bang was in thermal equilibrium, but there are good reasons to believe that it was quite close to being so. At thermal equilibrium the composition of matter is uniquely determined by the temperature, the masses of elementary particles, and the strength of nuclear interactions.

Ideally, the state of thermal equilibrium requires that there should exist as many particles as antiparticles. As far as is known, however, the present universe is composed almost exclusively of matter rather than antimatter. For the standard big bang model to account for this discrepancy, it must include an assumption that initially there was a very small asymmetry in the composition of matter and that particles slightly outnumbered antiparticles.

The elementary particles of matter found in nature can be divided into two families: leptons and hadrons. There are only a few kinds of leptons, the electron being the best known. Hadrons are more numerous; protons and neutrons—the elementary constituents of atomic nuclei—belong to this family. It now seems well established that all hadrons are composed of a much more limited number of entities called quarks.

During the first millionth of a second (10^{-6} second) after the big bang, the universe expanded and cooled so dramatically that most very massive particle-antiparticle pairs annihilated and in this process were transformed into photons. The matter that remained was composed mostly of quarks and antiquarks, electrons and positrons (antimatter electrons), and leptons called neutrinos. This is the earliest epoch that can be quite reliably described by the presently known and well-established laws of physics. What happened earlier, say in the first thousandth of a billionth of a billionth of a billionth of a second (10^{-30} second), is more speculative, but as will be discussed below, it is also the subject of some of the most exciting work in cosmology in the past five years.

Near the end of the first millionth of a second of expansion, quark-antiquark pairs annihilated and a small surplus of quarks combined to form protons and neutrons. Sometime during the next second or so electrons and positrons annihilated, leaving a small surplus of electrons to counterbalance the remaining protons and so preserve the electric neutrality of the universe. During the next two or three minutes the universe cooled sufficiently to allow protons and neutrons to combine and form light atomic nuclei. It turns out that the universe was not sufficiently dense and sufficiently hot for a long enough period to form elements heavier than helium. The big bang model predicts that the primordial matter should have contained about 25% helium and 75% hydrogen by weight. This prediction has been well confirmed by observations of elemental abundance in the present universe.

After this early activity the universe entered a long period of uneventful expansion—the primeval fireball. Its temperature was still very high and the energy density of radiation was much larger than the energy density of matter. The universe now grew much more slowly.

By the time the universe was about 100,000 years old the temperature had dropped to about 3,000 degrees kelvin, and the number of high-energy photons had decreased enough to allow electrons and nuclei to form atoms. This was the last time when radiation had a chance to interact with the prevailing "fog" of independent matter particles. Starting from the moment of formation of atoms, the universe became transparent to the radiation, and photons propagated freely. Photons making up the microwave background radiation that is observed today were emitted at that time. At about the same time the energy density of matter became equal to the energy density of radiation; subsequently the evolution of the universe was dominated by matter.

Over the intervening billions of years, radiation cooled down toward its present temperature of three degrees kelvin. Matter became more and more diluted. Slight irregularities or perturbations in density existed in the early universe, and as the universe expanded these irregularities grew. Eventually the excess matter in them accreted under mutual gravitational attraction, forming distinct objects. At the present time it is not clear if separate galaxies emerged first and condensed into clusters of galaxies, or if clouds of matter containing as much as several thousand

The standard big bang model attempts to reconstruct the early history of the universe from the singular state, but it has little to say about conditions before the first millionth of a second. By then the universe was cool enough to allow quarks to form protons and neutrons, which circulated with leptons and highly energetic photons in a kind of "soup." A second or so later particle energies dropped sufficiently for protons and neutrons to combine into nuclei of hydrogen and helium. After about three minutes and for the next 100,000 years the universe expanded uneventfully, dominated by the energy density of radiation. At that time the temperature became low enough to permit nuclei to capture and hold orbiting electrons, thus forming atoms. About the same time photons lost enough energy to bring the energy density of radiation down to, and then below, that of matter. Subsequently the evolution of the universe was dominated by matter.

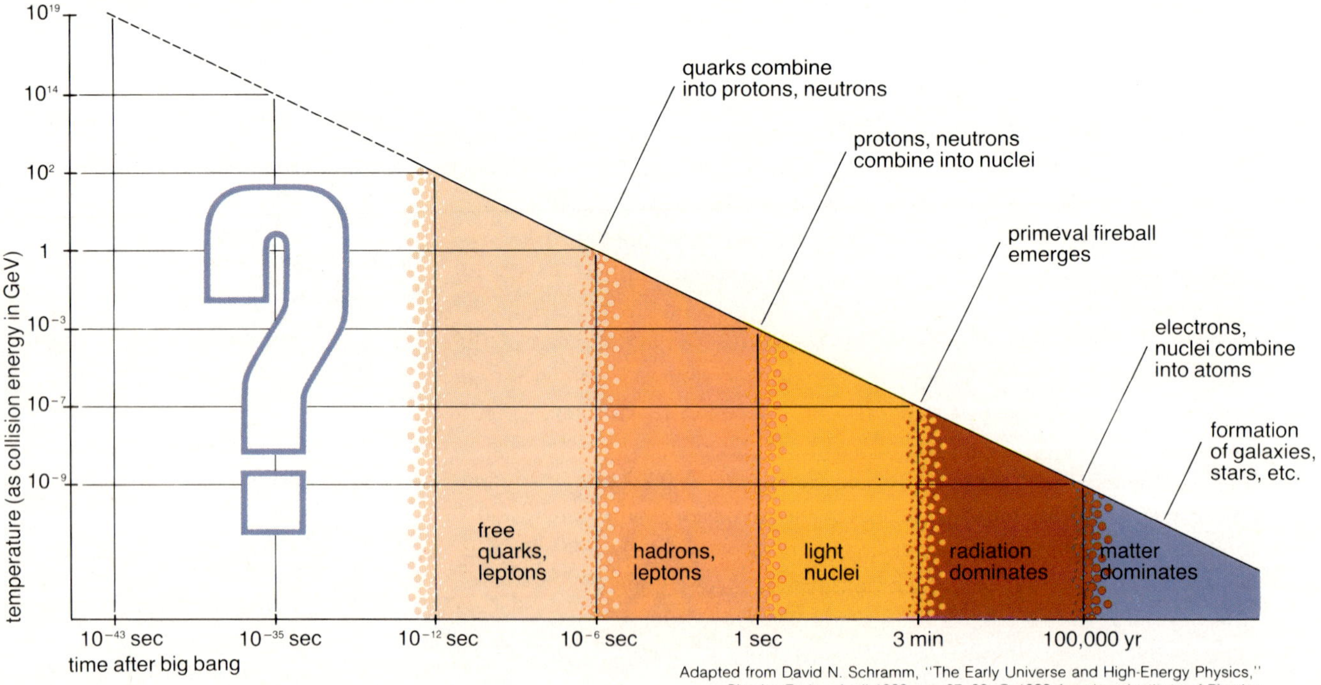

Adapted from David N. Schramm, "The Early Universe and High-Energy Physics," *Physics Today*, April 1983, pp. 27–33, © 1983 American Institute of Physics

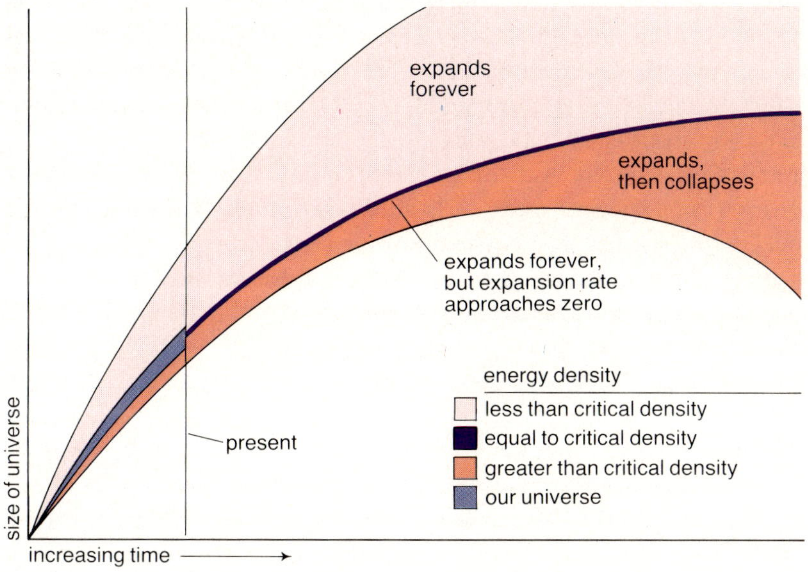

According to Einstein's formulation of general relativity, the fate of the universe depends on its mean energy density. If that density is less than a certain critical value, the universe will expand forever. If greater, the universe will stop expanding and collapse. If the mean energy density is just equal to the critical density, the universe will continue expanding, but its expansion rate will become slower and slower. Attempts to set limits on the mean energy density of the presently observed universe place its value somewhere between one-tenth and twice the critical density. This uncertainty has kept the future of the universe shrouded in mystery. Another intriguing question is why the mean energy density should be so close to the critical density. The big bang model does not answer this question but assumes the close correspondence as an initial condition.

galactic masses—corresponding to clusters of galaxies—formed first and later fragmented into galaxies. In either case, primordial gas finally condensed into stars. Thermonuclear reactions occurring in stars transformed hydrogen and helium into heavier elements. Stars more massive than our Sun evolved relatively quickly. Exploding at the endpoint of their evolution, they polluted the universe with heavy elements. In the throes of their explosive deaths, still heavier elements were synthesized. Dispersed clouds of hydrogen and helium containing a small admixture of heavy elements cooled down. The center of one of these clouds collapsed to form the Sun and the rest of the gas formed a disk that fragmented into planets, including the Earth.

When weight means fate

In the standard big bang model the future evolution of the universe depends on its mean energy density. According to general relativity, the mean energy density affects the structure of three-dimensional space. If it is higher than a certain critical value, space curves back on itself like a sphere; as a consequence the universe will eventually stop expanding and start contracting. If the mean energy density is lower than the critical density, space also curves but not back on itself; the universe keeps expanding forever. If the mean energy density is just equal to the critical density, the geometry of space is flat; the universe will continue to expand forever but just barely. While it is clear that determining the value of the mean energy density—in essence, weighing the universe—is crucial to predicting its ultimate fate, that value remains the most uncertain of the main cosmological parameters.

There are indirect ways to estimate the mean energy density. For example, one can base a value on the amount of luminous matter that can be observed. There are indications, however, that even our Galaxy contains much more matter than can be inferred from the amount of light it produces. Dynamical methods of mass determination involving the motions of galaxies in clusters also give much larger values than estimated from the observable matter. Nevertheless, precise determinations of the abundance of light elements, in particular deuterium and lithium, have been made, and these values place limits on the mean density of protons and neutrons in the universe. So even though the mean energy density is not known exactly, it is almost unanimously accepted that it could not be smaller than one-tenth of the critical density and not larger than twice the critical density. Yet, because of the uncertainty, the future evolution of the universe remains a great unknown.

Shortcomings of the standard model

The standard big bang model, coupled with a few simple assumptions, describes amazingly well the evolution of the universe. It fails, however, to answer some intriguing questions. It does not explain, for example, why the universe is so uniform on a large scale and so clumpy—with its planets, stars, galaxies, clusters, and large voids of empty space—on a small scale. It sheds no light on the origin of the small fluctuations that gave rise to galaxies and clusters of galaxies but rather postulates them as an initial condition.

When astronomers turn their instruments toward any two patches of sky separated by more than a few degrees, the photons of the microwave background radiation that they observe come from parts of the universe that were never before in causal contact; *i.e.,* close enough to affect each other. In principle, their energies could substantially differ, yet they turn out to be the same. In the big bang model it is simply postulated that initially radiation and matter were distributed uniformly.

The restrictions imposed on the value of the mean energy density of the present universe set it very close to the critical density. For this to be so, it must have been almost exactly equal to the critical density at the very early stages of evolution; even a small deviation would have become greatly magnified later on. Again, in the big bang model this fact is not explained but merely assumed as an initial condition.

The search for basic unities

As the evolution of the universe is pursued farther back in time, the contributions of particle physics becomes ever more significant. In fact, advances in both endeavors during the past decade show that the largest and smallest scales of the universe are intimately related. On the other hand, this synthesis has also pointed out additional difficulties in the standard big bang picture.

Four forces, or interactions, between the building blocks of matter are known to underlie practically all the complex physical phenomena in nature: the gravitational, electromagnetic, strong, and weak interactions. According to present understanding, all the interactions are mediated by the exchange of force, or field, particles. For example, two particles that carry electric charge, such as electrons, exert a mutual electromagnetic force by continuously exchanging photons. Photons are thus the elementary quanta of the electromagnetic field. The weak and strong interactions are also well described in terms of their own field particles, and so far only the gravitational interaction resists a quantum description. Its hypothetical field particle, the graviton, has not been discovered. (*See* table.)

The gravitational interaction is universal; all kinds of matter gravitate. Nevertheless, it is the weakest of all the forces, playing a dominant role only on an astronomical scale and in cosmology.

The electromagnetic interaction affects all charged particles. Although it is much stronger than the gravitational interaction, matter on a large scale is electrically neutral, and so the electromagnetic interaction plays almost no role in the large-scale structure of the universe and in cosmology. On atomic and molecular scales, the electromagnetic interaction is important, keeping atomic electrons in orbit around their nuclei and providing all the chemical bonds.

The strong interaction is responsible for the structure of all hadrons. In the quark model, every hadron including nuclear protons and neutrons is composed of quarks, and it is the strong interaction among the quarks that keeps them confined as hadrons. A less direct consequence of the strong interaction is the nuclear force that binds protons and neutrons together to form atomic nuclei.

The weak interaction is perhaps the most perplexing of all. As mentioned above, matter is composed of quarks and leptons. Quarks interact with each other by means of the strong interaction, quarks and charged leptons interact electromagnetically, and all of the fundamental particles engage in weak interactions. Weak interactions are responsible for the beta decay of radioactive nuclei, which seems to be the only noticeable effect of these interactions in the familiar natural environment. When processes governed by weak interactions are investigated more closely, it turns out that they violate some fundamental symmetry principles that were considered universal. For example, weak interactions distinguish between left- and right-handed orientations of frames of reference.

Looking at the list of different interactions one might wonder why nature requires such diversity, or put another way, why there should not be some underlying unity. Not so long ago electricity and magnetism were treated as separate interactions. Today it is known that at the elementary level electric currents are responsible for magnetic phenomena.

fundamental interactions

type	relative strength	field particle exchanged	particles affected	range
strong	1	gluon	quarks, hadrons	10^{-13} cm
electromagnetic	10^{-2}	photon	electrically charged particles	infinite
weak	10^{-13}	intermediate vector boson (W, Z)	all	less than 10^{-14} cm
gravitational	about 10^{-38}	graviton (hypothetical)	all particles with mass	infinite

Photos, courtesy, CERN

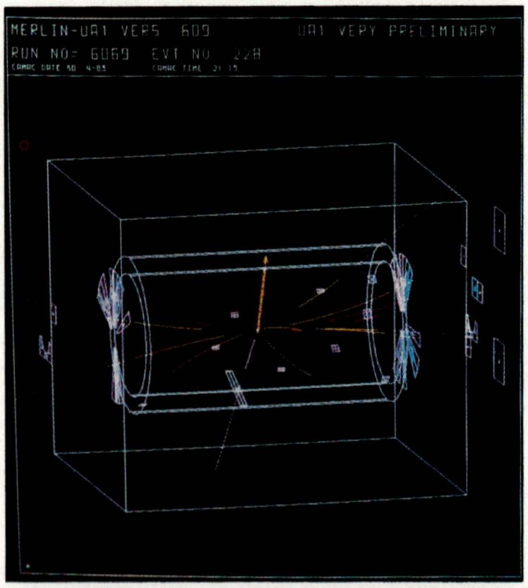

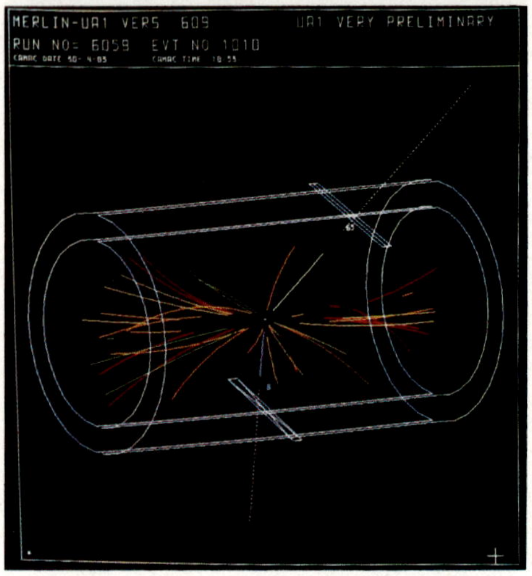

Track-detector images of proton-antiproton collisions at the CERN collider in Geneva provide evidence for the creation of W and Z particles, the field particles of the weak interaction. The purple track of a high-energy electron, visible amid a spray of collision debris (top), is a characteristic signature for the decay of a W particle. Likewise, the long blue and white tracks, denoting a high-energy electron-positron pair (above), are a telltale signature of a Z decay. The discoveries of the W and Z strikingly confirmed the electroweak theory, which suggests that the electromagnetic and weak interactions were indistinguishable at the extremely energetic conditions that existed in the very early stages of the universe.

The theoretical unification of electricity and magnetism by James Clerk Maxwell was a great achievement of 19th-century physics. It provided a consistent description of all known electromagnetic phenomena and led to new predictions, the most important being the existence of electromagnetic waves.

Another giant step in the search for unity in all the interactions in nature was made in the 1960s by theorists Steven Weinberg and Abdus Salam, who independently proposed a combined model of weak and electromagnetic interactions. In this model, at very high energies (about 100 billion electron volts, or 100 GeV) the weak and electromagnetic interactions are indistinguishable and are collectively termed the electroweak interaction. At lower energies the symmetry between interactions is broken, and the electromagnetic and weak interactions acquire their well-known properties. The electroweak theory of Weinberg and Salam, and later generalized by Sheldon Glashow, predicts the existence and masses of the field particles for the weak interaction. In 1983 these particles, the W and Z, were discovered in particle collisions that made available a total energy of 540 GeV, and estimates of their masses agreed well with the predictions.

Grand unified theories

Even before this great success of the unification program several models had been proposed that combine the strong, electromagnetic, and weak interactions. These models, known as grand unified theories (GUT's) predict that at extremely high collision energies the strong, electromagnetic, and weak interactions are indistinguishable. Because the energy required to achieve unification is clearly unobtainable given the current level of particle-accelerator technology, there is little hope that these theories will be accessible to direct laboratory tests in the foreseeable future. It is also impossible to find a present-day astronomical environment in which particles with such high energies could be created. As far as is known, only the very early stages of the universe, when it was 10^{-35} second old, could have provided a natural playground for extremely high-energy elementary-particle physics.

At collision energies higher than the critical value needed to achieve grand unification, estimated to be as high as a thousand trillion (10^{14}) GeV (11 orders of magnitude greater than the capabilities of the most powerful accelerators), all the interactions except gravity should have the same strength. Below this energy the symmetry between interactions is broken; the strong and electroweak interactions are distinct.

Spontaneous symmetry breaking has been known in physics for a long time. For an example, one can imagine the surface of a big lake early in winter. When the temperature of the water is above freezing, conditions at the surface at different points are the same. As the water slowly cools down, a small ice patch suddenly appears somewhere, but one cannot predict exactly where. Symmetry is now spontaneously broken, for not all points on the surface of the lake are indistinguishable. Soon other isolated ice patches may appear, distinct from the first. They expand until their edges meet, at which time the entire surface of the lake is transformed to ice.

In the grand unified theories spontaneous symmetry breaking is handled mathematically by including some additional fields, called Higgs fields, in the description. In the symmetric phase, at energies higher than the grand unification energy, all the Higgs fields have a value of zero. As the energy falls below the grand unification energy, at least one of the Higgs fields takes on a nonzero value. With this event symmetry is spontaneously broken, and the energy state of the system of particles being described drops rather abruptly to a stable state called the true vacuum. (In physics, the term vacuum does not mean an absolute void—even the vacuum of intergalactic space contains radiation and particles—but the lowest energy state that can be attained by a given collection of particles. Because physical systems have a natural tendency to seek and stay at their minimum available energy state, vacuum states are stable.)

When the grand unified theories are integrated into the standard big bang picture, they indicate that the universe cooled to energies below the grand unification energy 10^{-35} second after the big bang. At that time spontaneous symmetry breaking in the high-energy state ushered the universe into the true vacuum state, in which the strong and electroweak interactions gained their separate identities.

Although the application of GUT's to the standard big bang model has given a deeper insight into the evolution of the universe, it is not without problems. According to the special theory of relativity, it is not possible to send signals faster than the speed of light. This requirement puts a minimum limit on the time that must pass for one event to influence another, or be causally connected. In the big bang model, by the time spontaneous symmetry breaking occurs, the universe is already the size

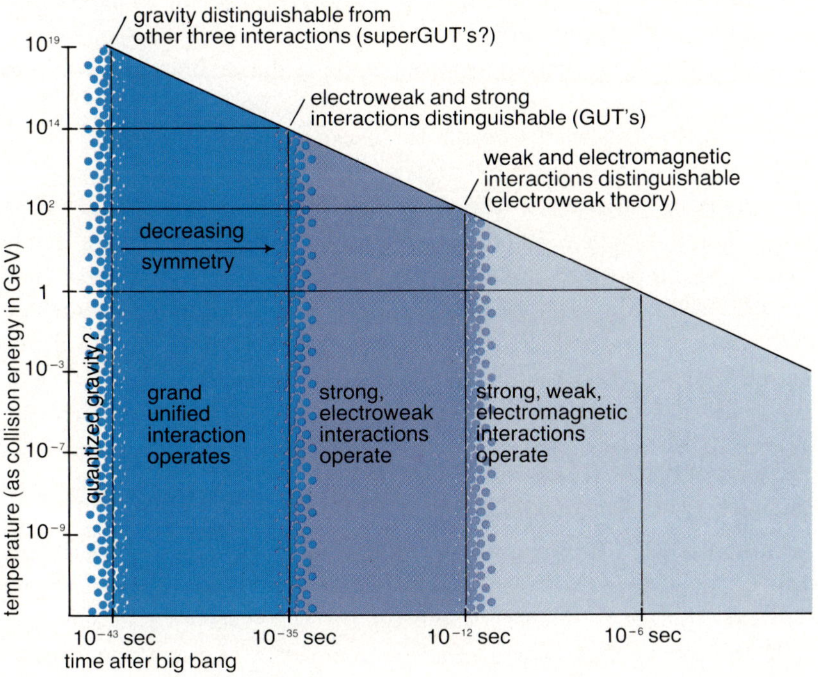

Recent advances in the search for unity among the fundamental forces of nature have helped extend the big bang picture into the past by some 30 orders of magnitude. The electroweak theory, confirmed by experiments that achieved particle collision energies believed to exist when the universe was only 10^{-12} second old, indicates that the weak and electromagnetic interactions were identical before that time. Grand unified theories (GUT's) carry this primordial unity back another stage, to an age of 10^{-35} second and a collision energy as high as 10^{14} GeV, when the strong, weak, and electromagnetic interactions were all one. Today gravity appears to be the final holdout; to bring it into the fold will require a quantum gravity or superGUT's theory that can describe conditions in the first 10^{-43} second after the big bang.

of a grapefruit, about ten centimeters (four inches) in diameter—large enough that vast numbers of adjacent regions are already out of causal contact. Since any coordination between these regions is impossible, spontaneous symmetry breaking can lead to a different physical state in every such region. As two adjacent regions of true vacuum state expand and meet, their incompatibility would create at least two kinds of massive defects along places of contact: pointlike defects, corresponding to magnetic monopoles, and surfacelike defects called domain walls. Magnetic monopoles are particles postulated to carry isolated north or south magnetic charges. In classical electromagnetic theory there are no magnetic charges, and magnetic phenomena are linked to the motions of electrically charged particles. Grand unified theories, on the other hand, predict free magnetic charges and magnetic monopoles. A monopole should be extremely heavy, about 10^{16} times the mass of a proton, or about the mass of an amoeba.

The difficulty with the prediction of these defects is that it violates present observational data. The mean density of the universe after its transition to the true vacuum state should be dominated by magnetic monopoles and domain walls and should be many orders of magnitude larger than the critical density, enough to halt expansion of the universe and bring about a collapse soon after its birth. Not only is this obviously not the case, but the existence of even a single monopole has yet to be confirmed.

Two schematic sequences on the facing page, based on differing models, follow the early evolution of the universe as it cools below the grand unification energy and depict the consequences of that evolution. In the sequence on the left, representing the big bang model extended by GUT's, the new universe is assumed to be hot, i.e., above the grand unification energy, and expanding. As the universe cools, spontaneous symmetry breaking generates independent regions of true vacuum state. Below the grand unification energy, these regions continue expanding but coalesce imperfectly, creating massive defects such as magnetic monopoles and domain walls where they meet. The resulting universe should be dominated by these defects to the extent that the mean energy density far exceeds the critical density. The inflationary model, on the right, requires that only one region of the universe be hot and expanding. As this region expands and cools past the grand unification energy, it becomes trapped in a false vacuum state and swells exponentially. About 10^{-32} second later, the inflated region makes a comparatively smooth transition to the true vacuum. Because a single inflated region could grow much larger than the entire observable universe, any defects produced during the transition would either be diluted by the inflation or be carried far beyond observation. Such a defect-free universe is compatible with present-day observation.

The inflationary scenario

In their search to resolve the outstanding questions and discrepancies in the standard big bang model, many investigators have become intrigued with a concept first proposed in 1980 by Alan Guth while working at Stanford University and subsequently improved by others. Called the inflationary model, it presents a scenario of the universe in its spontaneous symmetry-breaking phase that solves many of the problems in a natural way.

The inflationary model begins by focusing on a small region of the universe that is hot (above the grand unification energy) and expanding. As a result of expansion this region cools down until at a certain moment it reaches the critical grand unification energy. At this point it would become favorable for some of the Higgs fields, which have been equal to zero, to acquire a nonzero value and carry the expanding region to a new lower energy state. Instead, however, the expanding region remains trapped for a time in the previous energy state, because its Higgs fields are held at or near the zero value. This metastable state, termed the false vacuum, is analogous to the supercooled state of water: under the right conditions water will remain a liquid below its freezing point even though its solid form represents a lower energy state.

As the region continues to cool, the constant energy difference between the false and the true vacuum states drives a powerful expansion process. The region of space trapped in the false vacuum state suddenly balloons out, swelling exponentially at a rate that is proportional to the

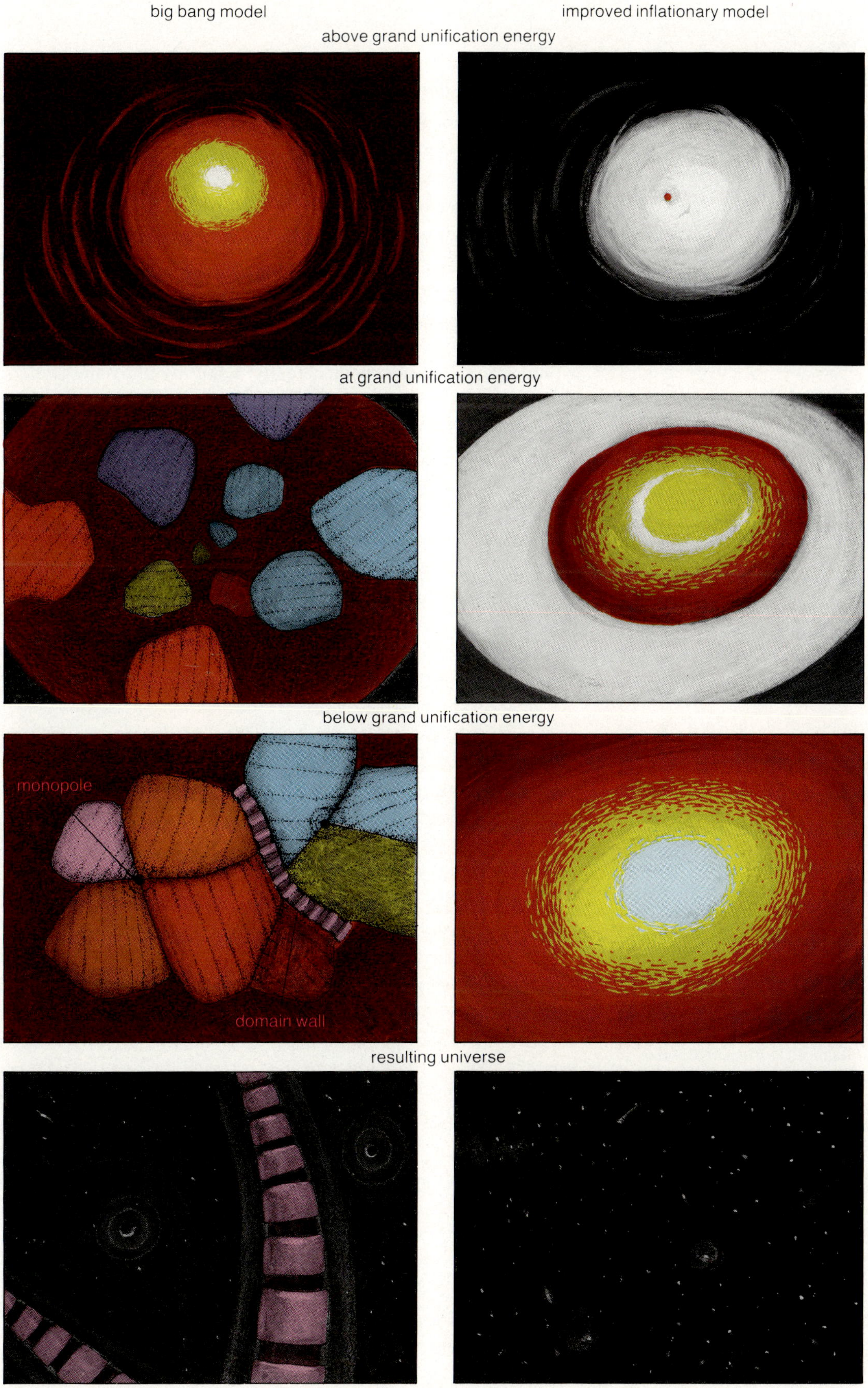
big bang model
improved inflationary model
above grand unification energy
at grand unification energy
below grand unification energy
monopole
domain wall
resulting universe

energy difference between the false and true vacuum states. If this phase lasts sufficiently long, about 10^{-32} second, the region could be inflated by an enormous factor, perhaps 10^{50} or more. A tiny part of this region, much smaller than the diameter of a proton, would expand to grapefruit size, large enough by itself at that point to evolve into the entire presently observable universe.

The quantum nature of the inflationary model provides for the inflated region to escape the false vacuum and bring exponential expansion to a halt. In Guth's original model the true vacuum state is separated from the false vacuum state by a small potential energy barrier. In quantum physics, random fluctuations can sometimes allow particles to penetrate a potential barrier without actually acquiring the energy to surmount it through a process called tunneling. Likewise, quantum fluctuations would occasionally cause the Higgs fields in a region of false vacuum to tunnel out of their zero-value condition and initiate "bubbles" of the true vacuum state. The difficulty with this version is that it allows for the formation of many such bubbles that are all out of touch. The result would be a region, containing our incipient universe within, that is filled with clusters of uncoalesced bubbles. Such a structure would give rise to extreme inhomogeneities in energy distribution, which are not presently observed.

The major difference between the original inflationary scenario and the improved version lies in their handling of spontaneous symmetry breaking. Each of the diagrams on the facing page depicts the energy density of the early universe as a function of two Higgs fields, a and b, at right angles to each other. All possible values of the energy density of the universe map out a three-dimensional surface having a central hump, the center of which corresponds to a state in which both Higgs fields have values of zero. The position of the ball on the surface corresponds to a given point in the evolution of the universe. In the original scenario (top), as the universe cools below the grand unification energy, its Higgs fields remain at their zero values while the universe settles into a false vacuum state, shown in the diagram as a central depression and separated from the true vacuum state by a hill, or energy barrier. For the universe to escape the false vacuum, quantum fluctuations must allow the Higgs fields in small regions of space to tunnel out of their zero-value condition and initiate bubbles of true vacuum. It can be shown, however, that these bubbles could never coalesce to form the universe observable today. In the improved version (bottom) the false vacuum state lies at the center of a rather flat plateau. After fluctuations nudge the Higgs fields from their zero values, the universe begins an initially rather slow and smooth transition to the true vacuum. Inflation takes place during this early stage, while the energy density remains roughly constant. By such a process a single region can grow large enough to contain the entire present-day universe.

A. D. Linde of the Lebedev Physical Institute in Moscow, Paul Steinhardt and Andreas Albrecht of the University of Pennsylvania, and Stephen Hawking of the University of Cambridge, among others, have circumvented this problem by developing a different type of arrangement for temporarily restraining the value of the Higgs fields—not with a barrier but with a very slow evolution away from their zero value early in the transition. A quantum fluctuation would merely start the evolution going instead of accomplishing it in one leap. Rather than inflation taking place in a completely trapped false vacuum state, it would occur at the earliest stages of the change from the false to the true vacuum. If this transition is made sufficiently slow, the region in question still expands exponentially and its size still increases enormously. The advantage of the improved model is that the true vacuum state does not arise from a collection of independent, imperfectly merging bubbles but from a single, smoothly evolving domain. Consequently the observable universe would be free of the irregularities expressed in the original model.

Finally, as the inflated region settles into the true vacuum state, the energy difference that has been powering the expansion is released, mainly as heat. Having cooled down dramatically during inflation, the region reheats to a temperature close to the grand unification value, and copious numbers of particles and antiparticles are produced. Its expansion drops to its preinflationary rate, and it then evolves according to the standard big bang model.

Postulating this era of exponential expansion helps to solve the most vexing problems of the standard big bang model. Magnetic monopoles, which are produced mostly at the beginning of the phase transition, are diluted so efficiently within the huge volume of inflated region that there should be practically no chance to find them in the relatively tiny part that makes up the observable universe. Consequently they do not contribute to the mean energy density of the universe. Likewise, the domain walls that form at the boundary surfaces of the inflated region are far beyond observation.

The dynamical equations describing the evolution of the universe relate the rate of expansion to the mean energy density of the universe and its curvature. During the exponentially expanding phase the mean energy density is dominated by the constant energy difference between the true and false vacuum states. The curvature term is inversely proportional to the square of the radius of the universe and is therefore exponentially decreasing, just as the curved surface of a spherical balloon becomes flatter as it is inflated. In fact, space becomes so flat for the observable universe that the inflationary model makes a prediction: the mean energy density of the universe is equal to the critical density; *i.e.*, the universe will continue expanding forever.

In the inflationary scenario the whole presently observed universe evolves from a preinflationary region that is 10^{50} times smaller in diameter than the corresponding region in the standard model. For its size it has ample time for the free exchange of information to smooth out all existing temperature differences. Therefore, it is no longer surprising that the observed temperature of the microwave background radiation is so uniform all over the sky.

The inflationary model has another very attractive feature: it allows the initial, preinflationary state to be quite chaotic. Whereas the standard big bang model requires quite rigid initial conditions to arrive at

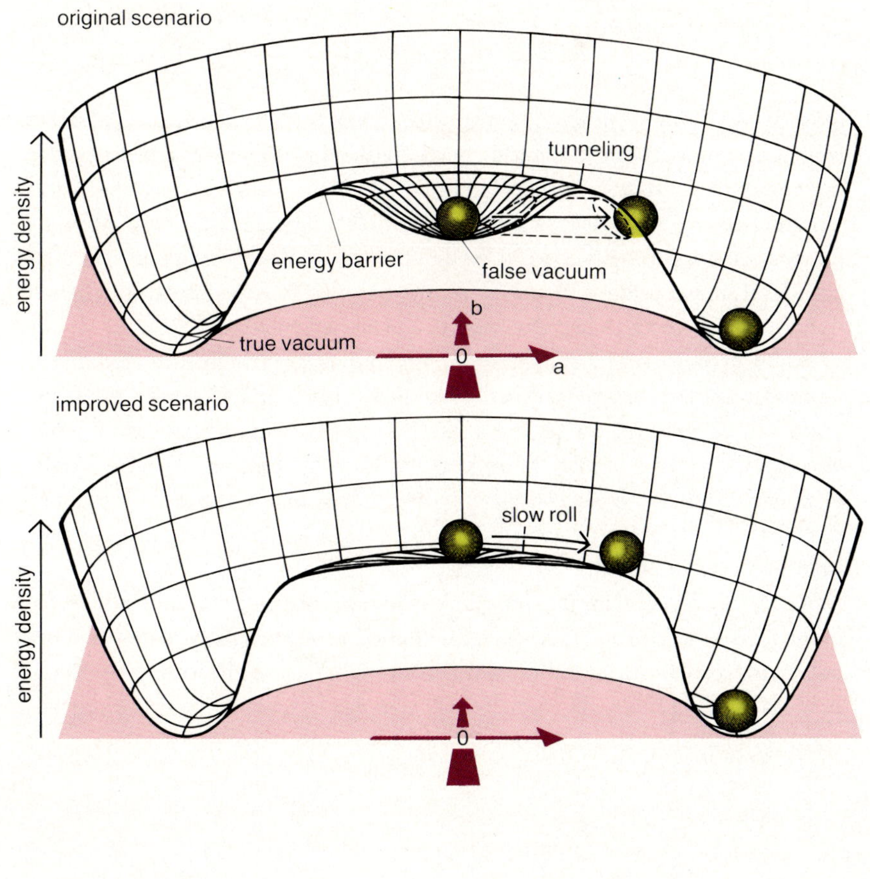

The big bang model and inflationary model are compared in graphs that trace the evolution of the universe in terms of three properties: radius (top), temperature (center), and energy density (bottom). In the inflationary model, before the universe cools to the grand unification energy, the volume that is to become the observable universe of today is many orders of magnitude smaller than that in the big bang model. By the end of the inflationary era, however, the sizes are identical. In the inflationary model the temperature of the universe plunges dramatically during inflation but then returns to a temperature close to the grand unification temperature as the energy that has been powering the expansion is released. Correspondingly, the energy density of the universe, which has remained roughly constant during inflation, drops abruptly.

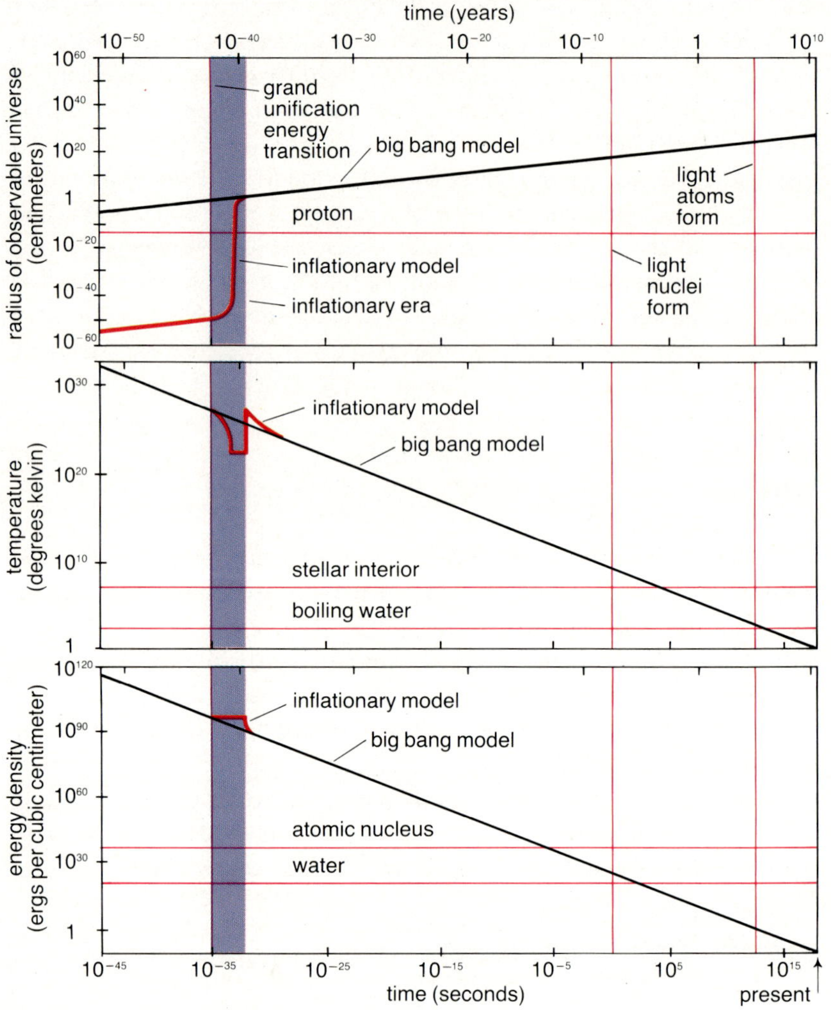

today's universe, the inflationary model requires only that at a very early stage of evolution there should be a small, expanding, and sufficiently hot region, which later inflates to encompass the present universe. Any irregularities in it would be smoothed out as a result of the exponential expansion. In the inflationary model the present universe not only is practically independent of initial details but also is wiped clean of them, like a person who has lost the memory of his infancy.

At its present stage of evolution, the universe is highly irregular on a small scale and intricately structured of matter and regions devoid of matter. It is generally believed that this structure is a result of a long and slow growth of small initial perturbations that were imposed on an otherwise homogeneous and isotropic universe. In the standard big bang model the perturbations must be introduced as initial conditions. In the inflationary scenario they are created in a natural way.

As the universe gradually undergoes the transition to the true vacuum state, ever-present quantum fluctuations could slightly accelerate or delay this transition at random in different parts of the universe. This process would create slight differences in density, which would then be magnified to a cosmic scale during inflation. Several theorists including Guth, Hawking, and A. A. Starobinsky of the Landau Institute of Theoretical Physics in Moscow showed that these perturbations could lead to the presently observed small-scale structure of the universe. There is, however, one weak point: the estimated size of the density perturbations in the inflationary model is far too large to agree with the observed uniformity of the microwave background radiation. This discrepancy might be an indication that the inflationary model is wrong, but it might only mean that more sophisticated grand unified theories are needed. It is, however, an unquestionable success of the inflationary scenario that it provides a physical mechanism for generating the density perturbations in the early universe, which therefore do not have to be specified as initial conditions.

Superuniverses and beyond

It is stimulating both scientifically and philosophically to consider how much the anthropocentric view of the universe has eroded with a deepening knowledge of man's place and origins. Less than 500 years ago the Earth lost its supposed status as the center of the solar system. Less than a century ago the Galaxy came to be seen as merely an "island universe" in a sea filled with galaxies, its place in space no more privileged than any other. Today, as a consequence of the synthesis of cosmology and particle physics, the inflationary scenario suggests that all of the galaxies our instruments can detect are just a small part of a domain of true vacuum—and perhaps not even a unique domain but only one hole in a superuniversal sponge.

What will be the next revelation? To probe the history of the universe before its inflationary period requires another major advance in particle physics, the final unification of the gravitational interaction with the grand unified theories. For the first 10^{-43} second after the big bang, quantum gravitational effects appear to be important, but as yet there is no adequate theory to describe them. If and when there is, will all of the initial conditions for the very earliest stage of the universe be found unnecessary? Encouraged by the inflationary model, some theorists have been tempted to speculate that the big bang itself was independent of initial conditions, appearing from literally nothing but a quantum fluctuation in a vacuum and a handful of physical laws.

See also *1985 Yearbook of Science and the Future* Feature Article: GIANT VOIDS IN THE UNIVERSE; *1980 Yearbook of Science and the Future* Feature Article: THE STRUCTURE OF THE UNIVERSE.

FOR ADDITIONAL READING

John D. Barrow and Joseph Silk, *The Left Hand of Creation: The Origin and Evolution of the Expanding Universe* (Basic Books, Inc., 1983).

Alan H. Guth and Paul J. Steinhardt, "The Inflationary Universe," *Scientific American* (May 1984, pp. 116–128).

Joseph Silk, *The Big Bang: The Creation and Evolution of the Universe,* (W. H. Freeman and Company, 1980).

CURVES, KNUCKLE BALLS, BREAKING PITCHES: THE KINEMATICS OF BASEBALLS

by Robert G. Watts

Curve balls do curve, and sinkers do drop suddenly. Pitchers have been using them for decades to baffle batters, and recent studies are revealing how they do it.

"A curve ball," according to Baltimore Oriole pinch hitter Terry Crowley, "comes straight in, then about four or five feet from home plate, it breaks straight down." Oriole catcher Joe Nolan claims that a curve ball does not appear to start "breaking" (changing direction suddenly) until it's about ten feet from the plate. Warren Spahn, an outstanding lefthanded pitcher for the Boston (and then Milwaukee) Braves, in the 1940s, 1950s, and 1960s, said that a slow curve breaks about a foot before it reaches home plate. A fast curve breaks even later. Rich Dauer of the Orioles, however, does not believe that a curve ball breaks suddenly. He describes a curve ball as "a big swoop. Like you drew it with a compass."

Sandy Koufax, a former star lefthanded pitcher for the Brooklyn (later Los Angeles) Dodgers, could throw a curve ball that many frustrated batters described as looking as though it rolled off a table. Legend has it that Walter Johnson (Washington Senators) could make his fast ball rise, appearing to momentarily defy the law of gravity.

The flights of curve balls and rising fast balls, as well as many other pitches—legal and illegal—are subjects of much baseball folklore. Does a baseball curve? If so, does it fly in an arc, as described by Rich Dauer, or does it change directions more suddenly, as many hitters believe? Is a rising fast ball possible?

Do curve balls curve?

It may be surprising to those who have hit (or swung at) a well-thrown curve ball that there has not always been universal agreement on whether a ball can be made to curve at all, let alone whether it breaks suddenly or curves in a smooth, circular arc. H. Allen Smith recalled that an old baseball scout claimed that there was no such thing as a curve ball and that everyone in baseball knew it. The story was contained in a letter to *The New Yorker* magazine on May 24, 1941, from R. W. Madden. Madden recounted a conversation with a man who claimed to be a major league

scout and who asked not to be named because "the guys in baseball have an interest in keeping the secret. It makes the game exciting to have curves . . . and some guys might be willing to say they doubted if there was such a thing as curves, only what would happen to 'em? I'll tell you. They'd be laughed out of town."

The response from readers to Madden's claim lasted for several months. Lucy Stanton in the June 7, 1941, *New Yorker* wrote that her father had seen a demonstration by Tommy Bond, a pitcher in Boston. Bond threw a curve ball toward three upright poles arranged in a straight line. The ball passed to the right of the first, to the left of the second, and to the right of the third. H. Allen Smith claimed that a similar test was made in Cincinnati in 1877. Will White of the old Cincinnati Red Stockings, "one of the league's best righthanders" accomplished the feat, witnessed by a large crowd.

Still doubts persisted. A university professor wrote in the *Cincinnati Enquirer* shortly after the 1877 demonstration that "it is not only theoretically but practically impossible for any such impetus to be conveyed to a moving body as would be required . . . to control the movement of what is termed a curve ball." Professor Lewis Swift of the University of Rochester represented the academic community on the other side of the debate, saying that the motion of a curve ball was "in accordance with the plainest principles of philosophy." Ray Lillibridge, writing in the June 7, 1941, *New Yorker,* called all of these early experiments into question, recalling that one man, after attending such a demonstration, suggested that "we were hypnotized, that the human eye could not detect which side of a stake an object the size of a baseball travelling at that speed had passed."

In that same year *Life* magazine picked up the gauntlet. The magazine's high-speed photographer, Gjon Mili, using high-speed stroboscopic lighting at night, positioned three cameras to photograph balls thrown by two pitchers, Cy Blanton of the Philadelphia Phillies and Carl Hubbell of the New York Giants. The conclusions drawn from the resulting photographs were that Blanton's "best curve is really a straight ball with a decided drop," and that "Hub threw everything in his pitching book" and "wound up with . . . two straight lines."

When *Life* conducted these experiments, both Blanton and Hubbell were in the declining years of their careers. Hubbell, who was 38 years old, finished that year with 11 wins and 9 losses and an earned-run average of 3.57. He lasted as a major league pitcher for two more years. This was not the same Carl Hubbell who had struck out Babe Ruth, Lou Gehrig, Jimmie Foxx, Al Simmons, and Joe Cronin in succession in the 1934 All-Star game. Blanton, 33, who won 6 and lost 13 with an earned-run average of 4.51, was in his next-to-last year. Nevertheless, although the editors failed at the time to recognize it, the pictures published in *Life* magazine are the first hard proof of the existence of the curve ball. In fact, Blanton, even in his next-to-last year in the major leagues, had both a good curve and a good screwball. The pictures taken by Mili are shown in Figure 1. Neither of these pitches is moving in a

ROBERT G. WATTS, *a former semiprofessional baseball pitcher, is Professor of Mechanical Engineering at Tulane University, New Orleans, Louisiana.*

(Overleaf) Illustration by Jane Meredith

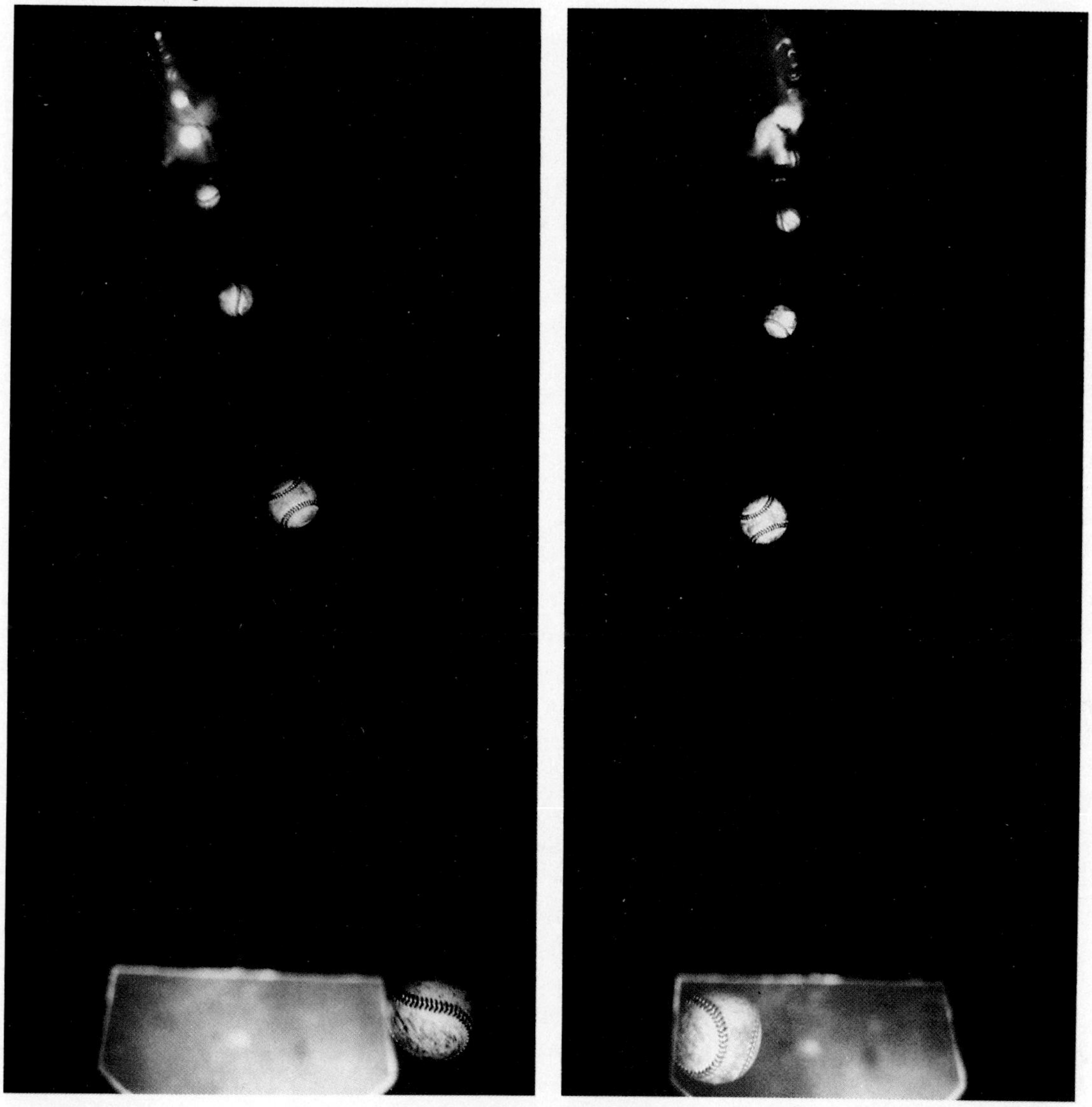

Figure 1. The first photographic proof that a curve ball curves appeared in Life *magazine in 1941. At the left a curve ball slants toward the right as it approaches home plate, while at the right a screwball curves in the opposite direction.*

vertical plane, as they would if they had been affected by gravity alone. The curve ball is clearly being forced toward the right and the screwball toward the left.

Since the photographic investigation published in *Life,* a number of articles have appeared in scientific journals and popular magazines seeking to answer definitively the question of whether a baseball curves—and whether it curves smoothly or breaks. In April 1942, less than a year after the *Life* article, a short paper appeared in the *American Journal of Physics* describing an experimental study by Frank L. Verwiebe. Five wooden frames with nets of fine cotton thread were placed between the pitcher's mound and home plate, and "straight balls," "right-handed outdrops," and "left-handed outdrops" were then thrown through the frames. The "outdrops" curved in the horizontal plane by 6.25 to 16.25 centimeters (2.5 to 6.5 inches). The curve was reported to be smooth with no sharp break observed. An interesting observation reported by

Verwiebe was that the balls also "dropped more sharply than would be the case for free fall alone."

In 1949 *Look* magazine published "Visual Proof that a Baseball Curves," and in 1953 *Life* magazine, this time using Gjon Mili's strobe photography of pitches thrown by Ken Raffensberger of Cincinnati, reversed itself, claiming that the curve ball curves but does not break.

In 1982, *Science* '82 magazine, together with three scientists, Robert Watts of Tulane University, Jim Walton of General Motors Corp., and Charlie Miller of Massachusetts Institute of Technology, set out to produce and study high-speed stroboscopic photographs of baseballs pitched by Scott McGregor, a pitcher for the Baltimore Orioles, and by Ray Miller, the Orioles' pitching coach. McGregor had a sore arm and was not throwing with as much enthusiasm as he would have under more nearly ideal conditions. Nevertheless, Watts and Walton reported in an address to the American Association for the Advancement of Science that McGregor's curve ball did, indeed, curve during a pitch of 60.5 feet (18.3 meters), the total deflection being about one foot.

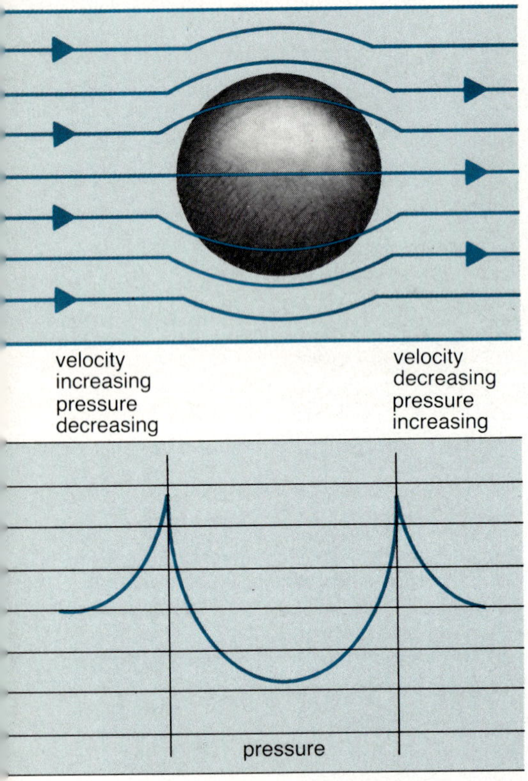

Figure 2. If air had no viscosity, it would flow smoothly past a smooth sphere. In such a situation the air must increase in speed as it moves toward the equator of the sphere and then decrease as it moves toward the rear. As a result the pressure decreases on the forward half of the sphere and increases on the back half.

Why a spinning ball curves

The fact that tennis balls curve because of the spin imparted to them was noted as early as 1671 by Sir Isaac Newton. Some two hundred years later John William Strutt, Lord Rayleigh, also in a paper describing the irregular flight of the tennis ball, credited the German physicist Heinrich Gustav Magnus with the first explanation of the lateral deflection of a spinning ball. From this the phenomenon derives its name, the Magnus effect. Magnus's explanation, however, was not published until 1853, more than a century after a similar explanation had been given by Benjamin Robins in his new book, *New Principles of Gunnery,* published in 1742. Both Magnus and Robins were interested in the trajectories of cannon and musket balls. Robins experimented with the effect of spin on the curvature of the paths of musket balls by firing them from a gun with a slightly curved barrel. With the barrel bent to the left the ball was forced into contact with the right side of the bore, thus imparting spin on the ball in the clockwise direction when viewed from above. The paths of the musket balls fired in this way were curved to the right.

The explanation given by Robins and Magnus was, in fact, not entirely correct. Their arguments went as follows: A spinning ball induces in the air around it a kind of whirlpool of air in addition to the motion of the air past the ball as the ball flies. This whirlpool effect slows down the flow of air past the ball on one side and speeds it up on the other. In accordance with a theorem of Daniel Bernoulli, an 18th-century Swiss mathematician, when the speed (or the kinetic energy) of a fluid increases, its pressure decreases. Thus, the side of the ball on which the air speed is lower experiences a higher pressure than the other side. The resulting imbalance in pressure (and force) causes the ball to move laterally toward the low-pressure (high-speed) side.

What actually happens is a bit more complicated. It involves the boundary layer, a thin layer of fluid (in this case, air) very close to the solid wall of the sphere (the ball). Particles of air that are in contact with the sphere move at the same speed as the surface of the sphere itself, while air particles that are located far from the sphere move more slowly. Between these extremes lies the boundary layer. In the case of air flowing past a baseball, this layer is only a few millimeters thick. Yet its presence is crucial to the ability of a pitcher to manipulate the paths of his pitches.

If air had no viscosity it would flow around the ball as shown in Figure 2. In this case, because of the acceleration of the air as it moves around the front half of the ball toward its equator, the speed of the air increases and the pressure decreases; the pressure increases again as the air slows down on the back half of the ball. Kinetic energy (energy associated with motion) increases at the expense of work done by pressure forces on the front half of the ball. The kinetic energy is given back as the rising pressure slows down the air on the back half. In the real case, however, air has viscosity and the picture is quite different (Figure 3). The viscous forces in the boundary layer near the surface of the ball sap energy from the flow on both the front and the rear of the ball. The loss of energy is so severe that, once a particle of fluid passes into the region where the pressure forces act against the flow, the fluid particles nearest the surface stop. Downstream of this place—called the separation point—a region of chaotic flow called the wake region forms. The pressure in this wake region is significantly lower than the pressure on the front of the ball. The existence of the wake creates a drag force called the pressure, or form, drag. If the wake is symmetrical, its effect will be simply to slow the ball down. If, however, the wake is not symmetrical, it will have the additional effect of redirecting the flow coming off the rear of the ball.

Figure 4 shows a remarkable photograph of flow around a spinning baseball. The ball is spinning clockwise, and the flow is from left to right. Separation is apparently delayed (occurs farther back) on the top surface and occurs prematurely on the bottom surface. Put rather simplistically, if a surface moves along with the flow, less energy is drained from

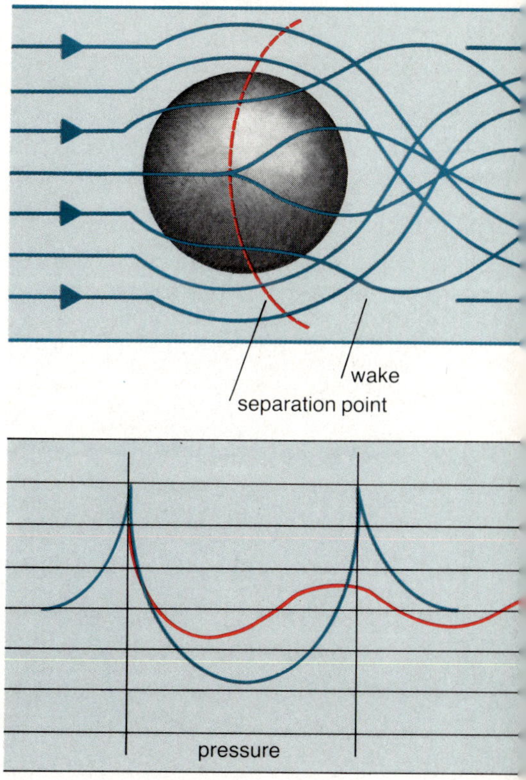

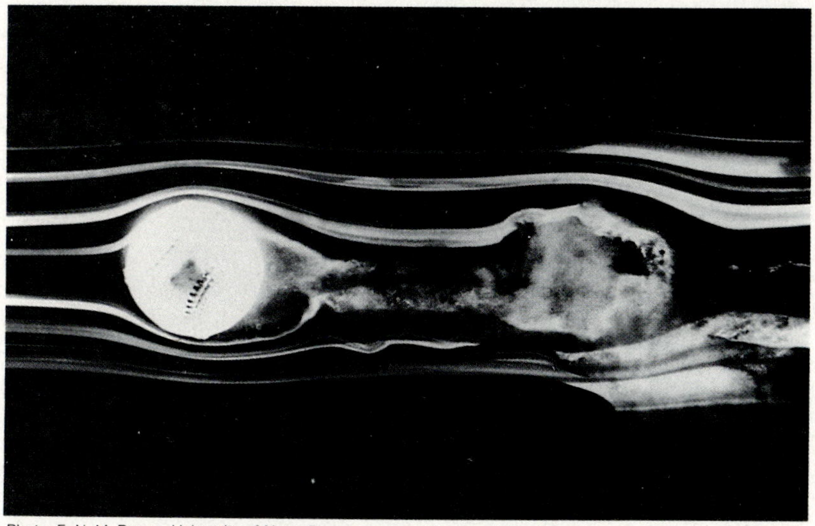

Photo, F. N. M. Brown, University of Notre Dame; courtesy, Thomas J. Mueller

Figure 3. By contrast with the hypothetical situation in Figure 2, in a real case air does possess viscosity (above). This causes the air to stall as it reaches the rear of the sphere, forming a turbulent wake (see text). In Figure 4 (left) a baseball is spinning clockwise, and the airflow is from left to right. The wake forms farther to the rear on the top than on the bottom, deflecting the airstream downward and creating an upward force on the ball.

A curve ball is thrown by pulling the fingers over the top of the ball (top) and then releasing it over the index finger (bottom). At the top right a pitcher puts a top spin on a curve ball by twisting his fingers over the top of the ball and turning his hand out during his follow-through.

(Left, top and bottom) David Stone—Berg & Associates; (right) ©Gordon Gahan—Prism and *Science 82;* stroboscopic lighting by C. E. Miller, MIT

the flow by the viscous forces, and separation is delayed. If the surface moves counter to the flow, separation occurs earlier. Since the shifted wake region deflects the flow downward, there must be a corresponding upward force on the ball.

Clearly, by throwing a baseball that is spinning about an axis, a pitcher can cause the ball to curve to the right or to the left, or to rise above or fall below its normal parabolic arc. He can do this, that is, if the forces that he can generate by spinning the ball are large enough.

Forces on a spinning ball

The most systematic experiment concerning the forces on spinning baseballs that is reported in useful detail in the scientific literature is one conducted by Lyman Briggs of the U.S. National Bureau of Standards. Briggs's work is especially important because it was reported in such a way that it can be used to investigate quantitatively the force on a spinning baseball. Briggs's idea was essentially to drop a spinning

Courtesy, the National Bureau of Standards

Lyman Briggs (right) of the National Bureau of Standards and Ossie Bluege (left) of the Washington Senators inspect the twists in a flat measuring tape that was attached to a pitched baseball. In this way they are able to measure the spin of the ball.

baseball across a horizontal wind tunnel in which the velocity of the air was known and to measure the deflection of the ball's path caused by the spin. The ball was at first mounted at the lower end of a shaft at the top of the wind tunnel, held in place by a suction-cup device. It was initially shielded from the wind by a small hollow cylinder. When the shaft was rotated by using a small electrical motor, the ball could be spun with its axis of rotation in the vertical direction. The rate of rotation was measured with a strobotac, a device that flashes a light off and on at a known, and adjustable, rate. The rate at which the flashing occurs was adjusted until a spot on the shaft of the spinning mechanism appeared to stand still. When this happened, the light was flashing at the same rate that the shaft was spinning. With the ball spinning at a known rate, the suction acting through the suction-cup device was turned off so that the ball fell out of its protective shield and dropped a distance of 6

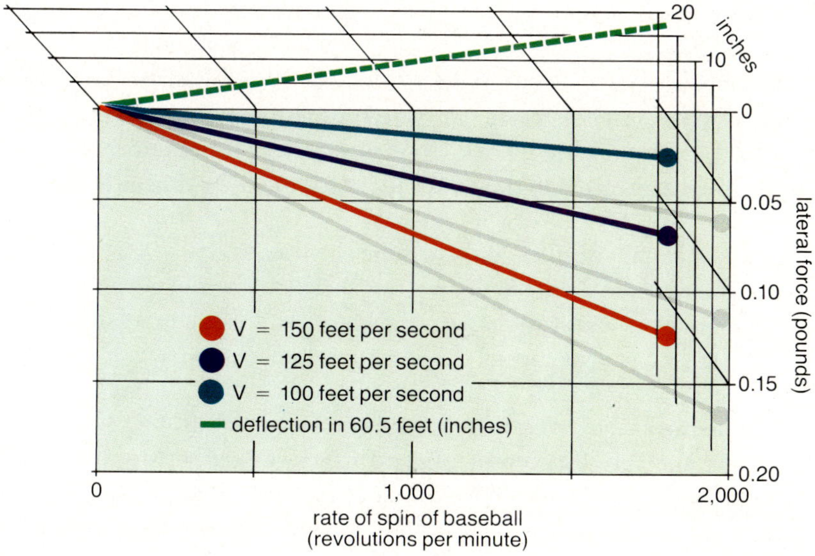

Figure 5. The lateral force on a spinning baseball increases with both the rate of spin of the ball and the square of its speed through the air. For example, a ball spinning at 1,800 revolutions per minute and having a forward speed of 125 feet per second experiences a lateral force of more than 0.1 pound. The distance that a ball curves in the 60.5 feet between the pitcher's rubber and home plate increases linearly with the spin. A ball spinning at 1,800 revolutions per minute is deflected by almost 18 inches.

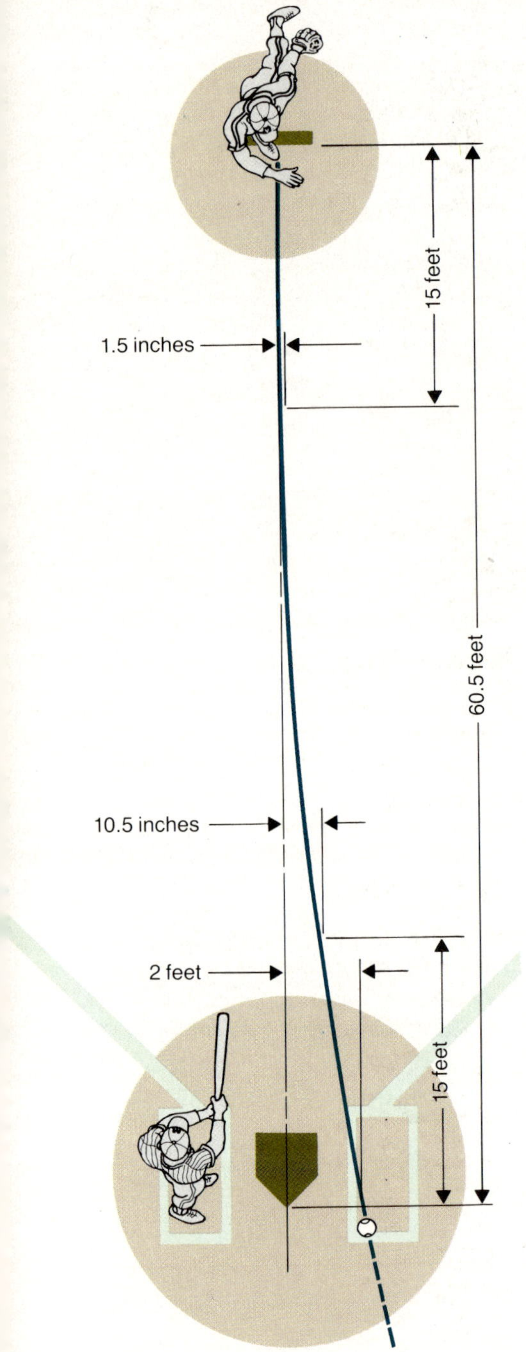

Figure 6. The "break" of a pitch is not entirely an optical illusion. Although the curve is a smooth circular arc, the ball moves to the side very little during the first 15 feet of its journey to home plate—in this case by 1.5 inches. In the last 15 feet it moves laterally by 10.5 inches, demonstrating that it is accelerating away from the batter.

feet (1.8 meters) across the wind tunnel. The air in the tunnel pushed the ball downstream, but, presumably because of the spin on the ball, the ball was also deflected to one side or the other, depending on the direction of rotation.

Using the measured lateral deflections from his wind tunnel experiments, Briggs was able to compute the lateral force on the baseball and also the distance a pitched ball would curve while traveling the distance of 60.5 feet from the pitcher's rubber to home plate. The results are shown, for various spinning rates and ball speeds, in Figure 5. One can see, for example, that a baseball thrown at 100 feet (30 meters) per second (about 65 miles [105 kilometers] per hour) and spinning at 1,800 revolutions per minute would suffer a lateral force of 0.08 pound (0.036 kilogram) and be deflected by about 17.5 inches (43.75 centimeters). The assumed speed is realistic, but the question arises as to whether a pitcher can rotate a curve ball that rapidly.

Briggs's approach to seeking an answer to this question was typically ingenious. He enlisted the help of two pitchers from the Washington Senators baseball team, Pedro Ramos and Camilo Pascual. After attaching one end of a flat tape to the ball and keeping the other end fixed, he asked the pitchers to throw curve balls. Then, by counting the number of twists in the tape, he determined that the pitches rotated 7 to 16 times on the way to home plate, corresponding to rotation rates of 700 to 1,600 revolutions per minute. It is worth noting that Carl Selin, in a doctoral thesis at the State University of Iowa in 1957, reported that curve balls thrown by the college pitchers whom he studied rotated at 1,750 to 1,970 revolutions per minute and had speeds in the neighborhood of 60 miles per hour.

As an historical note, Joseph Drury reported in *The Fireside Book of Baseball* that the aerodynamicist Igor Sikorsky performed wind-tunnel experiments that showed definitively that a spinning baseball curves. Although his results were apparently not published in the open scientific literature, Drury notes that Sikorsky indicated that a curve ball thrown at 80 miles per hour and spinning at just 600 revolutions per minute would curve 7.5 to 19 inches depending on the orientation of the seams. The larger value is obtained when the ball is released in such a way that all four seams meet the wind each time the ball rotates. Do major league pitchers release the ball optimally? A cursory examination of the photographs available to the writer indicates that they do not (Figure 1).

A zoo of legal—and illegal—pitches

When a constant force is applied to an object at right angles to the direction of motion of the object, the object will move in a circular arc. This is a simple principle of physics. Rich Dauer is correct: a curve ball is " . . . a big swoop. Like you drew it with a compass." The perception of the batter is, however, that the ball curves a great deal during its final few feet and very little during its first few feet, and he is perfectly correct. The points of view of the physicist and the baseball player (which are by no means mutually exclusive categories) can easily be reconciled by examining the peculiar geometry of a circular arc (Figure 6). The total deflection of the "curve ball" shown is 24 inches. During the first 15 feet of its flight the path moves to the left by only 1.5 inches. During the last 15 feet it moves 10.5 inches. Although the ball is moving in a circular arc, it is in fact accelerating away from the batter. But it does not suddenly change its curvature.

There are, however, pitches that do change curvature during their flights. Such pitches as knuckle balls, spitballs, and skuffballs are examples of pitches in which the pitcher manipulates the wake to produce a lateral force on the ball. The German aerodynamicist C. Wieselsberger discovered in the early 1900s that by placing a wire around the front part of a sphere one could under certain conditions delay the formation of the wake (Figure 7). The wire apparently introduces turbulence into the layer of air near the sphere, essentially increasing that layer's communication with the fluid farther out so that it can be swept farther around the sphere. Because of the presence of its seams, a nonspinning baseball offers to an airstream a sort of nonsymmetrical version of this picture. On one side of the baseball the presence of stitches may delay separation, while on the other side they may not (Figure 8). The fact that the wake is not symmetrical results again in a lateral force. Measurements published by Robert Watts in "Aerodynamics of a Knuckleball" indicate that lateral forces on a baseball as large as 0.15 pound can be achieved. The resulting lateral deflection would be 2 to 2.5 feet depending on the speed of the pitch. By throwing a ball so that it spins very slowly on its way to home plate, a pitcher can change the position of the wake during the flight of the ball. The magnitude and direction of the lateral force on the ball would change as the ball approaches the plate, causing the ball to curve first in one direction and then in another. This explains the erratic paths of such pitches.

There are a number of ways to throw a baseball so that it has very little spin. One method is to make the fingers slippery by applying such

Photos, Office National d'Études et de Recherches Aérospatiales (ONERA)

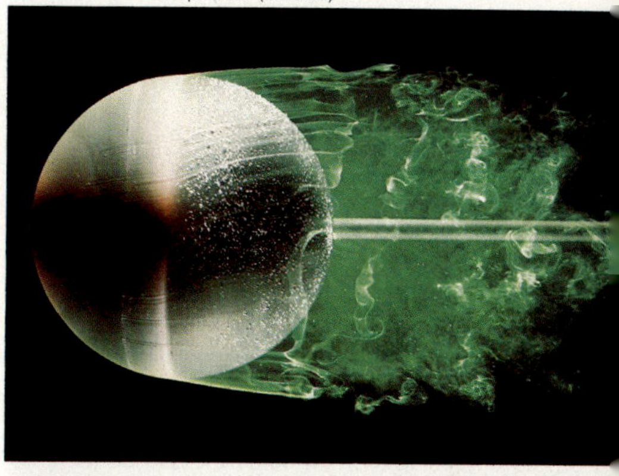

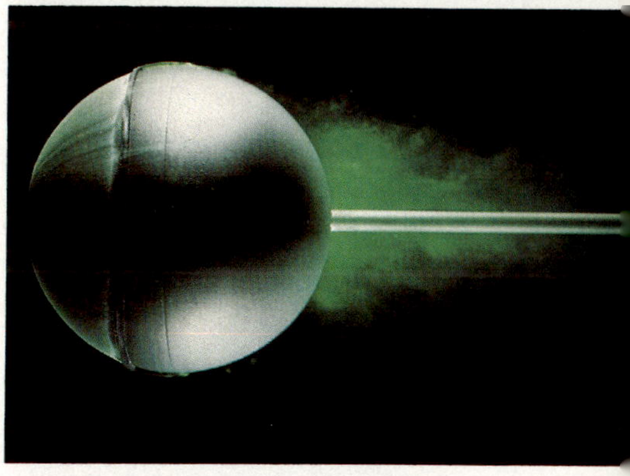

Figure 7. The flow of water past a sphere is made visible by dye in the water. At the top the separation point—where the fluid particles nearest the surface stop—forms near the equator, and a large wake is created. At the bottom a wire is wrapped around the sphere ahead of the equator. This causes the boundary layer of the water to become turbulent and delays separation. The wake is much reduced.

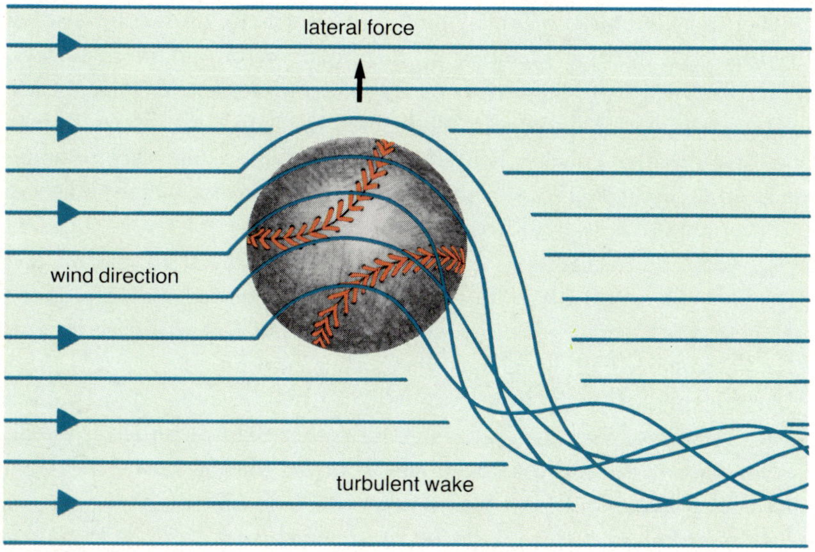

Figure 8. The seams of a nonspinning baseball can be oriented so that the wake forms farther back on one side of the ball than on the other. This nonsymmetrical wake results in a lateral force as large as 0.15 pound and a lateral deflection of 2 to 2.5 feet.

(Top, left and right) David Stone—Berg & Associates;
(bottom, left and right) ©John McGrail 1984; (below left) National Baseball Hall of Fame and Museum, Inc.

Knuckle ball (top) is gripped with the fingernails or first knuckles of one or more fingers; when the ball is released, the fingers are extended in order to inhibit the normal backspin of the ball. Spitball (bottom left), perfected in 1902 by Elmer Stricklett (below), is moistened and then squeezed out between the fingers. If properly thrown, it will drop deceptively as it approaches the batter. Bottom right, a pitcher scratches a ball with a needle hidden in a bandage. The resulting skuffball pitch curves toward the roughened spot.

material as petroleum jelly or saliva to the fingertips. (Many pitchers once chewed a substance called slippery elm for this purpose.) In releasing the ball, the pitcher then squeezes it out between his fingers the way one would squeeze a watermelon seed. According to Allison Danzig and Joe Reichler (*The History of Baseball*), the first pitcher to perfect the so-called spitball was Elmer Stricklett, in 1902.

The knuckle ball is a similar pitch. It is normally thrown by holding the ball with the first knuckles or fingernails of one or more fingers, rather than with the fingertips. As the ball is released, the fingers are extended, inhibiting the normal backspin of the ball. According to Danzig and Reichler the knuckle ball was introduced by Eddie Cicotte some years after the appearance of the spitball. It was, in fact, first known as the dry spitter. It is interesting to note that the spitball was outlawed many years ago because it was said to be too difficult to control and therefore was dangerous to hitters. Yet according to Ray Miller, pitching coach of the Baltimore Orioles, the spitball is in fact easier to control than the knuckle ball, which has remained a legal pitch.

Another illegal pitch is the skuffball. The ball is prepared by scuffing it on one spot so that one side is much rougher than the other. The ball is then thrown so that the rough spot remains on one side. The rough spot delays boundary layer separation on that side, shifting the wake and causing the ball to curve toward the roughened spot.

Batters and spinning balls

Pitchers are not the only baseball players who can manipulate the wake to their advantage. According to a report in the *Los Angeles Times* Steve Garvey, a veteran player in the National League, says that the ball carries better if the batter undercuts it a little. This can be checked by means of a calculation. One batter hits a baseball such that it leaves the bat at an angle of 20° from the horizon and at a speed of 200 feet per second. If a second batter puts just 0.5% of his hitting effort into rotational kinetic energy instead of translational kinetic energy by swinging slightly under the ball (undercutting), he can cause the ball to spin at 1,800 revolutions per minute. By trading 1% he can increase the rate of spin to 2,500. A ball with backspin will experience a lifting force, and so it might be expected to go farther before falling to the ground, as much as 50 feet farther in fact. The problem, as many hitters know, is to achieve the backspin without undercutting too much and popping the ball up. This is aggravated by the fact that the coefficient of friction between a wooden bat and a baseball is fairly low. This could be rectified if a hitter could put a sticky substance like pine tar on the part of the bat that strikes the ball. Hitters know this, of course, but it is not legal for them to do it.

In broadcasts of baseball games announcers often refer to long fly balls as "curving, curving, foul!" The question arises as to why they always curve foul—never fair. Could it be because of some strange way the wind blows in ballparks, always toward foul territory, regardless of whether in right or left field? The answer is that because of its oblique collision with the bat, a ball hit to right field has a clockwise rotation (viewed from above), while one hit to left field rotates counterclockwise. The lateral force induced by the rotation always takes the ball toward foul territory. Baseball players may not know why this happens, but they certainly know that it does happen. Their livelihoods depend on it.

FOR ADDITIONAL READING

Lyman Briggs, "Effect of Spin and Speed on the Lateral Deflection (Curve) of a Baseball; and the Magnus Effect for Smooth Spheres," *American Journal of Physics*, 1959.

Allison Danzig and Joe Reichler, *The History of Baseball* (Prentice-Hall, 1959).

Charles Einstein, ed., *The Fireside Book of Baseball* (Simon and Schuster, 1956).

Martin Quigley, *The Crooked Pitch: The Curveball in American Baseball History* (Algonquin Books, 1984).

Robert G. Watts and Eric Sawyer, "Aerodynamics of a Knuckleball," *American Journal of Physics*, 1975.

NONDESTRUCTIVE EVALUATION

COMING OF AGE

by Donald M. Forney and Dale E. Chimenti

Dedicated to finding trouble before it happens, scientists and engineers are applying radiation, ultrasonic beams, and magnetic fields in sophisticated new ways to assure the quality of critical parts and assemblies.

From the time people first acquired and traded material goods or fashioned implements to perform their work, they have felt a natural concern for quality. Depending largely upon their intuition and senses, craftsmen and consumers alike have searched for methods and tools to test for quality without damaging or otherwise lowering the value of the objects in the process—in other words, to test nondestructively. Some 2,200 years ago Archimedes' fortuitous discovery of the principle of specific gravity gave the Greek mathematician and scientist the means to prove for his friend King Hieron of Syracuse that a new gold crown was, in fact, not pure as claimed by its makers. At some time in antiquity master sword makers learned to strike a newly forged blade and listen to its ring as a measure of quality. These are early illustrations of what now is variously termed nondestructive testing (NDT), nondestructive inspection (NDI), or more generally, nondestructive evaluation (NDE).

Assuring quality in the modern world

Inspectors today employ a host of modern NDE methods to detect defects in component parts and to characterize physical properties of materials, none of which is observable with the naked eye. These methods follow one of two basic approaches. First, one can introduce some form of energy (primarily magnetic, electromagnetic, ultrasonic, thermal, optical, mechanical, or penetrating radiation) into a part and measure the change in that energy, which can be correlated qualitatively with the defects or properties being sought. Second, one can apply some type of superficial flaw indicator to the component and make a visual observation. Nondestructive methods of evaluation help assure quality in virtually every materials-processing and product-manufacturing industry. In addition, they form a cornerstone of performance-monitoring

Sygma

A DC-10 airliner spins out of control seconds before crashing near Chicago's O'Hare Airport on May 25, 1979. The consequent deaths of 274 people aboard and on the ground were traced to the failure of a cracked component in an engine pylon. Such a disaster underlines the importance of frequent, detailed inspections of all kinds of safety-critical parts and assemblies. (Overleaf) Microfocus X-ray radiograph reveals the internal parts of a wristwatch in intricate detail. A discerning observer may note the time as 10:40 and 17.5 seconds.

DONALD M. FORNEY is Chief of the Nondestructive Evaluation Branch at the Materials Laboratory, Wright-Patterson Air Force Base, Ohio. **DALE E. CHIMENTI** is a Physicist at the Materials Laboratory and an Adjunct Faculty Member in the College of Engineering, Ohio State University.

(Overleaf) Rolls-Royce, Ltd., Bristol, England

and maintenance-inspection programs for many industrial products and assemblies once those materials enter into service. Nevertheless, no NDE method is 100% accurate or reproducible. Rather, the type or size of the defects that can be detected and the consistency of their detection depend on the inherent physical sensitivity of the method, the precision of the test equipment, and the test procedures employed.

NDE technology was first catapulted into prominence in the modern industrial world by the unparalleled mass production required for World War II, in which hardware quality and reliability were critical objectives. By the end of the 1950s NDE had developed (primarily through trial-and-error methods) into a fledgling engineering field that promised not only a means of verifying the quality of manufactured products but also a route to safer product designs, more efficient production methods, and lower costs. That decade and its successor brought the jet age, the space race, sophisticated military weapons, and nuclear power plants. With these new technologies came heightened concerns about product safety and reliability and the growing costs of ever more complex manufacturing and maintenance procedures.

Safety concerns center on the possible catastrophic failure of a critical component, due to an undetected flaw or deficiency, that could endanger human life. When such failures happen, their consequences become newspaper headlines. Recent history includes several examples: the crash of a DC-10 jet aircraft on takeoff from Chicago's O'Hare Airport in May 1979, in which 274 lives were lost, was due to the failure of a cracked

Photos, Southwest Research Institute, San Antonio, Texas

component in an engine pylon; the collapse of a section of the Connecticut Turnpike's Mianus River Bridge in June 1983 left three dead; and the partial sinking of the "Alexander Kielland," an oil platform in the North Sea, claimed 123 lives in March 1980. This last catastrophe was later traced to the failure of a support leg bracing due to an undetected weld crack. Responding to such disasters, industry and government since the mid-1970s have begun to emphasize damage-tolerant designs and have specified more frequent and detailed inspections of all kinds of safety-critical systems, in which the failure of a part would jeopardize safety.

The manufacturing cost of many products, whether it be for an automobile or an orbiting satellite, is greatly affected by the customer's quality requirements. On the other hand, the introduction of cheaper manufacturing methods often depends on the availability of sensitive, accurate NDE techniques. For example, to fabricate an expensive, complex-shaped component, manufacturers of jet aircraft engines recently began using a new forging method that cut the cost per part in half. The new method, which eliminated most of the extensive machining operations of the old method, first required development of a computer-automated ultrasonic scanning system to perform the necessary quality-assurance tests. Today these cheaper engine components are helping to carry millions of passengers and tons of freight on aircraft in service around the world.

Once a product enters service NDE becomes even more economically significant. Rails and railroad rolling stock throughout the world, for instance, undergo extensive safety inspections. In the United States all

According to the Department of Transportation, most of the more than 571,000 bridges in the U.S. need regular safety inspections. Many of these structures present difficult conditions to inspectors attempting to examine support I beams and trusses. A boom over the side of the American Falls Dam Bridge (top) holds safety inspectors working in subfreezing weather on the underside of the structure. Inspectors (above) conduct an ultrasonic examination of bridge support beams.

New computer-automated immersion ultrasonic scanning methods are being used to evaluate the quality of jet engine disks that have been forged close to their final shape. At right, the ultrasonic probe located at the bottom of the vertical column moves under computer control in a precise pattern over the disk surface as it searches for flaws.

Rolls-Royce, Ltd., Bristol, England

railroad car wheels are scrapped, whether or not an actual flaw has been detected, when visual inspection reveals a predetermined degree of heat discoloration caused by brake-shoe friction. The cost of this conservative wheel-replacement policy approaches $200 million annually. A reliable NDE method to cull out the truly defective wheels from those safe for additional service would produce major cost savings. In another example, petroleum-industry inspection experts have pointed out that new or improved NDE methods to perform on-stream inspections while equipment continues to operate profitably has saved the industry substantial expense. Because of advances in NDE technology the time between oil refinery shutdowns required for a "general inspection" has increased from approximately one year in the late 1950s to three or more years today.

(Facing page) X-ray imaging is a nondestructive technique widely employed for verifying the quality of industrial products. For penetrating a thin vertical tail for an aircraft (bottom left) a comparatively small X-ray source (arrow) is adequate, whereas the thick walls of a 50-ton steam turbine shell (top) require a source of higher energy and intensity (arrow). Because elements attenuate neutrons and X-rays differently, neutron radiography frequently complements X-ray techniques. A neutron radiograph of a motorcycle carburetor (bottom right) reveals features not seen in an X-ray image. The high absorption of neutrons by hydrogen nuclei in the plastic float and rubber-composition fuel hoses results in darker images, while the comparatively low absorption by the aluminum housing makes its image virtually transparent.

The Big Five

More than two dozen nondestructive evaluation methods are presently in use or under study for industrial applications. Nearly all of them have appeared since the 1920s, and most since 1940. Five NDE methods, however, are used industrially far more often than any others: radiography, ultrasonics, eddy current, magnetic particle, and fluorescent liquid penetrant. Workers in the field popularly refer to this group as the Big Five.

Radiography was the first method of internal visualization to be adapted to NDE, and by the early 1930s—35 years after X-rays were discovered by Wilhelm Röntgen—it was being employed in some industrial inspections, based on the pioneering work of Horace H. Lester

(Top) E. I. du Pont de Nemours & Co., Wilmington, Delaware; (bottom left) McDonnell Douglas Corporation, St. Louis, Missouri; (bottom right) reprinted courtesy of G. Matsumoto, M. Teramura, K. Okubo, Y. Ikeda of Nagoya University and *Materials Evaluation,* vol. 42, no. 7, copyright © 1984 American Society for Nondestructive Testing at U.S. Army Watertown Arsenal in Massachusetts. During World War II millions of radiographs were taken to search for defects in a large variety of military-related components. Just as in the familiar medical X-ray this method sandwiches the component between a source of highly penetrating radiation and a piece of photographic film to make a film record of the internal condition of the object.

Three types of radiation are used in industrial radiography: X-rays, gamma rays, and neutrons. Both X-rays and gamma rays are forms of electromagnetic radiation, like radio waves or visible light, but much higher in energy. The collision of high-speed electrons with a metal target produces X-rays, the radiation coming from the lost kinetic energy of the rapidly decelerated electrons. Gamma rays are emitted naturally from the nuclei of decaying radioactive isotopes such as cobalt-60. Electrically uncharged neutrons, on the other hand, are ejected during nuclear fission or certain nuclear reactions induced by atomic particles striking a suitable target.

Ultrasonic waves launched into a component can quickly form a complex propagation pattern arising from reflections or conversions from one wave type to another at each surface or defect encountered. Using special visualization techniques and glass models that match ultrasonic behavior in steel counterparts, researchers at British Rail in the U.K. are studying wave propagation in railroad rails. At the right an ultrasonic transducer launches a compressional wave toward a "defect" hole five millimeters (0.2 inch) in diameter. At the far right, as the original compressional wave continues beyond the hole, part of it is reflected as a radial compressional wave while part is converted into a trailing shear wave. Both of these returns will be sensed as flaw echoes by the transducer.

When a beam of radiation penetrates an object under examination, the intensity of the beam is reduced, or attenuated, in proportion to the thickness being penetrated and the material's opacity to that form of radiation. Voids, changes in thickness, or regions of differing material composition will produce projected shadows and highlights of themselves on film. Since neutrons interact directly with the nucleus rather than with an atom's electrons, as do X-rays and gamma rays, neutron radiography serves to augment the two more common radiographic methods. For instance, neutrons are strongly absorbed by hydrogen, boron, and cadmium, while considerably less by aluminum, iron, and copper or by the heavy metals such as tungsten, lead, and uranium. Therefore, the height of a water column in a lead standpipe can easily be determined by neutron radiography, a feat not practical with X-rays or gamma rays.

Ultrasonics ranks high on the list of NDE methods and is in wide use today. Like radiography it is capable of detecting defects hidden in the interior of a part. Ultrasonics, however, is more sensitive to minute flaws than radiography and can be used to inspect from one side only. The

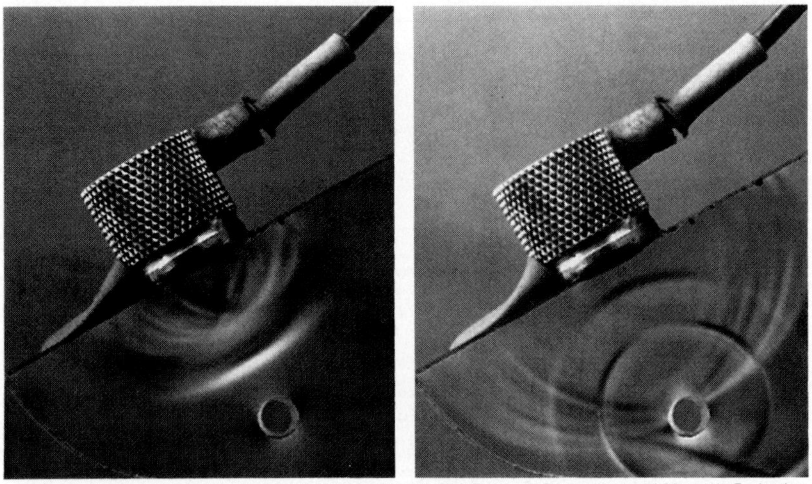

Photos reprinted courtesy of K. G. Hall of British Rail Research Department and *Materials Evaluation*, vol. 42, no. 7, copyright © 1984 American Society for Nondestructive Testing

method consists of launching high-frequency sound waves (at millions of cycles per second) into the material being inspected and either detecting echoes from a discontinuity—such as a crack or a foreign substance—encountered by the wave (pulse-echo mode) or measuring the reduction in signal strength as the attenuated wave exits the material (through-transmission mode).

In virtually all procedures the ultrasonic wave is launched and received with piezoelectric transducers made of such material as lead zirconate titanate. These crystals vibrate mechanically at high frequency when an oscillating electric field is applied to them; conversely they produce an oscillating current when mechanically vibrated. Coupled to a component by a thin grease film or a water path, the vibrating crystal induces elastic waves of the same frequency, which propagate into the material under test. Either compressional or shear waves (vibration along or perpendicular to the direction of propagation) can be produced. Compressional waves have higher sensitivity to some defects; other defects can be found more easily with shear waves.

The first recorded experiments with the propagation of sound in solids were conducted by Jean-Baptiste Biot in France in 1808. In 1929 S. J. Sokolov in the U.S.S.R. experimented with ultrasonic waves for detecting defects in metals, as did A. Trost, O. Mulhauser, and R. Pohlman in Germany during the 1930s. It was, however, the invention of the ultrasonic "reflectoscope," a forerunner of modern pulse-echo equipment, by Floyd A. Firestone of the University of Michigan in 1940 that led to the development of modern ultrasonic NDE equipment and, with it, most of the fundamental research that has followed.

The eddy-current method of testing was first investigated systematically in the early 1930s in the U.S. by Cecil Farrow at Republic Steel Corp., but the in-depth analytical and experimental work in the 1940s by German scientist Friedrich Förster provided most of the early scientific basis for the method. Eddy current is at its best in locating surface or near-surface flaws in electrically conductive materials or, to a lesser extent, in evaluating material properties like hardness and in sorting alloys by type. It exploits the secondary currents (eddy currents) that are induced electromagnetically within the surface region of a conductive part under test when a wire coil, itself excited by an alternating current, is positioned close to the part. When the coil probe passes over a crack in the part, the eddy currents are distorted, producing a measurable signal in the electronic circuit monitoring the coil. The most important factor in an eddy-current test is the frequency of the electromagnetic field generated by the coil, since frequency substantially determines flaw detectability. Lower frequencies allow deeper field penetration, but at the expense of reduced sensitivity to small surface flaws. On the other hand, higher frequencies increase surface-flaw sensitivity while rendering subsurface defects virtually undetectable. The effectiveness of an eddy-current inspection is also influenced by such factors as coil design, electronic instrumentation, test procedures, signal analysis methods, and the skill of the inspector.

Surface and near-surface flaws in electrically conductive materials can be detected by eddy-current inspection. In this method a test coil energized by an alternating current induces a magnetic field in the part, which then drives eddy currents. In the diagram below a crack distorts the eddy currents, producing a measurable signal in an electronic circuit monitoring the coil.

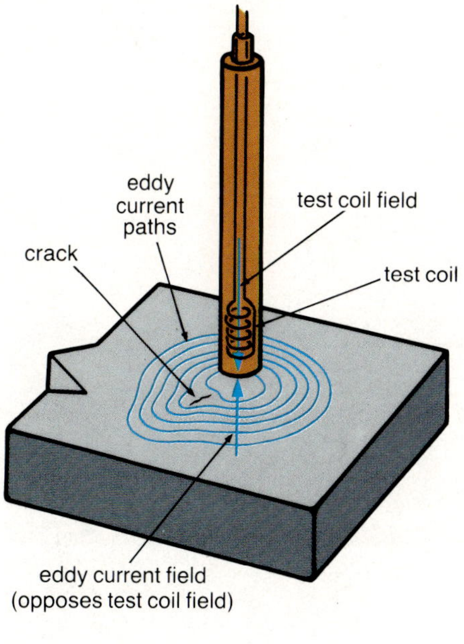

Photos, Magnaflux Corporation, Chicago, Illinois

Magnetic-particle inspection (MPI) is used to find defects in new steel billets. Billets are moved automatically through the inspection process (above) at rates exceeding 40 pieces per hour. After being magnetized, the billets are flooded with fluid carrying fluorescent iron particles (above center) and then viewed under ultraviolet light for flaws (above right); in this case the fluorescence is an indication of seams formed during the billet-rolling operation.

NDE methods based on the way surface cracks disturb magnetic fields in ferromagnetic (strongly magnetizable) materials such as iron and many steels began to appear in the late 1920s. Elmer A. Sperry and H. C. Drake in the U.S. experimented with so-called magnetic leakage fields that they observed emerging from surface cracks in sections of magnetized railroad rail. Soon after their initial success they founded the Sperry Rail Service, which since has inspected more than 12 million kilometers (7.5 million miles) of rails. In 55 years Sperry inspectors using improved versions of Sperry and Drake's method have found almost four million flaws requiring rail section replacement in the U.S. and Canada. In 1928 Alfred V. de Forest of the Massachusetts Institute of

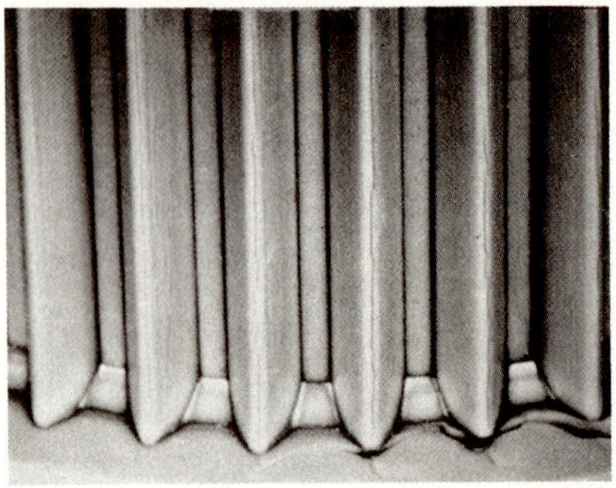

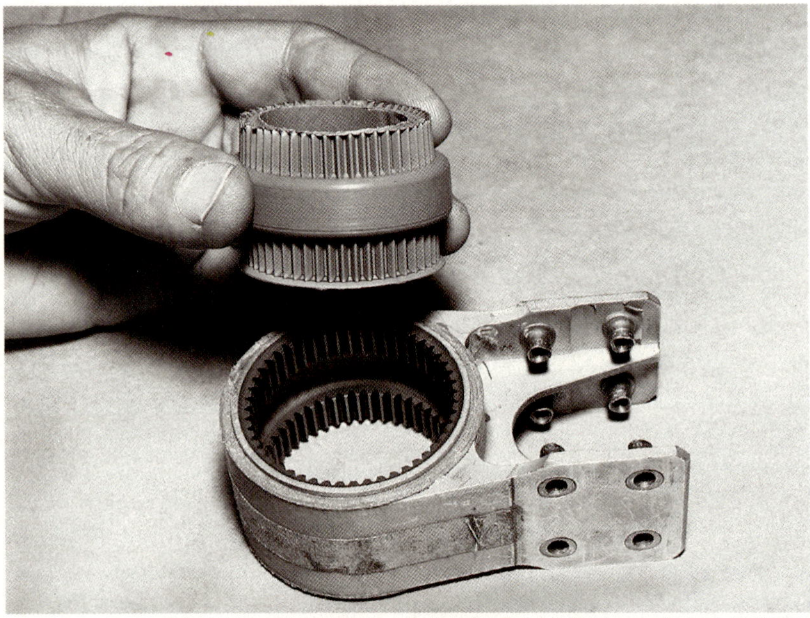

Photos reprinted with permission from H. J. Weltman, W. T. Kaarlela, J. D. Reynolds, and J. E. Halkias, and *Materials Evaluation*, vol. 42, no. 7, copyright © 1984, American Society for Nondestructive Testing

(Left) McDonnell Douglas Corporation, St. Louis, Missouri

Fluorescent penetrant inspection (FPI) is an effective means of evaluating both small and large components for surface flaws or cracks. A large titanium alloy forging (left) is moved into a penetrant booth for FPI processing and examination. A corner of a part (below) shows the fluorescence typically indicative of a crack.

Technology demonstrated a magnetic-particle method in which fine iron powder is attracted to a surface flaw in a magnetized ferromagnetic part; the two sides of a crack behave like opposing magnetic poles, exerting a force that retains the particles at the crack. Today the magnetic-particle inspection (MPI) method is optimized for each particular application, including magnetization/demagnetization procedures and type of iron powder used. The powder may be dry or suspended in a light oil, brightly colored or coated with fluorescent material. MPI is in wide use as a simple, economical NDE method for industries that deal with ferromagnetics, from metal production to field maintenance.

The use of such nonmagnetic structural materials as aluminum, magnesium, and stainless steel (primarily for aircraft construction) rose sharply with the approach of World War II. The need for an NDE method applicable to these materials, and as successful as MPI became for ferromagnetics, was obvious. Following several years of experimentation in their Cleveland, Ohio, laboratory, Robert C. and Joseph L. Switzer introduced the fluorescent penetrant inspection (FPI) process in 1942. Rapidly embraced by manufacturers of military aircraft, the method was used to inspect propellers, engine components, castings, bearings, and cutting tools. Refined considerably since that time, FPI is relatively simple, economical, and especially well suited to the inspection of complex-shaped components for defects that open to the surface. A part is immersed in a penetrant liquid bath—an oil or a glycol-based fluid with good wetability—and the liquid seeps into open surface cracks. After the excess penetrant is rinsed away and the parts are dried, the penetrant trapped in cracks is drawn out by the application of a developer, which acts like a blotter. The result is a slightly enlarged replica of the crack, which fluoresces under ultraviolet light. FPI finds use across a broad spectrum of manufacturing and maintenance NDE from large structures to minute electronic microcircuits.

(Facing page, bottom) An MPI variation called magnetic rubber inspection uses a fast-curing liquid rubber containing dispersed black magnetic particles. After the liquid is introduced into the area to be inspected, an applied magnetic field causes the magnetic particles to migrate through the liquid and concentrate in vivid dark lines at cracks in the test surface. Once cured, the solid reversed replica is removed from the part. A reversed replica of an aircraft flap actuator gear (bottom right) reveals cracks at the roots of several gear teeth when viewed under low magnification (bottom left).

A move toward higher technology

The need to toughen quality control while lowering costs has motivated industry scientists and engineers to invent new NDE procedures and systems for the production floor. Much of the emphasis is on microprocessor-controlled inspection systems, some combined with robots to provide the speed, precision, and untiring attention to detail not achievable with manual methods. Such systems are frequently linked to a fast central computer that quickly processes inspection data and makes recommendations to accept or reject a part based on preprogrammed minimum-quality standards. As an example, most aerospace hardware manufacturers that produce large aircraft structures made of graphite-fiber composites are using computer-automated ultrasonic scanning systems to inspect for such flaws as adhesive voids, delaminations (separation of bonded layers), and trapped foreign matter. Compared with manual methodsthese automated systems typically improve the accuracy of flaw detection, speed data analysis, and reduce inspection time by as much as 90%.

Improvements to basic NDE techniques are also allowing more accurate and economical inspections. A good illustration is the continuous-motion, fluoroscopic inspection used today by most manufacturers of large-diameter pipe intended for transporting oil or natural gas. An X-ray beam checks the quality of welds along the length of the pipe as it moves past the inspection unit at five meters (more than 16 feet) per minute. This method, which produces immediate, electronically enhanced video images, is faster and cheaper (by $30 per meter of pipe in the mid-1980s) than conventional film radiography. Most commercial airline passengers are familiar with a lower power, less sensitive application of this technique for airport security inspections of carry-on baggage.

Compared with NDE in factories, maintenance inspection of products in the field presents the greater challenge, because these inspections are not as easily automated. Furthermore, many advanced materials have rather different, perhaps more severe, flaw-detection requirements during service, and once assembled into the finished product and put into service, components may be less accessible than during manufacture.

For example, the inspection of offshore drilling platforms and pipelines in the hostile underwater environment, motivated by government regulatory pressure and the industry's own economic and safety concerns, clearly poses one of the most difficult problems in the NDE field. W. J. Harris of University College, London, pointed out in a recent study that the worldwide commercial market for the nondestructive evaluation of offshore structures will approach $500 million annually by the mid-1980s, thus spurring development of new specialized methods and equipment. The major structural problem in offshore structures is the accumulation of fatigue damage and extensive cracking in welded joints in the legs or support members of the platform, caused by flexing forces from the relentless ocean currents.

At present, underwater inspection relies almost completely on divers applying "dry-land" methods adapted to underwater use and, to a lesser

Several "dry-land" NDE methods, including ultrasonic, magnetic-particle, radiographic, and magnetic-leakage field testing, have been adapted for inspecting offshore structures. Below, a diver positions a specially designed ultrasonic scanner in preparation for an automatic inspection of a support-leg butt weld on an offshore drilling platform in Southeast Asia. Marine growth has been grit-blasted from the inspection site, a task that alone can take more than four hours per weld. Carried by a drive motor along a guide belt (the black strap in the upper right portion of the photo), the scanner traverses the full length of the weld around the hollow leg, which is 1.2 meters (4 feet) in diameter, at a speed of one meter per minute. Such an inspection, including preparation, can take more than eight hours per weld.

RTD Rotterdam and Shell Amsterdam

Martin Marietta Baltimore Aerospace Division, Baltimore, Maryland

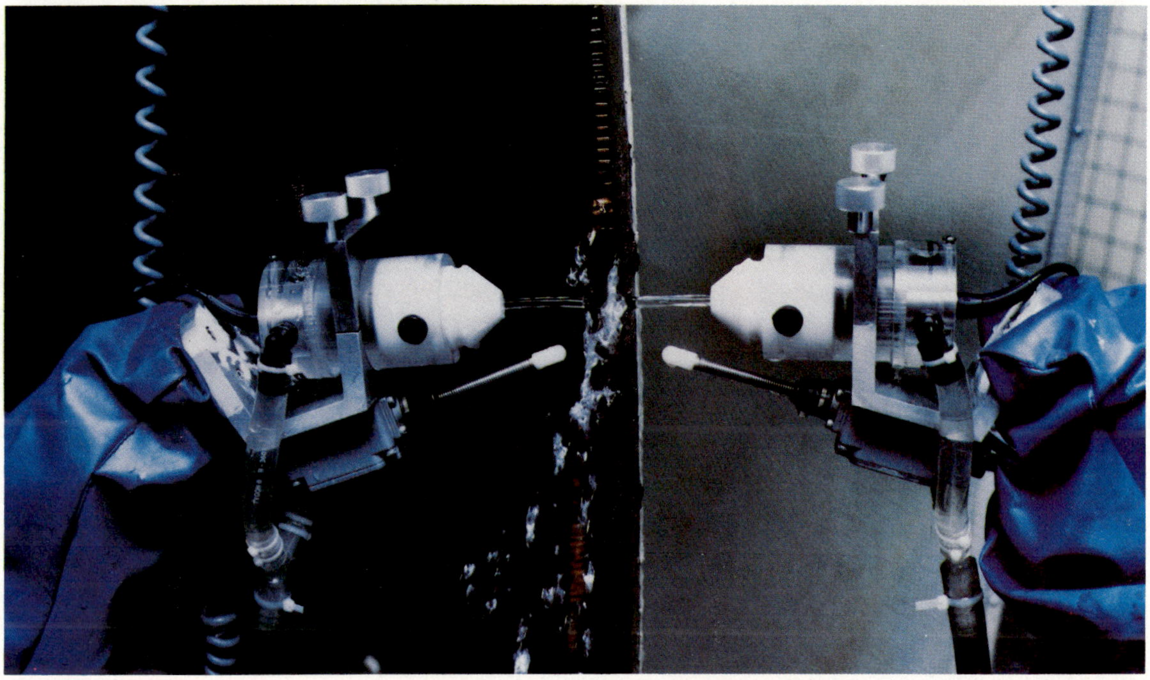

Fully computerized two-robot ultrasonic NDE system (above) uses ultrasonic signals sent and received through specially designed water jets to detect separated bonds in large graphite-fiber-composite fan reversers for jet engines. The transducer-containing robots maintain mutual alignment on either side of the part under inspection while automatically adjusting positions to follow the part contour. (Below) Automobile-wheel castings move continuously through a high-resolution X-ray video fluoroscope at production rates. Images are recorded on magnetic tape for analysis, storage, and playback of inspection results.

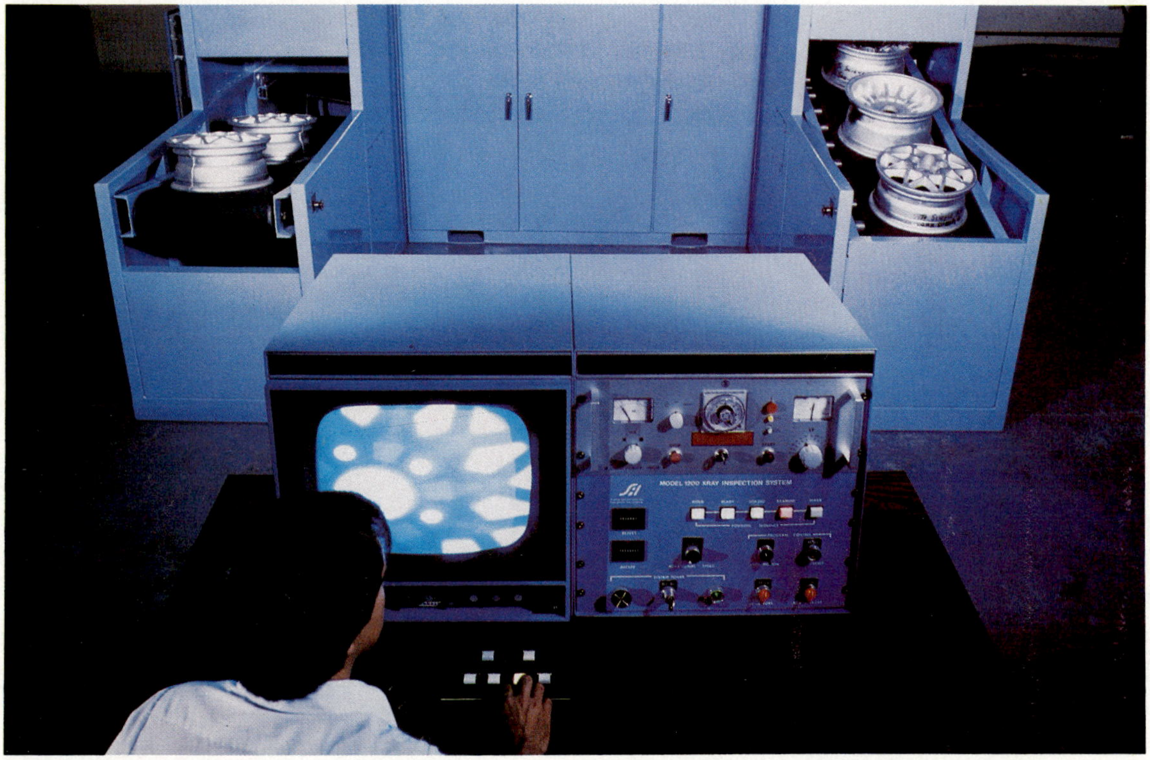

Science Applications International, San Diego, California

extent, on remotely controlled observation and manned submersible devices. These procedures prove to be both physically and mentally taxing. All of the methods require preliminary stripping away of marine growth over the inspection area, an operation that can take hours per welded joint. Harris concluded that any future NDE strategy for offshore structures should include a permanently installed diagnostic monitoring system, for example, vibrational analysis or acoustic emission, to give early warning of accumulating damage. Acoustic emission consists of low-level ultrasonic signals, detectable under certain conditions, that are generated by a fatigue crack during its growth. He was quick to add, however, as do many engineers, that the solutions to several difficult technical problems must be found before such a system would be economically feasible.

Another case deals with the construction and operation of nuclear power plants. These complicated and potentially dangerous systems place high demands on NDE; this rigor is due in part to conservative safety engineering mandated by nuclear regulatory agencies around the world. The large, water-filled, reactor pressure vessel, including inlet and outlet nozzles and associated piping fabricated with numerous welded joints, receives the greatest inspection emphasis in order to detect any slow-growing stress-corrosion cracks that may have developed in the welds. To minimize the time spent by technicians in the radioactive area, most in-service inspections of the exterior surface (primarily pulse-echo ultrasonics and some liquid-penetrant techniques) are conducted using specially designed, remotely operated, articulated mechanical scanning devices, which can be quickly installed and removed. Ultrasonic test data are transmitted electrically to a central computer, whereas penetrant test results are sent over optical fibers to a video camera and cassette recorder. Inspections in the extremely adverse environment on the inside of the water-filled pressure vessel use immersion ultrasonic methods and must be conducted entirely by remote control, including device positioning, inspection scanning, and data collection.

Internal pipe crawler physically inspects the inside surfaces of pipes by remote control, obviating the need for an inspector to enter the pipe. The device can perform visual or ultrasonic inspections in pipe with a diameter of 50 centimeters (20 inches) or more. It crawls in a caterpillarlike fashion by means of an internal piston in conjunction with positioning of its "legs."

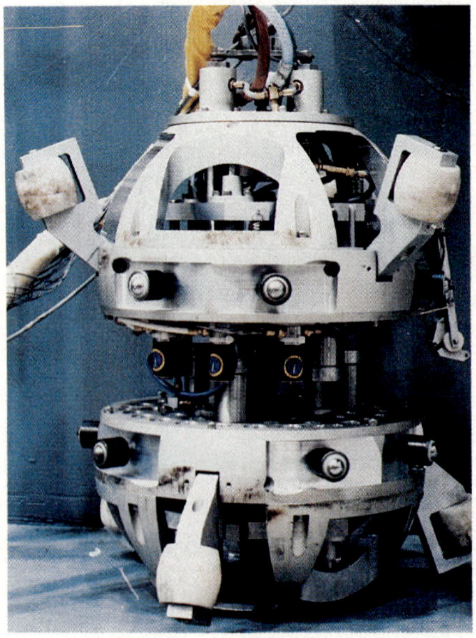

Southwest Research Institute, San Antonio, Texas

Research in progress

Beginning as a shop-floor quality assurance tool in the 1920s, nondestructive evaluation has evolved into a mature field of engineering, bolstered by an active research community supported with public and private funding. While nearly every aspect of NDE has been investigated in the laboratory, most current research activities are concentrated in eddy current, ultrasonics, and radiography.

A particularly fruitful topic of research has been in eddy-current or, more generally, electromagnetic methods. Theoretical efforts to calculate the electromagnetic field of a coil near a metal component have improved coil design. Moreover, new flaw-detection techniques emerging from the laboratory are no longer dependent solely on conventional multiturn wire coils. For instance, tiny yttrium-iron-garnet crystals that resonate in a magnetic field have been employed by Bert Auld and his associates at Stanford University to improve detection of very small surface cracks in fracture-critical aircraft components—parts whose failure will likely

Photos, Southwest Research Institute, San Antonio, Texas; drawing based on information obtained from Southwest Research Institute

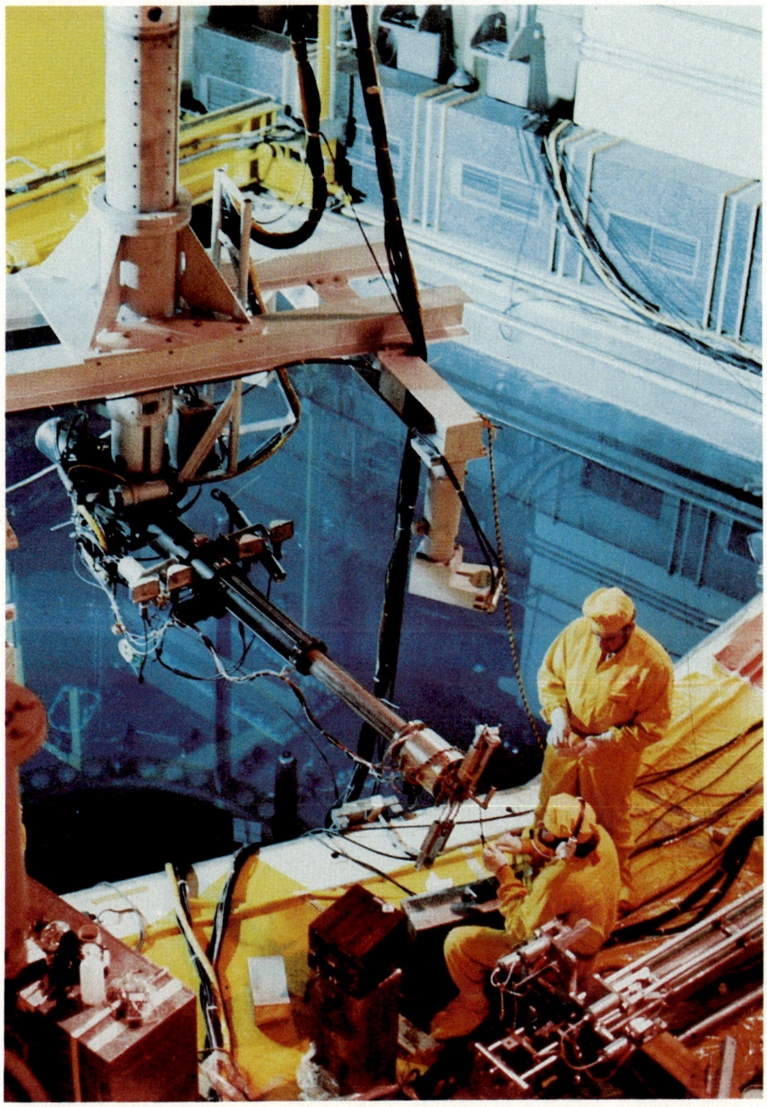

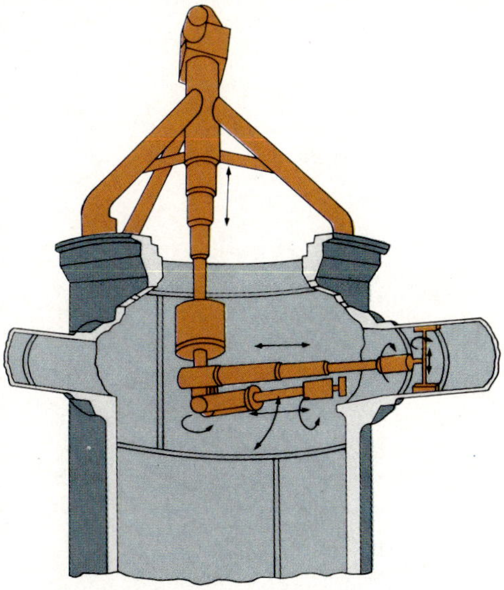

Ultrasonic inspection head mounted on the adjustable boom of a remotely operated mechanical inspection system (above left) is calibrated before being lowered into a water-filled nuclear reactor pressure vessel. An underwater television camera and lighting system for monitoring the inspection can be seen attached to the boom. After alignment on the top flange of the pressure vessel (top right) the boom is lowered into the vessel to begin weld inspections. Diagram (above) illustrates the system's remotely controlled movement capabilities, which allow inspections at any interior location using the appropriate head.

cause a crash. In another development the read/write heads originally developed for computer floppy-disk systems may find application as sensitive detectors of flaws near corners or edges, where conventional coils operate less effectively. Discrete defects are proving to be not the only anomalies detectable with eddy current. When metal components are fabricated, it is possible, perhaps desirable, that unrelieved mechanical forces or stress will be trapped in the finished part. Recently investigators have demonstrated that eddy-current maps of residual surface stress can show the location and intensity of these trapped forces. And research currently under way promises the ability to deduce from eddy-current measurements the depth of a fatigue crack, a vital quantity for the analytical prediction of a component's resistance to fracture.

In ultrasonics one principal thrust is the understanding and mathematical modeling of the complex ways that ultrasonic waves disperse, or scatter, from isolated defects in components. As difficult as this task

Photos, Battelle Pacific Northwest Laboratory, Richland, Washington

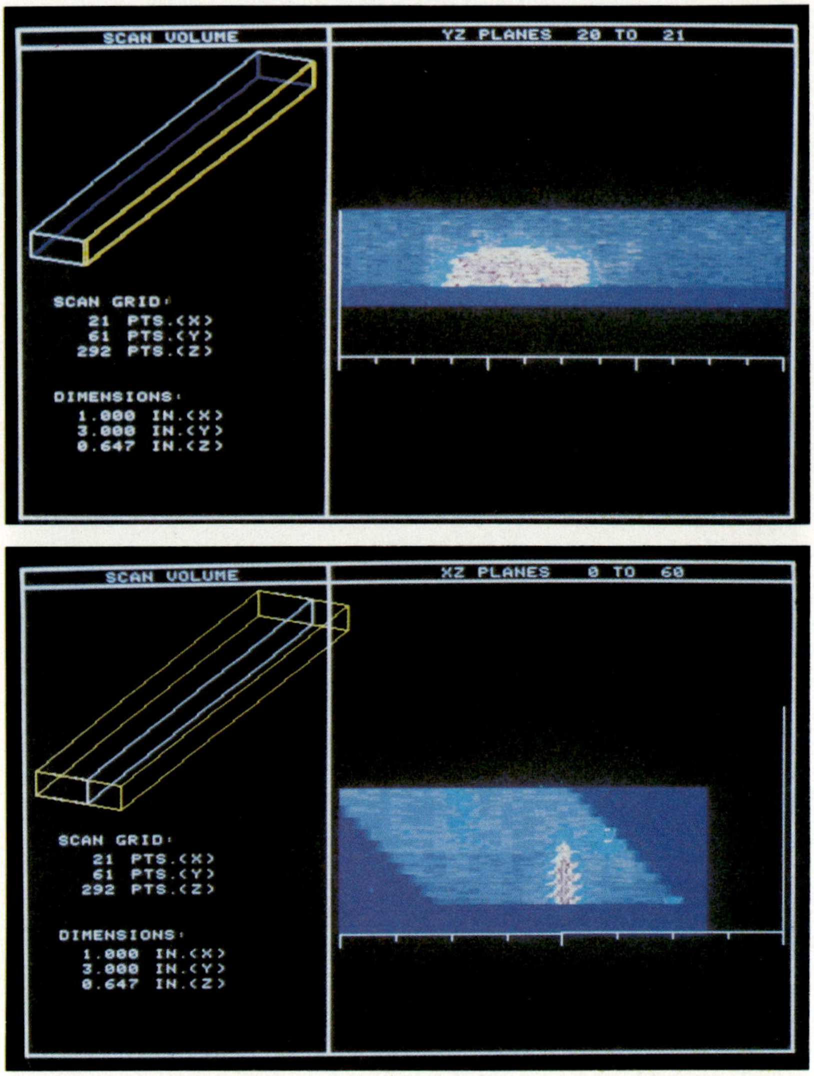

Images of a crack extending from the inside wall surface of a stainless steel pipe were constructed from ultrasonic flaw signals by means of ultrasonic imaging. The method treats the sound wave as a coherent signal, recording its amplitude and phase. A computer then reconstructs the image by projecting it mathematically back to its source, much as a hologram does for visible light. The crack image can be displayed as a view from the side (top) or edge-on (bottom).

is, considering the complicated equations used to describe the wave-flaw interaction, it is eclipsed by an even more challenging aspect of an NDE measurement: the accurate determination of size, shape, and other measurable characteristics of a flaw from a small sample of the scattered wave, known to specialists as the "inverse problem." The situation is similar to casting a stone into a quiet pond and then trying to deduce the size, shape, and composition of the stone by analyzing the resulting surface ripples at one point along the bank.

In 1977, drawing on techniques first used by German physicist Max Born in the 1920s to model the scattering of colliding subatomic particles, James A. Krumhansl and collaborators at Cornell University, Ithaca, New York, adapted the Born approximation to ultrasonic NDE; later they suggested a way to solve the inverse problem. Since then, variants of their method have been extended to include cracklike flaws, multiple defects, and defects very near component boundaries, whose strong reflective properties could mask the small flaw signal. Although the technique

E. L. Ginzton Laboratory, Stanford University, Stanford, California

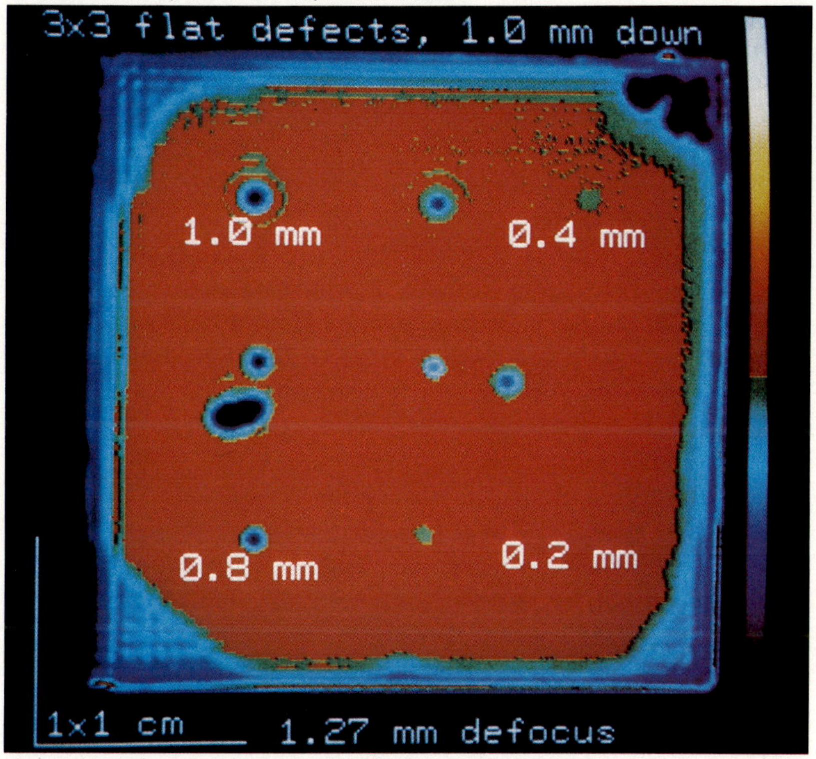

Color-enhanced acoustic micrograph reveals small flat defects near the surface of a test sample; the known sizes of the defects are also indicated. In a conventional ultrasonic scan the defects would be swamped by the far larger reflection from the surface of the component. The transducer of the acoustic microscope sends a sound wave through a couplant fluid like water to the surface of the part, where the wave becomes converted to a surface acoustic wave (SAW). Hugging the surface, the SAW skims along, scattering from defects in its path. After a short distance the energy of the wave leaks back into the couplant fluid and returns, now perturbed by the defects it has encountered, to the acoustic microscope transducer. The signal is then processed to produce color-enhanced images.

works for a wide variety of flaws and host materials, it breaks down for complex-shaped flaws or defects falling outside an expected size range. Current work is aimed at reducing these limitations.

However useful the size and composition might be to characterize a defect, it is often much more desirable to have a complete picture of a discontinuity, permitting the unequaled reasoning powers of the human mind to bear on the problem. Significant progress has been made over the past several years in generating reconstructed images of internal and obscure surface flaws. By combining new physical insights with improved experimental procedures, and aided by the ready availability of computers, researchers have succeeded in producing, under controlled conditions, flaw images of remarkable clarity.

The pursuit of technology to create undistorted images of features directly from ultrasonic reflections reaches back more than a decade, following Pal Greguss's experiments in 1965 at the Technical University in Budapest. In one approach to this problem a transducer is scanned in two dimensions across a defect site, collecting information about the strength of the returning echoes and their arrival times. This information is then processed by computer, according to equations governing wave propagation, to produce an image. Coherent ultrasonic imaging goes beyond standard ultrasonic NDE in that it preserves all the scattered-wave information from the scan. An oscillatory signal, such as a sound wave, is characterized by two quantities: its amplitude, or strength, and its phase, or the time shift between a reference point and the arrival of a wave crest. Conventional methods use only the amplitude of the reflected sound wave observed from a single point. Ultrasonic imaging, or acoustical holography as some refer to it, captures and processes both the amplitude and phase of the reflected sound wave within the scan frame, permitting a faithful reconstruction of the defect image. In this instance NDE has borrowed from the work of Nobel laureate Dennis Gabor in London and holography pioneers Emmett Leith and Juris Upatnieks at the University of Michigan.

The requirement above for a moving transducer can be relaxed if the single device is replaced by an array of transducer elements. By suitably timing the signals sent to each element, one can point the sound wave around the scan area electronically rather than mechanically. A group of researchers at Framatome in France have used a 128-element linear array, fabricated into a probe 6.4 centimeters (2.5 inches) wide, to produce images of internal flaws.

The rich variety of phenomena associated with ultrasonics is well illustrated by the example of acoustic microscopy. A burst of ultrahigh-frequency sound enters a couplant fluid through a hemispherical acoustic "lens" that focuses the sound waves to nearly a point. When this focal point is made to coincide with the surface of an object being examined, the reflected ultrasound will yield microscopic images of near-surface structure. Invented by Calvin F. Quate of Stanford University, this technique is capable of discerning surface and subsurface defects with a spatial resolution that can exceed that of optical microscopes and rival that of electron microscopes. At the expense of such extreme resolution an inspector can choose to cover more area or to probe more deeply into the surface by reducing the acoustic frequency. To make an image, the acoustic focal point is scanned in a raster pattern across the surface being examined. Since this procedure can be performed very rapidly, acoustic micrographs are quickly built up on a video monitor screen. The remarkable abilities of acoustic microscopy have proven useful in tasks as diverse as the inspection of microelectronic circuits and the elucidation of the internal structure of biological cells.

X-ray computed tomography offers a powerful method for obtaining high-quality images of subtle internal features of objects. A small solid-fuel rocket motor (above) is first imaged with a preview scan (facing page, top), an undistorted reconstructed projection—akin to a conventional X-ray—from which locations can be selected for "slice" reconstruction. One such slice through the motor (facing page, bottom) shows details of the case, liner, and solid propellant. Voids or density changes in the propellant (enclosed in white boxes) stand out in considerable detail.

One of the most dramatic achievements in imaging component features and flaws is X-ray computed tomography (CT, or CAT scanning). Based on the Nobel Prize-winning work of Briton Godfrey Hounsfield and American Allan Cormack, CT was introduced in the early 1970s as a medical diagnostic tool capable of an image quality far higher than standard X-ray procedures. To achieve this improvement CT produces an image in a fashion radically different from that of conventional radiography. Instead of recording the X-ray attenuation perpendicular to a plane through the object, CT measures the intensity of X-ray transmission through the object along many ray paths in a fanlike pattern lying in the plane to be imaged. After this procedure has been repeated at many angles in the plane, a computer processes the data, calculating the point-by-point X-ray attenuation in the image plane and displaying the result as a cross-sectional "slice." CT overcomes a principal shortcoming of conventional radiography, the superposition of features lying at different depths when a three-dimensional object is viewed in two dimensions.

Photos (below and facing page), Aerojet Strategic Propulsion Company, Sacramento, California

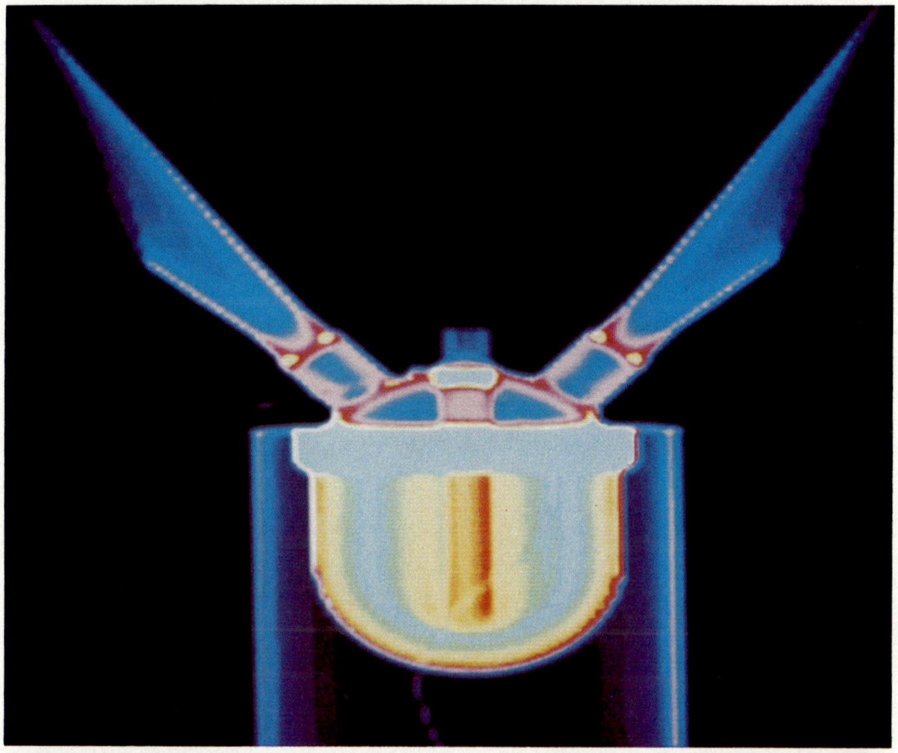

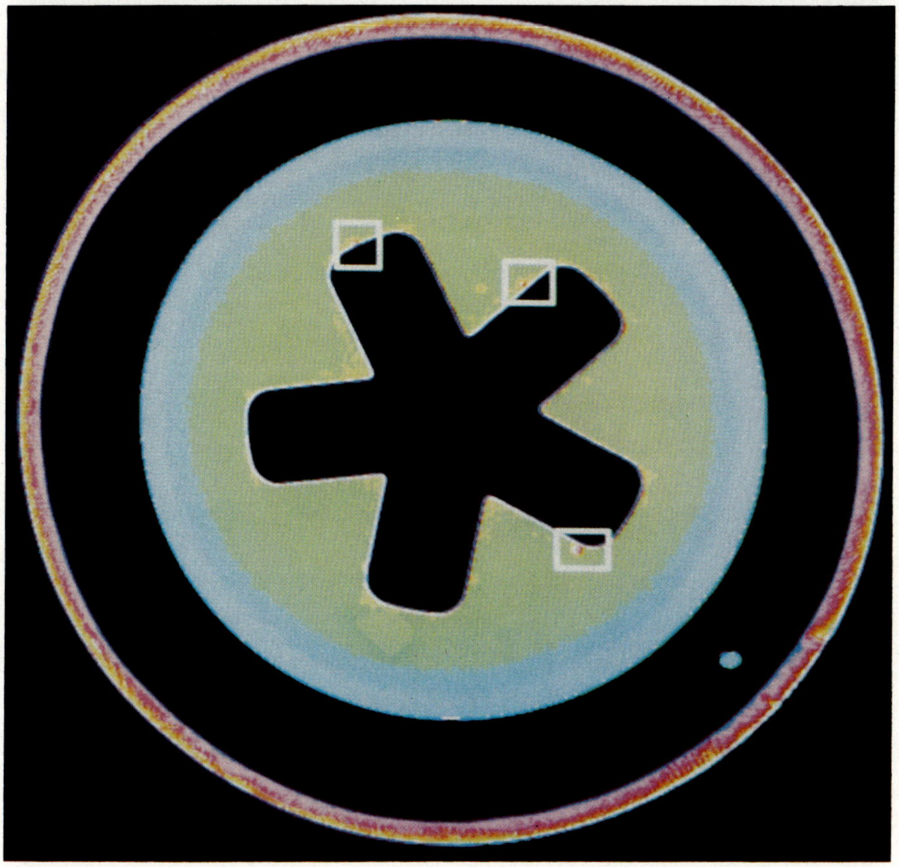

Photos, Wayne State University, Detroit, Michigan/Therma-Wave, Inc., Fremont, California

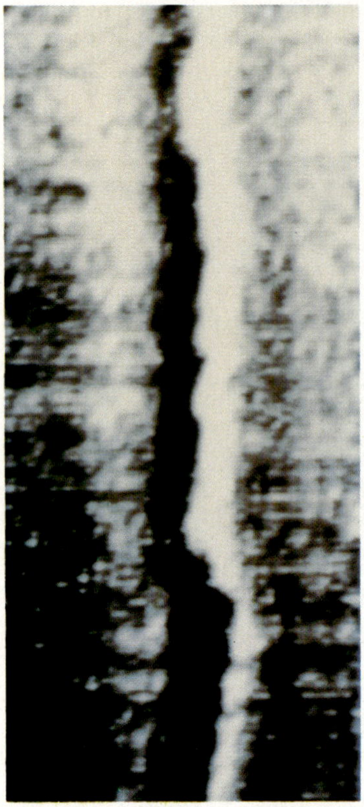

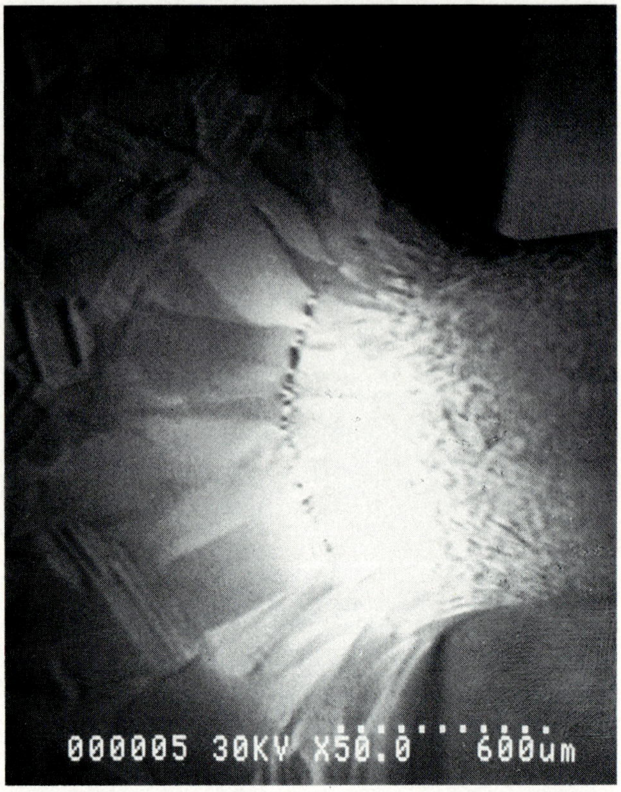

Thermal-wave imaging takes advantage of the variations in heat conduction across the surface of flawed materials to isolate and image near-surface defects. The image of a fatigue crack on the inside surface of a bolt hole (above) is made by scanning a tightly focused laser beam across the surface. A second laser beam, crisscrossing the path of the first just above its beam spot on the part, samples the air temperature at that point. Depending on the way in which the heat deposited by the first laser is conducted away, the temperature of the minute air volume above the beam spot will vary. (Above right) A weld in a valve of an artificial heart is depicted in a thermal-wave micrograph. The facetlike structure in the left of the image is due to crystalline grains, and the thin ring just at the neck of the one-millimeter (0.04-inch) tubular portion is an incipient subsurface crack in the weld zone. Conventional optical micrographs show none of these details.

Several new CT instruments under development since 1978 are being tested for industrial inspection applications. These systems possess more intense X-ray sources and are capable of resolving finer detail than typical medical units. They will be able to detect cracks or fissures with openings as small as 0.1 millimeter (0.004 inch) and resolve contrast differences of 0.03%. Once obscured by complex structure or by material properties unsuited to the application of other techniques, subtle internal features of objects that range in size from microcircuits to satellite rocket motors now leap into view. Investigators are also applying the principles of computed tomography to neutron beams to make cross-sectional images of fuel bundles for nuclear reactors.

Departing from conventional NDE techniques altogether are methods that exploit the variations in heat conduction across the surface of flawed materials. Imaging with "thermal waves" consists of periodically depositing heat, such as from a repeatedly interrupted laser beam, at a point on the surface of a part under test and then measuring by one of several noncontact methods the consequent temperature change at that point. As with ultrasonic imaging both the amplitude and phase of the thermal wave are recorded. In this case "phase" refers to the time delay between switching on the laser and the maximum of the subsequent temperature rise. Since the heat can propagate efficiently as deeply as one millimeter (0.04 inch) into the part, it will interact with superficial features having thermal properties different from those of the host material. By scanning the focused laser beam across the surface, one can generate a digital image that reveals near-surface thermal discontinuities in the part. Experiments have demonstrated the ability of the technique to image microscopic flaws as well as metallurgical features such as crystalline phases and grains. Although still experimental, like most of the research activities described in this section, prospects for eventual routine application of thermal-wave imaging seem bright.

Yet to come

For the future, exciting challenges await researchers in NDE as the difficulty of the problems increase and the sophistication of the technology available to solve them keeps pace. Each of the many advanced materials now under development may pose its own unusual set of perplexing inspection problems. Arrayed against these puzzles are innovative approaches, some derived from allied disciplines, others unique to NDE. Included among several possibilities is integration of robot position sensors with NDE probes, enabling a robot to find, orient, manipulate, assemble, and inspect components in a single operation. New ways to generate and detect familiar forms of energy—like ultrasonic waves with lasers instead of transducers—are being investigated. Scientists have known for some time that an intense pulse of laser light striking an opaque solid will deposit enough heat to cause the material to expand rapidly at that point. The ensuing mechanical disturbance propagates away from the heated zone as a sound wave, which can interact with defects inside the part. As ultrasonic reflections reach the generating surface, they can be detected with laser optical interferometry. Laser-ultrasonic NDE could lead to rapid inspection of large, complex-shaped parts or components that are radioactive, very hot, or in motion, since no physical contact with the surface is required.

It may well be that no single technique will suffice for very complicated materials. In that case flaw information from intrinsically different methods might best be combined and interpreted by a heuristic, or self-correcting, method of analysis, such as that used in computer systems equipped with artificial intelligence. It is even possible that one day some structures may carry the necessary NDE sensors, designed for remote interrogation and embedded in their interior at manufacture, to assure their continued fitness for service. Sensors built into the structural members of a passenger aircraft, for example, could allow inspectors to detect preliminary signs of metal fatigue during periodic maintenance.

In the last 25 years remarkable progress has been made in this maturing engineering field. Yet, most NDE specialists would agree that the really exciting times are yet to come.

THE DEVELOPMENT OF SKYSCRAPERS

THE CHANGING SKYLINE

by Paul D. Spreiregen

From masonry spires to glass boxes to the varied shapes of postmodernism, architects have sought to express themselves and their times with buildings that rise high into the sky.

Deep in the souls of those who build structures on the Earth lies the desire to raise them to great heights. Such aspirations are the stuff of which architecture is made. For, while everyday buildings serve everyday needs, architecture goes further; it enables people to go about their tasks while stirring their visions to larger prospects. And this fascination with the vertical in architecture is, as much and maybe more than any of architecture's many avenues of interest, an especially illuminating focus.

In the American Southwest the pueblo or village-building Indians made ladders to descend into below-ground ceremonial rooms. The ladder's two side poles were made from long slender trees and extended without rungs well upward into the sky, describing an arc suggestive of flight. Such harmonious combinations of the pragmatic and the spiritual—in a vertical form—can be found in the architecture of all cultures.

A historical perspective

Among the historic cultures of the eastern Mediterranean and the Near East, history's most prodigious architectural vertical building was an early achievement. The Great Pyramid of Khufu (Cheops) in Egypt, dating from 2700 BC, reached a height of 146 meters (482 feet), a record not seriously challenged for almost four thousand years. Roman and, soon afterward, Byzantine builders only approached the 60-meter (200-foot) mark. The 12th-century Moorish builders of Seville in Spain built a minaret, "La Giralda," which reached 90 meters (298.5 feet). The 80-meter (265-foot) "Porcelain Tower," erected in the 15th century in Nanking, is representative of Chinese capabilities and ambitions. Medieval architects in Italy surpassed the 300-foot mark in erecting bell towers (campaniles) in town centers. The campanile in St. Mark's Square, Venice, is nearly 98.5 meters (325 feet) high, while that in

© John G. Ross—Photo Researchers

The Great Pyramid of Khufu (Cheops) in Egypt (above) reached a height of 146 meters (482 feet); completed about 2700 BC, it was not exceeded in height for almost four thousand years. Later vertical achievements (opposite page, top to bottom) included "La Giralda" in Spain, built in the 12th century to a height of 90 meters (298.5 feet); the 80-meter (265-foot) "Porcelain Tower," erected in China during the 15th century; and the Eiffel Tower in Paris, created to an original height of 298 meters (984 feet) in 1889.

the Campo in Siena measures 101.5 meters (335 feet) to its topmost element. The largest and tallest of the Renaissance domes, on St. Peter's Basilica in Rome, reached 123 meters (405 feet).

Gothic builders set their sights higher than any others until those of the past century. The timber tower of Old St. Paul's Church in London, completed in 1315, is recorded to have been 149.1 meters (489 feet) high, but that is difficult to verify. It burned after standing for two and a half centuries. French cathedral builders conceived spires of 400–500 feet but seldom attempted to realize them. Salisbury Cathedral's 13th-century tower stands at 125 meters (414 feet), and Strasbourg's 15th-century tower at 139 meters (455 feet). Cologne's 14th-century design for 151-meter (500-foot) twin towers and Ulm's 15th-century design for a single 160-meter (528-foot) tower were realized only late in the 19th century during a period of Gothic revivalism. These belated but still remarkable achievements coincided with two others that celebrated not religious faith but a historical figure and modern engineering. The Washington Monument, 168 meters (555 feet) high, was completed in Washington, D.C., in 1884; the Eiffel Tower, originally 298 meters (984 feet) high, was completed as the symbol of the Paris Exposition of 1889.

Only the modern skyscraper, making its first appearance in the United States soon after the Civil War, came to rival and then exceed these historical accomplishments. And only after steel-frame skyscraper construction was made feasible in the early 1900s did it exceed them assuredly. All of the skyscraper's forerunners, from the Great Pyramid to the Eiffel Tower, are remarkable achievements in engineering and construction. And all, their particular purposes being varied, share in their denial of one function. None was built for habitation. That role remained for the skyscraper. It was a role necessitated by the functions and conditions of the 19th-century industrial and commercial city. And it was a possibility afforded only by the scientific technology developed in that century.

PAUL D. SPREIREGEN is an architect and planner in Washington, D.C.

(Overleaf) © Peter Arnold

Early industry required concentrations of workers near sources of power and convenient to routes of long-distance transport. That led to the development of multistory factory buildings in Europe and the U.S. The buildings were typically three or four stories in height, sometimes up to six. Such a configuration suited water- and then steam-powered machines as well as spatial arrangements for production.

During the 19th century manufacturing, warehousing, and shipping were undertaken in proximity to one another in cities, in addition to one relatively new function—administering the manufacturing, warehousing, and shipping. That was the work of buying and selling, of financing and contracting, of accounting and insuring, of documenting and negotiating. Such services had to be close to one another for convenience in transaction and to draw on a large worker population. With all of these functions needing to be in close proximity, a new form of city and of city center gradually came into being.

Thus were planted the seeds of the skyscraper, a type of building whose initial purpose was the efficient accommodation of administrative, clerical, and service workers but which soon came to be used for hotel and, later, apartment dwelling. In few places in the world is this chapter in architecture—and architecture's fascination with the vertical—more dramatically visible than Chicago, with New York playing a key supporting role. It was in Chicago during the post–Civil War boom years when the city was so free of precedent, so open to possibility, so unfettered of habit, so heedless of building style or custom, that the new form of architecture found its main pathway.

First efforts

Numerous technical resources were available to skyscraper designers when large office buildings began to be built in the 1850s in the United States. Masonry walls could be built to great heights. Water could be pumped vertically by means of a series of pumping stages. Coal gas could be distributed in a building for lighting purposes and steam or hot water for heating. A reliable and safe passenger elevator had been developed. Available building materials were abundant as were skilled workers and unskilled laborers.

Construction techniques were quite advanced, including mechanical or steam hoists for building tall structures. Iron and then steel frames were used to build bridges and for structures to span large spaces, such as railroad stations and exposition halls. Cast-iron columns to support floors were in use. There was a well-developed system of fabricating cast-iron frames for buildings of up to perhaps six stories. The production of steel, to replace cast iron, awaited the introduction of the Bessemer and open-hearth processes, both in wide use shortly after 1850.

In 1857 the first power passenger elevator was installed, in the Tiffany Building in New York City. The lack of passenger elevators was the single factor restraining building height from rising above six stories. In the 1860s low walk-up office buildings were still being built in New York. But by the early 1870s, after the Civil War and with New York

(Top) © Adam Lubroth—Art Resource; (center) The Bettmann Archive; (bottom) Milton & Joan Mann

City booming, the first buildings that can properly be called skyscrapers appeared; they were structures of more than six stories in which the elevator was the principal means of vertical transport. The nine-story Tribune Building designed by Richard Morris Hunt rose to 79 meters (260 feet), including its tower; the 10½-story Western Union Building, designed by George B. Post, rose 70 meters (230 feet).

Both of these skyscrapers posed two of the major problems of tall building design, one structural and the other aesthetic. The structural problem was that of load-bearing walls. Made of masonry, the walls of the lower stories had to be quite thick to support the weight of the walls and floors above. That prevented the use of large windows in the lower floors, where they were needed for light and ventilation. The thickness of the walls also reduced usable floor space. The aesthetic problem centered on how to treat such an architectural mass. Was it a series of layers? Was it a shaft? Was it a base, shaft, and cap? The professional architects of the time turned to the architecture of the past for answers. The answer was not there.

Though the first skyscrapers rose in New York City, it was in Chicago that the key ingredients of opportunity for architects presented themselves, a result of post–Civil War growth. Conveniently, the city's central area had been cleared for building by the great fire of 1871. By the 1880s Chicago's booming growth generated both the need and resources that resulted in the Chicago skyscraper. The major contribution was the development of a metal frame that served as the building's entire structural support. The route to the full acceptance of that system was marked by a series of structural systems—masonry and cast iron, cast iron and steel, and finally steel alone.

The Montauk Building of 1882, designed by Daniel Burnham and John Wellborn Root, was supported by an iron frame sheathed in masonry on its exterior. The two designed the 16-story Monadnock Building of 1891, its distinction being that it remains the world's tallest masonry building. Its exterior brick walls are 1.8 meters (6 feet) thick at the foundations tapering to 12 inches at the top story. Shaped like an elongated slab, it has no stylistic allegiances or historic references. Its shape and appearance are an unapologetic consequence of its reality.

To a building engineer, William Le Baron Jenney, goes the credit for the solution to the problem of designing the fully supportive metal frame. His skill came into full development in the creation of the 12-story Home Insurance Building in Chicago, 1883–85. Chicago's 16-story Manhattan Building, 1889–91, was another demonstration by Jenney of the logic of the fully supporting metal frame.

The structural advances credited to Jenney have important parallel accompaniments, contributed by many Chicago engineers and builders. Soil engineering and foundation design are principal among them, and they have not been given their proper due. Chicago's soil was sandy and required a system of foundations that would distribute loads. The ingenious foundation techniques developed in Chicago included systems whereby the side walls of buildings were cantilevered out so that side

Modern skyscrapers became possible with the development of a safe and reliable passenger elevator. The drawing below is from a patent granted in 1852 to Elisha Graves Otis, who lined the elevator shaft with ratchet bars that would catch the elevator car and prevent it from plunging to the ground if the lifting cable should snap.

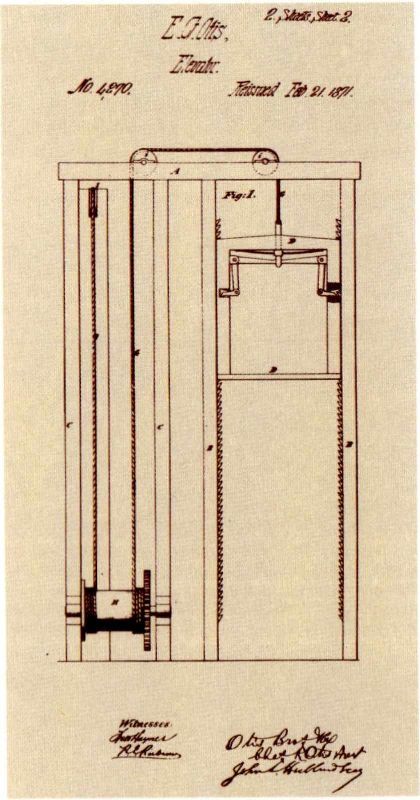

United Technologies/Otis Elevator

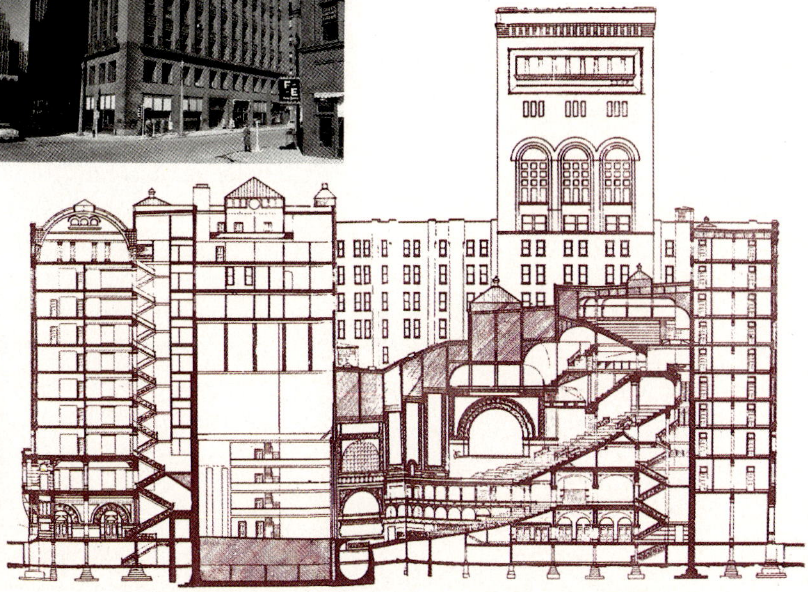

foundation loads were reduced, thus preventing disturbance to neighboring buildings. In the three-part Chicago Auditorium building (with attached hotel and office), designed by Dankmar Adler and Louis Sullivan, a technique for weighting the building's separate foundation components during construction was used. The differential weighting brought about a calculated foundation settlement that avoided a later and more damaging settlement in the completed building.

The flowering of these techniques into architectural form can be seen in the early 1890s in the hands of several architects. The 14-story Reliance Building, designed by Daniel Burnham and Co., and the 13-story Chicago Stock Exchange, designed by Adler and Sullivan, were both completed in 1894. Both are eminent examples of the Chicago steel-frame skyscraper. Both are composed principally of uniform floors with a maximum of glass exterior—as much as column thickness, spacing, and masonry fire protection would allow. Both express an architectural base, that of the Adler and Sullivan building being three stories and the Burnham design two. Both also have architectural "tops" in the form of projecting cornices, the Adler and Sullivan design being more pronounced.

Facade design was liberated by the steel frame. One could hang on it whatever one wished, including an architectural exterior to suit an owner's or a designer's fancy. Such opportunity has often been and continues to be one of architecture's greatest hazards. But it was the glory

During the 1880s and 1890s Chicago became the center of skyscraper construction. Counterclockwise from the top left to the bottom right are notable examples from that era: a portion of the steel and wrought-iron frame of The Fair department store under construction in 1891; the Monadnock Building of 1891, at 16 stories still the world's tallest masonry structure; the Home Insurance Building of 1883–85; and a longitudinal cross section of the Auditorium Building (1887–89), which contained a hotel at the left, a theater in the center, and an office tower at the right. The Wainwright Building (1891) at the upper right is in St. Louis.

(Top left and bottom right) Chicago Architectural Photographing Co.; collection of David R. Phillips; (top right) Bill Hedrich—Hedrich-Blessing; (bottom left) courtesy, The Montauk Co.; (bottom center) Chicago Architectural Photographing Co.

of the Chicago group's creativity that one of its members, Louis Sullivan, illuminated and demonstrated a new path in architecture's ancient and eternal search—the search for a fully integrating form, conjoining function, structure, materials, and much else in harmonious resolution. Three examples of Sullivan's finest skyscraper works still stand. They are the Wainwright Building in St. Louis, Missouri (1891), the Guaranty Building in Buffalo, New York (1895), both recently restored, and the Bayard Building in New York City (1898). All have steel frames, two-story architecturally articulated bases, and terminating projecting cornices. The first two are clad in a combination of red brick and terra-cotta, the third in glazed tile.

Sullivan's great contribution to skyscraper architecture lies in the way in which he expressed its overall form and in its detailing, the latter including decorative embellishment. In this he established himself as a master, comparable to the artists of any time or place. It was an America of optimism and a calm self-confidence that he proclaimed. In his system of architectural adornment he celebrated that area of architecture which is sometimes architecture's triumph, too often its folly. Sullivan derived his system of ornamentation from nature, utilizing such forms as plant stems and leaves, and he did so without sentimentality, apology, or nostalgia. But even as he did, skyscraper architecture, indeed U.S. architecture generally, was becoming preoccupied with matters of appearance.

The 58-story Woolworth Building in New York City, completed in 1913, reflects the design experimentation that was taking place at that time. Its design was inspired by the Gothic style of the Middle Ages, while other skyscrapers of the time were patterned after such models as Italian bell towers and the classic structures of ancient Rome.

Brown Brothers

The skyscraper as expression

Architecture fully approached is more than appearance. Unhappily, one of those moments when architecture is overly fascinated by appearance occurred just as the skyscraper was finding its form. The event was the World's Columbian Exposition of 1893, in Chicago, celebrating the 400th anniversary of Columbus's discovery of America. The architectural style of the exposition buildings was neoclassicism, specifically that of imperial Rome.

The effect of the exposition on U.S. architecture was the almost immediate and nearly total dismissal of all that Sullivan's work stood for. The skyscraper became an opportunity for superficial architectural exhibitionism. This occurred in all developing U.S. cities but most prominently in New York. Gothic verticality inspired the 58-story Woolworth Building in New York City, designed by Cass Gilbert. Italian bell towers, much enlarged, were an obvious model, from the Bromo-Seltzer Tower in Baltimore to the Customs House Tower in Boston. Greek and Roman temples capped the tops of skyscrapers, as did exaggerated mansard roofs.

The range of styles and forms is well documented in a book published in connection with the international competition of 1922 for the design of the new Chicago Tribune Building. Every conceivable western architectural historic style is represented. Included are concepts of gigantic classical columns and equally huge statue-buildings of heroic American Indians, both punctured with windows. As significant as was this spectrum of styles were the designs, many from Europe, in which the skyscraper was posed in its most direct self-expression, without borrowed or con-

(Left) The Bettmann Archive; (right) Brown Brothers

Exuberance and variety characterized skyscrapers built in the early 1930s. The Chrysler Building in New York City (far left), at 319.4 meters (1,048 feet), was the world's tallest building when completed in 1930 but was surpassed one year later by the Empire State Building (left), also in New York; at 381 meters (1,250 feet), it remained the world's highest building for 39 years. The Philadelphia Savings Fund Society Building in Philadelphia (below), completed in 1932, demonstrates another design style of the time.

trived covering. Thus, while stylistic preoccupation dominated attention, a more fundamental architectural impulse was seen, the impulse to find form through the inherent nature of the task and technology at hand.

The height of stylistic emphasis occurred in the 1920s, mostly in New York City. There the proliferation of opportunities and an atmosphere of artistic open-mindedness enabled creative architects to experiment with many systems of expression. Stylistic inventiveness came to the skyscraper in such forms as Art Nouveau, Art Deco, and Art Futuro. They were an expression of a need for identification with the culture of past values, of progress, of the machine, and of the promise of the future.

Of those buildings that best expressed the exuberance and originality of that era the most representative are two in New York City, the Chrysler Building, 1930, designed by William Van Alen, and the Empire State Building, 1931, designed by the firm of Shreve, Lamb and Harmon. The Chrysler Building broke the height record of the Eiffel Tower by 18 meters (60 feet); the Empire State Building, in turn, exceeded the Chrysler Building's height by just over 60 meters (200 feet). Its height record, 381 meters (1,250 feet) in 102 stories, prevailed for 39 years. Among other notable achievements of the time are the Philadelphia Savings Fund Society Building, 1932, designed by George Howe and William Lescaze and the McGraw-Hill Building in New York City, 1932, designed

Courtesy, Philadelphia Savings Fund Society

(Left) Jan Moline—Photo Researchers; (right) © Ezra Stoller—ESTO

Reflecting the design philosophy of "form follows function," skyscrapers of the 1950s refrained from exterior ornamentation and introduced the extensive use of glass in the exterior walls. Among the outstanding examples are the United Nations Secretariat Building in New York City (above), completed in 1950, and the Seagram Building (above right), also in New York and completed in 1958.

by Raymond Hood, a co-designer of the Gothic-clad Chicago Tribune Building a decade earlier. McGraw-Hill is architecturally the leanest of all of these, and representative of an urge to simplicity. The most distinctive of all is Rockefeller Center in New York City, dating from the late 1920s and realized in large part during the 1930s. The work of a number of architects, it is both an elegant composition of rising slabs and towers and a much-loved series of urban spaces at ground level.

The postwar years: "form follows function"

The Depression and World War II put an end to these exuberant skyscraper achievements. But activity was resumed throughout the world in the postwar decades. Indeed, the four decades since the end of the war can be reckoned as the great age of the skyscraper. It has been dominated by a philosophy of design that derives form from function and that refrains from ornamentation. This era produced skyscrapers pristine in purity of overall form, with articulations limited to those necessitated by mechanical, structural, or other functional considerations. Its visible opulence was limited to its display of fine materials—thin cut stone, thin metal panels, and glass.

This aesthetic had been proclaimed even before the postwar skyscraper building period began, largely through the writings and drawings of the

Swiss-born French architect Le Corbusier and the German architect Ludwig Mies van der Rohe. Le Corbusier visited the United States in 1935. He had been impressed by free-standing grain silos in the western prairies and equally alarmed by the close-packed skyscrapers of New York and Chicago. He commented that U.S. skyscrapers should be much larger and should be spaced far apart, allowing the advantages of concentration while leaving much of the ground free. Le Corbusier had few opportunities to design skyscrapers, but those few had wide influence, such as the Ministry of National Education and Public Health in Rio de Janeiro in the late 1930s and, most of all, the United Nations Secretariat Building in New York City in the late 1940s and early 1950s.

The UN Secretariat Building's all-glass side walls proved an enormous influence on architecture, as many later buildings followed its lead. During the building boom of the 1950s the glass-sided tower was the dominant form. The firm of Skidmore, Owings & Merrill, then with offices in New York and Chicago, produced such influential designs as Lever House, the Chase Manhattan Bank, and the Union Carbide Building, all in New York City, and the Inland Steel Building in Chicago. Their commissions were from prestigious clients. Their buildings were generally more expensive than multitenant commercial office structures built at the same time. All slab or towers in shape and generally slender, they were not always as efficient in the amount of usable interior floor areas which they yielded as were their less glamorous commercial counterparts. This was due to the amount of space consumed by their mechanical "cores"—arrangements of elevators, stairs, restrooms, and other facilities in the center of the floors.

A clear influence on the work of Skidmore, Owings & Merrill was Mies van der Rohe. His principal and most influential skyscrapers in this period were the twin Lake Shore Drive Apartments in Chicago (whose all-glass walls created interior cooling problems in the summer) and the Seagram Building in New York City. Lever House, Chase Manhattan, and Seagram popularized the notion of an open public plaza as an urban setting for the skyscraper. The problem of controlling sunlight on large glass surfaces was met by the development of tinted and insulated glass, which reflects sunlight and prevents excessive heat loss or gain.

The predilection toward the all-glass 1950s skyscraper spread throughout the world. But even at the height of its time other possibilities were being explored. The Price Tower, in Bartlesville, Oklahoma, the only realized design of Frank Lloyd Wright that can be called a skyscraper—and a modest one at that, being less than 60 meters (200 feet) high—was a direct challenge to the glass prism. It was also a different concept of the skyscraper, combining offices and living quarters. It is a delicate vertical, standing alone in a flat prairie town. The Finnish architect, Alvar Aalto, offered still another avenue. In the late 1950s he designed a 22-story masonry-clad residential high-rise, Neue Vahr, in Bremen, West Germany. In that and in a modest group in Tapiola, Finland, Aalto made an important contribution to skyscraper silhouette. The plans of the Bremen and Tapiola buildings resemble an open fan, but the roofs

The Price Tower in Bartlesville, Oklahoma, demonstrates an alternative to the dominance of the glass box during the 1950s. Completed in 1955 and the only realized design of Frank Lloyd Wright that can be considered a skyscraper, it combines offices and living quarters and stands alone in a flat prairie town.

Lawrence Smith—Photo Researchers

This masonry-clad apartment building is one of a group of seven designed by Finnish architect Alvar Aalto and completed in the late 1950s and early 1960s. The buildings, located in Tapiola, Finland, have flat roofs and plan forms like open fans. From afar their skyline silhouettes are passive, but up close they are dynamic.

Paul D. Spreiregen

are flat. Seen from afar, the structures are flat in profile but well proportioned—passive yet complementary to the skyline, and not drawing undue attention. But if one looks upward at one of these buildings from nearby, its fanlike skyline silhouette becomes apparent, and the building takes on a more animated personality.

In rebuilding their war-damaged cities European architects were, out of necessity and culture, especially sensitive to the relationship between their new buildings and the surrounding environment. The site of the Torre Velasca in Milan, Italy, also a 1950s product and designed by the Milanese architects Banfi, Belgiojoso, Peressutti and Rogers, is very close to the cathedral of Milan. Some 18 stories of a square-plan concrete office tower rise out of a narrow site in an area of narrow streets. Above that the building widens, like the battlements of the city's 15th-century Sforza Castle, for nine additional stories of apartments. It was designed to be seen from afar as a prominent and complementary feature of the Milan skyline. Its details, while fully of its time, are also as much of its place. Not far from the Torre Velasca is the Pirelli Skyscraper, designed by Gio Ponti and a group of associates. It is a slender and elegant glass and masonry slab, over 120 meters (400 feet) high and free-standing. Because of its slenderness and in order to resist lateral wind forces its interior columns are extremely thick at their base. As with contemporary U.S. skyscrapers there is little outward expression of that fact.

The twin towers of the World Trade Center in New York City (opposite page, top), completed in 1972, rise 409 meters (1,350 feet) and are exceeded only by the Sears Tower in Chicago. The towers are, effectively, structural tubes formed by rigid outer frames that consist of closely spaced steel columns and beams (opposite page, bottom) to resist lateral wind forces and twisting.

Further explorations in form and technology

During the decade of the 1960s architects of skyscrapers continued to explore designs other than the glass box. The Toronto (Ontario) City Hall, designed by the Finnish architect Viljo Revell with the Canadian firm of John B. Parkin, is a sculptural work; two curved slabs create a ground space for a low building. The twin cylindrical Marina City Towers in Chicago, designed by Bertrand Goldberg, accent and enliven the city's downtown area with their openness. The Ford Foundation Building in New York City, designed by Roche and Dinkeloo, is an L-shaped office tower built around a remarkable interior enclosed court, an element that was to become increasingly popular. Eero Saarinen's contribution to skyscraper design is the CBS Building in New York City. Its mechanical core is at the center of the almost square floor plan. Between it and the exterior the floors are column-free for maximum flexibility.

Along with those accomplishments the 1960s marked the advent of a new assault on the limits of skyscraper height and size. The World Trade Center, designed by the architect Minoru Yamasaki and the New York firm of Emery Roth, was built in lower Manhattan starting in 1962. Its 110 stories, in twin towers, contain ten million square feet of floor space. Its height of 409 meters (1,350 feet) exceeded that of the Empire State Building by 30 meters (100 feet).

Though proving somewhat of an architectural disappointment because its huge size in relation to the lower Manhattan skyline did not add to the grace of that cityscape and because its facade columns did not express their bend-resisting function, the World Trade Center is important for several reasons. First is its sheer size. It has seven times the floor area of the Empire State Building. From an engineering standpoint it raised several questions inherent in what might well be called "superscrapers." These include the problems of structure, of vertical transportation, and of safety. Structurally, the two towers act like masts, subject to lateral displacement, or sway, induced by wind forces. The towers, once swayed, will then sway back, passing through their normal vertical position and then on to an opposite position, like the arm of a metronome. Such a sway tendency must, of course, be damped in order to prevent a continued oscillation. The solution was to treat the entire building structure as a tube, square in cross section, with the exterior walls consisting of close-spaced columns. These massed columns, acting with those of the central interior core, offer maximum resistance to the bending movement of the towers. Unfortunately, that structural fact is not conveyed in the appearance of the buildings, even though it is an engineering accomplishment of no small magnitude.

For vertical transportation a new system of elevators was devised for the World Trade Center. A series of express elevators takes passengers from the ground floor directly to several upper "sky lobbies," from which they take local elevators to their destination floors. This has become a workable solution to the problem of transporting people long distances in the "superscrapers." In future large buildings the express elevators may themselves be several stories high.

The question of safety involved a number of considerations, including fire prevention, fire fighting, and, of highest priority, the evacuation of occupants. This problem in high buildings has led to the establishment of a special discipline in fire safety and emergency management. Systems of computerized fire-detection monitors have reached the stage of advanced development and are installed in all buildings with large populations. (The recently completed IBM Building in New York City, designed

Hedrich-Blessing

© W. B. Finch—Stock, Boston

by Edward Larrabee Barnes, has its central control console in its lobby, where it is presented to the public as a frank architectural element.) Such systems can detect fires and initiate fire-control responses automatically. The control systems include: informing people in threatened areas to evacuate by predetermined routes into "safe areas"; directing supplementary water supplies to an activated sprinkler system; closing ducts and openings that might channel fire; and altering the building's air flow directions in order to confine smoke and prevent fire augmentation.

The new technology utilized for the World Trade Center made skyscrapers of comparable dimensions possible. The two most notable achievements are the John Hancock Center and the Sears Tower, both in Chicago, both contemporaries of the World Trade Center, and both the designs of Skidmore, Owings & Merrill. The Hancock Center is a tapered shaft whose sway resistance is achieved by large cross braces that are expressed on the building's exterior. The Sears Tower, currently the record holder for height at 441 meters (1,454 feet) employs one of the most advanced structural systems yet devised. "Bundled tubes" are connected at several levels by "belt trusses" that have a binding effect, thus increasing the strength of individual tubes by causing them to act in concert and strengthen one another. As the tubes are bent by the wind, the side of one tube under compression is aided in its resistance to being bent by the tension stress in the side of the adjoining tube. One pushes, the other pulls, and so both in concert help each other.

The cross braces of the Hancock Center and the bundled tubes of the Sears Tower both were products of a new assault on heights. At this writing a still more ambitious assault was being mounted, its author the Boston engineer William J. LeMessurier. For a new generation of "super-scrapers" LeMessurier proposed the use of thick high-strength concrete shafts interlocked with lacelike diagonal steel struts.

The next generation of larger skyscrapers is likely to utilize a so-called hollow truss system, a gigantic structural frame of large diagonals, verticals, and horizontals, acting together as a vertical cantilever and resisting lateral flexure. The Bank of China Building in Hong Kong, designed by I. M. Pei and containing 70 stories, is an example of this.

The Hancock Center experienced one of those unforeseen technical problems that sometimes bedevil skyscrapers. High-level rain and mist froze against its upper surfaces, like "icing" on an airplane wing. The ice layers built up, falling when their weights were no longer supported by friction, such as when they were warmed. The falling ice sheets were potentially lethal, requiring the isolation of areas of ground level plaza and streets. But no skyscraper difficulty received more attention than the John Hancock building in Boston, designed by I. M. Pei. Its problem was that many of its windows fell out. Various theories were examined—sway, air suction currents, glass stiffness, glass holding joints. It is believed that the cause, never publicized, was a failure in the bonding of the double glass window panels. This may have been caused by an alteration in the chemical composition of the outer glass panel, which did not adhere sufficiently to the edge sealant element.

(Left) © George Kufrin—CLICK/Chicago; (right) adapted from information obtained from Skidmore, Owings & Merrill

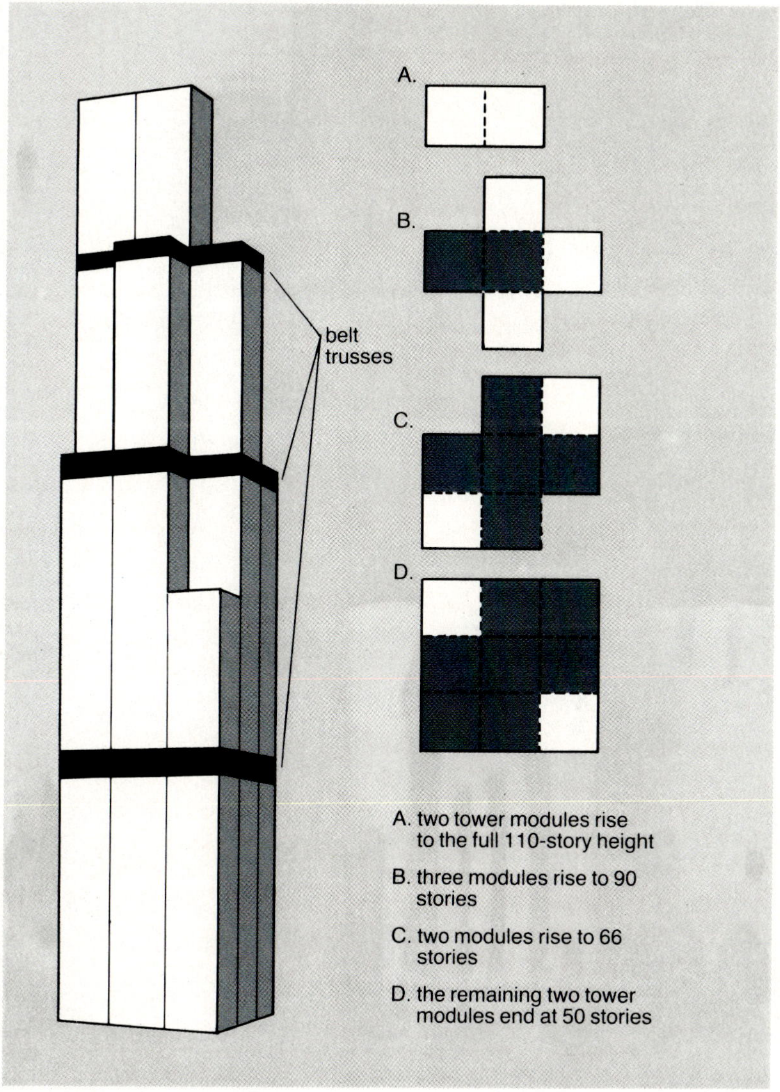

The John Hancock Center in Chicago (opposite page, top) utilizes large diagonal cross braces (opposite page, bottom) to help achieve sway resistance. For the Sears Tower in Chicago (above left), at 441 meters (1,454 feet) the world's tallest building, such resistance is gained by using "bundled tubes" (above), assemblages of small tubes that are joined with one another and that terminate at various heights (see text).

A return of expressionism

The last decade of skyscraper design has offered new possibilities in aesthetic expression. Skyscraper design again has come to be regarded as an opportunity for very-large-scale visual art. As at the turn of the century, when skyscraper technology made it possible to design any kind of exterior, the new level of technological capability achieved in recent years has allowed freedom in external appearance.

Among the foremost architects to begin working in these new modes was Philip Johnson, with his partner John Burgee. Johnson and Burgee can be credited with not breaking but smashing the mold into which skyscraper design had found itself. Their new generation of works, starting in the 1970s, include: the IDS Center in Minneapolis, Minnesota, with its great internal atrium and multiple functions; Pennzoil Place in Houston, Texas, twin glass prisms with angled tops; and, the one that attracted the most attention and generated the most discussion, the 37-story AT&T

(Left) © Guy Gillette—Photo Researchers; (right) Richard Payne—AIA

The AT&T Building in New York City, completed in 1983, is clad in pink granite. Its top recalls the broken-pediment motif of the late Renaissance. One among several challenges to the dominance of the glass box skyscraper of the postwar decades, and one of the most publicized buildings of any time, it is also regarded by many as an aberration.

Building in New York City. It is a pink granite tower on a five-story arcaded base. Its top, an angled pediment with a circular opening, became the focus of critical discussion. Johnson and Burgee's subsequent work, much of it based on historic reference, has continued to challenge and upset the established architectural tenets, at the same time creating a new one that has become referred to as postmodernism. Their more recent works include the Republic Bank and Transco Towers in Houston and the Pittsburgh Plate Glass Industries headquarters in Pittsburgh, Pennsylvania.

These two designers are not alone. Just as challenging has been the work of Michael Graves, in the Portland (Oregon) Public Services Building and the Humana Headquarters in Louisville, Kentucky. And one must also mention the work of Cesar Pelli (the Museum of Modern Art tower and Battery Park City, both in New York City, and the Four Leaf Tower in Houston) and that of Helmut Jahn (One South Wacker Drive and the Chicago Board of Trade addition, both in Chicago, and

(Left) © Peter Mauss—ESTO; (center) courtesy, Murphy/Jahn Architects; (right) courtesy, I. M. Pei & Partners

the design for the Bank of the Southwest Tower, in Houston). The work of I. M. Pei, particularly the Texas Commerce Tower in Houston, the 50-story Oversea-Chinese Banking Corporation in Singapore, and the already mentioned 70-story Bank of China Building in Hong Kong, is also deserving of examination.

As is the case with all contemporary architecture, energy efficiency has become of primary concern for the skyscraper. In that regard it is ironic that the glass tower, because it is glass, conveys the impression that it wastes energy by allowing heat to be transported between the inside and outside. In fact, the glass surfaces of modern skyscrapers are composed of a combination of tinted and insulated glass. The tinting reduces the passage of ultraviolet rays, thus reducing summer cooling loads; the insulation is achieved by arranging the glass in layers with vacuum spaces in between them, thus reducing heat transmission to the outdoors. Thus many all-glass buildings are more energy efficient than buildings with thick stone walls.

Skyscrapers planned and built in the 1980s reveal a variety of styles. From left to right, they include the masonry-clad Humana Headquarters in Louisville, Kentucky; the planned 82-story Bank of the Southwest Tower, with an innovative system of internal bracing in which triangular steel frames will interconnect columns on opposite sides, in Houston, Texas; and the glass-sheathed Bank of China in Hong Kong, which in 1985 was the tallest building outside of New York City and Chicago.

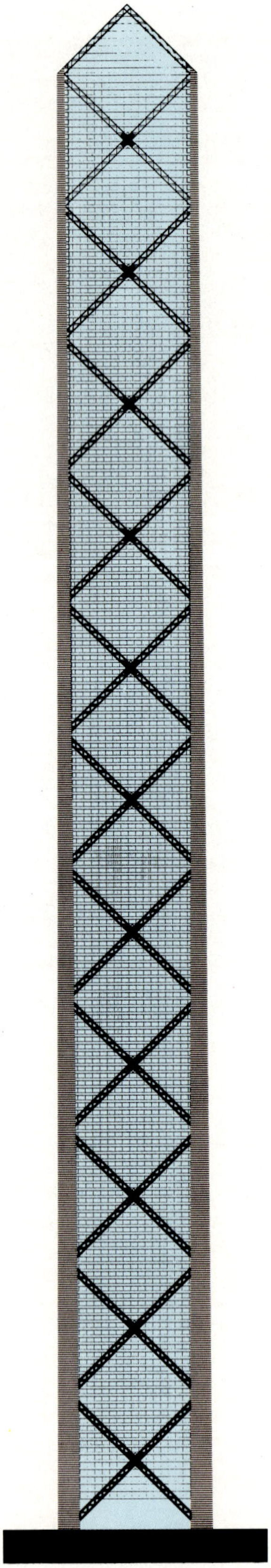

"Superscrapers" under consideration for the future include (far left) a 200-story building some 800 meters (2,640 feet) high that would be supported by massive posts of high-strength concrete at each corner tied together with steel diagonal cross braces. At the left is a proposed 150-story tower comprising five distinct vertical segments of 30 stories each.

(Left) courtesy, Le Messurier/SCI; (right) courtesy, Harry Weese & Associates

The skyscraper and urbanism

The skyscraper and its future must be considered within the context of urbanism. The evolving relationship between the two must be reckoned a factor in the future form of both.

Of the world's 100 tallest buildings half were constructed in the past decade and one quarter since 1980. In the minds of developers—public and private—skyscrapers of 120, 130, 150, and even 210 stories are now under consideration. Even a 500-story skyscraper has been contemplated. Whether such structures will be built does not appear to depend upon engineering or financial capability; both are available. Instead, the question goes beyond the skyscraper as a building enterprise unto itself. It involves the shape and content of the urban habitat.

Born as a by-product of the evolution of the modern city, the skyscraper as a building form has come to characterize the modern city. Forces of centralization compelled this. Those countries in which investment capital largely determines the form of urban development built the biggest skyscrapers and the greatest number of them.

It is curious but understandable that the American skyscraper, built almost exclusively in an urban setting, has been predominantly conceived as a solid mass with only incidental regard for its potential contribution to urban design. Many skyscrapers fill every inch of their site spaces at ground level. The only public spaces at the bases of the Chrysler Building and the Empire State Building are their surrounding public sidewalks. It is possible to pass both at ground level and be totally unaware of the fact that they are among the world's highest buildings. One sees only their lower floors and is likely to perceive them as a two-dimensional street wall rather than as a three-dimensional building mass. One has to be quite far from both buildings to see anything of their silhouettes or their three-dimensional forms. In reaction to this the forecourt or plaza came into being, first gratuitously as in Lever House and the Seagram Building, then as a result of zoning ordinances which made them advantageous to building owners by allowing more building area in upper stories.

Such skyscrapers that can be seen as whole compositions from the ground are often the results of fortuitous sites, two cases in point being the Chicago Tribune Tower and the more recent Lake Point Tower not far away. One is hard pressed to identify skyscrapers that were intentionally sited to be wholly visible compositions from nearby ground positions.

The first major exception to this shortcoming was Rockefeller Center in New York City, conceived from the start as a composition of urban spaces at ground level as much as a composition of slabs aloft. But not even in the area to the west of the original Rockefeller Center, on the Avenue of the Americas, was this urban design lesson repeated. Nor did

Cervin Robinson

Site plan (below) reveals the location in New York City of the IBM Building (above) and its enclosed "Bamboo Court" (left). They are part of a complex that also includes the Bonwit Teller store and the Trump Tower and that provides an indoor pedestrian system between Madison and Fifth avenues.

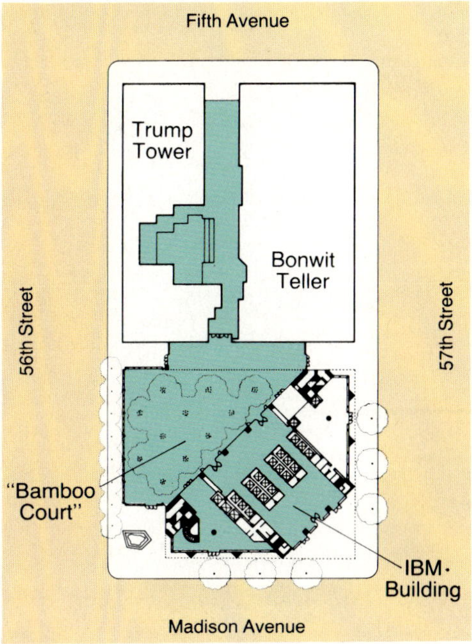

(Left) © Peter Mauss—ESTO; (right) adapted from information obtained from Edward Larrabee Barnes Associates

© 1985 Elliott Varner Smith

Certain skyscrapers play key roles in their urban settings. Two examples are the Toronto (Ontario) City Hall (opposite page) and the Embarcadero Center in San Francisco (above).

an attempt by the architects of New York City to reestablish this idea through urban policy in the 1970s succeed. The closest New York City has come to repeating Rockefeller Center is a group between Madison and Fifth avenues. This consists of the IBM Building and its "Bamboo Court" (Edward Larrabee Barnes), the Bonwit Teller department store, and the Trump Tower (Swanke, Hayden, Connell). That group is interconnected, indoors, by a through-the-block pedestrian system.

When one attempts to identify those skyscraper developments that are as important as urban designs as they are as individual buildings, a limited though hopeful picture emerges. A near contemporary of Rockefeller Center was the Nebraska State Capitol in Lincoln, designed by Bertram Goodhue (1932), as was the University of Texas Main Building in Austin, designed by Paul Cret (1933). The Tower of the Forrestal U.S. Naval Hospital in Bethesda, Maryland, also designed by Cret (1940), can also be counted. Beyond, one can include the Office Tower portion of the Congress House in Brasília, Brazil, designed by Oscar Niemeyer; the Toronto (Ontario) City Hall, designed by Viljo Revell and John B. Parkin; Reskustorni in Tapiola Town Center near Helsinki, Finland, designed by Arno Ervi; the Economist Group in London, designed by Peter and Alison Smithson; and Place Victoria, designed by Luigi Moretti and Pier Luigi Nervi and Place Ville Marie, designed by I. M. Pei, both in Montreal. Those in the U.S. include Kips Bay in New York City, designed by I. M. Pei; the Society Hill Towers and town houses in Philadelphia, also by Pei; Waterside Plaza in New York City, designed by Davis and Brody; and the Golden Gateway and the Embarcadero Center, both designed by John Portman and both in San Francisco. The conclusion is that American accomplishments are extremely small in number in relation to the opportunities.

It is not that U.S. developers (public or private) are indifferent to this opportunity or neglectful of responsibility. Quite the contrary, if one considers such public plazas as the aforementioned Lever House and Seagram Building, as well as the Boston City Hall plaza, or such interior atrium spaces as the IDS in Minneapolis or the State Office Building in Akron, Ohio, a rather hopeful picture emerges. But these worthwhile and commendable contributions beg a bigger question. That question is better answered in those countries in which effective systems of metropolitan and regionally scaled urban planning are practiced. Among them are Canada, Switzerland, France, the United Kingdom, and West Germany. Such systems are able to coordinate private developments with the broader urban interests of the community. The resulting high buildings—mainly offices or residences—are better situated as visual elements of the urban landscape.

Such urban planning and development requires considerable public investment in infrastructure, particularly transportation, utilities, open space, and natural resource protection. Particularly noteworthy in this regard is the recent development of greater Toronto; Vancouver, British Columbia; Stockholm; and many cities in West Germany. In those cities high-rise buildings have been placed so as to allow concentrations without crowding. The big buildings furnish concentrations alongside smaller ones that maintain a human scale.

© Peter Arnold

Future prospects

There can be no doubt that larger and taller skyscrapers will be built. Technology and capital, the two cornerstones of great buildings, make it inevitable. Beyond them is a third factor, the skill of a society in shaping its habitat. The techniques for doing that are as developed as the techniques for financing and building. But they are not as evenly practiced, especially in the United States, where many of the great opportunities still lie ahead. But should that skill improve one wonders if the present trend in skyscraper appearance—the present tendency to aesthetic exuberance if not exhibitionism—may yield to a quieter and better mannered skyscraper expression. If so, it will be one in which there is less need for exuberance, because the skyscraper's form and hence appearance will derive from a harmonious, mutually respectful, and balanced relationship between site and structure. It will be much like the unforgettable power of seeing the spires of Chartres Cathedral rise from the plains of the Beauce, white spires against a horizontal of golden wheat.

There is an ancient building custom, probably Nordic in origin, that has been revived in recent years. When the highest point in a building's structure is reached, the leader of the building team ascends to the top and fixes a fresh-cut evergreen on it, yet surmounting it. The recent revival of this custom occurred at the very time that the new generation of great skyscrapers was rising. May it be taken as a symbol of recognition of deference and respect, of acknowledgment that all our design works are guests in the greater design we call nature. Its rules of harmony have always been our best guide.

FOR ADDITIONAL READING

Wayne Attoe, *Skylines: Understanding and Molding Urban Silhouettes* (John Wiley & Sons, 1981).

Carl W. Condit, *The Rise of the Skyscraper* (University of Chicago Press, 1952).

Carl W. Condit, *The Chicago School of Architecture* (University of Chicago Press, 1984).

ENR Symposium, "Super Tall Buildings: The Sky's the Limit," *Engineering News-Record* (November 3, 1983, pp. 44–67).

Paul Goldberger, *The Skyscraper* (Alfred A. Knopf, 1981).

Charles Jencks, *Skyscrapers—Skycities* (Rizzoli, 1980).

Wolfgang Schueller, *High-Rise Building Structures* (Wiley Interscience, 1977).

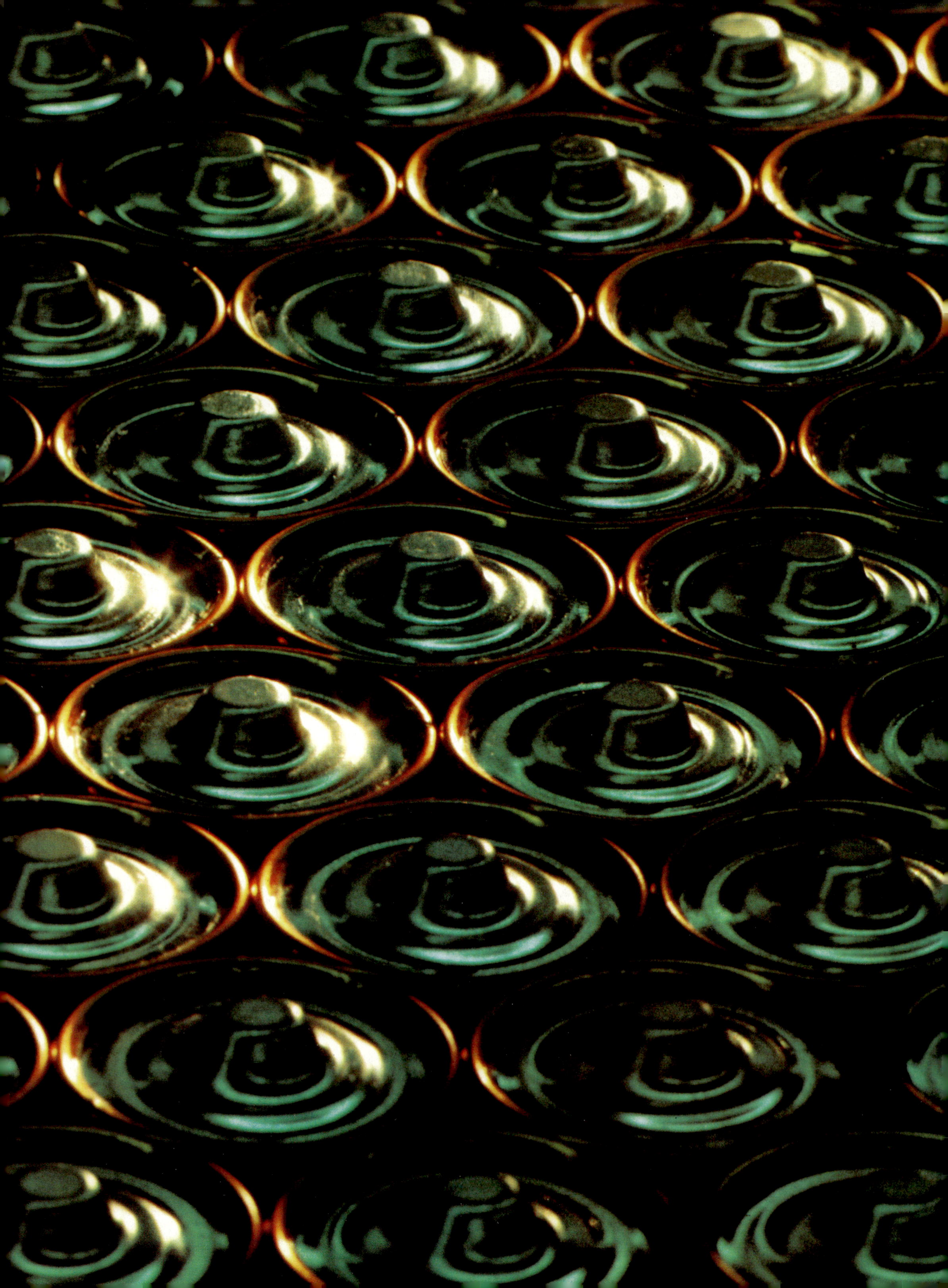

BATTERIES
A TECHNOLOGY RECHARGED

by Colin A. Vincent

Once identified mainly with flashlights and automobiles, batteries are now finding thousands of new applications. Today's lighter, more powerful systems are but the first fruits of a revitalized research and development program.

Although batteries may have been one of the scientific marvels of the early 19th century, their manufacture seems hardly a high-technology glamour industry of the 1980s. Nevertheless, the growing popularity of portable electrical products—from digital watches to audio and video recorders to cordless power drills—as well as the increasingly specialized needs of the semiconductor industry, the military, and the world's energy suppliers have made the battery the object of fierce competition among manufacturers and a topic of serious research. Today smaller, lighter, more powerful, and cheaper batteries, some based on entirely new chemical systems, are being developed for applications that did not even exist a decade ago.

A battery, or electrochemical cell, is a device that enables the energy liberated in a chemical reaction to be converted directly into electricity. A primary battery, for example, an ordinary flashlight dry cell, contains a fixed amount of reagents; when these have been consumed by the discharge process, the cell cannot be used again. Because certain electrochemical reactions can be made to go in reverse, it is possible to devise secondary batteries that can be charged; that is, used to store electrical energy that has been generated by some other means. Such batteries, which include the common lead-acid automotive battery, can be put through many charge-discharge cycles.

In an electrochemical cell the overall chemical reaction may be considered to take place in two steps, one of them occurring at each of the cell's two electrodes. During discharge electrons are released at the negative electrode, or anode, flow through the external circuit, and are consumed in the process taking place at the positive electrode, or

COLIN A. VINCENT *is Reader in Chemistry at the University of St. Andrews, Scotland, and principal author of* Modern Batteries: An Introduction to Electrochemical Power Sources*.*

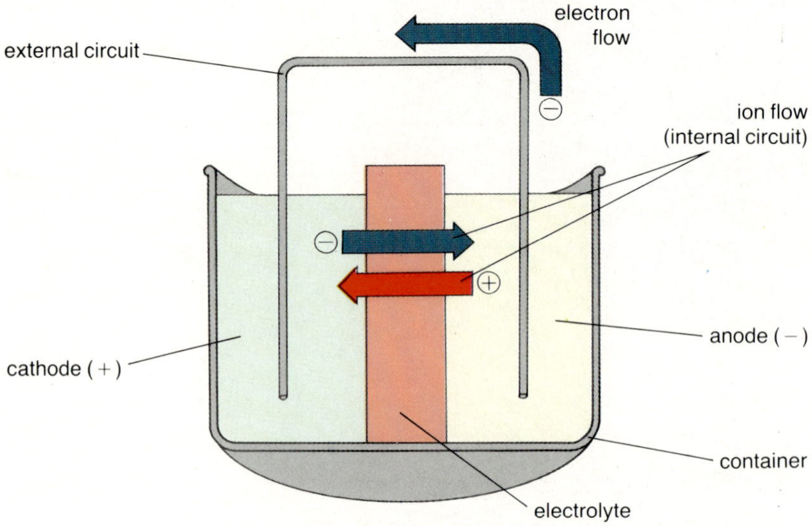

A battery consists of two separate electrode materials, the anode and the cathode, interconnected by a substance, the electrolyte, that contains mobile ions. During discharge, electrons released at the anode flow through an external circuit to the cathode. The circuit is completed internally by the migration of positive and negative ions through the electrolyte toward the electrodes.

cathode. To complete the circuit, the two electrodes are interconnected by a substance containing mobile ions (charged atoms or molecules), known as the electrolyte.

The total electrical energy that a battery can deliver depends on the particular chemical reaction associated with its operation, the size of the battery, and its physical construction. Electrical energy is commonly measured in watt-hours; there are 3,600 joules in one watt-hour. The smallest commercially available batteries, producing about 0.1 watt-hour, are used for such applications as data retention in the integrated circuits of computer memories. Standard silicon-chip packages with built-in lithium-based batteries guarantee the preservation of data for ten years in the absence of external power. In contrast, high-capacity batteries, consisting of many cells in series and parallel arrays, have total energies ranging from 3 million watt-hours for submarines to 100 million watt-hours for storage devices being planned for the electric power industry.

Increasingly diverse demands for portable electric power have spurred the marketing of batteries having a wide variety of shapes, sizes, chemistries, energy densities, and discharge characteristics (below). Uses range from the millions of portable radios, televisions, tape players, calculators, and other consumer devices sold yearly (below center) to advanced-technology applications. Odex 1 (below right), an experimental walking robot being evaluated for its industrial and military potential, operates on a 24-volt lead-acid rechargeable battery.

(Left) Courtesy, Matsushita Battery Industrial Co., Ltd.; (center) James Pozarik; (right) courtesy, Odetics Inc.

The expanding role of commercial batteries

During the period between the first description of an electrochemical power source by the Italian scientist Alessandro Volta in 1800 and the large-scale introduction of the dynamo some 70 years later, batteries provided the only source of electricity. Today, because of the much lower cost of electricity from generating stations powered by fossil fuels or nuclear energy, the use of batteries has become centered on two principal applications. First, batteries provide the most convenient portable source of electric power. Examples include the well-known button cells in electronic watches and pocket calculators, the lead-acid batteries in cars and trucks, and rather more specialized power units such as those for arming artillery shells or powering orbiting communications satellites when their solar-cell arrays are in the Earth's shadow. Second, batteries provide emergency power for fixed installations such as hospitals, telecommunications facilities, and public buildings and allow the temporary storage of energy generated by solar arrays.

The consistent growth in the battery market throughout the world—by a factor greater than 20 since the end of World War II—is a result of commercial demand arising from new applications and concurrent advances in battery technology. For more than a century commercial battery production has been dominated by two types of battery: the Leclanché, or zinc-manganese dioxide, primary cell and its variants (which include so-called carbon-zinc and alkaline dry cells) and the lead-acid secondary battery. The popularity of the former—annual sales exceed $1 billion in the U.S. alone—is due largely to cheap raw materials and ease of manufacture. The emergence of microelectronics and the consequent evolution of large-scale integrated circuits were the stimuli for developing miniature batteries (0.1–2 watt-hours) with a much higher ratio of available energy per unit volume than that of conventional cells. A typical electronic watch circuit requires some 0.5–2 microamperes for operation and hence a total consumption of 0.015–0.060 watt-hours (15–60 milliwatt-hours) per year. A suitable battery must supply this energy reliably over a period of two to ten years. It must, in addition, maintain

United Technologies Mostek

Two tiny lithium button cells, seen in the uncased assembly at the right, allow this memory chip to preserve data entered into it for ten years in the absence of external power.

William Schroeder (below left), recipient of an artificial heart driven by compressed air, enjoys the mobility provided by a shoulder-slung portable compressor, which runs on rechargeable cadmium-nickel oxide (nicad) batteries. (Below) A bank of more than 1,600 lead-acid storage batteries housed in a stationary facility the size of a basketball court protects data from power failure at a Boeing computer center in the U.S.

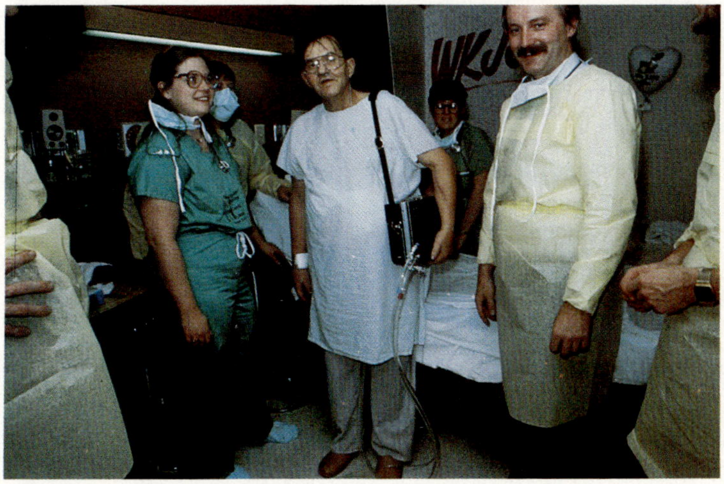

(Left) William Strode—Humana Inc./Black Star; (right) © 1984 Steve Firebaugh

(Left) Adapted from "Battery Research is Back in Business," Colin A. Vincent, *New Scientist*, vol. 101, no. 1403, pp. 34–38, March 29, 1984; (right) Medtronic

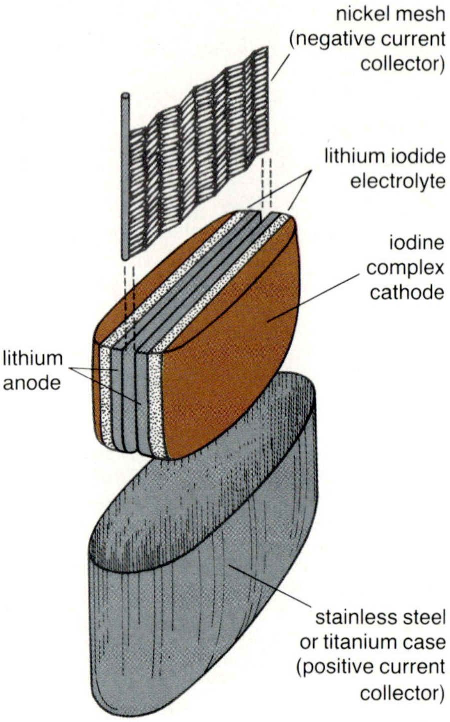

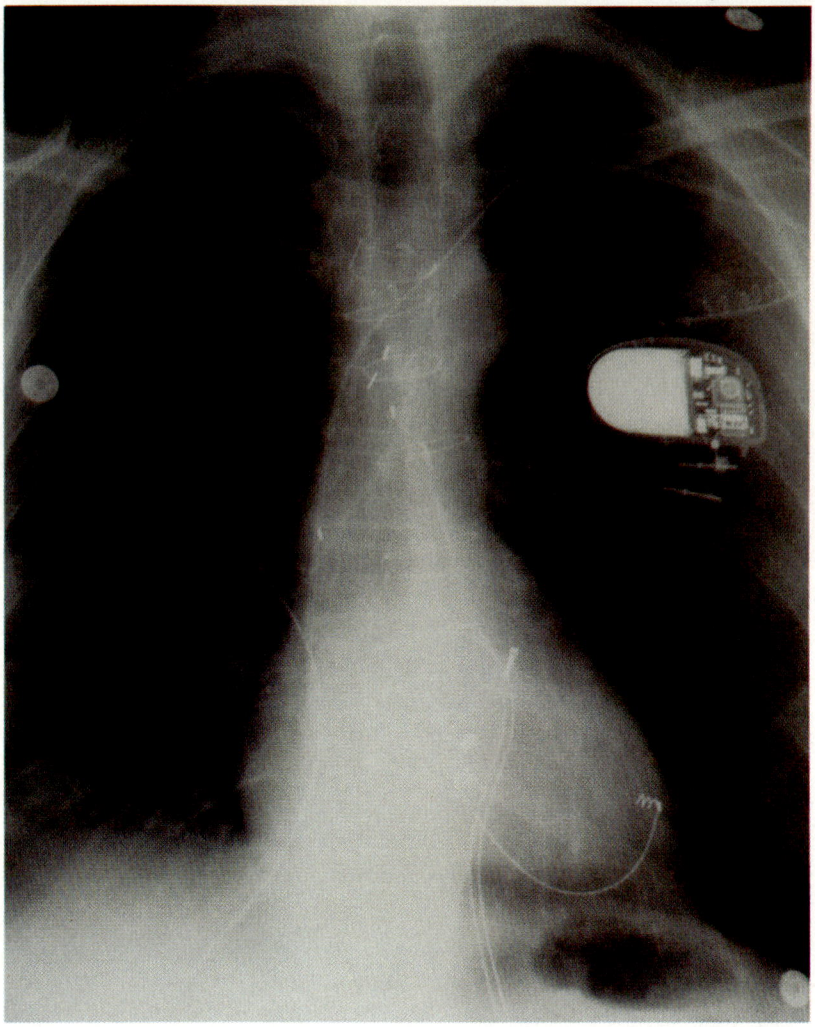

X-ray of an implanted cardiac pacemaker (above right) shows its position in the body as well as inner details of the device itself. Power for the pacemaker's microcircuitry is supplied by a long-lived lithium-iodine battery, which occupies the left side of the case. The all-solid-state pacemaker battery (above) employs lithium metal for the anode, a complex of iodine for the cathode, and a thin layer of crystalline lithium iodide for the electrolyte.

a nearly constant voltage during its working life, be capable of supplying high currents for brief periods to operate alarms or to backlight liquid-crystal displays, and occupy a volume not much larger than a cubic centimeter, about the size of two aspirin tablets. Batteries with outputs of 2–100 watt-hours are required for the flood of portable equipment being developed for home use (*e.g.*, cordless power tools, kitchen and garden implements, headphone radios and tape players, portable computers, and electronic toys) and for military applications (night-vision devices, communications sets, and missile guidance). Most modern applications require higher gravimetric or volumetric energy densities (energy per unit mass or volume) and superior electrical characteristics than are generally available from batteries based on earlier technologies.

Lithium primary batteries

The basis of recent advances in battery technology lies in a combination of research and development in electrochemistry, engineering, and materials science. Of the new materials being exploited, none is more important than lithium. This metal is attractive as a battery anode because of its low density (it is the lightest metal element) and the large chemical energy associated with its oxidation; *i.e.*, the loss of one of its atomic electrons. However, lithium reacts vigorously with water so that lithium cells must use nonaqueous electrolytes.

The successful development of the all-solid-state lithium-iodine battery is an excellent example of the fusion of ideas from different branches of science and technology. In this cell the anode is lithium metal, and the cathode a complex of iodine. The electrolyte is a thin layer of crystalline lithium iodide through which current is carried by the migration of lithium-ion vacancies in the crystal lattice. Solid-state cells of this type are hermetically sealed and may be used over a wide range of operating temperatures, pressures, and accelerations. They have excellent storage stability, about ten years, but the relatively high electrical resistance of the solid electrolyte restricts them to applications in which current drain is low. Their high reliability and the absence of problems associated with electrolyte leakage or gas generation makes these cells ideal for powering implanted medical devices. The battery powering a typical modern "demand" pacemaker, which functions only when it detects an irregular heartbeat, has a volume of about six cubic centimeters and a lifetime of a decade.

The term ambient-temperature primary lithium cell covers a wide variety of power sources all of which employ lithium as the anode but use different cathode materials and electrolytes. The most common electrolytes for this new generation of miniature and small primary cells are based on polar organic liquids containing dissolved lithium salts. Cells with solid cathodes such as carbon monofluoride or the oxides, sulfides, and oxosalts of manganese, chromium, and other metals are fairly easily fabricated into button and cylindrical shapes that can produce as much as three watt-hours of energy. They have high volumetric energy densities and deliver a very constant output voltage throughout their lifetime. Cells with soluble cathodes use gases such as sulfur dioxide dissolved in an electrolyte, or liquids such as thionyl chloride, which act both as cathode material and as solvent for the electrolyte. Despite the fact that the reactive lithium anode is in contact with the cathode material, the formation of a protective film on the lithium prevents significant self-discharge.

Cells based on sulfur dioxide have an outstanding performance and are capable of a high rate of discharge. They can operate at temperatures as low as −50° C (−58° F) and as high as 70° C (158° F) and have a very good shelf life, losing less than 10% of their available energy after six years of storage. Such cells are generally fabricated as cylinders with energies as high as 75 watt-hours. One problem is that, if abused (*e.g.*, by exposure to high temperature or prolonged short circuit), lithium-sulfur dioxide cells may explode or vent a toxic gas. Until recently applications have been limited mainly to military devices such as portable communications sets and prototype pilotless, electrically powered aircraft, where high power density is a great advantage. Civilian use has been restricted due to safety considerations. The need to fabricate gastight seals, incorporate pressure-relief safety vents, and maintain a high standard of construction makes these batteries comparatively expensive.

Lithium cells based on liquid thionyl or sulfuryl chloride, containing lithium aluminum chloride ($LiAlCl_4$) as electrolyte, have among the highest energy densities of any practical battery system. For large, low-discharge-rate cells, values of 500 watt-hours per kilogram (225 watt-hours per pound), or 0.9 watt-hour per cubic centimeter, have been achieved. A typical lithium-thionyl chloride battery the size and shape of a D-cell (54 cubic centimeters) has a practical energy output of 34 watt-hours, compared with about 5 watt-hours for a standard carbon-zinc Leclanché cell and 12 watt-hours for an alkaline cell of the same size. A wide range of these primary lithium batteries are manufactured, from miniature button cells to 60,000-watt-hour units used to provide standby power in missile silos.

New-generation aqueous primary batteries

In addition to the commercial introduction of lithium-based systems, progress is also being made in the development of aqueous primary cells. After the first patent for a dry cell, that is, one in which the liquid electrolyte is immobilized in a paste or gel, by Carl Gassner of Germany in 1886, continuous advances in the performance of zinc-manganese dioxide cells approximately doubled the energy density every 20 years until 1960. Since then, increasing use of synthetically prepared manganese dioxide (in preference to naturally occurring ores), thin coated-paper separators, improved electrolyte formulations, and new cell-sealing techniques have boosted performance still further. Nevertheless, in its standard carbon-zinc form this cell is best suited for intermittent or low-current use and in situations when a constant output voltage is not important.

An increasingly popular variant, the alkaline cell, can operate satisfactorily at higher rates. This cell has an electrolyte based on potassium

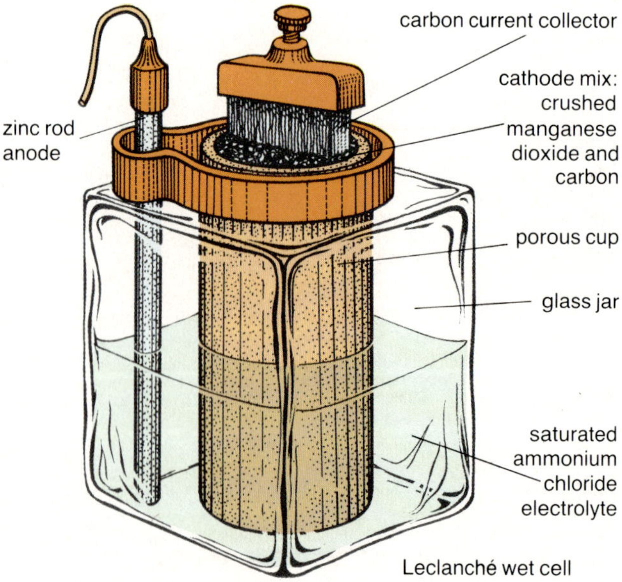

Leclanché wet cell

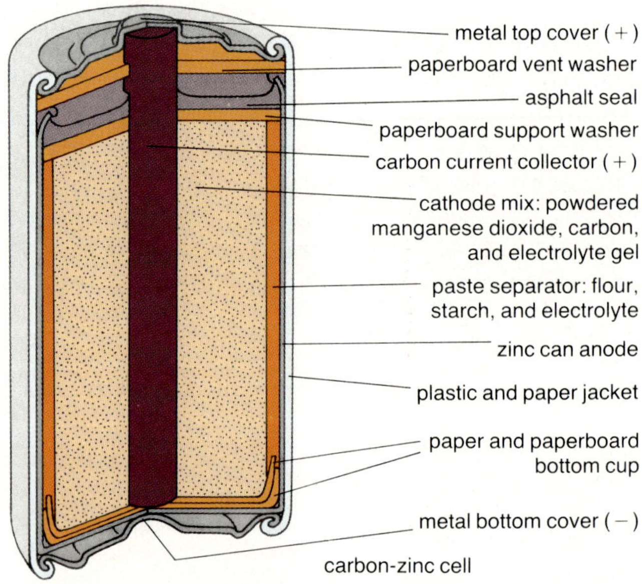

carbon-zinc cell

hydroxide, a strong corrosive, which necessitates the added expense of a steel case and a well-engineered sealing assembly. Two other aqueous primary systems are also commercially important: the silver (zinc-silver oxide) cell and its closely related precursor, the mercury (zinc-mercuric oxide) cell. Both cells have a high volumetric energy density and electrical characteristics at least as good as lithium cells. Although widely available, mainly as low-drain button cells, the cost of materials is high, making them comparatively expensive.

The primary objective of miniature battery design is to pack as much energy as possible in a small container of fixed size. Using atmospheric oxygen as the cathode eliminates the need to store cathode material; the freed space can be filled with additional anode material, which translates into an increase in energy density. Cells based on this principle, called metal-air batteries, have been known for more than a century, and for many years low-cost cells with air cathodes have powered electric fences, signaling circuits, and similar low-power devices. More recently researchers have developed a thin, high-performance oxygen electrode consisting of a Teflon-bonded catalyst composite covered with a gas-permeable waterproof membrane. The result is a zinc-air button cell of unrivaled volumetric energy density, especially suitable for devices requiring fairly high current drain such as hearing aids.

Today's trends in the worldwide primary battery market are fairly clear. In the five-to-ten–watt-hour realm of the flashlight battery, consumers are moving from carbon-zinc to alkaline cells because the superior performance and shelf life of the latter makes them generally more cost-effective. Lithium systems are likely to take over from mercury and silver miniature batteries for a number of reasons. Not only do lithium batteries have electrical discharge characteristics as good as those of the best mercury and silver miniature cells, but they have higher gravimetric energy densities, operate at lower temperatures, and demonstrate supe-

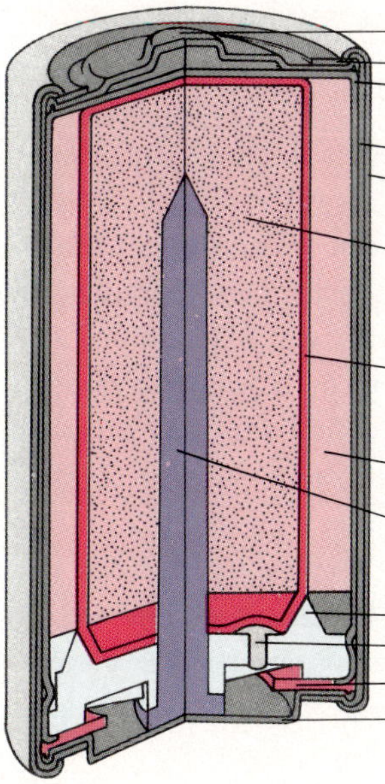

The original form of the Leclanché zinc-manganese dioxide cell, invented about 1865, is compared with two modern variants, the carbon-zinc and alkaline cells. Although their basic chemistries are virtually identical, continuous advances in materials, design, and manufacturing techniques have greatly boosted the performance of today's versions. The design of the carbon-zinc cell is simply a reengineered form of the original Leclanché cell, with the liquid electrolyte replaced by a gel that is mixed with the manganese dioxide cathode, the porous cup by a paste separator, and the glass container and zinc anode rod by a zinc can that serves both purposes. The alkaline cell uses the carbon-zinc design turned inside out: a core of powdered zinc and alkaline electrolyte separated by fabric from an outer layer of manganese dioxide and electrolyte.

(Top) Courtesy, Matsushita Battery Industrial Co., Ltd.; (center) U.S. Army Harry Diamond Laboratories; (bottom) courtesy, COMSAT

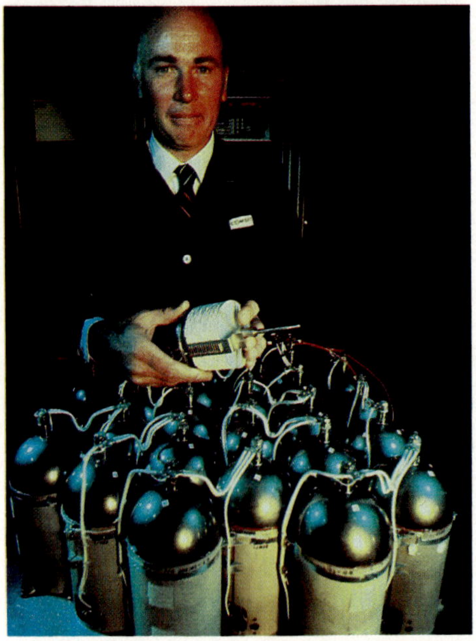

Three very different batteries illustrate the great diversity of forms: (from top to bottom) a "paper" lithium primary cell the size of a postage stamp and only 0.5 millimeter (one-fiftieth of an inch) thick; a lead-lead oxide artillery-projectile battery, activated in a fraction of a second by the spin imparted to the projectile as it travels through the gun barrel; and a cluster of rechargeable nickel-hydrogen cells used to power Intelsat V communications satellites.

rior shelf life. Individual lithium cells commonly supply about 3 volts, compared with about 1.5 volts for the aqueous cells, thus reducing the number of cells needed in series combination for certain applications. Furthermore, lithium is a relatively abundant material; as noted above, costs of lithium cells will fall as production rises. On the other hand, the high cost of silver and toxic nature of mercury and its compounds do not improve the competitiveness of cells based on these metals.

Nevertheless, primary batteries remain a costly way of purchasing energy. For example, it is over 1,000 times more expensive to buy electricity in the form of a carbon-zinc cell than from a public utility; for a zinc-silver oxide button cell the factor is over 100,000. Such economics underlie an important pursuit of battery research: the replacement of primary batteries with compatible rechargeable units of acceptable characteristics and performance.

Electrochemical energy storage

The mainstays of electrochemical storage systems are the Planté lead-acid battery, which possesses 90% of the market, and the alkaline cadmium-nickel oxide, or nicad, battery, which accounts for most of the remainder. About 75% of the lead-acid market is for automotive batteries for starting, lighting, and ignition, and car owners who have experienced the changeover to low- and no-maintenance batteries will be familiar with the improvements of the past 20 years. In particular the use of low-antimony, calcium-lead grids for electrodes has practically eliminated water loss during charging and greatly reduced the rate of self-discharge. Another achievement has been the development of small, sealed, maintenance-free lead-acid and nicad cells for such applications as cordless power tools, portable televisions, and personal computers.

Because of increased concerns with global energy resources and the environment, attention is focusing on electrical energy storage on a scale many orders of magnitude greater than present practice. While renewable natural energy sources such as solar, wind, wave, and tidal power offer decreased dependence on fossil fuels and reduced pollution, a major problem is that they are intermittent. To exploit them efficiently it is necessary to store energy during favorable generating periods for later demand.

A growing requirement for energy storage also exists within the traditional electric power industry. For nuclear reactors and certain other power plants to operate efficiently they must run at a fixed output level; *i.e.*, the cheapest electricity is produced under conditions of a constant level of demand. Unfortunately demand during a 24-hour cycle invariably plunges during the night and peaks during the day. The principle of load leveling is to smooth out the demand by operating above minimum demand levels at night and storing the surplus to help meet the peak demand later in the cycle. A number of load-leveling methods are currently in use. For example, in mountainous, temperate regions pumping water into elevated reservoirs for later hydroelectric generation can be economical despite the high capital costs.

Batteries, it has been suggested, will soon find increasing application in load leveling. Being modular, they enable an energy storage facility to be expanded, reduced, or moved to a new site. A particular advantage is that small units, free from environmental pollution, can be placed in urban locations near industrial load centers and so lead to significant savings in transmission costs, especially if the distribution network is reaching its maximum capacity during peak demand periods.

Electrically powered cars and trucks represent another potential large-scale application for storage batteries. Electric road vehicles have a long history stretching back to the late 1830s when the world's first "electric carriage" was built in Scotland. Before the end of the 19th century a French battery-powered car had pushed the world land speed record above 100 kilometers (62 miles) per hour, and by 1916 annual production of electric vehicles in the U.S. had reached 10,000. Presently, however, despite the manufacture of a large number of industrial and leisure electric vehicles, no major production line exists for commercially competitive general passenger or freight vehicles. The main reason for the failure of the electric car to compete with its internal-combustion-powered rivals lies in the comparatively low energy density of conventional storage batteries such as the lead-acid system, which restricts the range and payload of practical vehicles. For example, to deliver the same energy supplied by a liter (about a quarter gallon) of gasoline would require a lead-acid battery weighing some 40 kilograms (88 pounds).

The performance needed from batteries intended for storing electricity from intermittent renewable energy sources, for load leveling, and especially for practical electric vehicles greatly exceeds what is likely to be achieved by improving conventional systems. In contrast to the familiar car battery, which is rarely deeply discharged, batteries for the above

Gaston Planté's original lead-acid cell of 1860 (below) is contrasted with a modern 12-volt maintenance-free automotive battery (below left) and a 6-volt predecessor from the early 1950s (inset). Electrode grids made from calcium-lead rather than antimony-lead alloys allow maintenance-free batteries a capless, sealed design, since the need to add water is eliminated. The new grids also reduce self-discharge, permit a longer storage life, and resist corrosion and breakage.

(Left and center) Delco Remy Division/General Motors Corporation; (right) SAFT

(Left) Black & Decker; (right) Hewlett Packard

The development of lightweight rechargeable batteries for office and household products has been a significant achievement of battery research. The portable personal computer at right, which features 16-line liquid-crystal display, operates for as long as 16 hours per charge on three sealed, D-size lead-acid cells. Five nicad batteries power the handheld cordless vacuum cleaner, below, for 15 minutes on a full charge.

applications must withstand a deep, possibly complete discharge followed by recharge on a daily basis for as many as 2,000 cycles in order to give a five-to-ten-year lifetime. For stationary storage, energy density is not of overriding importance, but the cycle efficiency, defined as the energy delivered during discharge compared with the energy needed for charge, is critical. In the case of electric vehicle batteries the requirements are particularly demanding—to a long cycle lifetime and reasonably high cycle efficiency must be added a high gravimetric energy density and high peak power output.

Advanced aqueous secondary batteries

Conventional lead-acid batteries designed for deep-cycle applications have energy densities of the order of 25–30 watt-hours per kilogram (11–14 watt-hours per pound). Improvements may eventually double this value. On the other hand, a number of advanced batteries now under development have already achieved 100–140 watt-hours per kilogram (45–65 watt-hours per pound). Refinements in these devices and the introduction of new systems are likely to result in batteries that deliver 250 watt-hours per kilogram (115 watt-hours per pound) by the end of the century.

New-generation rechargeable batteries at an advanced stage of development fall into two principal groups depending on their operating temperature: ambient and high-temperature. Ambient systems based on aqueous electrolytes and operating at or close to the temperature of their surroundings achieve their high performance by employing light, reactive chemical components. A good example is the zinc-chlorine cell being developed both for electric vehicles and for load leveling. A key problem with this system is the storage and manipulation of the highly corrosive chlorine gas. In one ingenious solution the chlorine, produced during charging, is converted into the solid hydrate $Cl_2 \cdot 6H_2O$, which is stable below 9.6° C (49.3° F) and which may be readily stored in a refrigerated chamber. In this battery the electrolyte, zinc chloride dissolved in water, must be circulated through the electrode stacks using pumps; its disadvantages therefore lie in its complexity and in the energy required to operate the pumps and refrigeration equipment.

Bromine is the cathode material in a related battery. This element has a lower energy density than chlorine, but its comparative ease of handling and storage—in an organic compound rather than by refrigeration—makes it an attractive alternative. Although the lifetime of experimental cells has been shown to be very good, performance improvements—especially in cycle efficiency—continue to be pursued.

Metal-air rechargeable systems have the potential for development into practical batteries with superior energy densities. High-rate air electrodes have been subject of much research over the past 25 years in connection with fuel cells, and zinc-air and iron-air rechargeable batteries are under development in a number of countries. The aluminum-air system, which has an exceptional energy density, is regarded as a long-term possibility. Problems associated with changes in the form and structure (morphology) of the metal anode have been largely solved by the use of metal slurries and circulating electrolyte, so that the limiting factor in the production of practical cells lies in the behavior of the air anode. Development of a successful high-rate reversible air electrode presents one of the most formidable challenges in battery research.

Another system that relies on the pumped flow of materials from external reservoirs is the redox battery; for example, the $Fe^{3+}/Fe^{2+}-Cr^{2+}/Cr^{3+}$ (iron-chromium) system. In this case both of the chemical reagents and both of the products are soluble in water. During discharge the reactant solutions, which also contain the electrolyte, flow through a cell that is divided into two compartments by an ion-exchange membrane. Oxidation and reduction occur at chemically inert electrodes in the separate compartments, and the spent solutions are stored in external tanks. During charging the whole process is reversed. This type of system has a number of important advantages. Since reagents and products are al-

(Left) Photo Hutin; (right) Peugeot S.A.

Frenchman Camille Jenatzy's battery-powered "La Jamais Contente" (left), an early electric car, achieved a record speed for land vehicles of 105 kilometers (65 miles) per hour in 1899. The prototype 205 Electric (above), from Peugeot, runs on 12 six-volt nickel-iron secondary batteries that deliver twice as much energy as lead-acid units of identical weight. The newly developed batteries allow the car to travel about 160 kilometers (100 miles) in normal city driving on a ten-hour charge.

Robert Cumming

Under study by electric utilities for load-leveling applications, the ambient-temperature zinc-chlorine secondary battery at right relies on a complex system of refrigeration equipment and pumps to store and manipulate its highly corrosive cathode material, chlorine gas.

ways in solution, there are no problems with morphology changes in the electrodes and hence no limitations on discharge depth or cycle lifetime. In addition, because of the way that the redox system operates, capacity and power output are very flexible. Despite difficulties related to the mechanical complexity of the design and the energy wasted on circulation pumps, a number of prototypes are being studied for storing energy from solar arrays and wind-turbine installations in remote locations.

Other aqueous secondary batteries have been developed for specialized applications. For example, the excellent characteristics of the zinc-silver oxide system have been exploited in developing a rechargeable unit with the highest energy density of any commercially available, although the high cost of silver has restricted it mainly to military and space applications. Early zinc-silver oxide batteries were used in torpedoes;

U.S. Department of Energy

Ease of handling makes the zinc-bromine battery an attractive alternative to its zinc-chlorine relative, despite its lower energy density. In this system the bromine cathode is stored in an organic compound rather than by refrigeration.

more recently they powered the Lunar Rovers of the Apollo Moon missions and presently are running payload-handling equipment on the U.S. space shuttle. Another successful space-oriented storage battery is the nickel-hydrogen system, used extensively in communications satellites.

NASA

Two 16.8-volt zinc-silver oxide batteries supply electric power to the manned maneuvering unit, shown in use for the first time by a U.S. space shuttle astronaut in 1984. The cells, each delivering 752 watt-hours of electricity, can operate the gas-propelled backpack for a six-hour spacewalk. Other successful space-oriented applications for zinc-silver oxide systems include the Lunar Rovers of the Apollo Moon flights and payload-handling equipment aboard the shuttle.

High-temperature secondary batteries

High-temperature batteries operate in the range of 300°–450° C (570°–840° F). For electrolytes they use molten salts or solids having high ionic conductivity, enabling them to deliver high power levels. Further, by taking advantage of light, reactive metals such as lithium or sodium as anodes, such batteries achieve excellent energy densities. The two principal contenders for electric-vehicle and load-leveling applications from this group are sodium-sulfur and lithium-iron sulfide batteries.

The former is unusual in that it employs liquid electrodes (molten sodium and sulfur) and a solid electrolyte, most commonly sodium beta-alumina that has been heated and compressed to form a ceramic tube. During cell discharge positively charged sodium ions migrate through the solid electrolyte into the positive compartment where they combine with sulfide ions being produced there. As the cell is recharged, the sodium ions move back through the beta-alumina tube to be reduced to metallic sodium at the negative electrode. Sodium-sulfur systems, often called beta batteries, are under development in a number of countries. Considerable advances have been made in cell technology, thermal management, and reliability, and numerous vehicular and stationary batteries have been built and tested. Full commercial production awaits the development of more durable ceramic tubes and seals and a longer cycle life without serious loss of capacity.

Alternative systems based on the liquid-electrode/solid-electrolyte configuration but running at somewhat lower temperatures have been proposed in an attempt to alleviate the materials problems encountered above 300° C. In the sodium-antimony trichloride cell, sulfur is replaced by $SbCl_3$ dissolved in a melt of sodium chloride (NaCl) and aluminum chloride ($AlCl_3$) at 220° C (430° F). In another version of the sodium-sulfur cell, the electrolyte is fabricated from many thousands of hollow glass fibers that conduct sodium ions.

The second major high-temperature battery program involves cells with solid electrodes—namely, a lithium-aluminum or lithium-silicon alloy as the anode and iron sulfide (FeS or FeS_2) as the cathode—and a molten salt electrolyte. The latter is generally a low-melting mixture of lithium chloride and potassium chloride or of lithium chloride, fluoride, and bromide. As with all high-temperature batteries, one of the main tasks is to find suitable materials for cell components that can withstand the extremely hostile environment of the cell. In this case the problems are particularly severe since the reactivity of lithium precludes the use of common ceramics such as silica and alumina, while the corrosion of steel current collectors in FeS_2-based cells necessitates the use of expensive molybdenum alloys. The most successful separator material tried in prototype cells has been a boron nitride fabric, but it is so expensive that it accounts for as much as 75% of the total cost of cell materials. A number of alternatives are being studied. Electric-vehicle batteries producing 40,000 watt-hours and based on FeS have been built and tested, and a design study has been made for a 100,000-watt-hour stationary electrical storage facility.

Ambient-temperature lithium secondary batteries

A practical, ambient-temperature rechargeable lithium battery is one of the principal goals of modern battery research. A possible approach to producing cathodes that will cycle reliably without degradation involves the use of an "intercalation" reaction. The principle of such a process is that, during battery charging, positively charged lithium ions from the electrolyte are reduced to neutral lithium atoms and inserted into spaces within the crystalline lattice of a host material while causing little or no structural changes to the latter. On discharge the lithium leaves the insertion-host lattice and reenters the electrolyte. Investigators have had considerable success in devising practical insertion-host cathodes based on such materials as titanium sulfide (TiS_2) and the vanadium-oxygen compound V_6O_{13}. The main problem still plaguing such cells is the poor cycling behavior of the lithium anode. In primary lithium cells the lithium reacts with components in the electrolyte solution to form a protective film that gives these cells their long shelf life. This same lithium-electrolyte reactivity, however, greatly restricts the cycling characteristics of lithium anodes, leading to failures in secondary cells. Major research programs are concentrating on new electrolytes or modified lithium substrates in an attempt to solve this problem.

Recent research has yielded a number of new polymer-based materials that may play a large part in the future development of batteries of this type. Of particular interest are polyethers containing dissolved lithium salts that can function as electrolytes. Their advantage over conventional liquid electrolytes is that they can be formed as thin, coherent films, so

Courtesy, Chloride Silent Power Limited

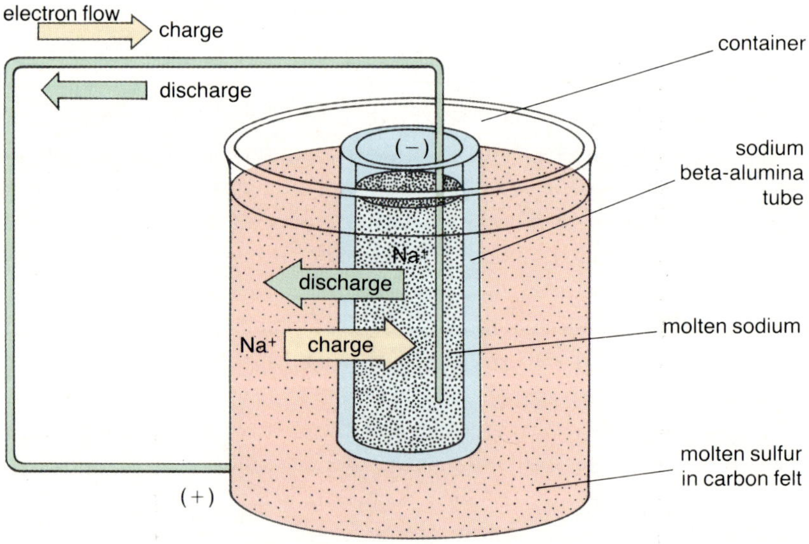

The rechargeable sodium-sulfur, or beta, battery (right) employs a molten sodium anode and a molten sulfur cathode separated by a solid ceramic electrolyte of sodium beta-alumina. During discharge positively charged sodium ions (Na+) migrate through the beta-alumina tube to the cathode where they combine with sulfide ions being produced there. During charging sodium ions move back through the tube to the anode where they are reduced to metallic sodium. The solid electrolyte tube shown above, fabricated for a beta battery, is 33 millimeters (about 1.25 inches) in diameter and has a wall thickness of 1.5 millimeters (0.06 inch).

allowing the configuration of batteries with very high energy densities. Experimental lithium cells employing such a polymer electrolyte and running at just over 100° C (212° F) are being studied.

The possibility of using electrochemically doped, electronically conducting polymers (*e.g.*, polyacetylene) as electrode materials and hence of producing an all-polymer battery has attracted much attention, but implementation of practical systems still appears to be some way off. Presently polyacetylene-based systems suffer from two major problems, one connected with the overall chemical stability of the material, especially in an oxidizing environment like the atmosphere, and the other involving difficulties experienced in regaining all the electrical energy used to charge polyacetylene-based electrodes when they have been left on standby for long periods or when high currents are required. Pursuit of solutions to these problems lies at the forefront of research, as does the development of other electronically conducting polymers with superior characteristics. Nevertheless, it seems likely that all-polymer batteries will first appear in low-power applications, *e.g.*, self-powered credit cards containing silicon chips, before being tried in high-power devices.

The future

At the present time future trends in commercial batteries are not entirely clear. Most developmental work in small primary cells continues to be on lithium systems. Although such cells now account for only a small percentage of the total primary-cell market, significant gains seem certain as availability increases and costs fall with high-volume manufacture.

The status of new-technology storage batteries is rather different; it is still unclear which, if any, of the competing systems is likely to succeed commercially. On the other hand, a practical large-scale electrochemical energy storage system is widely held to be of key importance in future energy-management plans. Investment in research into new battery materials will remain an essential component of any strategy to maintain economic growth in a world in which fossil-fuel reserves are rapidly vanishing.

FOR ADDITIONAL READING

V. S. Bagotzky and A. M. Skundin, *Chemical Power Sources* (Academic Press, 1980).

J. O'M. Bockris, Brian E. Conway, Ernest Yeager, and Ralph E. White (eds.), *Comprehensive Treatise of Electrochemistry,* vol. 3: *Electrochemical Energy Conversion and Storage* (Plenum Press, 1981).

Jean-Paul Gabano (ed.), *Lithium Batteries* (Academic Press, 1983).

Johannes Jensen, Peter McGeehin, and Ronald M. Dell, *Electric Batteries for Energy Storage and Conservation* (Odense University Press, 1979).

David Linden (ed.), *Handbook of Batteries and Fuel Cells* (McGraw-Hill Book Company, 1984).

Colin A. Vincent, with Franco Bonino, Mario Lazzari, and Bruno Scrosati, *Modern Batteries: An Introduction to Electrochemical Power Sources* (Edward Arnold Ltd., 1984).

PENDULUMS IN THE TEST TUBE

by Irving R. Epstein

Not long ago a simple inorganic reaction that oscillates periodically would have been thought to violate a natural law. Today such reactions are being systematically designed and studied as analogues of periodic behavior in living organisms.

Show a child a wobbling top, a physicist a pendulum, or a physiologist a beating heart, and they nod with understanding and a certain degree of pleasure. To most people and particularly to most scientists, periodic oscillation is a normal and pleasing mode of behavior for the objects of nature. For the majority of chemists, however, until quite recently the notion of a test tube full of liquid that could periodically change its color from red to blue or green to yellow and back again was an extremely perplexing one.

New advances in chemistry have made chemical oscillation not only credible and accessible but a source of potential insight into a variety of problems in chemistry, biology, the earth sciences, and other disciplines. Although the first chemical oscillators were discovered accidentally, it is now possible to design reactions that display not only oscillatory behavior but other, more exotic phenomena as well.

Early developments

The first evidence for oscillation in a homogeneous chemical reaction was obtained accidentally by W. C. Bray at the University of California at Berkeley in 1921. While studying the way iodate (the IO_3^- ion) catalyzed the decomposition of hydrogen peroxide in solution to oxygen and water, Bray noticed that a brown color characteristic of iodine periodically appeared and disappeared and that bubbles of oxygen were produced in bursts at regular intervals. Bray's discovery encountered more skepticism and indifference than enthusiasm in the chemical community, and it was largely ignored for the next 40 years. This response, or lack of it, was due largely to a widespread but mistaken belief that the laws of thermodynamics forbid chemical oscillation just as surely as they prevent the operation of perpetual motion machines.

A second chemical oscillator, also discovered by serendipity, was found by B. P. Belousov in the Soviet Union in 1958. Again, there was initial resistance to the notion of such behavior, but Belousov and especially his compatriot A. M. Zhabotinsky continued to probe this reaction, in which an organic acid, usually citric or malonic acid, loses carbon dioxide and gains a bromine atom when it reacts in solution with bromate ion (BrO_3^-) in the presence of a metal catalyst such as cerium or manganese. As the reaction progresses, the solution changes periodically from colorless to pale yellow and back. In part because developments in the theory of nonequilibrium thermodynamics had shown that chemical oscillation was neither a physical impossibility nor a scientific fraud and in part because Zhabotinsky published recipes that allowed the oscillations to be reproduced easily in the laboratory, a number of chemists soon became aware of and fascinated by the surprising Belousov-Zhabotinsky (BZ) reaction.

Throughout the 1960s the BZ system was known to a relatively small group of chemists as an entertaining but enigmatic lecture demonstration. Its possible relevance to the built-in biological clocks of humans and other organisms was recognized, and it was the subject of much discussion at a conference in Prague in 1968 on "Biological and Biochemical Oscillators."

IRVING R. EPSTEIN *is Professor and Chairman, Department of Chemistry, Brandeis University, Waltham, Massachusetts.*

(Overleaf) Sequence of photos taken over a 12.5-second period follow the color changes in the Briggs-Rauscher oscillating chemical system. The reaction, a hybrid of the Belousov-Zhabotinsky (BZ) and Bray systems, changes from colorless to gold to dark blue and back again. A magnetic stirrer, added to keep the reactants well mixed, is responsible for the funnel-shaped vortex in the center of the flask.

(Overleaf) William Shea, Brandeis University

In 1972 Richard Field, Endre Koros, and Richard Noyes at the University of Oregon proposed a reaction mechanism for the BZ system. To a chemist a reaction mechanism represents a complete understanding of the way in which a particular chemical reaction takes place. It consists of a series of "elementary steps"; that is, chemical equations each of which describes an actual encounter between molecules. For example, the overall equation for the combustion of octane, an important process when gasoline is burned in a car engine, is:

$$\underset{\text{octane}}{2C_8H_{18}} + \underset{\text{oxygen}}{25O_2} \rightarrow \underset{\text{carbon dioxide}}{16CO_2} + \underset{\text{water}}{18H_2O}$$

The equation does not imply, however, that the reaction occurs when 27 molecules collide simultaneously to form the appropriate products. Such many-body collisions are wildly improbable. Far more plausible is a scheme in which an initial encounter between one molecule each of octane and oxygen results in the transfer of a hydrogen atom from the octane to the oxygen. Reactive intermediates are thus formed, which react by way of further two-molecule collisions until the final products are formed. Much research in combustion science today is devoted to elucidating just what those further elementary steps are.

The successful formulation of a mechanism that accounted for oscillations in the BZ reaction meant that, in at least one system, chemical oscillation could be understood in the same terms that describe the behavior of any other chemical reaction. Chemical oscillators might be complicated, but they were not magic. The time seemed ripe for an attack on the general problem of what conditions were necessary and sufficient for a chemical system to produce oscillation.

From accident to design

The 1970s produced a number of studies aimed at elucidating the nature of chemical oscillation. Chemists discovered conditions that were necessary—but not sufficient—for a particular mechanism to be capable of oscillation. They were able to show that if no more than two molecules were involved in each elementary reaction or molecular collision, then oscillation required the existence of at least three independent chemical species in the system. Several simple models of chemical oscillators were formulated and analyzed mathematically.

In spite of this progress, by 1980 it had become apparent that no one yet truly understood chemical oscillations and that the number of examples available on which to test general theories was disturbingly small. Aside from the accidentally discovered Bray and BZ reactions, the only known chemical oscillators either were minor variants, in which one or more components were replaced by chemically similar compounds, or were preparations of biological origin. These latter systems generally consisted of a cell-free extract prepared from a living or once-living system, such as yeast or beef heart. On addition of a nutrient such as glucose, oscillations could be observed in the concentrations of oxygen or other molecules involved in various biochemical pathways.

Accident, of course, was not a reliable route to new oscillating reactions. Variants of existing systems gave little new insight into the origins of oscillatory behavior. The biologically based oscillators, though numerous, were far too complex, containing many, often unknown substances, to serve as useful models for constructing or testing theories of chemical oscillation. A deeper understanding thus seemed to hinge on a deliberate search for new chemical oscillators, preferably inorganic ones in order to keep the chemistry as simple and understandable as possible. How might one systematically design a procedure for creating such systems?

The results of nonequilibrium thermodynamics provide one starting point. A reaction reaches chemical equilibrium when it begins to run at the same rate in its forward and reverse directions so that the net amounts of reactants and products do not change. At this point the reaction may be considered finished; for the specified conditions under which the reaction occurred, the maximum amount of reactants has been converted to products. In order to exhibit sustained oscillation, a chemical reaction must be maintained far from equilibrium. One way to do this is to employ an open system, one which exchanges matter, energy, or both with its surroundings. Living things support their oscillatory functions, *i.e.*, life itself, by operating in this fashion. They regularly take in reactants high in available energy—nutrients—from the environment and excrete low-energy products—wastes—to the environment. If either the input or the output flow is stopped, the oscillations soon cease and the organism dies.

An analogue of this arrangement is to be found in a tool long employed by chemical engineers, the continuous-flow stirred tank reactor (CSTR). Streams of different reactants are pumped into the reactor, which is essentially no more than a beaker with a stirrer and a water jacket around it to maintain constant temperature. The reactants mix and undergo

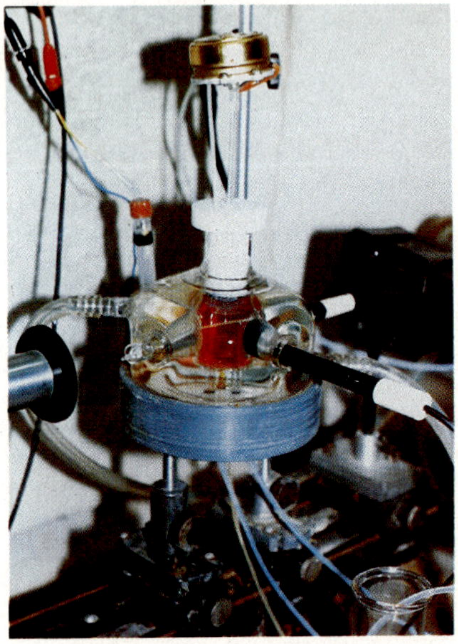

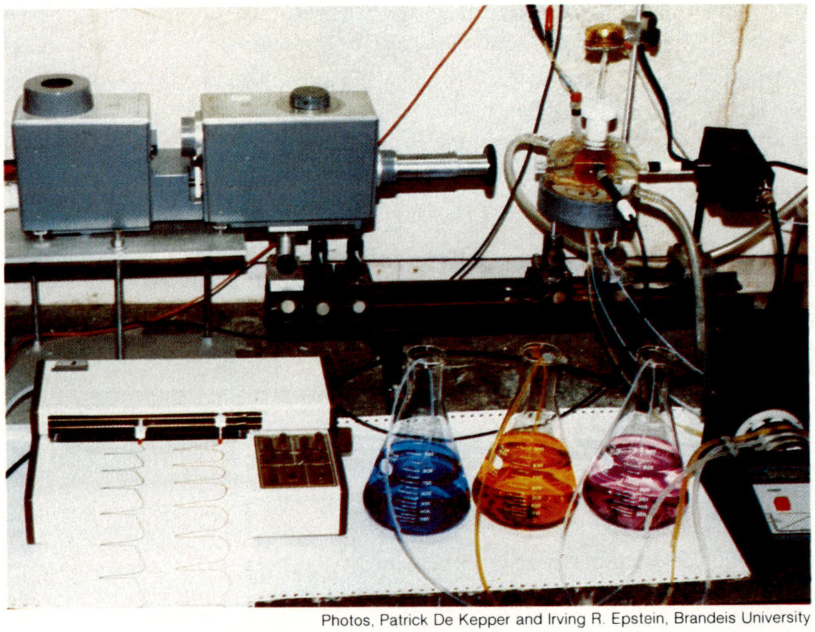

Experimental arrangement pictured at right was used by De Kepper, Epstein, and Kustin at Brandeis University in 1980 for running the first systematically designed oscillating reaction. The continuous-flow stirred tank reactor (CSTR) can be seen in the upper right portion of the photo and in the close-up above. The reactants—solutions of arsenite, iodate, and chlorite ions held in the three flasks—are pumped continuously into the base of the reactor. A light source and photocell on opposite sides of the reactor monitor the color changes, while electrodes mounted in the reactor sense oscillations in the concentration of particular ions. Periodic changes in color and concentration appear on the chart recorder at the lower left.

Photos, Patrick De Kepper and Irving R. Epstein, Brandeis University

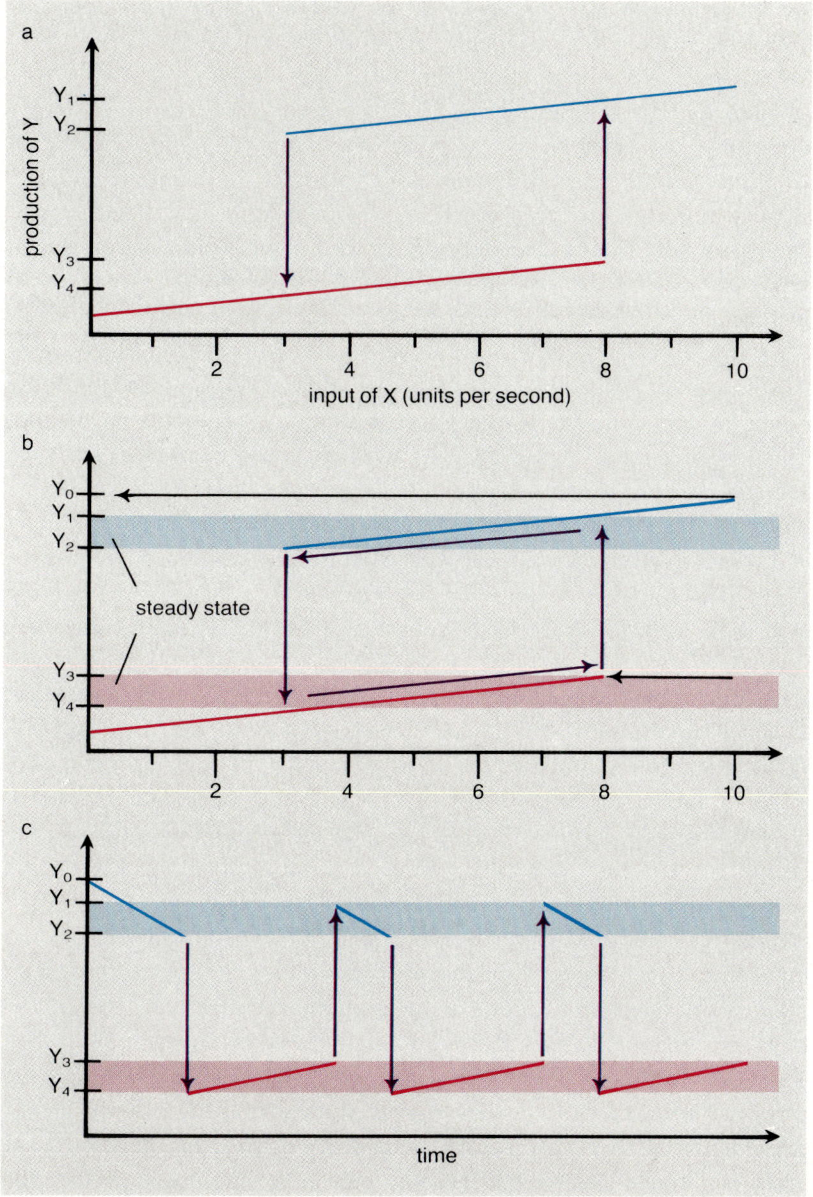

The rationale for designing an oscillating chemical reaction in a CSTR, *as described by way of a hypothetical example in text on page 148, is shown in three graphs. Graph a illustrates how a system with two stable steady states can exhibit hysteresis. As one increases the flow of colorless reactant X into the red-colored reactor, the production of blue compound Y first increases slowly (red curve). When the input of X reaches some critical value (here eight units per second), the production of Y rises dramatically (from Y_3 to Y_1), causing the reactor to turn blue. If one then decreases the input of X, the production of Y falls slowly (blue curve), even when X is decreased below the first critical point. Only when a second critical point (three units per second) is reached does the system jump back to the red state (Y_2 to Y_4). Graph b shows what happens to the system when one adds a substance Z that acts selectively to change the available level of X in the reactor. Whereas adding Z decreases the amount of X available to both states of the system, the decrease is much greater in the blue state (horizontal arrow at Y_0) than it is in the red state (horizontal arrow at Y_3). If one fixes the input of X at a high value (say ten units per second) and the system is in the blue state, Z drives down the level of X until the system jumps to the red state; then Z allows the level of X to climb again until the system returns to the blue state. Consequently the system cycles through the path traced in purple. Graph c follows the concentration of Y over time in the bistable system containing the proper amount of Z. The reaction oscillates from one state to another, causing the color of the reactor to alternate between red and blue.*

reaction in the reactor and then exit through an outflow tube so that the volume of the system is maintained constant. The composition in the reactor may be monitored, for example, by shining light through it and analyzing the absorbance at a particular wavelength or by connecting electrodes placed in the solution with an external recording device.

The CSTR takes care of the requirement that the system operate far from equilibrium, since reactants are always being refreshed and products removed. But one still needs the right chemistry for oscillation. One fairly obvious necessary condition is that the system possess some sort of feedback mechanism. That is, the product of some reaction in a sequence must influence the rate of some earlier step in the sequence in which that compound or a precursor is produced. Experience suggests

(Facing page) Major chemical oscillators known by the mid-1980s can be grouped into five main families. Solid lines connect related systems within families, while broken lines between families relate systems that share certain characteristics. The BZ reaction, discovered in 1958, belongs to the metal-catalyzed branch of the bromate family; the Bray reaction, found in 1921, and the Briggs-Rauscher system, devised in 1973, are classified in the iodate family. Aside from the above reactions, their minor variants, and oscillators derived from biological systems, all of the listed oscillators have been discovered since 1980.

that the simplest type of feedback and one that occurs in nearly all known chemical and biochemical oscillators is autocatalysis.

Autocatalysis is the acceleration of the rate of a reaction by one of its products. The simplest example of an autocatalytic chemical reaction is given by the equation $A + X \rightarrow 2X$, in which X is produced at a rate proportional to the concentrations of both the starting material A and the product X. Such behavior is almost ubiquitous in biology; it is the basis of the exponential (Malthusian) growth of unchecked populations, in which X is a particular organism and A is its food supply. It is a relatively rare occurrence in chemistry, but autocatalytic reactions do exist.

In studying autocatalytic reactions in a flow reactor, one observes that they can show bistability, the existence of two different stable steady states for the same set of reaction conditions. This peculiar behavior, which is analogous to that of a ball rolling along a surface with two depressions in it, can occur only in open chemical systems. It provides a starting point for the systematic design of a chemical oscillator.

A bistable system shows hysteresis. That is, as some control factor such as the concentration of an inflowing reactant or its flow rate is varied first in one direction then in the other, the system undergoes transitions between the two stable steady states. For example, such a system might consist of a solution of a red-colored reactant A and a colorless reactant X, which are fed into a reactor and react to produce a blue compound Y. When the inflow of X is low, the steady-state level of Y is low, and the solution in the reactor remains red. As the input of X is gradually increased, the level of production of Y also grows, and the solution becomes slightly purplish. When the input of X reaches a certain critical level, say eight units per second, a sharp transition occurs, the steady-state level of Y increases dramatically, and the solution suddenly becomes blue. If the input of X is increased still further the solution remains in the blue state. Now, if one begins to reduce X, the solution remains blue even if the input goes below the critical value of eight units per second. Only when a second critical point, say three units per second, is reached does the reverse transition occur and the system jumps from the blue to the red state. Between the two transition points, from three to eight units per second, the solution is either blue or red depending upon whether the system arrived at the present conditions from higher or lower inputs of X. Thus the state of the system depends upon its past history.

A mathematical model developed by Jacques Boissonade and Patrick De Kepper of the Paul Pascal Research Center in Bordeaux, France, leads to an interesting prediction about such bistable systems. In the example above, suppose a substance Z can be found that reacts with the substance in the reactor to change the level of X. If Z reacts sufficiently slowly and if the effect of Z is much greater when the system is in one of the steady states than in the other, then addition of the proper amount of Z to the bistable system will produce a feedback that will cause the system to jump periodically from one branch of steady states to the other; *i.e.*, to oscillate. The Boissonade-De Kepper prediction is an integral part of a systematic method to search for or design new chemical oscillators.

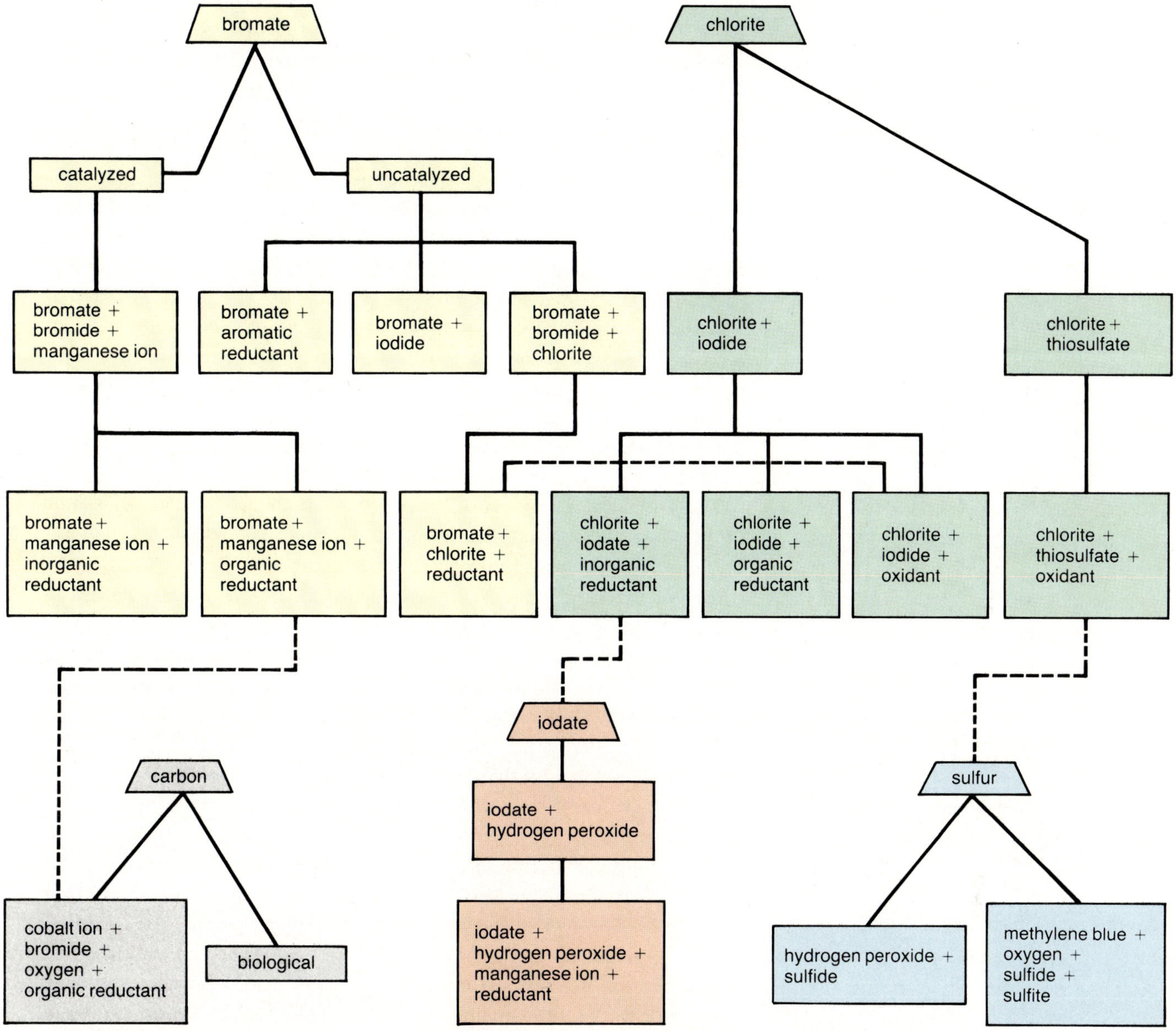

The procedure starts with finding an autocatalytic reaction. Although autocatalysis is not common, several of these reactions are well-known as lecture demonstrations since they give rise to sudden, dramatic color changes. After selecting an autocatalytic reaction, one runs it in a flow reactor to find conditions under which the system is bistable. If bistability is obtained, one then seeks a suitable feedback substance. When the input concentrations are adjusted, hopefully the system will oscillate.

This technique was first applied successfully in 1980 at Brandeis University, Waltham, Massachusetts, by De Kepper, Irving Epstein, and Kenneth Kustin. The autocatalytic reaction involved arsenite (AsO_3^{3-}) and iodate ions, which generate iodide (I^-) as the product. The feedback substance was chlorite ion (ClO_2^-), which actually reacts autocatalytically with the iodide produced by the first reaction. Thus, two autocatalytic reactions are coupled together, and the color of the system alternates between brown and colorless. Later studies led to the discovery of more than two dozen oscillators involving the chlorite ion.

More oscillators

The first, accidentally discovered chemical oscillators and the later, systematically designed chlorite oscillators are related to each other chemically in a significant way. Each is based on the chemical reactions of an oxyhalogen ion: iodate (IO_3^-) in the case of the Bray reaction, bromate (BrO_3^-) for the BZ reaction, and chlorite (ClO_2^-) for the new systems. That is, in all of these systems the key to oscillatory behavior is an ion consisting of a halogen atom—an element from column VIIA of the periodic table that has all but one of the possible places in its outermost electron shell filled—and several atoms of oxygen.

Why should oxyhalogens be so favorable to chemical oscillation? Could elements from other parts of the periodic table be used as well?

The crucial property of the halogen elements appears to be their ability to support a large number of oxidation states. In other words, iodine, bromine, and chlorine are relatively stable whether they are associated with one electron more or with one, three, five, or seven electrons less than they possess as neutral atoms. This versatility allows them to shift rapidly from one form to another, a requirement for participating in oscillatory processes in which intermediate species are produced and consumed with each cycle.

One might then expect that new types of oscillators should be sought among those elements that also possess multiple oxidation states. Since 1983 success has been achieved with two nonhalogen elements, carbon and sulfur.

The ubiquity of oscillation in living systems suggests that there exist literally thousands of organic, carbon-based oscillators. Owing to their complexity, however, it has not proven possible to isolate the chemical origin of oscillation in any of them. Scientists at E. I. du Pont's Experimental Station in Wilmington, Delaware, have discovered oscillations in a comparatively simple organic system, the cobalt- and bromide-catalyzed air oxidation of two common organic chemicals, benzaldehyde and cyclohexanone. These reactions, which are similar to industrial processes that produce billions of kilograms of organic chemicals each year, constitute the first two well-characterized organic oscillators. The periodic change in the oxidation state of the cobalt ion causes the system to alternate between pink and greenish-brown.

A similar breakthrough in sulfur chemistry has produced two new oscillators. The first, developed by Richard Field and Maria Burger at the University of Montana, involves the reaction of sulfide (S^{2-}) and sulfite (SO_3^{2-}) ions with dissolved oxygen in the presence of the dye methylene blue. The solution oscillates between blue and colorless. The second sulfur-based oscillator, discovered by Miklos Orban of Eotvos University in Budapest, Hungary, is somewhat simpler in composition, containing

Photo sequence (facing page, top to bottom) taken at half-minute intervals follows the formation of two-dimensional bull's-eye patterns in a thin layer of a modified BZ formula on the bottom of a petri dish. Blue spots appear spontaneously on the red field of the BZ mixture and spread into circles. Shortly thereafter the centers of the circles regain their red color, and new blue spots appear in the same places as the previous ones. As these chemical oscillations continue, concentric rings are built up.

Photos, A. T. Winfree

only sulfide ion and hydrogen peroxide. Since this system alternates between acidic and basic (alkaline) states, the oscillations can be followed in a variety of colors by adding any of a number of color-changing acid-base indicators.

Patterns in space and time

The first studies of chemical oscillation were directed toward eliciting simple, periodic, homogeneous behavior. The systems were required to be well-stirred and uniform. Ideally each cycle of oscillation would be exactly the same as the previous cycle and would contain one maximum and one minimum in each concentration. More recently, by varying the conditions and looking carefully, chemists have discovered fascinating phenomena that are intimately related to the oscillations studied earlier but far more complex.

Perhaps the most striking example is the spontaneous development of spatial patterns in an initially uniform, unstirred medium. If the iron-based oxidation-reduction indicator ferroin is used instead of cerium in the BZ reaction, then the system can oscillate between a red and a blue state. With certain concentrations of the reactants, however, the system is not oscillatory but excitable. Stirred and left undisturbed, it remains in one state, say red, but a small disturbance like a drop of acid can cause it to turn blue and remain that way for several seconds before returning to the stable red state. When this excitable mixture is poured in a thin layer at the bottom of a petri dish and left alone, a remarkable thing happens. After a few minutes one or more blue spots appear. Each of these enlarges and its center turns back to red, so that the spot becomes a ring of blue. As time goes on, a new blue spot forms in each red center and grows into a ring. Soon the dish is filled with sets of concentric blue rings on a field of red. When rings get large enough to touch one another, they annihilate and disappear, leaving behind the red background.

How can such a thing occur? Intuition suggests that any nonuniform pattern that develops in an initially homogeneous medium will soon fade away as a result of diffusion, just as the random motions of molecules tend to disperse a drop of ink placed in a glass of water. But in an excitable and potentially oscillatory system in which autocatalytic reactions are possible, something else may occur. A small fluctuation, perhaps a spontaneous local increase in the concentration of a reactant, may cause a tiny bit of solution to change from red to blue. Since the reaction is autocatalytic, as diffusion spreads the blue material, it causes nearby areas to turn blue as well as the red-to-blue reaction accelerates. The spot enlarges. Soon, however, the central region where the disturbance started returns to its red state. A ring has formed. As the system continues to oscillate, further rings are generated.

Must such behavior be restricted essentially to two dimensions on the bottom of a petri dish? Early experimenters found that if the layer of solution was too thick, the patterns would not form. Later work showed, however, that beautiful and intricate three-dimensional patterns can de-

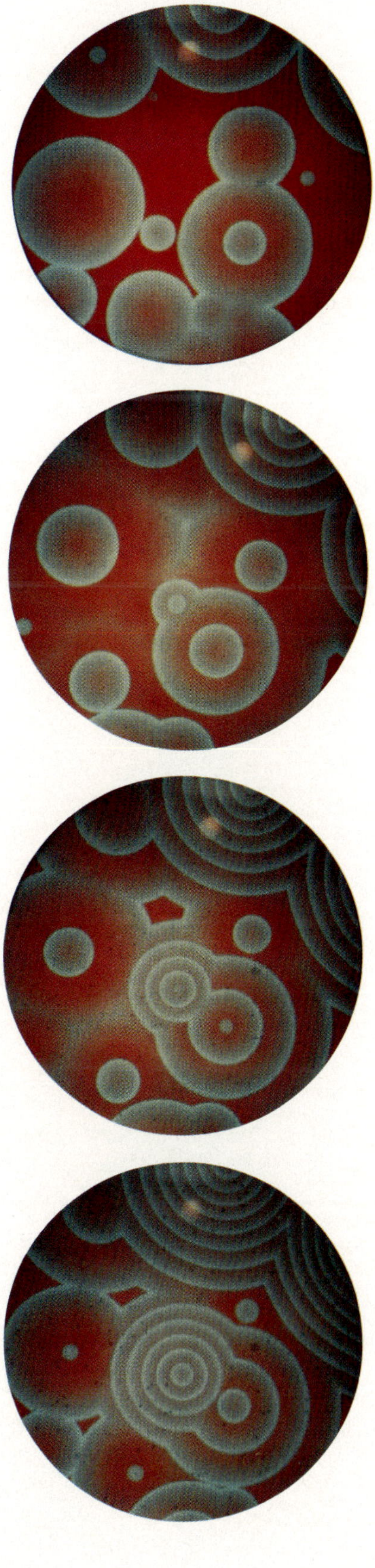

velop under appropriate conditions. Theoretical studies suggest that there may be a hierarchy of structures involving increasingly more complex sets of intertwined rings, spirals, and scrolls.

Bistability between two different steady states played an important role in designing new chemical oscillators. There are other, more exotic forms of multistability. Some chemical systems can exist in either a steady or an oscillatory state, or in either a high-frequency or a low-frequency oscillatory state, depending on their past history. Investigators have even found tristability, the existence of three different states stable under the same set of conditions. All such multistable systems have the inherent property of chemical memory. By observing the current state of the system, one can infer something about its previous state.

People tend to think of oscillations, whether chemical or mechanical, as both simple and periodic; they need not be either. Complex periodic oscillations consist of two or more different kinds of swings or peaks per cycle. For example, a pendulum might be made to give one long swing, two short ones; one long, two short; and so on in a perfectly repetitive pattern.

Similar behavior has been observed in a few chemical oscillators. One of the most striking cases is the reaction between chlorite and thiosulfate ($S_2O_3^{2-}$) ions in a CSTR. At low flow rates one finds simple large-amplitude oscillations. When the flow rate is increased beyond a critical value, the pattern changes to one large- and one small-amplitude peak per cycle. Further increases in flow rate produce successive transitions to modes of oscillation in which the single large peak is combined with even larger numbers of small peaks in each cycle. Finally, above another critical flow value, the system returns to simple oscillation, but now with only small peaks. Thus, chemical oscillators can form patterns not only in space but in time as well.

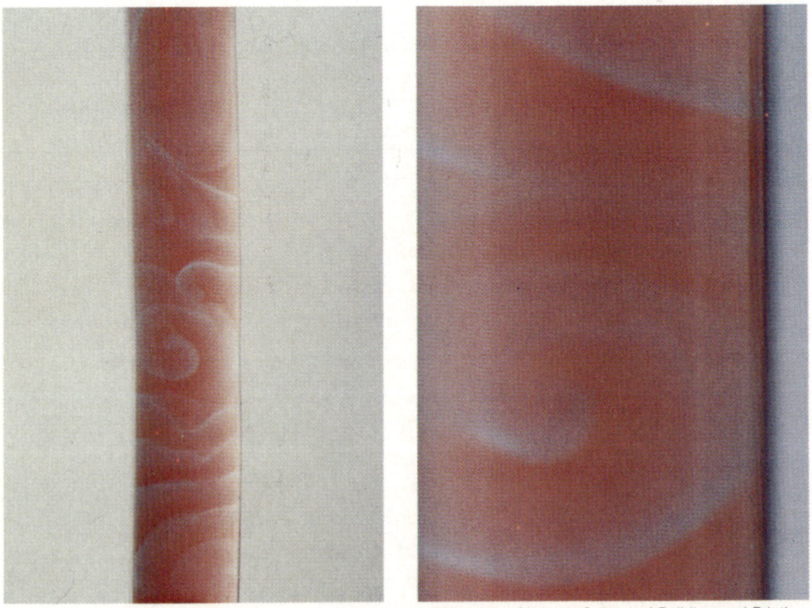

Photos, Martha Thomson, Glasgow College of Building and Printing, and Arthur E. Burgess, Glasgow College of Technology

Another type of nonsimple behavior is aperiodic oscillation or, as it is sometimes called, chemical chaos. Imagine a pendulum whose swings vary in length, never repeating the same sequence. Although the swings appear random, there is an underlying determinism; given the length of one swing and a knowledge of the system's makeup, one can predict exactly the length of the next swing. Such oscillation is aperiodic, but nonrandom. The chemical analogue of such a system would exhibit a nonrepetitive pattern of highs and lows in the concentrations of its components.

Recently this type of behavior has intrigued mathematicians as well as scientists working with physical systems in a variety of seemingly unrelated fields such as electronics, hydrodynamics, and meteorology. Many of these systems can be described by sets of differential equations that possess solutions known as strange attractors. The solutions correspond to the sort of nonrandom, aperiodic variation described above. They have two other properties of interest, which give rise to their name. They are attractors in the sense that if one starts off the physical system in a state different from that of the attractor, the state changes in such a way as to bring the system ever closer to the attractor. Ultimately the system will exhibit the behavior characteristic of the attractor.

The attractor itself is strange in the sense that if one starts off the system twice, once from each of two almost identical sets of conditions, within a short time the system evolves to two very different sets of conditions. In a system having a strange attractor it is impossible to "perform the same experiment twice." Regardless of the effort, one can never prepare the system with exactly the same starting conditions, and the small initial differences will ultimately grow into major discrepancies.

Chemical oscillators have been shown to possess strange attractors, and in some cases the structure of the attractor has been characterized.

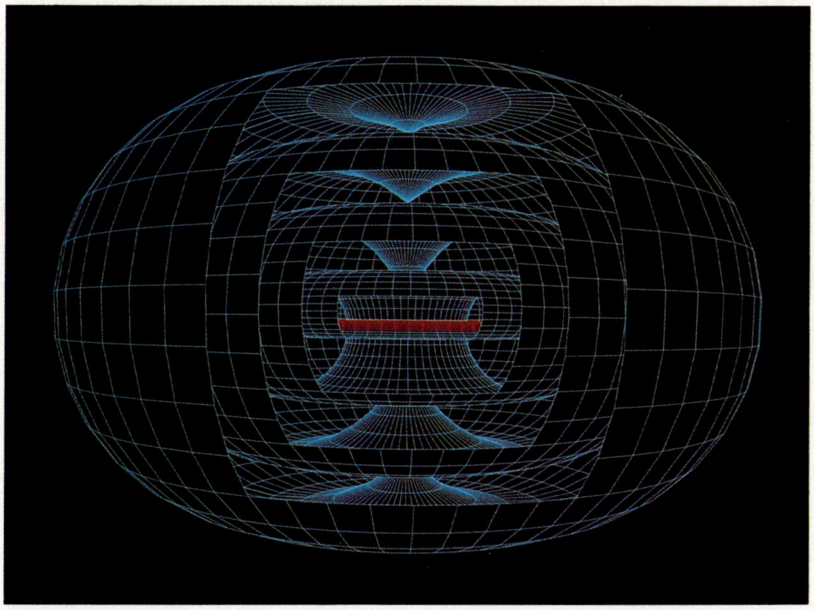

A. T. Winfree

Under appropriate conditions a test tube filled with an initially homogeneous solution of BZ reactants can produce intricate three-dimensional patterns (facing page, left). The most common pattern is the scroll wave (facing page, right), a solid spiral. One variation, in which the ends of the scroll close to form a doughnut-shaped ring, is shown in a computer-generated image (this page, left); sections of the scroll surface have been removed, creating "windows" to inner detail.

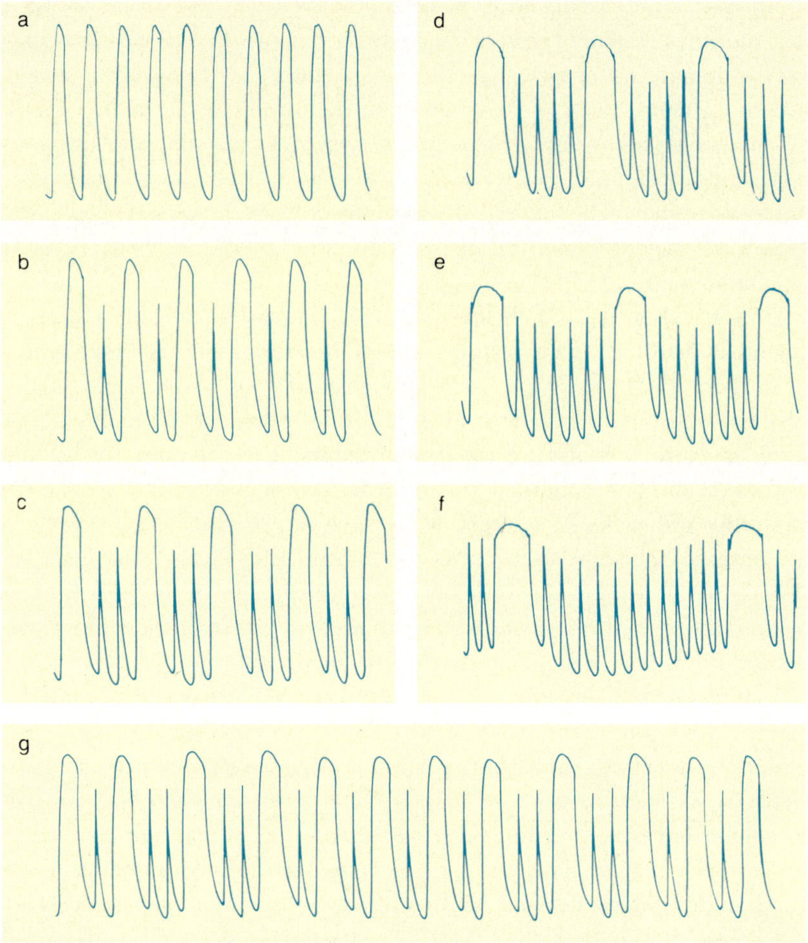

Periodic oscillations in a reaction between solutions of chlorite and thiosulfate ions in a CSTR *have been found to exhibit a variety of patterns when the input flow of reactants is varied. At low rates ion concentrations show simple large-amplitude oscillations over time (a). As flow rates are raised, the system makes successive transitions to complex periodic modes in which large peaks alternate with increasing numbers of smaller peaks (b–f). Between adjacent modes flow rates can be "fine-tuned" to give aperiodic, or chaotic, patterns that are apparently random mixtures of the two modes. For example, between the periodic oscillations shown in b and c can be found the aperiodic oscillation shown in g.*

Adapted from "Oscillating Chemical Reactions," Irving R. Epstein, Kenneth Kustin, Patrick De Kepper, and Miklos Orban,

The first and most thoroughly studied chaotic chemical system was the BZ reaction, but chaotic behavior has been observed in several chlorite reactions and in a biochemical oscillator, the horseradish peroxidase system, as well. Current research is aimed at elucidating the underlying mechanisms of chemical chaos, the processes by which a periodic oscillator becomes chaotic as some element, *e.g.*, a reactant concentration, is varied, and the different types of chaotic behavior.

Expanding applications

As chemical oscillators have become better understood, the principles that underlie them have found application in fields other than chemistry. Three of these fields, all rather closely related to chemistry, are geology, meteorology, and biology.

For almost a century chemists have been fascinated by the periodic precipitation patterns known as Liesegang rings. Two separate solutions, each containing one of a pair of ions—such as silver (Ag^+) and dichromate ($Cr_2O_7^{2-}$)—that react to form a nearly insoluble compound, are placed at opposite ends of a tube filled with a gel. As the ions diffuse through the gel toward one another, a precipitate forms—in this case, silver chromate (Ag_2CrO_4). Contrary to expectations, however, the precipitate does not deposit uniformly, but in a series of sharp, nearly evenly spaced bands. Recent theories account for this phenomenon by invoking the competitive autocatalytic growth of different-sized particles of precipitate coupled to diffusion of the reactants, an explanation analogous to that offered for pattern formation in the BZ system. Scientists are presently using similar theories to model the formation of band and zone patterns in a wide variety of rock formations, despite the vast difference in time and length scales between the geological and chemical systems.

The existence of strange attractors may have profound consequences for the prospects of long-range weather forecasting. One of the earliest models of a strange attractor, proposed by meteorologist E. N. Lorenz of the Massachusetts Institute of Technology, describes the flow of air in the atmosphere. Suppose that the weather is a system that possesses a strange attractor. If forecasters knew the equations that govern the changes in the weather, could they, even in principle, develop good long-range weather forecasting? The answer appears to be no because of the nature of the strange attractor. One never knows the current weather with infinite accuracy. Temperatures are measured only to within a degree, or perhaps a tenth of a degree, wind velocities to within a kilometer per hour or so. If two initial states differing by a tenth of a degree can evolve within days or weeks to situations in which temperatures differ by tens of degrees, then long-range forecasting is an impossiblity.

The field on which studies of chemical oscillators are likely to have their greatest effect is biology. For example, nerve impulses involve the transmission of electrical signals from neuron to neuron across gaps called synapses. The mode of transmission involves messenger molecules called neurotransmitters and the change in permeability of the cell membrane to various ions in response to changes both in neurotransmitter concentrations and in the electrical potential across the membrane. This membrane potential undergoes rhythmic fluctuations, called action potentials. Neurobiologists are beginning to formulate molecular models to account for membrane oscillations based on the idea suggested by

The activity-rest cycle in animals (below), the daily opening and closing of flowers, the 90-minute dreaming cycle of the human brain, and the second-by-second beat of the heart are but a few examples among thousands of the oscillatory behavior apparent in living organisms. Many biologists suspect that most if not all biological oscillation is controlled at the molecular level by chemical oscillators.

Photos, Alan A. Moss

Concentric rings of silver chromate precipitate (above) can be formed by placing a crystal of silver nitrate at the center of a glass plate coated with gelatin containing a dilute solution of potassium dichromate. Recent theories for this phenomenon, called Liesegang rings, give an explanation analogous to that offered for pattern formation in the oscillating BZ system. The banding patterns in agate (above right) and many other minerals closely resemble that of Liesegang rings and are believed to arise by way of the same general mechanism.

(Right) B. M. Shaub

Nobel laureates Alan Hodgkin and Andrew Huxley in 1952 that the membrane contains "voltage-gated" channels for different types of ions. These channels open or shut as the membrane potential passes through a critical value.

One of the most exciting applications of these ideas to problems in biology is a recent proposal by Arthur T. Winfree of Purdue University, West Lafayette, Indiana, that sudden cardiac death syndrome can be understood in terms of the response of a chemical or electrical oscillator to a disturbance. Winfree argues on the basis of theorems in topology that there must be a particular, "singular" stimulus of exactly the right strength and timing that is capable of disrupting the regular oscillation of the heart's natural pacemaker. When this singularity develops it initiates a three-dimensional rotating wave of electrical activity, an action potential, that passes through the cardiac tissue, much as the colored wave of chemical activity passes through a petri dish of BZ reagent. The result of this wave is fibrillation, a rapid, irregular contraction of the heart muscle, which, if it persists, is fatal. The stimulus could come from outside the body or, more likely, from a random impulse from one of the many neighboring nerve cells. Should Winfree's hypothesis prove correct, it might provide important guidance in designing drugs and other therapies to lessen the risk of a sudden cardiac death.

One may speculate whether the mechanisms that produce spatial structure in simple chemical reactions are related to those that are involved in the formation of biological patterns. For example, the similarities between the rings formed in the BZ and other "simple" chemical systems and those seen in the aggregation of the slime mold *Dictyostelium discoideum* are truly striking. *D. discoideum* and other cellular slime molds, demonstrate a remarkable life cycle that has been studied for clues to the way the cells of higher organisms communicate and become organized. Part of their time is spent as separate, free-living amoebalike cells. When food becomes scarce, they signal to each other chemically and subsequently aggregate into a multicellular structure that culminates in a spore-filled fruiting body. In the earliest stage of aggregation, cells of *D. discoideum* release the substance cyclic adenosine monophosphate (cAMP) at about five-minute intervals and in a way that establishes centers of concentration. Each cell has receptors on its surface that allow it to sense and move toward higher concentrations of cAMP. As aggregation progresses the periodic release of cAMP organizes the moving cells into spiral or concentric circular patterns about the concentration centers. This coupling between temporal oscillation and motion toward a chemical attractant is similar to the interaction of oscillatory chemistry and diffusion in the nonliving BZ system. One may also ponder whether the interaction between diffusion and excitable chemical reactions might provide a means whereby a single, apparently uniform cell could divide into increasing numbers of identical daughter cells, which at some point become irrevocably destined for skin or bone or brain cells.

Future prospects

It is hard to predict the next decade for a field that was essentially nonexistent 15 years ago. Some directions seem nearly self-evident while others are highly speculative.

Work is likely to continue on developing new oscillators and on understanding both existing systems and those still to be discovered.

A fatal irregular beating, or fibrillation, of the heart muscle is the only manifestation of an as yet inexplicable category of heart disease called sudden cardiac death syndrome. Both theory and clinical observation suggest that the condition may be caused by a three-dimensional rotating wave of electrical activity—similar to the waves produced in a BZ reaction—that disrupts the regular oscillation of the heart's natural pacemaker. One model of such a wave, a computer-generated square array of counterrotating vortices, is shown below. At the lower left a drawing depicts the electrical activity recorded from the ventricles of a living human heart during open-chest surgery. The tip-on perspective, as if viewed from directly below the heart, reveals a five-cycle-per-second rotating wave pivoting around the apex of the ventricles.

(Left) adapted from information obtained from A. T. Winfree; (right) A. T. Winfree

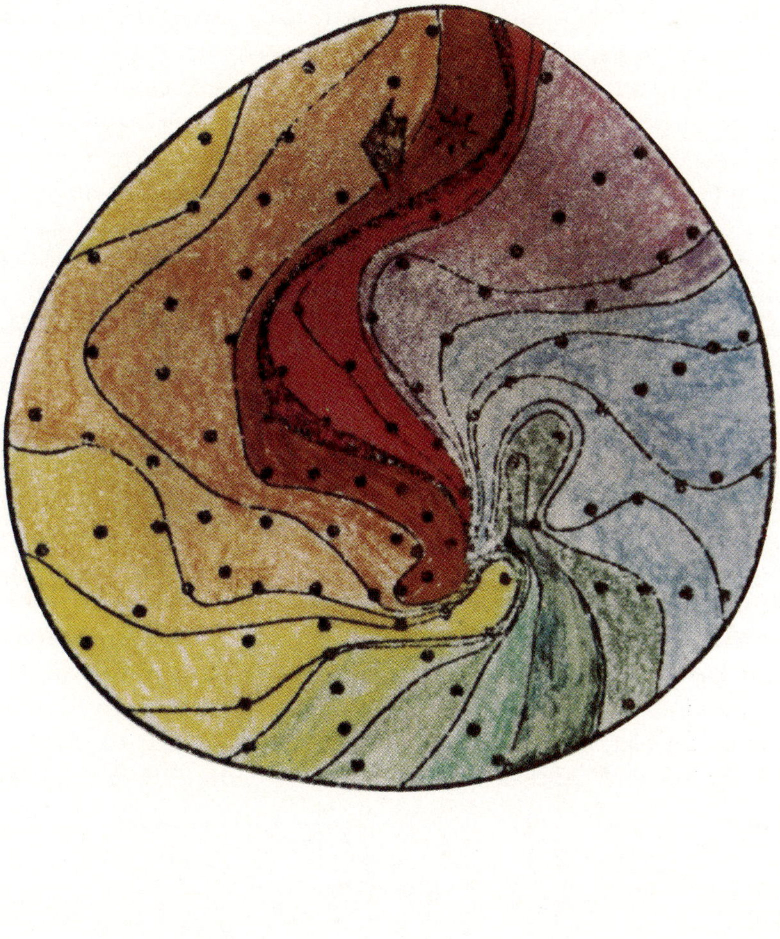

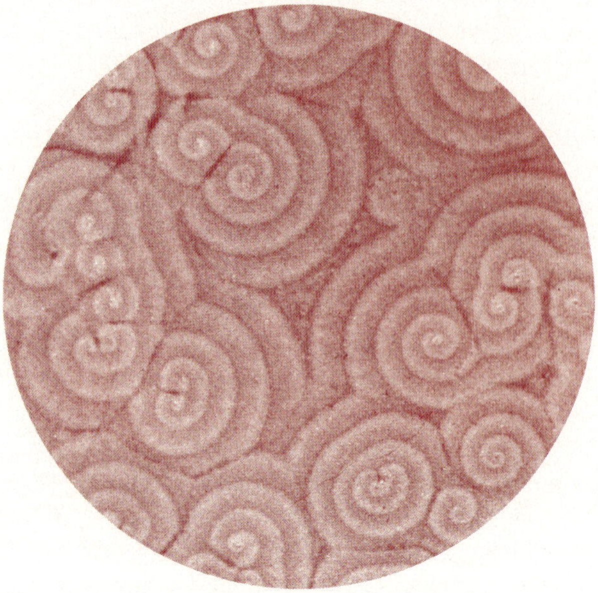

Spiral waves of oscillating chemical activity in a petri dish of BZ reactants (above) bear a striking similarity to the organized waves of cell movement during aggregation of the slime mold Dictyostelium discoideum *(above right). Both phenomena may hold clues to the way cells of higher organisms communicate during morphogenesis, the complex process by which the cells and tissues of an embryo become patterned into a complete organism.*

(Left) A. T. Winfree; (right) Peter N. Devreotes, Department of Biological Chemistry, The Johns Hopkins University School of Medicine

Biological oscillators are based primarily upon the chemistry of organic compounds and transition metals, particularly systems in which enzymes are involved. Now that the first nonbiological organic oscillators have been found, many more should follow. Scientists have begun to design oscillators based on such transition metals as chromium, cobalt, and vanadium, which support a wide range of oxidation states. Increasingly sophisticated experimental and computational techniques should yield rapid progress in explaining reaction mechanics. While it took 15 years from the discovery of the BZ reaction to develop a mechanism for it, the chlorite-iodide oscillator, of comparable complexity but discovered in 1981, was explained mechanistically within three years.

More attention will probably be devoted to the practical applications of chemical oscillation. Chemical engineers have known for years that industrial reactors which produce large quantities of chemicals occasionally go into oscillation spontaneously. They have developed methods of suppressing this behavior, since an uncontrolled oscillating reactor may escape its usual operating range and either explode or shut down. It makes sense, however, to ask whether certain processes might not be more efficient, in the sense of yielding more product or a higher ratio of desired to undesired material, if they were operated in a controlled oscillatory mode rather than the usual steady state.

The role of chemical oscillation in catalysis is a concept worthy of investigation. Catalysts, substances that speed up a reaction without being changed themselves, are essential factors in almost every process of industrial or biological significance. In order to perform its work, each catalyst molecule or particle must oscillate between two or more different states. In most cases, however, oscillations of the total catalyst concentration have not been observed, either because they occur too fast to be seen without special techniques or because the different portions of the catalyst oscillate independently of one another so that, on the average, the fraction in any given state is independent of time. Might it be possible to induce the molecules of catalyst to oscillate in phase with one another? And could such coherence increase the efficiency of the catalyst?

Finally, it is likely that studies of chemical oscillators will draw closer to biological and biochemical systems. This approach may take several forms. Experiments with chemical oscillators coupled together either by physical transfer of material from one reactor to another or by sharing a common component in the same vessel have given rise to a complexity of behavior not found in the individual uncoupled system. Further studies of this sort may shed light on the workings of biochemical pathways or structures that may be viewed as a set of linked oscillators.

Investigations of this type attempt in a sense to simulate and gain insight into biological complexity by starting from simple chemical systems and making them increasingly more complicated. An alternative method is to begin with a biological system and dissect it into pieces to which the techniques developed for studying simpler oscillators may be applied. For example, studies are under way to probe whether a single enzyme, phosphofructokinase (PFK), will generate oscillations in a flow reactor. If so, this would be strong evidence that PFK is responsible for the oscillatory behavior of glycolysis, a process used by nearly all living organisms to generate energy-rich adenosine triphosphate (ATP) from glucose.

During the past decade chemical oscillation has become a well-established and well-studied phenomenon. During the coming years scientists will ascertain whether those fascinating beakers of solution that turn from red to blue and back again can help answer such questions as why the heart beats so regularly and why sometimes, unexpectedly, it stops.

FOR ADDITIONAL READING

Britton Chance, E. Kendall Pye, Amal K. Ghosh, and Benno Hess (eds.), *Biological and Biochemical Oscillators* (Academic Press, 1973).

Patrick De Kepper, Kenneth Kustin, and Irving R. Epstein, "A Systematically Designed Homogeneous Chemical Oscillator," *Journal of the American Chemical Society* (April 1981, pp. 2133–2134).

Irving R. Epstein, Kenneth Kustin, Patrick De Kepper, and Miklos Orban, "Oscillating Chemical Reactions," *Scientific American* (March 1983, pp. 112–123).

Richard J. Field and Maria Burger (eds.), *Oscillations and Traveling Waves in Chemical Systems* (Wiley, 1985).

Richard J. Field, Endre Koros, and Richard M. Noyes, "Oscillations in Chemical Systems. II. Thorough Analysis of Temporal Oscillation in the Bromate-Cerium-Malonic Acid System," *Journal of the American Chemical Society* (December 1972, pp. 8649–8664).

Christian Vidal and Adolphe Pacault (eds.), *Nonequilibrium Dynamics in Chemical Systems* (Springer-Verlag, 1984).

Arthur T. Winfree, *The Geometry of Biological Time* (Springer-Verlag, 1980).

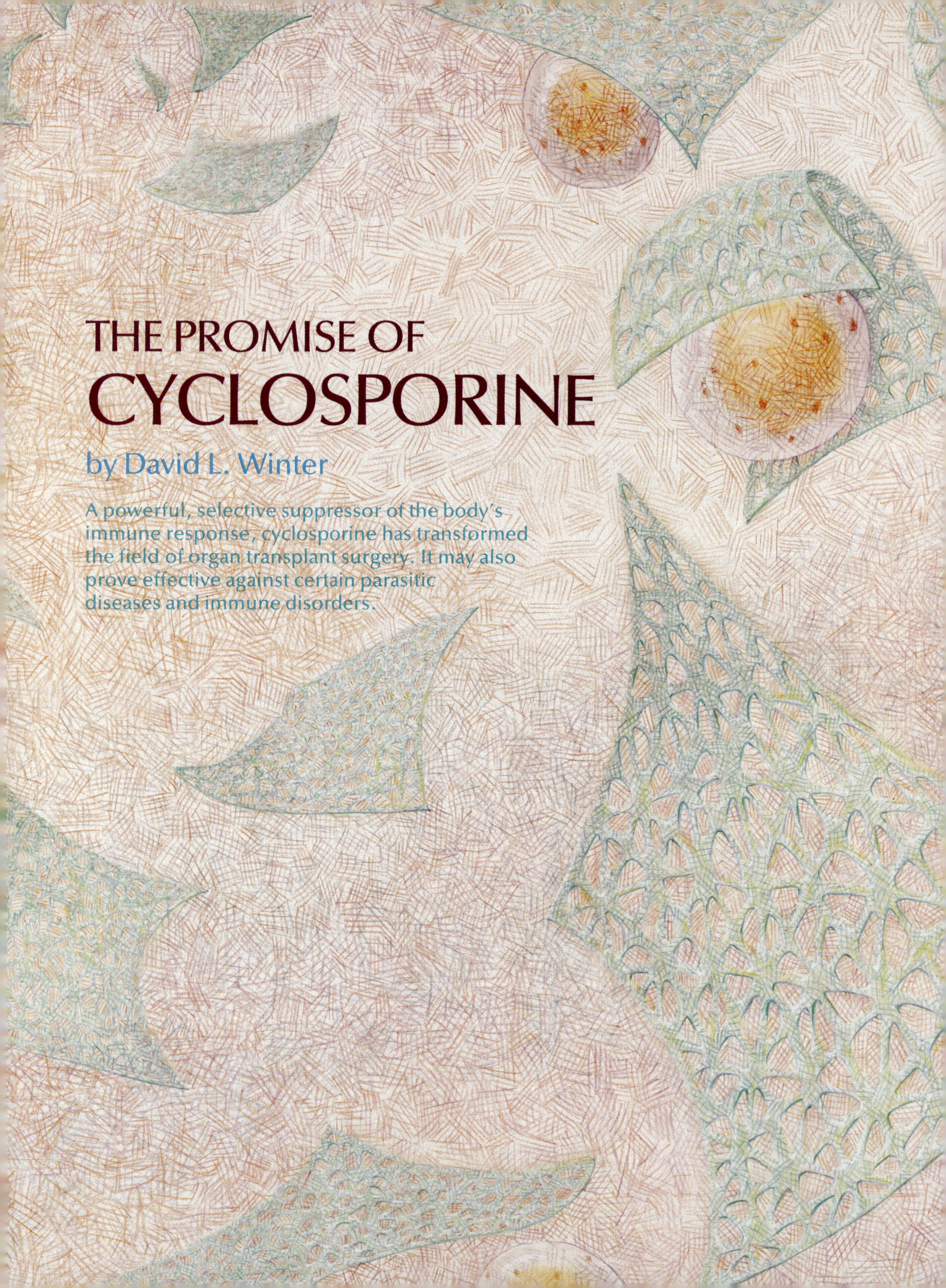

THE PROMISE OF CYCLOSPORINE

by David L. Winter

A powerful, selective suppressor of the body's immune response, cyclosporine has transformed the field of organ transplant surgery. It may also prove effective against certain parasitic diseases and immune disorders.

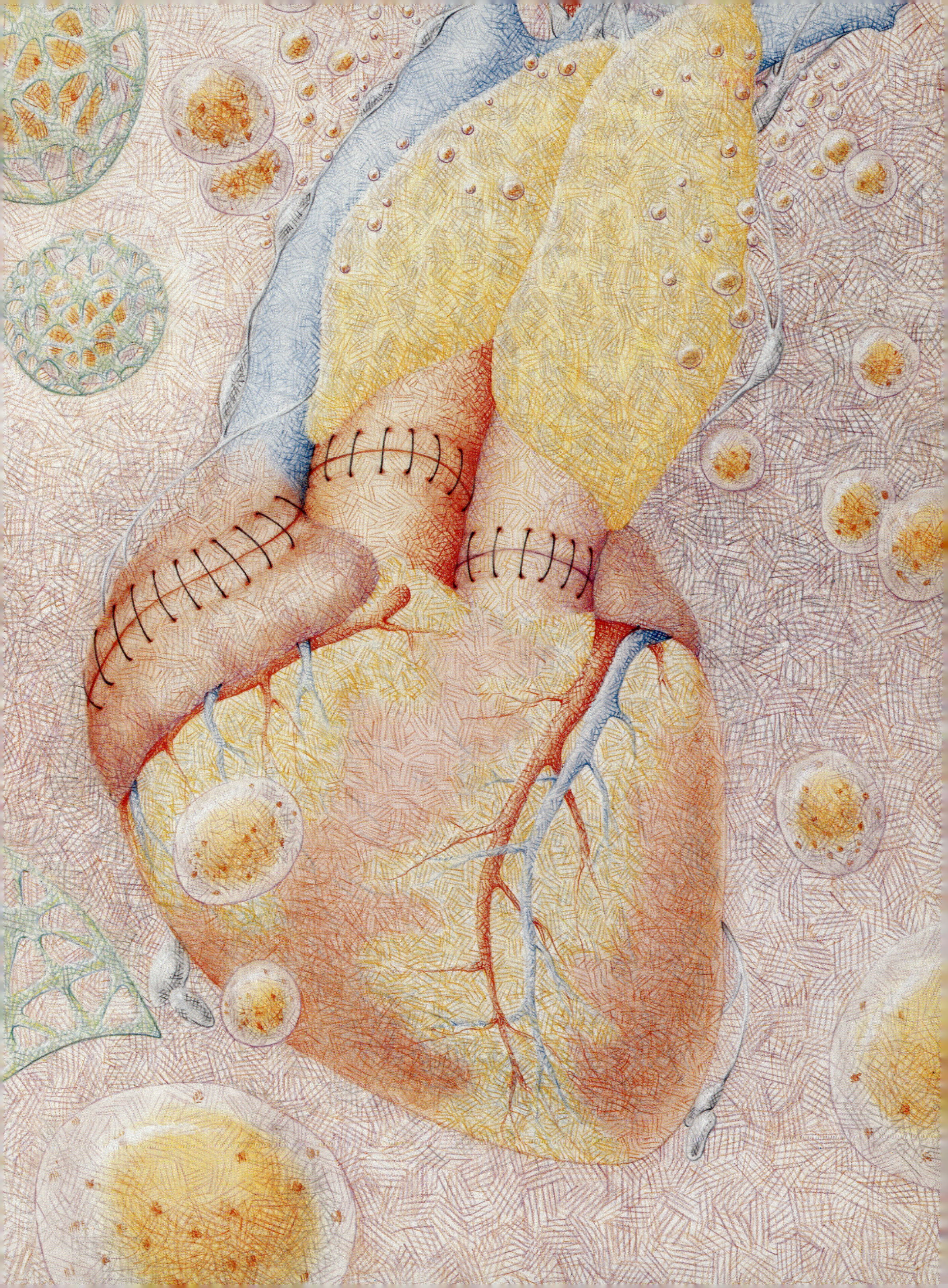

The story of cyclosporine, a highly unusual drug, had an unusual beginning. From an arid plateau high above a Norwegian fjord a vacationing scientist from Sandoz Limited in Switzerland removed several soil samples as a matter of company practice and took them back to work. There, microbiologists studied the samples for signs of any unusual forms of bacteria or fungi. From one sample a new strain of fungi imperfecti was isolated and given the name *Tolypocladium inflatum* Gams. Little did anyone suspect at the time that this discovery, made in 1970, would lead within 15 years to a revolution in the field of organ transplant surgery and to promising therapies for certain parasitic diseases and disorders of the body's immune system.

From a crude extract of *T. inflatum* Sandoz researchers obtained small amounts of material that was subsequently screened for antifungal and antibacterial activity. This material, which contained many chemical substances, was quite active in vitro (in the test tube) against several strains of fungi; unfortunately the strains were of no medical importance. Although the material showed very little toxicity, its medical usefulness clearly seemed limited. It was put on a shelf and not explored biologically any further. Sandoz chemists, however, continued to study this mixture of compounds because it showed some interesting chemical properties. By 1971 the mixture had been further purified to the extent that it contained only a few substances, and by 1972 enough purified material was available for additional biological screening.

At that time Jean Borel, a Sandoz biologist, reexamined the material and found that it exhibited a unique combination of properties: it suppressed the immune system but did not kill cancer cells. Since no compound previously studied had shown a separation of these activities, characterizing the material became an attractive goal. By 1973 the substance had been reduced to a single compound, and biological studies picked up.

By 1974 the unique properties of the compound now called cyclosporine began to show themselves. By means of a combination of studies in vitro and in vivo (in living organisms) it became clear that cyclosporine selectively inhibited the immune system's white-blood-cell lymphocytes, mainly a subset called helper T cells and, in some concentrations, another subset called effector T cells. It was also noted that cyclosporine did not kill bone marrow cells in doses that were immunosuppressive. This observation was particularly important since most immunosuppressive compounds are toxic to all rapidly dividing cells including cancer cells, lymphocytes that are proliferating after encountering a foreign substance, and the bone marrow's stem cells, the source of all blood cells. Most importantly, when the drug was withdrawn, the lymphocytes resumed their normal activities; thus cyclosporine was not toxic to them. In studies involving the mouse, rat, guinea pig, rabbit, dog, and monkey, cyclosporine was found to behave in the same way.

The chemical structure of cyclosporine was established in 1975. The following year investigators began long-term toxicity studies as well as experimental transplants of heart, kidney, liver, pancreas, and other

DAVID L. WINTER *is Director of Medical Research for Sandoz Pharmaceuticals, East Hanover, New Jersey.*

Illustrations by Leon Bishop

Photos, courtesy, Sandoz Ltd., Basel

The fungus Tolypocladium inflatum, *which synthesizes cyclosporine as a metabolic by-product, was discovered in soil taken from a site above a Norwegian fjord (left). A scanning electron micrograph (below) reveals its microscopic structure of branched tubular filaments, a typical fungal characteristic.*

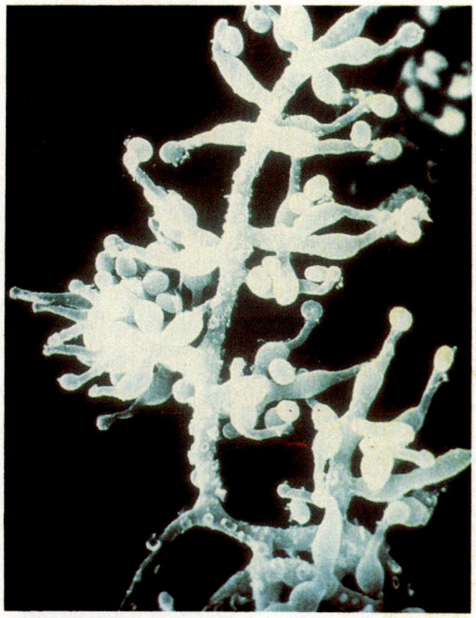

organs in different animal species. The results were impressive, and in 1978 the cyclosporine was tested for the first time in humans.

The first clinical trials, carried out in England, involved transplants of kidney and bone marrow and were soon followed by studies in several centers in the United States. By 1983 transplant surgeons participating in the trials were reporting how cyclosporine had transformed their field of medicine with dramatic increases in the survival rates of kidney and liver transplant patients. A full application was submitted to the U.S. Food and Drug Administration, and in November 1983 cyclosporine was approved for general use in kidney, liver, and heart transplantation. In Europe the drug has approval for use in bone marrow transplantation as well. Cyclosporine is available under the trade name Sandimmune.

The immune system

When foreign protein, or antigen, enters the body, it evokes a complex immune response that consists of both cellular and humoral (chemical) mechanisms. Cellular immunity involves the participation of many cell types, including lymphocytes, which are primarily responsible for recognizing the antigen and ensuring that the response is directed specifically against it, and accessory cells such as macrophages, monocytes, polymorphonuclear leukocytes, and tissue mast cells, which may initiate, amplify, or control the response. Humoral immunity relies on chemicals either actively secreted by cells during the immune response or already circulating within the body. These substances include antibodies, complement, and other mediators of cell function called lymphokines.

Lymphocytes originate from precursor stem cells, which are located in the bone marrow. These precursors migrate in the bloodstream either to the thymus, where they mature into T cells, or to other, nonthymic tissues, where they become B cells. In addition, some stem cells transform into cells called natural killer (NK) cells. All three cell types are considered lymphocytes.

T cells, which mature in the thymus under the influence of various thymic hormones, play an important role in the immune response. While maturing, they acquire recognition structures, or receptors, on their surfaces. The receptors, which consist of proteins, enable T cells to recognize specific molecular characteristics called antigenic determinants. An antigenic determinant is a structurally unique portion of a molecule, such as a foreign protein, that has a shape complementary to that of the receptor. When specific T cells combine with their complementary determinants, they undergo a series of changes. It is the nature of these changes that further characterizes T cells into at least five functional subsets. While T cells themselves do not produce antibodies, they enable other cells, B cells, to do so and in addition have other functions.

Cytotoxic T cells attack foreign cells as well as those cells of the body whose surface receptors are altered and thus appear foreign. Amplifying T cells help cytotoxic T cells to differentiate and proliferate. Helper T cells cooperate with other lymphocytes called T-dependent B cells so that they are able to respond to specific antigens. Helper T cells may also react to an encounter with antigens by releasing lymphokines, substances including interleukin-1 (IL-1), interleukin-2 (IL-2), interferon, and chemotactic factors (chemical attractors of cells) that activate lymphocytes, macrophages, and a variety of other cells. Delayed-hypersensitivity T cells release lymphokines and produce delayed hypersensitivity, a slowly evolving inflammatory reaction to a foreign substance that indicates previous exposure to it. Suppressor T cells inhibit various responses of other T cells and also some B cells.

B cells, which mature outside of thymic tissue, also develop recognition structures on their surfaces. Like the T cell receptor, this structure is specifically designed to fit a particular antigenic determinant. When the B cell encounters this determinant, it enlarges and generates additional B cells bearing the same receptor. The rapidly proliferating B cells become mature plasma cells and secrete the surface receptor, which is now called antibody, into the body fluids. Those B cells that respond to antigen only in cooperation with T cells are termed T-dependent, while other B cells, capable of responding alone, are termed T-independent.

The body's immune system responds to a foreign substance with a variety of cellular and humoral, or chemical, components that all have specialized, interlinked functions. Cellular immunity involves the participation of many types of cells including B and T lymphocytes and accessory cells. Humoral immunity relies on antibodies, blood complement, lymphokines, and other types of molecules either already circulating in the body or secreted by cells during the immune response.

Natural killer cells, a third category of lymphocytes, are able to kill some tumor cells in vitro without the need for antibody. Their role in vivo, however, is not understood.

The accessory cells, like lymphocytes, originate from the stem cells of the bone marrow but develop in a different fashion, becoming either mononuclear cells (monocytes and macrophages) or polymorphonuclear cells (granulocytes). The mononuclear cells either circulate in the bloodstream or remain as specialized cells in many body tissues. They have two functions that are vital to the immune response. As macrophages their prime function is to ingest, digest, and destroy foreign materials. Their second function is to process foreign antigen and present it to lym-

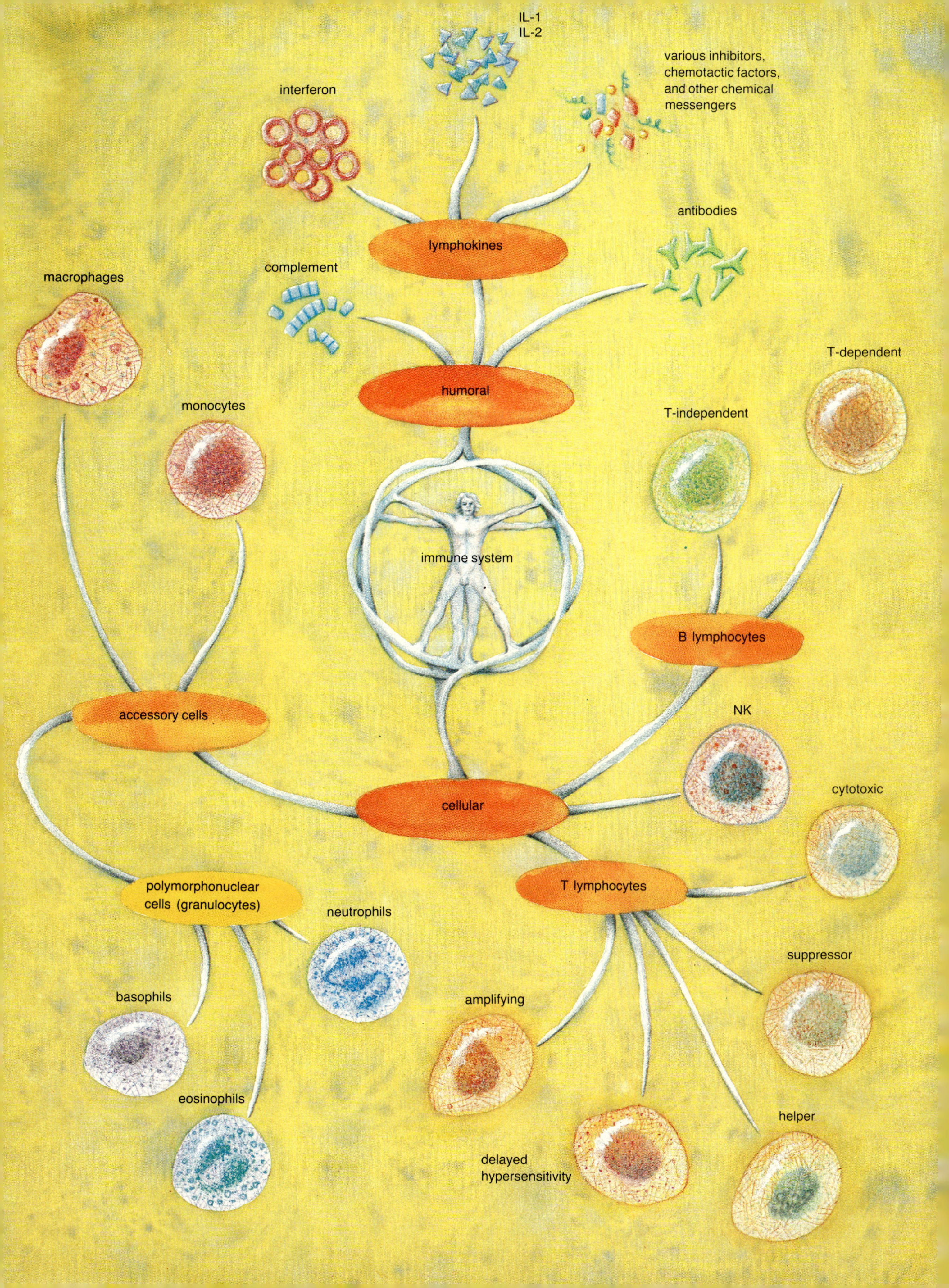
IL-1
IL-2
various inhibitors, chemotactic factors, and other chemical messengers
interferon
antibodies
lymphokines
complement
macrophages
humoral
T-dependent
T-independent
monocytes
immune system
B lymphocytes
accessory cells
NK
cellular
cytotoxic
polymorphonuclear cells (granulocytes)
T lymphocytes
neutrophils
suppressor
basophils
amplifying
eosinophils
helper
delayed hypersensitivity

phocytes in a form that will evoke a strong, specific immune response.

The polymorphonuclear cells also have a great ability to ingest and destroy foreign materials. On the basis of the way they take up stains when being prepared for microscopic examination these cells are divided into neutrophils, basophils, and eosinophils.

Neutrophils are the most abundant type of polymorphonuclear cell in the blood and migrate to antigens faster than macrophages. They constitute the first line of defense against foreign organisms and, in cooperation with other substances, promote the acute inflammatory response to foreign invasion. Basophils in the blood and their related counterparts in tissue, the mast cells, produce such substances as histamine, heparin, and bradykinin, which play a major role in allergic reactions. The eosinophils also play roles in allergic conditions and in the response to parasitic infections.

(Facing page) The body's rejection of a foreign transplant can be considered to take place in three stages. In the first stage (top left) T cells bearing specialized recognition structures on their surfaces recognize and bind to the antigenic determinants of the transplant. By way of a complex route involving cellular and humoral components, this recognition triggers the proliferation of helper T cells and their secretion of lymphokines. Serving as chemical messages, the lymphokines initiate the second stage of rejection (center) by stimulating the transformation of certain populations of T cells from their quiescent state into rapidly proliferating cytotoxic T cells whose function is to destroy the transplant. In the third stage (bottom right) cytotoxic T cells infiltrate the transplant and try to destroy it. (Below) The recipient of a liver transplant is monitored in an intensive care unit for infection and acute rejection of the foreign organ. During the first few days or weeks after transplantation a patient may experience one or more powerful rejection episodes.

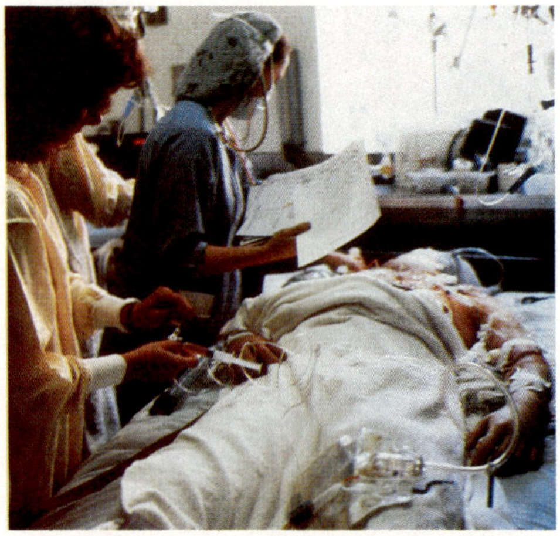

Ergun Cagatay, *Life* Magazine © 1982 Time Inc.

Transplant rejection

Transplant rejection is a type of immune response in which humoral and cellular mechanisms are directed against the foreign antigens of transplanted organs or tissues. The severity of the response and its timing depend on the antigenic dissimilarity of the transplant to the recipient, or host, and the ability of the host to mount a defense. The genetic disparity between host and donor is the major determining factor in rejection. The more dissimilar, the greater the response.

Depending on the genetic similarity of the donor and host, transplantation may evoke no immune response, a violent response, or a response somewhere in between. Four types of transplants are defined on the basis of the donor-recipient relationship.

An organ or tissue given back to the individual from which it was taken is called an autograft. An example is bone marrow removed from a patient, stored, and then reintroduced. An isograft is a transplant between genetically identical individuals; for instance, a kidney from an identical twin that is transplanted into his sibling. An organ or tissue taken from an individual of a species and transplanted into a genetically different member of the same species is called an allograft. Today the allograft is the most common type of transplant between humans. A fourth type, called the xenograft, consists of a part of the body from one species that is transplanted into a different species. A baboon heart given to a human host is an example of a xenograft.

Autografts and isografts, because they are genetically identical to the host, survive indefinitely. An allograft, however, provokes a powerful response from the recipient's immune system, which seeks to expel what it regards as a foreign substance. The response of a host to a xenograft is generally an even more violent response, again in an effort to reject the foreign "invader."

The rejection response of the host can be divided into at least three stages. The first stage appears to be the recognition of antigenic determinants by the surface receptors of T lymphocytes. This recognition triggers the proliferation of a subpopulation of T lymphocytes, helper T

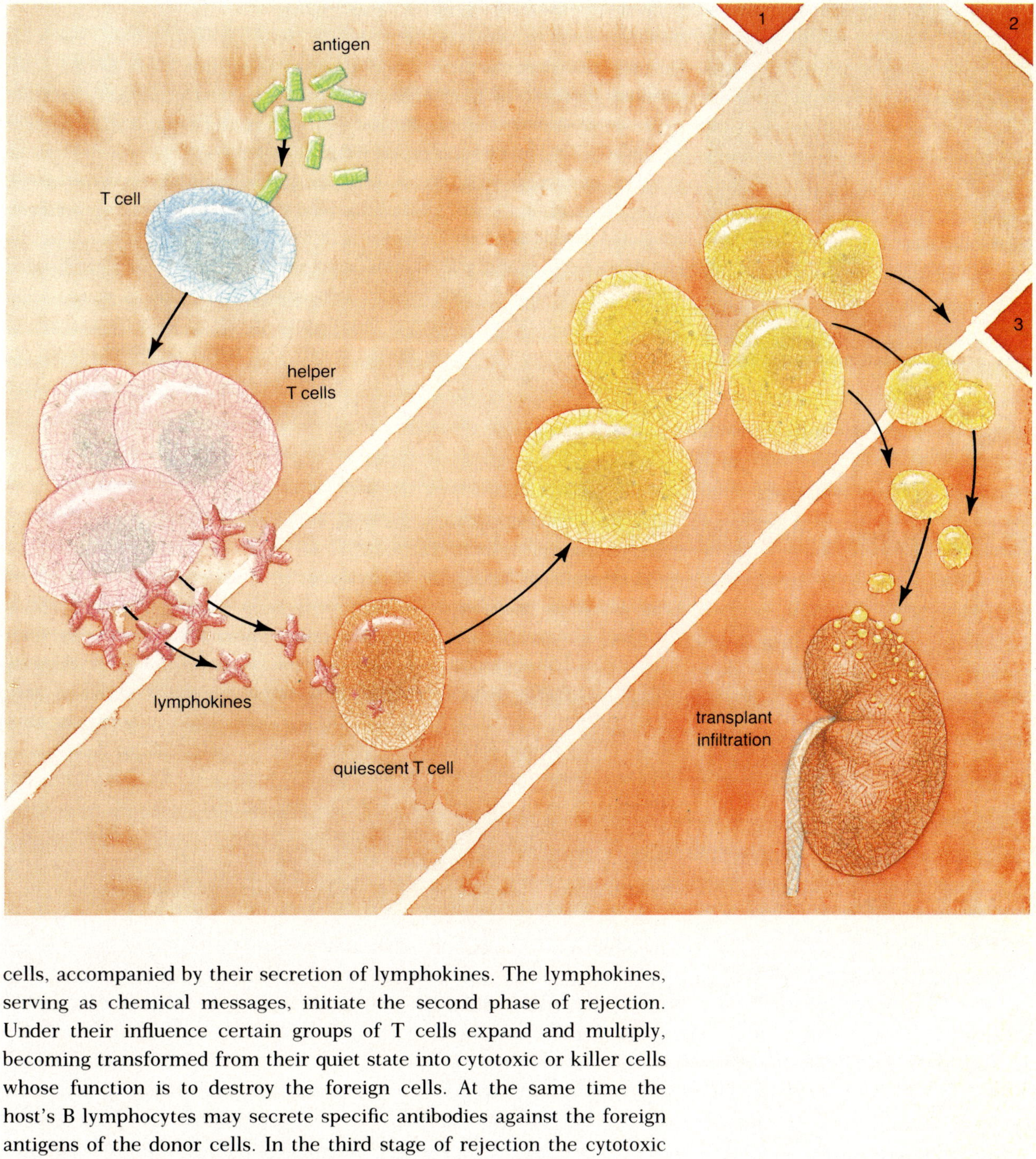

cells, accompanied by their secretion of lymphokines. The lymphokines, serving as chemical messages, initiate the second phase of rejection. Under their influence certain groups of T cells expand and multiply, becoming transformed from their quiet state into cytotoxic or killer cells whose function is to destroy the foreign cells. At the same time the host's B lymphocytes may secrete specific antibodies against the foreign antigens of the donor cells. In the third stage of rejection the cytotoxic lymphocytes infiltrate the transplant and attempt to destroy it.

Acute transplant rejection, which occurs within days or weeks after transplantation, appears to be primarily a cellular response that originates from T cells. In the more delayed types of rejection, in which the transplant gradually deteriorates over a period of months or years, humoral mechanisms appear to play a more important role. In all cases, however, T cells and B cells together mediate the graft rejection.

(Top) Courtesy, Sandoz Ltd., Basel; (bottom) adapted from information obtained from Sandoz Ltd.

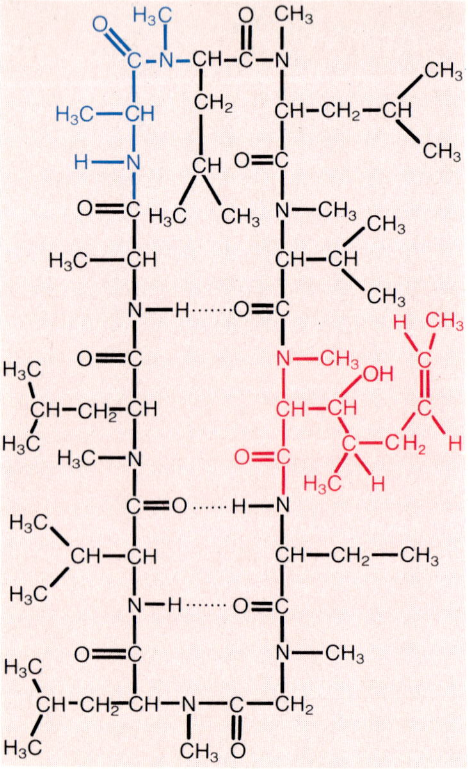

Two portraits of cyclosporine, as purified crystals (top) and as a molecular structure (above). Cyclosporine's structure is unusual in two respects: one of its 11 amino acids (highlighted in the diagram in red) had never been found before in living organisms, and at one position the amino acid alanine (blue) appears in its right-handed form rather than in the left-handed form observed in most organisms.

Cyclosporine and immune suppression

Cyclosporine, sometimes called by its older name of cyclosporin A, is a ring-shaped polypeptide (short protein) consisting of 11 amino acids linked together in a chainlike fashion. It is produced as a metabolite, or by-product, of the fungus *T. inflatum*. Its empirical formula is $C_{62}H_{111}N_{11}O_{12}$, and its molecular weight is 1202.64.

The molecular structure of cyclosporine is unusual in two respects. One of its amino acids, containing nine carbon atoms, had never been isolated previously from living organisms. And at one position in the molecule the amino acid alanine is found in the dextro or right-handed form rather than in the levo or left-handed form that is observed in most living organisms. All the other amino acids in its structure are in the usual levo form. Cyclosporine is effective when taken by mouth; its ability to be absorbed by the intestine, an uncommon characteristic for a small peptide, may be due to the fact that seven of its amino acids have attached methyl groups (CH_3—). Moreover, it is not inactivated by gastric enzymes or chemical conditions in the digestive tract. Cyclosporine is neutral and insoluble in water but highly soluble in organic solvents and lipids. Patients swallow the drug dissolved in vegetable oil.

Cyclosporine inhibits the normal rejection response by specifically and reversibly affecting the behavior of T lymphocytes. Whereas it does not appear to interfere with their recognition of antigen, it suppresses the steps immediately following: the proliferation of helper T cells, their release of lymphokines, and the induction of cytotoxic T cells. Because the drug inhibits helper T cells, it also suppresses the secretion of antibodies from T-dependent B cells. On the other hand, it seems to have little effect on the formation of antibodies by T-independent B cells.

Data from in vitro and in vivo studies suggest that cyclosporine differentiates between subsets of T-cell populations, allowing suppressor T cells to remain active but inhibiting the proliferation of helper and cytotoxic T cells. While most immunosuppressive agents work by killing lymphocytes, cyclosporine only inhibits their activation and multiplication. It also appears to have no effect on lymphocytes that have already matured into cytotoxic T cells. The drug's specificity for only certain components of the immune system seems to account for its ability to prevent transplant rejection while still allowing the body to fight cancer and infectious organisms.

How cyclosporine works on a molecular level is still a matter of conjecture and the focus of research. It may act locally on the cell membrane where receptors are found. Recent studies indicate that lymphokines such as IL-1 and IL-2 probably play a major role in the immunosuppressive action of cyclosporine. On the basis of experiments with mice the compound appears to have a twofold action: it selectively inhibits activated helper T cells from producing or releasing IL-2, which is essential for the proliferation of T lymphocytes, and it inhibits macrophages from releasing IL-1, which is essential for helper-cell activation and IL-2 release. In addition, cyclosporine may prevent T cells from developing surface receptors for IL-2.

The body absorbs cyclosporine from the gastrointestinal tract incompletely and variably. Taking the drug with food has no consistent effect in influencing its absorption, its peak concentration in the blood, or its rate of elimination. Cyclosporine reaches its highest concentrations in the blood about 3½ hours following ingestion. Compared with an intravenous infusion, the amount of cyclosporine absorbed from an oral solution is about 30%.

Once cyclosporine is in the body, most of it appears in the cells of the blood: approximately 4–9% in lymphocytes, 5–12% in granulocytes, 41–58% in erythrocytes (red blood cells), and 33–47% in plasma. In plasma about 90% of the drug is bound to proteins. After administration is stopped, the concentration of cyclosporine in the blood drops by half roughly every 19 hours. It is eliminated primarily in the feces; only 6% is excreted in the urine. Cyclosporine is extensively metabolized by the liver, but there is no single major chemical pathway. Only 0.1% of the dose is excreted in the urine unchanged. Some 15 different metabolites have been characterized in human urine, but the changes to the original molecule are minor.

The changed outlook for transplant patients

Because of its impressive effects in experimental animals and the lack of toxicity to bone marrow cells, cyclosporine was first tested in human recipients of kidney and bone marrow transplants. In these studies it showed itself to be a very powerful immunosuppressant and antirejection

How cyclosporine affects T lymphocytes on a molecular level is still the subject of investigation. It does not appear to block the antigen-recognition process itself but interferes with the steps immediately thereafter. The activation of helper T cells by antigen is much more than a simple one-to-one encounter. Macrophages participate in recognition by presenting the antigen to T cells in a more response-provoking form, and they also release the lymphokine interleukin-1 (IL-1), which stimulates the antigen-triggered T cells to release another lymphokine, interleukin-2 (IL-2). IL-2 in turn induces receptive T cells to proliferate and give rise to populations of cytotoxic T cells. On the basis of animal experiments cyclosporine appears to interfere with several links in this chain of events: it inhibits antigen-triggered T cells from producing or releasing IL-2 and inhibits macrophages from releasing IL-1. It may also prevent T cells from developing surface receptors for IL-2.

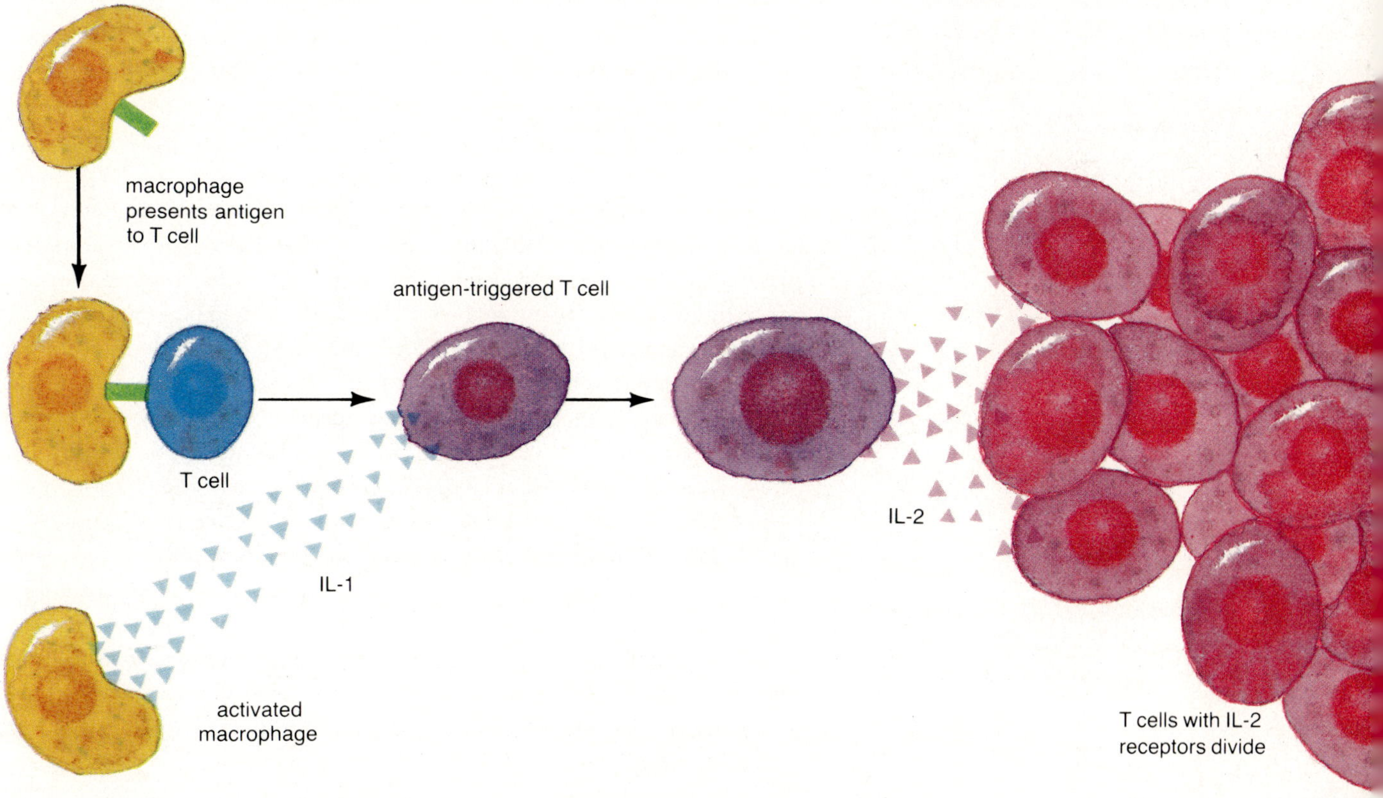

Photos, © Hank Morgan—Rainbow

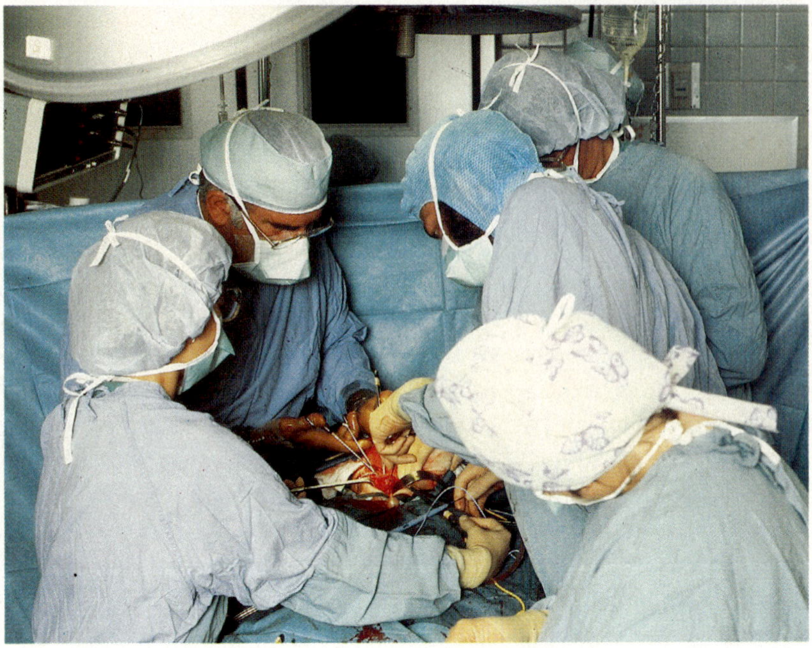

Cyclosporine's dramatic improvement in the survival rate of patients following kidney transplant surgery (right) has made it the drug of choice for this kind of operation. For patients whose transplants bear some antigenic similarity to their own tissues, usually because of blood relationship, the probability of rejection may be lowered still further. (Below) A mother comforts her young son, the recipient of one of her own kidneys.

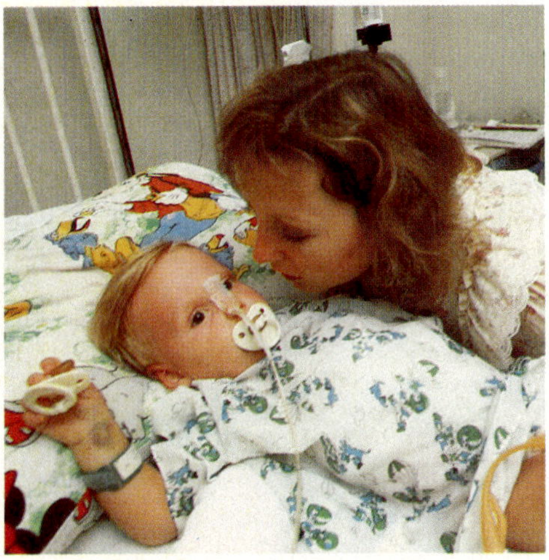

agent. This potency, however, led to unexpected side effects. Some patients experienced kidney and liver damage; others acquired lymphomas, cancerous tumors arising from lymphocytes or other stem-cell derivatives, which can appear when the immune system is severely depressed. In the belief that these problems were caused by excessive doses of cyclosporine coupled with the use of other immunosuppressive agents, new transplantation studies included a different therapeutic regimen: cyclosporine was used either as the sole immunosuppressive agent or with low doses of conventional immunosuppressants. This new approach reduced serious side effects while it maintained the improvement in transplant survival rates over standard therapy.

Controlled studies of kidney transplant patients were conducted at major institutions in the United States, Canada, and Europe between 1978 and 1982. Patients were chosen at random to receive either cyclosporine or azathioprine, a conventional immunosuppressant. The results from all trials were quite consistent: the transplanted kidneys of patients receiving cyclosporine survived significantly longer. In addition, the patients experienced fewer rejection episodes and, in many cases, fewer infectious complications and shorter hospital stays. In combined averages from five pivotal studies patients on cyclosporine had a one-year transplant survival rate of 80%, compared with 61% for those on azathioprine therapy. In these studies all patients received kidneys taken from cadavers.

Today cyclosporine is widely considered the drug of choice for kidney transplants. It appears to be most effective in transplants from cadavers, in patients who have received few blood transfusions, and in patients who are receiving retransplants after previously rejecting other kidneys. By early 1985 well more than 8,000 kidney transplant patients had received cyclosporine.

Before the advent of cyclosporine liver transplantation had been all but discontinued. By 1978 only one transplant center in the world still continued this operation, thought to be highly experimental, and was achieving one-year survival rates of only 32%. Initial studies with cyclosporine raised one-year survival rates as high as 70%. This dramatic improvement prompted numerous centers to begin liver transplantations. No longer considered experimental, the procedure has saved the lives of many young children who, born with liver abnormalitites, could not otherwise have survived. It is in this area of transplantation that cyclosporine excels. Today throughout the world all liver transplant surgery is performed with cyclosporine.

Like liver transplantation, heart tranplantation had been virtually abandoned prior to the success of cyclosporine. After initial enthusiasm for the procedure in the early 1970s, poor survival rates caused all but one transplant center to stop heart transplant surgery. After clinical studies of cyclosporine began, the average survival rates of heart transplant patients rose markedly, from 55–59% to about 80%. These figures encouraged many centers to resume heart transplant surgery or start it for the first time. In some centers heart transplantation has become an accepted therapeutic procedure.

Cyclosporine has also made possible the simultaneous transplantation of heart and lungs, a strategy that failed dismally when first tried in the late 1960s with immunosuppressants available at the time. Since 1981 heart-lung transplantations have been successfully performed; as of early 1985 two-thirds of all heart-lung recipients were alive and well. This operation, reserved for those patients whose heart and lungs are both failing, has opened the possibility for life where none previously existed.

In Europe cyclosporine is approved for the transplantation of bone marrow into patients with leukemia, some forms of anemia, or certain immune disorders. In this procedure healthy donor marrow is injected into the recipient after that person's marrow has been destroyed with radiation and chemotherapy with the expectation that the new marrow will grow and replace the old. Cyclosporine appears to speed the growth of the new marrow, to aid in its engraftment, and to reduce the problems of graft-versus-host disease, in which lymphocytes present in the donor marrow react against the host tissue, in effect rejecting the recipient. In 1985 an application for the use of cyclosporine in marrow transplantation in the U.S. was pending.

Currently cyclosporine is being tested as an antirejection agent in pancreas transplantation. In this still experimental procedure a pancreas from a cadaver is implanted in a diabetic patient with the intent of replacing those insulin-secreting cells that have been lost in the diabetic's own pancreas. Thus far, studies have shown that cyclosporine inhibits the rejection of transplanted pancreases.

In 1984 cyclosporine was tried in a highly experimental operation in which the fatally defective heart of a 12-day-old baby was replaced with the heart of a young baboon. The infant died after 20 days from heart and kidney problems following an attempt by her immune system to reject the xenograft. This outcome offered a sharp reminder of modern medicine's still limited understanding of all the immunologic factors that are important in determining the success of a transplant.

Side effects

In animal studies cyclosporine had produced only minor side effects, a particularly noteworthy characteristic for an immunosuppressive agent since the most common problem of such agents has been bone marrow suppression. Nevertheless, the animal work failed to warn of a number of side effects that appeared once the drug was tried in human beings.

Kidney damage is the most significant side effect of cyclosporine and is, of course, a special problem in managing patients with kidney transplants. This damage is manifested by elevated levels of creatinine and blood urea nitrogen (BUN) in the blood, two indirect indicators of abnormal kidney function. Most transplant patients, however, do run elevated creatinine levels, and the long-term effects are not yet known. Creatinine and BUN levels can be brought down and controlled by reducing the dosage of cyclosporine.

Liver damage afflicted some early transplant patients. As the therapeutically effective dose of cyclosporine has been reduced, however, the incidence of liver problems has decreased sharply and is no longer considered a major concern. Hirsutism, or increased hair growth, appears in about one out of four patients taking cyclosporine and can be a cosmetic problem for some people. Overgrowth of the gums, high blood pressure, and tremor have been noted in 10–30% of patients.

Lymphomas, noted in early cyclosporine trials, have dramatically decreased in incidence with a change in therapeutic approach. As of 1985 fewer than 1% of cyclosporine-treated transplant patients worldwide have developed either a lymphoma or some other disorder that involves abnormal proliferation of lymphocytes.

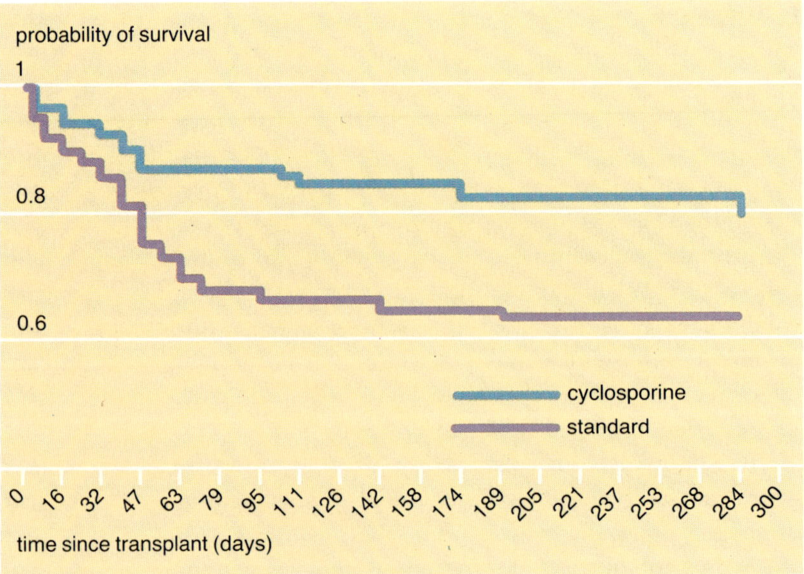

Controlled studies of kidney and liver transplant patients, chosen at random to receive either cyclosporine or the conventional immunosuppressant azathioprine, were conducted at major medical institutions in several countries. The results from all trials were quite similar: the transplanted organs of patients receiving cyclosporine survived significantly longer. Graph at right plots the probability of survival over time for one such study of kidney transplant recipients taking cyclosporine or azathioprine.

Adapted from information obtained from Sandoz Ltd., Basel

Present and future research

Although the body's immune system is adapted to recognize and destroy foreign antigens, in some situations it recognizes part of itself as foreign and produces autoantibodies, antibodies directed against the body's own proteins. What causes this abnormality is unknown, and the process itself is poorly understood. When the autoantibodies lead to structural damage, the resulting conditions are termed autoimmune disorders or diseases.

Presently cyclosporine is not approved for use in any autoimmune disorder. Nevertheless, encouraged by the results of tests on animals having autoimmune diseases that serve as models for their counterparts in humans, medical researchers in several countries have started clinical studies on patients with diseases thought to be autoimmune in origin or to have an autoimmune component. These include insulin-dependent diabetes mellitus; posterior uveitis of noninfectious cause, a potentially blinding inflammatory eye condition; rheumatoid arthritis; multiple sclerosis; primary biliary cirrhosis, a disease of the liver; systemic lupus erythematosus; and amyotrophic lateral sclerosis, or Lou Gehrig's disease. Although the positive responses to cyclosporine in some early trials have inspired exciting speculation, there still is no definitive evidence that such therapy will be successful. Only through the conduct of appropriately controlled trials will a proper assessment be obtained.

Cyclosporine potentially represents the first in a new class of compounds. As an immunosuppressant its activity is unique and will certainly stimulate the search for other immunospecific agents. One can look forward to immunologic compounds that will be highly specific in their actions compared with the broad suppressive power of previous agents.

Cyclosporine itself is also distinct in that it appears to produce no fewer than five different effects. In addition to inhibiting T cells, it shows antifungal activity, kills certain parasites—including the agents of malaria and schistosomiasis—in vitro and in experimental animals, reduces inflammation, and in some way causes damage to the kidney. In future investigations of cyclosporine and the analogues that will be derived from this core molecule, researchers will attempt to separate or maximize some of these specific powers. For example, a cyclosporine analogue may be found that will destroy parasites without suppressing any component of the immune system; likewise, a compound that emphasizes anti-inflammatory activity while reducing or eliminating kidney damage. The prospect of such achievements is extremely exciting to the scientist working on cyclosporine today as well as to the practitioners and beneficiaries of modern medicine.

See also *1985 Yearbook of Science and the Future* Feature Article: ANTIBODIES: NATURE'S "MAGIC BULLETS."

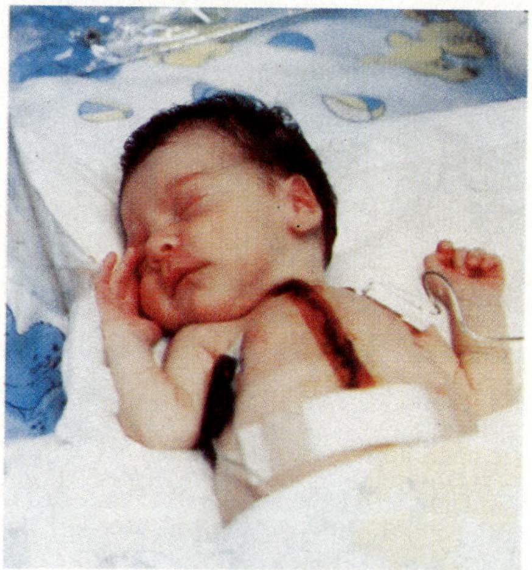

In 1984 a 12-day-old infant, known pseudonymously to the public as Baby Fae, received cyclosporine as part of an experimental procedure in which a baboon heart, a xenograft, was substituted for her own fatally deformed heart. The infant died after 20 days despite the best that modern medicine could offer.

PHOBIAS:

FEAR OUT OF CONTROL

by Thomas R. Insel

Heights, spiders, cats, open and closed spaces—these are among the phenomena that can disrupt the lives of susceptible people by generating in them fears of crippling intensity.

George Tooker, "The Subway," 1950; egg tempera on composition board, 18 x 36 inches. Collection of Whitney Museum of American Art, Juliana Force Purchase, acq. no. 50.23

Imagine yourself in a familiar place away from home, perhaps a supermarket or a restaurant. Although you have been there countless times before, this time is like none other. Without warning and for no apparent reason, terror suddenly overwhelms you. Your heart pounds, you seem short of breath, you find yourself perspiring, your mind races, and, most of all, you feel you have to escape. This is a panic attack, and it is the source of a common yet severe form of phobia called agoraphobia.

What causes such an attack? Does it result from an unconscious conflict? Is it triggered by something in the environment that recalls a childhood trauma? Or is it the result of a chemical event in the brain? There are several psychological theories concerning the origin of phobias, but recently the success of behavioral and medical treatments have focused scientific attention on "contextual" and biomedical aspects of the problem.

Phobia, in the most general sense, means fear. The word can be traced to the Greek god Phobos, who was, appropriately enough, the son of Ares, god of war, and Aphrodite, goddess of love. Phobos symbolized not only fear but terror and flight. To this end, ancient Greek warriors would paint the image of Phobos on their shields to intimidate enemy soldiers.

Phobias, in a scientific context, involve not only the emotional component of fear but also a physiological component (such as a racing heart in the presence of the phobic stimulus) and usually a behavioral component, manifested as avoidance of the phobic situation. It is important to remember that fear is a normal emotion, essential to survival. Phobias, by contrast, are senseless or disproportionate fears that interfere with normal functioning. For those who have never experienced a phobia the term "fear" may be too weak—a phobia involves dread, terror, or panic. Routine activities, such as eating in public or using elevators, become so terrifying that jobs or important relationships are lost. A person with agoraphobia may be so afraid to leave home that he or she may spend years as a housebound invalid.

Fear

One approach to understanding phobias is to study normal fear. This can include the study of species other than man, for, as Charles Darwin pointed out more than a century ago, fear is an emotion that is shared by all mammals, and its expression is largely biologically determined. Darwin described the expression of fear this way:

> The frightened man first stands like a statue motionless and breathless . . . the heart beats quickly and violently . . . the skin instantly becomes pale . . . perspiration immediately exudes from the skin . . . the hairs also on the skin stand erect . . . the mouth becomes dry . . . one of the best marked symptoms is trembling of all the muscles of the body and this is often first seen in the lips.

The facial expression of fear is also surprisingly universal. When photographs of the faces of frightened people are shown to strangers from diverse cultures, not only can the emotion be identified correctly but also fear can be consistently distinguished from the expressions of anger or surprise.

__THOMAS R. INSEL__ is a psychiatrist and psychopharmacologist in Bethesda, Maryland. He also works as a research scientist for the National Institute of Mental Health.

Courtesy, National Portrait Gallery, London; oil painting by John Collier

Charles Darwin more than a century ago pointed out that fear is an emotion shared by all mammals and that its expression is in large part biologically determined.

In general, fear develops under three interrelated influences. The first is genetic. Animal behaviorists have developed strains of rodents and primates that are genetically more fearful or less fearful. Although the genetic locus for this behavioral difference has not been identified, recent research has demonstrated that more fearful animals show significantly more distress when isolated from their families even in the first few days of life. Cross-fostering, that is, raising fearful offspring with genetically less fearful parents, does not eradicate the fearfulness. Thus there is some genetic component that appears to persist in spite of environmental influences. Humans also show genetic or temperamental differences on a dimension of fearfulness, with timidity at one end of the spectrum and an active risk-taking personality at the other. However, scientists still do not know the importance or magnitude of the genetic component of fear in humans.

A second major influence on the emergence of normal fear is related to the developmental tasks in the first few years of life. With each new "passage" there seems to be an associated fear. As an infant begins to

(Left) Candee Productions; (right) Jean-Loup Charmet, Paris

As children grow older, their fears change. "Childhood separation anxiety" (above), evidenced by a clinging to mother or another familiar person when strangers are present, is followed by fears of animals and imaginary creatures (right).

crawl, he or she may avoid heights, steering away from what has been called the "visual cliff." At about eight months of age "stranger anxiety" normally develops, with desperate crying and clinging to mother when an unfamiliar person is present. From ages two to four many children who previously had been attracted to animals become fearful of them. Thereafter, fears of the dark, storms, and imaginary creatures take over. Beginning school is a particularly stressful "passage" that may be accompanied by a number of new fears and can be associated with "school phobia," when a child becomes terrified of school itself. Most of these normal, developmental fears dissipate by nine or ten years of age.

A third influence on the development of fear stems from the environment. Some people are buffeted by more dangers and traumas than others. The behaviorist John B. Watson demonstrated more than 60 years ago that traumatic experiences can become associated with neutral, coexistent cues so that fears have a tendency to spread to innocent elements in the environment. In a classic experiment with "little Albert," Watson found that a boy who enjoyed playing with a pet rat became "fearful" of the rat after he was exposed to a frightening noise each time he reached for his pet. This observation led to a "conditioning theory" of fear, which proposed that fear was largely a learned response.

In the past few years scientists have begun to realize that the environment affects fears not only by association but also through more subtle influences such as parental modeling of fearful behavior. For instance, a parent may be fearful and foster dependence and avoidance. Other parents may communicate that fear is unacceptable, thus provoking confusion and insecurity. And some parents may push their children into fearful situations, thereby generating feelings of vulnerability and helplessness. Such effects are rarely simple, however, as they depend not only on what the parent does but on how the child interprets the parent's behavior. Two children may react to the same parental actions in entirely different ways. Susan Mineka at the University of Wisconsin recently examined parental influences on fear by studying two groups of rhesus monkeys: a group of parents that had been raised in the wild, where they had become fearful of snakes, and a group of their young offspring that had been raised in the laboratory in peer groups away from their parents. The young monkeys showed no innate avoidance of snakes, but Mineka found that when they were permitted to observe their parents' fearful responses, even for as little as eight minutes, they also began to avoid snakes, By means of such experiments the very rapid and powerful effects of parental modeling can be studied in detail.

The anatomy of fear

Although many organs of the body, from the heart to the sweat glands, participate in the fear response, it is the brain that orchestrates the entire reaction. The brain consists of billions of nerve cells, or neurons, many of which are clumped together into small structures. Neurons function somewhat like telephone wires, carrying an electrical signal from the body of the cell out to tiny branching processes. The message from one neuron to another is not transmitted by an electrical signal but by a chemical one. Neurons meet at tiny junctions called synapses. Within each synapse one neuron that has been electrically stimulated releases a specific chemical, a neurotransmitter, which binds to a specific receptor on the second neuron. This action leads to local chemical changes that contribute to a new electrical impulse. There are billions of synapses in the human brain and scores of different neurotransmitters to convey messages across them.

Courtesy, Susan Mineka, University of Wisconsin, Madison

Monkey reveals fear of the snake in the foreground by retreating to the rear of its cage and leaving its food untouched.

Courtesy, Thomas R. Insel

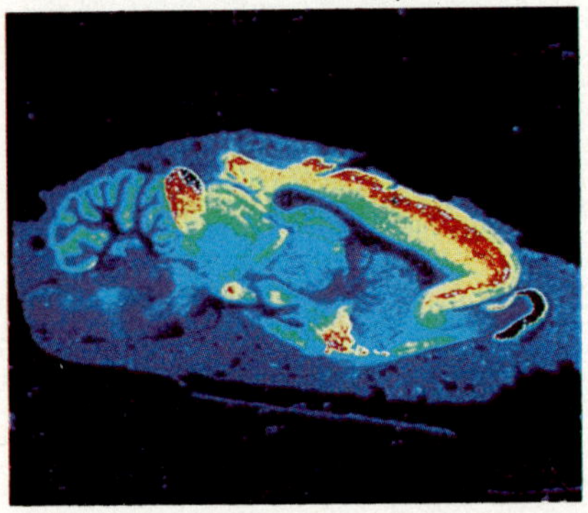

Autoradiogram of a rat's brain reveals the areas in the limbic system, labeled in red and yellow, that are receptors for such drugs as Valium and also for neurotransmitters (chemicals in the brain) that affect fear.

Which brain structures and which neurotransmitters mediate fear? Researchers still do not know all of the brain systems that are important for the experience of fear, but by electrically stimulating certain brain structures neuroscientists have been able to identify certain areas that elicit fearlike behavior in animals. The principal regions of the brain involved are the amygdala, the hippocampus, and the hypothalamus, all part of a primitive circuit in the brain called the limbic system. This circuit has been described by neuroanatomist Paul Maclean as the middle layer of a "triune," or three-layer brain, sandwiched between an inner, ancient reptilian core brain that mediates primitive postural responses and an outer, "newer" shell called the neocortex that is particularly developed in the human brain and is responsible for uniquely human functions such as language.

Using a technique known as autoradiography, scientists recently have been able to show that within the limbic system there are receptors for certain drugs and for many of the neurotransmitters that affect fear. One such neurotransmitter that has been extensively studied is norepinephrine. Neurons containing norepinephrine or norepinephrine receptors have been localized to a discrete system in the brain. Although this norepinephrine system accounts for fewer than 1% of the neurons in the brain, it may be essential to the fear response because the stimulation of certain norepinephrine neurons elicits fear or alarmlike behavior in nonhuman primates. Another brain system that may be important for the mediation of fear has been suggested from studies with the benzodiazepines, a class of drugs that includes diazepam (Valium) and chlordiazepoxide (Librium). These drugs, which potently decrease anxiety and in some animal studies reduce fearlike behavior, appear to work by binding to specific receptors in the brain. Benzodiazepine receptors are found throughout the limbic system and may elucidate one of the mechanisms by which the brain mediates fear. Undoubtedly many others will be revealed in the next few years.

In summary, fear is a normal human emotion, tied to a legacy from man's biological past yet influenced by a number of developmental and environmental factors. In regard to phobias the question arises as to whether they are just extreme fears or qualitatively different from normal fear. The answer to this question requires an explanation of the different kinds of phobias.

Varieties of phobias

Phobias are usually classified into three general types: simple phobias, social phobias, and agoraphobia. Although any given individual may have all three varieties, there are some distinctive features that make their separation important.

Simple phobias, perhaps the most common and usually least troublesome, involve persistent fears of a particular object or situation. Common examples of simple phobias are: acrophobia, fear of heights; ailurophobia, fear of cats; arachnophobia, fear of spiders; claustrophobia, fear of closed spaces; xenophobia, fear of strangers; and zoophobia, fear of animals.

(Left, top and bottom) Dan Morrill; (right) Arthur Tress

Simple phobias are persistent fears of a particular object or situation. Among the most common of them are fear of heights (above), of cats (top left), and of spiders (bottom left).

Heights and spiders may evoke fear in most nonphobic people, but there is a quantitative difference between them and sufferers from phobias. For instance, an acrophobic would not only fear looking over a precipice (which might arouse a fear of falling in most people) but would experience equal discomfort inside a building behind a closed window through which there was no danger of falling. As long as a simple phobic can avoid his or her "phobic stimulus," there is little problem with fear and in most cases little disruption of normal functioning.

A social phobia is somewhat different. Again there is avoidance, but in the case of a social phobia the fear is one of humiliation or embarrassment in a public setting. Common examples are fears of eating or writing or speaking in public. Often this problem is chronic, emerging in the early twenties in people who are shy and isolated by nature. While it may not be readily apparent that this could be a very distressing problem, one should remember that people who cannot sign their names in public would have great difficulty managing business affairs, and public speaking phobics may be unable to function in some professions.

A third form of phobia, agoraphobia, was originally described by the German neuropsychiatrist Karl Westphal in 1872. Agoraphobia means literally "fear of open places," but a more accurate definition is a dread of being unaccompanied in public places. In most cases this leads to progressively restrictive confinement to a "safe" haven, usually home. Agoraphobia to one degree or another afflicts more than 1% of the United States population. Women are affected twice as often as men. A U.S. psychiatrist, Donald Klein, demonstrated that in most cases of agoraphobia there is a preceding history of panic attacks—spontaneous, sudden episodes of severe anxiety much as described at the beginning of this article. Panic attacks may be associated with any or all of the physical symptoms of fear but may also include the curious psychological experience of "depersonalization," in which a person feels disconnected or disembodied from himself. Agoraphobia may be secondary to the panic attacks. What the agoraphobic fears is not the public setting as such but the experience of another panic attack in such a place.

People with agoraphobia are prone to severe depressions and very frequently have relatives who suffer from agoraphobia, depression, or alcoholism. In addition, they may have a history of severe responses to separations in childhood. In the past few years studies have accumulated evidence suggesting that, in contrast to people with normal fears or simple phobias, agoraphobics may have some biomedical abnormality.

PET (positron emission tomography) scanner (below) reveals images of cerebral blood flow (below right) in a person with a panic disorder. The left half represents brain blood flow, and the right half reveals the difference in blood flow between the left and right hemispheres of the brain. This difference is expressed as a percentage of the left hemisphere blood flow in the left half of the calibrated scale. The quantity of the blood flow in the brain corresponds to the right half of the scale. The most abnormal distribution of blood flow, as indicated by the bright yellow spot in the right hemisphere, occurs in the parahippocampal gyrus, a region where the limbic system meets the outer layer of the brain.

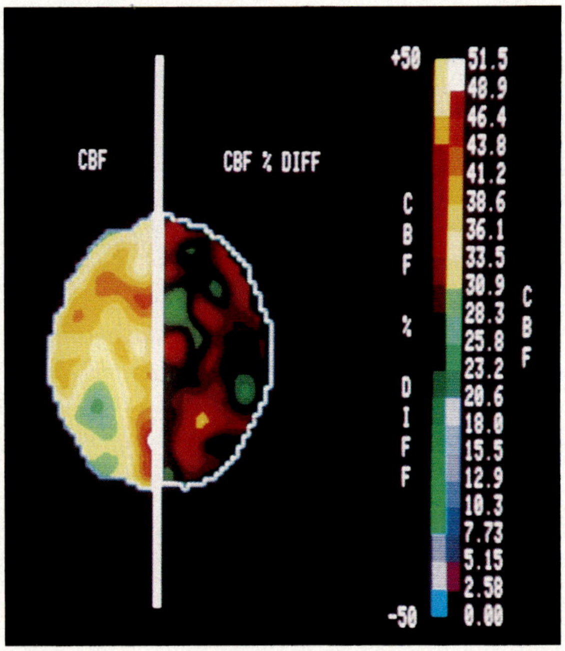

(Left) Dan McCoy—Rainbow; (right) courtesy, Marcus E. Raichle, Washington University School of Medicine

For instance, panic attacks can be induced by certain chemical changes in the brain. Lactate, a chemical that changes the acid-base balance in the brain, and yohimbine, a drug that increases the activity of the brain pathways containing norepinephrine, are both capable of evoking panic attacks in agoraphobics but not in nonagoraphobic volunteers. These findings point to a neurochemical abnormality, perhaps involving the norepinephrine system, that may make agoraphobics particularly vulnerable to panic attacks.

Recently scientists have developed new techniques such as the PET (positron emission tomography) scan to look inside the brain in order to study the pattern of brain blood flow or energy use. Using this method in a study of people with panic attacks and agoraphobia, researchers at Washington University in St. Louis, Missouri, reported an abnormal distribution of blood flow in an area called the parahippocampal gyrus, one of the regions where the limbic system meets the outer layer of the brain, the neocortex.

These pharmacological and anatomical findings so far have not provided a single simple understanding of why some people have panic attacks that progress to agoraphobia. They have, however, suggested that earlier theories that assigned all of the symptoms to an unconscious conflict may have been creating a psychological explanation for a problem that, in part, is a physiological one.

One other form of phobia deserves to be discussed separately. Over the centuries in every culture there have been individuals who suffer from a fear of contamination, a fear that may be described as a dirt or germ phobia. Such people feel contaminated almost constantly, and their only respite is to wash themselves, often for several hours each day. Contamination phobics usually avoid public rest rooms, refuse to touch doorknobs or money, and may carry a handkerchief as a safety barrier to the world. People with this problem are generally unlike those in the three previous phobia groups. Indeed, these people have a problem that is not simply a phobia but is called an obsessive-compulsive disorder. They recognize that their sense of contamination is irrational, but the feeling of being dirty is so strong that they cannot resist the compulsive urge to wash. Unlike a true phobia, in this obsessive-compulsive disorder there is no way to fully avoid what the person fears; "contamination" is everywhere, it is invisible, and it is inescapable. The cause of this strange and painful disorder is not known, but recent research has provided new medications and new behavioral treatments that help to control the obsessive fears and the compulsive behavior.

Art Pahlke

People who suffer from a fear of contamination, which may be described as a dirt or germ phobia, wash themselves frequently, sometimes for several hours each day.

Origins of phobias

How phobias develop is not clear. One curious observation is that phobias seem often to focus on the same stimuli. Snakes or heights, for instance, are very common fears among phobics, whereas guns or knives, objects potentially as dangerous, rarely become the focus of a phobia. Why should some objects or situations become phobic so often? One theory is that we are biologically predisposed to respond with greater fear to

Arthur Tress

An unusual aspect of phobias is that they rarely are focused on such potentially dangerous objects as guns or knives.

objects that would have been a threat to our mammalian ancestors. A Norwegian psychologist, Arne Öhman, tried to test this theory by studying the physiological responses to pictures of various objects. In support of the theory he found that nonphobic volunteers showed an increased intensity of responses to pictures of objects that frequently become the focus of a phobia compared with pictures of objects that, though potentially as dangerous, rarely are associated with phobias. Perhaps humans are genetically programmed to respond with more intense fear to such "prepared" stimuli.

Just as certain genetic strains of animals appear more fearful, certain families have a high incidence of phobias. This is particularly true for agoraphobia as well as for certain kinds of simple phobia. The question is then asked whether this is a genetic effect or the result of the modeling of fear by parents for their children. The answer requires a study of offspring from phobic families adopted by nonphobic foster parents. This cross-fostering strategy has been used to study the genetic aspects of other psychological disorders, but it has not been employed for phobias. An alternative strategy, comparing identical and fraternal twins for concordance of phobic symptoms, could provide similar insights because a genetic disorder should be shared more often by identical twins, whereas environmental factors should affect both kinds of twins almost equally.

In a study of Norwegian twins Sven Torgersen found a much higher concordance rate for panic disorder and agoraphobia in identical twins as compared with fraternal twins, supporting the hypothesis that this form of phobia is, in part, genetically determined.

But genetics is only part of the story. Phobias are frequently precipitated by the loss of a relative or close friend. In some cases it appears as if the phobic person substitutes a focused fear of a tangible object for the terrible insecurity and anxiety resulting from the loss. It is possible that most phobias originate with a traumatic event in a genetically predisposed individual.

Occasionally, simple phobias seem to result from a conditioning process in which an innocent object becomes phobic by being associated with a fearful event. Isaac Marks, in his book *Fears and Phobias,* describes a case of a four-year-old girl who was playing with a pet dog. The girl's baby sister sitting nearby was knocked over by the dog, receiving a cheek wound. The wound later became infected, and the baby died. The mother, unaware of the dog's role, blamed the little girl for the baby's death. Several days later the little girl began to dislike dogs and, sometime thereafter, developed a dog phobia that lasted 40 years. While such a scenario might be expected in every case of phobia, a history of a discrete event precipitating the phobic symptoms is the exception rather than the rule. More commonly a series of stresses precedes the emergence of a phobia.

New treatments for phobias

Early in this century Sigmund Freud described the psychoanalytic treatment of phobias. In a classic case history, "Analysis of a Phobia in a Five-Year-Old Boy," Freud explained the case of Hans, the son of a colleague. Hans had developed a fear of horses that, Freud concluded, was a displaced "Oedipal fear" of conflict with his father and also a disguised wish to have his father dead. This was the essential psychoanalytic theory of phobias: the fear results from a conflicted, unconscious wish. The psychoanalytic treatment attempted to shed light on the unconscious conflict by the process of free association and dream interpretation. In recent years this classic psychoanalytic approach has evolved to more interactive psychodynamic treatments that rest on the same assumptions of cause but usually involve more give-and-take with the therapist and focus more on the experience of vulnerability than on childhood traumas. All such treatments are basically "talking therapies"; that is, the patient talks about his phobia, and the therapist and patient together try to understand its cause. Although there are many exceptions, for the most part an understanding of the factors leading to the phobia emerges only after the phobia has been mastered.

This leads to the question of how one masters a phobia. The answer seems simple and yet is extremely difficult in practice. The best treatment for simple and social phobias is prolonged and repeated exposure to the phobic stimulus. With patience, fortitude, and an encouraging, supportive therapist, most people with phobias can overcome their avoidance and gradually habituate to spiders, elevators, eating in public, or whatever else they have allowed to control their behavior. The key is exposure. But how can exposure to a phobic stimulus be a treatment when that stimulus was the very thing that precipitated such extreme fear? Part of the answer is that phobic persons may begin to fear their fears; the longer they avoid something the more they dread it, until the anticipation of the fear becomes as bad as the original problem. Exposure reduces this secondary fear by returning to the original stimulus and showing a phobic person that fear can be confronted and not just avoided. In addition, the therapeutic approach is a gradual one, involving an encouraging helper who can teach the phobic person to rely on numerous safety cues in the environment.

There is also a role for medication in the treatment of phobias. In the case of the simple and social phobias medication may be used to reduce immediate anxiety, but it is not a substitute for exposure. For people who have panic attacks and agoraphobia, the situation is somewhat different. Two kinds of medications that were previously used mainly to relieve depression, namely the tricyclic antidepressants and the monoamine oxidase inhibitors, also reduce panic attacks. These medications do not work simply by making people feel euphoric so that they forget about their problems. Antidepressants are not euphoria-producing drugs; in fact, they have little effect on mood except in people with severe degrees of depression. More important, in people with panic disorder who are not depressed, these compounds reduce panic attacks. Researchers are attempting to determine whether the biochemical mechanism for this effect is related to the mechanism for the antidepressant effects. Neither mechanism has been fully elucidated, but these medications do share some common effects on brain neurotransmitters. For instance, they modify the activity of the norepinephrine system, suggesting to many

Arthur Hardy, a physician, helps an agoraphobic woman overcome her fears of dealing with strangers.

TERRAP, Menlo Park, Calif.

scientists that some change in the regulation of this system may mediate both the symptoms and the treatment of depression and panic disorder. But whatever the mechanism, medication does not constitute the full treatment for people with agoraphobia. An active behavioral approach that helps the person to leave home and enter public areas is essential once medication has controlled the panic attacks.

Obsessive-compulsive disorders are usually more difficult to treat. In addition to exposure, behavior therapy must include firm restrictions on washing. In some of the most successful treatment centers obsessive-compulsive washers may not be allowed to clean themselves at all for 48 or 72 hours while having to confront the very things they most dread. The combination of exposure and "response prevention," as this treatment is called, leads to rapid and sustained recovery in as many as 80% of those who agree to the treatment. As with panic disorder and agoraphobia, antidepressants also are helpful to many people with this problem. And in very severe cases, psychosurgery has been tried with some success.

Future prospects

Phobias have been recorded as a mysterious part of the human experience for many centuries. Man has always had a theory to explain them. During a more religious era they were believed to be the work of the devil. In Victorian times Freud provided a sexual explanation. Today, research is primarily directed at a neurochemical cause. In many ways the theories reveal as much about the historical era as they do about the disorder.

In the current view the phobias are a spectrum of disorders that range from mild, discrete fears to an incapacitating, pervasive syndrome. Just as with normal fear, there are both biological and environmental factors that lead to phobic symptoms. While our understanding of the origin of phobias is still limited, our capacity to treat both mild and severe phobias has increased greatly in the past few years.

The future will certainly take us inside the brain to view both chemically and anatomically the pathways of fear. A major focus for research in the next few years will be the study of different elements of the fear response, defining which neurotransmitters are involved and how they work in an integrated fashion to lead to behavior. By elucidating these elements, perhaps we may learn for the first time how phobias originate from normal fears and how current treatments relieve phobic suffering.

FOR ADDITIONAL READING

Jeffrey Gray, *The Neuropsychology of Anxiety* (Oxford University Press, 1981).

Donald F. Klein and Judith Rabkin (eds.), *Anxiety: New Research and Changing Concepts* (Raven Press, 1981).

Isaac M. Marks, *Fears and Phobias* (Heinemann Medical, 1969).

Isaac M. Marks, *Living With Fear* (McGraw-Hill, 1978).

Manuel D. Zane and Harry Milt, *Your Phobias* (American Psychiatric Press, 1984).

INDOOR AIR:

NEW BATTLEGROUND AGAINST POLLUTION

by Anthony V. Nero, Jr.

Progress has been made in combating outdoor air pollution, but another threat to human health of at least equal importance—polluted air within buildings—has been largely ignored.

During recent years scientists and policymakers have paid substantial attention to airborne substances that can threaten human health or environmental quality. Studies of these pollutants have led to the enactment of measures to control them. Controls on emissions from automobiles or industries that burn gasoline, kerosene, or coal have noticeably improved the quality of the air. A growing effort is being devoted to isolation of industrial activities and waste dumps that can release complex chemicals into the air and water. And a substantial and relatively effective regulatory structure is in place to control releases of radioactivity to the general environment.

But, in terms of human health, scientists may have missed the main point: that people typically spend 80 to 90% of their time indoors. Recent research has revealed that concentrations of many pollutants can be higher indoors than out and that the factors contributing to indoor pollution are virtually unaffected by controls on outdoor air pollution. This perspective is not new just to the public or government officials. Even environmental scientists and engineers specializing in air pollution have been startled to discover that the highest personal exposures to combustion emissions occur not in urban smog but in homes with unvented combustion appliances. Concentrations of organic chemicals in homes and offices are often a hundred or a thousand times higher than they are outdoors, and airborne radioactivity in homes is more significant by far than that released from nuclear power plants.

The question arises as to how this situation came about. The reason is simply that substantial amounts of these pollutants are emitted from indoor sources into what are actually very small microenvironments, the atmospheres inside buildings. Understanding indoor air quality requires substantial efforts to characterize the sources of each pollutant class,

__ANTHONY V. NERO, JR.,__ is Co-Leader of the Building Ventilation and Indoor Air Quality Program, Lawrence Berkeley Laboratory, University of California at Berkeley.

(Opposite page) Illustration by John Zielinski

ZIELINSKI

Rene Burri—Magnum

A substantial effort has been made in recent years to control outdoor air pollution such as that generated by heavy industry.

the removal of pollutants by ventilation and air movement, and the interactions of airborne substances with each other and with the building and its contents.

Nature and origin of indoor pollutants

Air pollution arises indoors from two major sources: materials or appliances that are part of the building, or directly connected with it, and outdoor emissions that are carried indoors with outdoor air. The exchange of outdoor and indoor air provides ventilation of building interiors, whether by infiltration through the walls of buildings, passage through windows or doors, or forced movement by a fan and, perhaps, duct system. In a typical home, ventilation is accomplished largely by means of infiltration or open windows and serves to exchange the indoor and the outdoor air about once an hour. This provides at least one limit to the buildup of pollutants emitted from indoor sources. At the same time outdoor pollutants may be carried in, but it appears that indoor pollutant concentrations are dominated by contributions from indoor sources; indoor concentrations are substantially greater than those outdoors when indoor sources are present but are less when they are not.

The indoor pollutants most closely related to outdoor pollution are emissions from the combustion of fuels such as kerosene, natural gas, and wood. To varying degrees these produce oxides of nitrogen (primarily NO and NO_2) and of carbon (CO and CO_2) as well as particles, oxides of trace substances such as sulfur, and complex organic substances (some of which are present on the particles emitted). Indoor levels of these emissions are determined by the fuel composition, the details of the combustion process, and the degree to which the emissions are vented to the outdoors. Central home heating systems that burn oil or natural gas vent virtually all the combustion products outdoors unless there is a defect in the system. But many space heaters and gas ranges, as well as some other appliances, are not vented or merely have a local exhaust fan (which often goes unused).

As examples, a kerosene heater of the modern "clean-burning" type can cause indoor concentrations of nitrogen oxides approaching one part per million if used with low ventilation rates. The outdoor long-term limit for NO_2 is about 0.06 part per million. Under similar conditions indoor concentrations of CO_2, not ordinarily considered to be an environmental pollutant with health signficance, can reach thousands of parts per million, which is comparable to the upper limit applied to industrial workers. Use of a gas range also noticeably affects indoor concentrations. But this has less potential for causing high concentrations than a space heater, simply because more fuel usually is burned for heating a home than for cooking. Villages in many less developed countries, however, strongly contradict this generalization. In such areas cookstoves burning available solid fuels—wood, dung, etc.—in poorly ventilated conditions cause very high concentrations of gaseous and particulate combustion emissions. Such conditions can cause visibility reductions across a 5-meter room that are comparable to those from a 2-kilometer (1.25-mile) thickness of Los Angeles smog. Exposures to these remarkable concentrations may have a significant effect on the health of village women; just by cooking they receive exposures to a number of substances (such as benzo-a-pyrene, a known and potent cancer-producing agent) that substantially exceed those received by heavy smokers of cigarettes.

Cigarette smoking is a form of combustion that can also influence indoor air quality, thus affecting the health, not only of the smoker, but of his companions. It is well established that cigarettes introduce into the indoor atmosphere substantial amounts of organic material. The chemical content of airborne cigarette smoke is extremely complex, and its effect on the health of those who inhale it is not well understood. Of course, cigarette smoking itself can have significant effects (such as lung cancer and emphysema) on the health of those who smoke. Whether the same types of effects are caused to a lesser degree among those who occupy the same space with cigarette smokers may never be known with certainty.

Thus combustion in general and cigarette smoking in particular can be a rich source of airborne organics, sometimes with substantial health implications. But combustion is certainly not the only source of organics. This is particularly true indoors, where organic substances may arise from a variety of materials and products in common use and may often reach levels that far exceed those regarded to be a concern outdoors. Major classes of sources include building materials and furnishings, various types of liquid and aerosol products, specific types of machines or appliances, and the combustion sources mentioned above. Together they emit a wide range of substances with a variety of known or suspected effects on building occupants if they occur at relatively high levels. For example, many organics are known to cause nonspecific ill effects—such as eye or respiratory irritation, headaches, nausea, or general weakness—among at least a portion of the people exposed to them. In addition, some organics are known to have carcinogenic (cancer-inducing) effects quite apart from those that may be attributed to cigarette smoke. However, the question of which organic substances appear in indoor air and what health effects they may have is far from fully understood. Even characterizing the wide variety of sources of these substances is a bewildering problem, since they include many of the most common features of people's daily lives.

For example, in homes the structure and many of the furnishings may give off significant quantities of formaldehyde, merely because of the use of pressed-wood products (which generally contain formaldehyde-emitting resins) as building materials and as major elements of modern furniture. The use of urea formaldehyde foam insulation has also been shown to raise the levels of formaldehyde within the home. In modern office buildings many of the common finish materials as well as the adhesives used with them are potentially significant sources of organics. A second class of potentially great importance includes products that are used indoors for specific purposes: cleaning fluids, paint strippers, and insecticides. Such products are used to a significant degree in all types of buildings, and the resulting airborne concentrations are generally not known other than the elementary fact that, because indoor environments are so small, associated concentrations of organics are likely to be considerably higher than in outdoor air. Finally, as an example of a distinct class of sources, photocopying machines employing special process fluids may be considered.

The total significance of organics from these sources and from the others indicated above is only beginning to be examined. But there is a significant possibility that they constitute a major element in the so-called "sick building" syndrome, wherein occupants of modern buildings often

Ethiopian woman uses a cookstove to bake bread. Available materials such as wood or dung are generally used as fuel in these stoves; when cooking is done under conditions of poor ventilation, the result is a high level of air pollution.

Brian Seed—CLICK/CHICAGO

find themselves to have an unusually high incidence of nonspecific illnesses that appear to be associated with the buildings themselves.

It is not altogether surprising that building interiors should be significant causes of exposure to combustion gases and organics. But it is disconcerting to find that significant exposures to radiation take place in homes. This results from the naturally occurring radioactivity that comes from the ground. All of the Earth's crust has trace amounts of uranium. The amounts are, as expected, much higher in areas that are suitable for uranium mining, but even ordinary soil and rock contains uranium and, of more significance to indoor atmospheres, radium. When radium decays, it produces radon, a gas that is chemically inert and can therefore be transported with the air in the soil and thus reach the outdoor or indoor atmosphere. The decay products of radon are isotopes of polonium, bismuth, and lead that are chemically active. Once they are formed from radon in the air, if inhaled they can attach to the lining of the lungs, either directly or along with airborne particles to which they have become attached while in the air. When they decay, these isotopes produce radiation that affects the lungs. It is this type of radiation exposure that has been shown to be the cause of elevated rates of lung cancer among uranium miners. For that reason their exposure to radon decay products has been severely limited.

Art Pahlke

Smoking cigarettes introduces substantial amounts of organic material into an indoor atmosphere.

Indoor concentrations of radon are generally but not always lower than those to which uranium miners are exposed. However, ordinary indoor concentrations are still high enough to cause a risk of lung cancer that is thought to be significant—higher, for example, than the average risk caused by the presence of cigarette smoke in the air. Thus it is quite typical to have a chance of one in five hundred of contracting lung cancer during one's lifetime merely because of the radon (and its associated decay products) present in one's home. Such a risk is small enough that it would be difficult to detect directly in studies of lung cancer incidence among different populations. It is large enough, however, to constitute a significant part of the total lung cancer rate. In fact, one evident complication is the potential synergism between cigarette smoking and radon decay-product exposures: these two insults to the lung may act together to cause more lung cancer than would be expected from merely adding their individual effects.

Occasionally, indoor concentrations of radon are so high that they cause exposures comparable to or greater than those that uranium miners (now) receive. This is no trivial matter because the associated risk far exceeds those that have been associated with other environmental exposures. At the highest levels encountered in ordinary homes, during a 50-year period of use such a home would have approximately a 50% chance of causing one of its occupants to contract lung cancer. Even homes at a tenth of this risk level must be regarded as candidates for remedial action.

The radon that is present indoors arises from three major sources: directly from the ground or rock surrounding the building foundation, from building materials (such as concrete or brick) that use crustal materials as

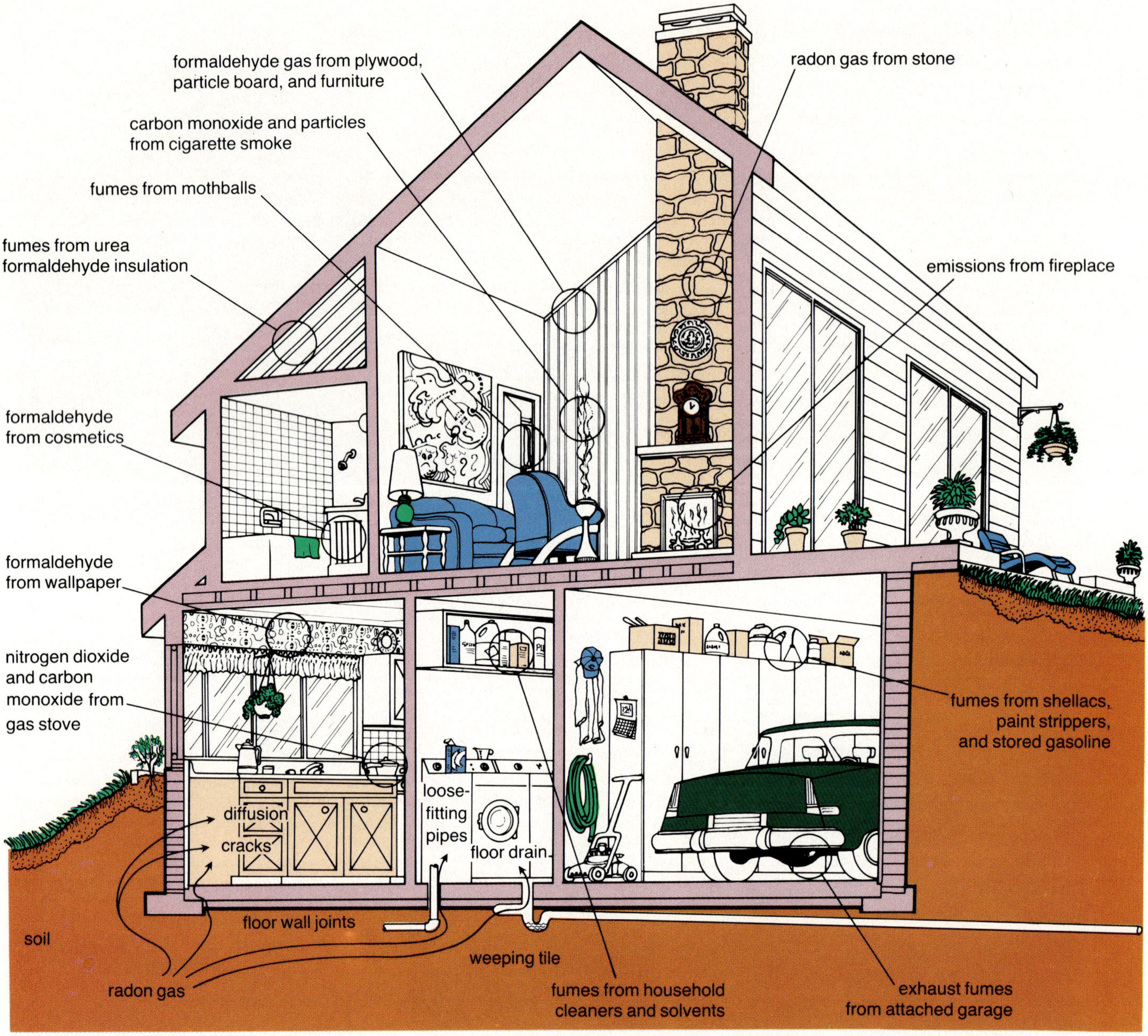

A wide variety of air pollutants can be found in a typical house. Some enter through the foundation, while others are generated by combustion within the house or are emitted as fumes from various building materials and furniture.

Illustration by John Draves

components, and from domestic water supplies drawn from underground sources. The relative contribution from each of these sources depends on the location, the type of building, and, to some extent, on occupant behavior. Typically, the largest contribution in single-family houses is from the ground, merely because such houses have relatively direct contact with a large amount of radium-bearing material. In multistory buildings the indoor concentrations tend to be lower and to arise primarily from the building materials and from outside air. There are, however, circumstances in which building materials can cause high concentrations. This occurred in a substantial part of the housing stock in Sweden because of the use of concretes made with alum shale, which was later found to cause very high indoor concentrations of radon because of its very

William Moriarty—Rainbow

Insulation is installed under a window frame in order to improve the energy efficiency of the building. Contrary to a widely held belief, the "tightening" of a building that results from such measures does not increase significantly the level of indoor air pollution.

high radium content. But even in Sweden, as in other countries where single-family houses constitute a major part of the housing stock, the ground constitutes the major source—perhaps 80%—of indoor radon.

There are classes of indoor pollutants other than the ones just described. The most obvious are "viable particles," including bacteria and spores, and special substances such as lead and asbestos that are present in buildings because of specific materials used in the past.

Factors influencing concentration levels

One of the main incentives for the recent increase in research on indoor air quality has been the awareness that energy conservation programs could have a significant effect on pollutant levels. Specifically, if buildings are "tightened" and ventilation rates reduced in order to decrease the amount of energy devoted to heating or cooling the air that comes in from the outdoors, the levels of indoor-generated pollutants could rise simply because less ventilation is being provided to flush them out. In homes tightening typically entails measures to reduce infiltration of outside air through walls and around windows and doors. Much concern has been voiced about such measures, but it is a fact that indoor pollutants are a problem regardless of energy conservation programs. The major determinant of whether or not a building has excessive levels is the emission rate from various indoor sources, and ordinary measures to reduce infiltration in homes, such as weather-stripping and caulking the windows, have a relatively minor effect. Concentrations of any given pollutant vary by factors of hundreds to thousands within the building stock, whereas energy-saving measures ordinarily reduce ventilation rates by only 10 to 30% and thus increase indoor concentrations by roughly the same amount.

The question then arises as to why concentrations vary so widely. In the case of combustion products one reason is clear: the presence or absence of an unvented appliance or smoker has a decisive effect on the emission rate and, therefore, on indoor levels. Similarly, the presence or absence of furnishings or walls with formaldehyde-emitting materials can have a substantial effect on the rate at which this substance enters the indoor atmosphere. Such differences in emission rates tend to be more important than differences in ventilation rates, as has been demonstrated in recent research on radon. In a number of studies both the radon concentration and the ventilation rate were measured in groups of houses, including both ordinary houses and some that had unusual energy-conservation features. The surprising result was that, although there were great differences in concentations among the houses, they had no apparent correlation with the ventilation rates. Moreover, the ventilation rate had a considerably smaller variability than the radon concentration, suggesting a wide range in the rate at which radon finds its way into indoor air. This conjecture was confirmed by detailed studies of the mechanisms by which radon enters homes from the ground. Most of the radon entering homes does not move indoors by molecular diffusion but rather because the house is often slightly depressurized with respect to the ground. This depressurization occurs because of winds and—more interesting—because house interiors tend to be warmer than the outdoors, giving rise to a ''stack'' effect that draws air into the house through its understructure or lower walls. These are the same factors that cause infiltration, and it is not surprising that some of the infiltrating air can be drawn from the ground and can carry much more radon into a house than could enter by diffusion through the understructure.

Studies of the behavior of radon and of the influence of ventilation rates on indoor concentrations have led to a conclusion that has also been found to apply to other substances. It appears that, in general, the indoor pollutant concentrations observed—whether for radon, combustion products, or organics—are determined by three factors: the source strength or emission rate, the ventilation rate, and various kinds of interactions that pollutants have with one another or with the building. Taken

Art Pahlke

Some new houses have been built with such attention to energy efficiency that the rate of infiltration by air from the outside has been reduced by factors of five to ten below the average. Such homes are likely to suffer from high levels of indoor air pollutants.

Photos, courtesy, Anthony Nero, Lawrence Berkeley Laboratory, University of California

Portable laboratory (left) contains complex instrumentation (below) that is used to monitor a wide range of pollutant concentrations in air drawn through sampling tubes from several places inside a building.

together, the variability in these factors accounts for the wide range of concentrations found in homes and other buildings. And, given this wide range, one cannot expect a 20% change in ventilation rate to make the difference between an ordinary concentration and one that is deemed excessive. Similarly, except in cases where for some reason the ventilation rate is extremely low, one cannot look to increases in ventilation as the primary means to solve or prevent indoor air quality problems. With a ventilation rate in the normal range the most obvious means of control is to focus on the sources of indoor pollutants.

There is, however, one class of building concerning which specific attention needs to be given to the ventilation rate as well as to other means of controlling pollutant concentrations: new homes that are built to have infiltration rates far below normal. Most homes in the United States have infiltration rates in the range of 0.5 to 1 air change per hour. Thus, the amount of air entering hourly from the outdoors is a substantial fraction of the house volume. New homes typically have infiltration rates at the lower end of this range, as low as 0.4 per hour, and older homes, with careful application of tightening measures, can also achieve relatively low rates. However, European homes already tend to be considerably tighter than this. And in the United States small numbers of homes are being built that employ such special construction features as inclusion in the wall of a plastic barrier with no penetrations, effectively providing a complete seal against infiltration. Such techniques can achieve infiltration rates not just 50% lower than average but lower by a factor of five or ten. Such houses have the potential for raising pollutant concentrations by a large factor, therefore requiring special attention to air quality.

Before turning to the question of control, more needs to be said about the interactions among indoor pollutants. Those who are interested in outdoor air pollution have long been aware of the critical influence of chemical and physical reactions on outdoor concentrations. For example, the worst components of smog are not the emissions that come directly from automobile exhausts or power plants but rather the substances that

Courtesy, Anthony Nero,
Lawrence Berkeley Laboratory, University of California

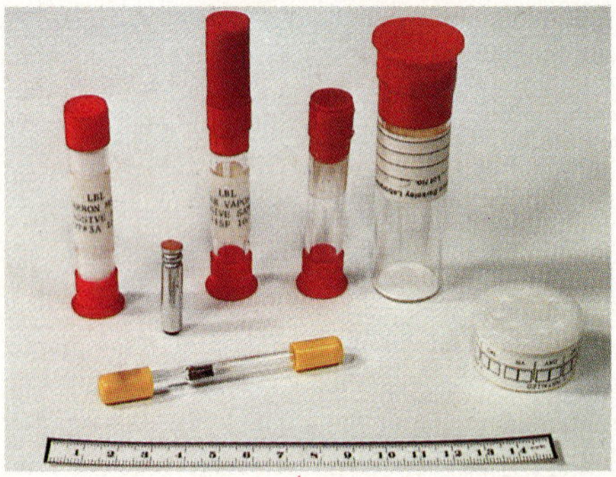

Passive monitoring devices (compared for size with a 15-centimeter [6-inch] ruler in the foreground) are placed in a building for about one week and then sent to a laboratory for analysis. The four tall upright instruments are used to measure concentrations in air of (from left to right) carbon monoxide, water vapor, nitrogen dioxide, and formaldehyde. The short device, together with the sampler lying on its side, measures the ventilation rate, while the instrument to the far right detects concentrations of radon.

result from photochemical reactions involving those emissions. Similarly, reactions can be important for each class of indoor pollutants. For example, in many circumstances NO_2 appears to be removed from indoor air as much (or more) by reactions as by ventilation. Virtually nothing is known about the products of chemical reactions involving NO_2, and the products themselves may be more significant than the primary emissions. Also, organic substances, including formaldehyde, may have a complex behavior indoors, but such questions are only beginning to be examined. And, finally, it is known that the chemical activity of the decay products of radon, in particular their tendency to attach to airborne particles or to walls and other surfaces, has a substantial influence on the amount of those products that is present in air and on the degree to which they cause a radiation dose. Indeed, the behavior of NO_2, formaldehyde, and radon decay products all highlight the importance of interactions with walls.

It is clear that our understanding of the behavior of indoor pollutants is far from complete. Much of the variability observed in concentrations is associated with differences in sources, although ventilation rates also vary substantially. The role of reactions is harder to characterize. Ordinarily, one would not expect significant differences in reaction rates, but it is apparent that such differences occur in some cases. Overall, the major influences on concentrations appear to be, in order, sources, ventilation, and chemical and physical reactions. This provides some basis for action to control each of the important pollutant classes.

Control measures

Strategies for achieving or maintaining acceptable indoor air quality involve two rather different components. The first concerns the technical means to control concentrations of pollutants. These techniques consist of reduction in emission rates, changes in ventilation rates or in the manner in which ventilation is supplied, and the use of air cleaning systems. The second component is more complex; it involves the practical methodology for identifying situations in which attention to pollutant concentrations is needed and for choosing (and applying) the appropriate solutions.

Control methods themselves are not difficult to formulate, given an understanding of the causes of the high concentrations. In most cases the most effective means of reducing levels is reduction of the pollutant emission or entry rate, mainly by the provision of venting for combustion devices, by attention to the manufacturing process for materials that emit organics, and by blocking the radon that rises from the ground. In some cases specific forms of ventilation are used effectively. These include a hood that ventilates above a gas stove and ventilation beneath the understructure of a home to exhaust the radon that would otherwise be drawn indoors. Each of these methods provides a relatively small amount of ventilation to bring about a significant reduction in the entry rate of the pollutants and, therefore, in their indoor concentrations.

An alternative control technique is to increase the general ventilation rate, an approach that has the advantage of applying to virtually any

(Left) Art Pahlke; (right) courtesy, Gas Research Institute

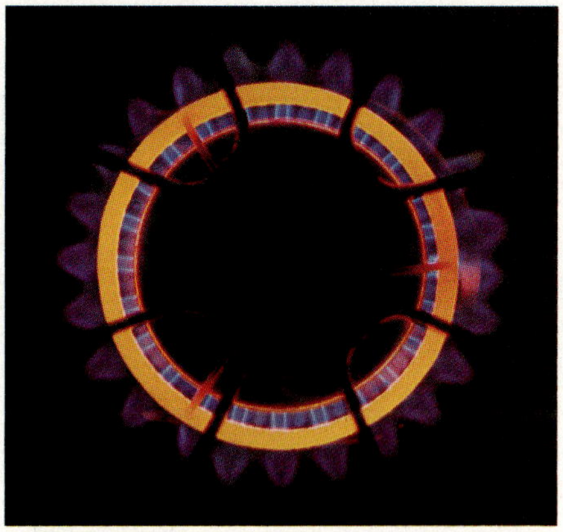

Experimental burner on a gas stove (above) has been designed to reduce the nitrous oxides produced during cooking and thereby achieve a cleaner indoor environment. Another effective control device is a hood (above left) placed over the stove to provide ventilation.

pollutant that arises indoors but that also requires the conditioning of substantial amounts of air during heating and cooling seasons. This difficulty can be largely avoided by the use of mechanical ventilation systems that recover heat. One class of systems uses an air-to-air heat exchanger to transfer energy between an exhaust airstream and the stream of air entering from the outdoors. Such systems are available in forms that are built into a duct system as well as in window-mounted configurations similar to air conditioners. In either case the principle is the same: the outgoing airstream is used to condition the incoming one, thereby avoiding the expenditure of energy to do so. Alternatively, some systems simply exhaust air from the building and recover heat for heating water or other purposes. However, the use of such a system without an active supply duct can depressurize the house considerably, which in some cases may increase the radon entry rate significantly.

Finally, for some purposes devices that actually clean the air may be appropriate. For example, a number of techniques may be used to remove airborne particles. Among them are filtration and electrostatic precipitation, two of the principal methods for cleaning stack gases from industrial facilities that burn fossil fuels. Commercially available devices that employ these techniques are relatively effective for removing particles (including smoke and pollens) from indoor spaces. However, the small, inexpensive devices that have become popular in the United States do not appear to be very effective.

As for gaseous pollutants, removal techniques are available in principle, but the commercial availability of effective devices is limited, probably because of considerations of convenience and cost. Taken as a whole, air-cleaning techniques are generally not as effective as reduction of the source or provision of more ventilation, but they can be useful in some cases.

Thus indoor pollutants can generally be controlled to levels that are acceptable to most people. (The portion of the population suffering from unusual sensitivities to airborne agents is another matter, often requiring special measures, such as the total exclusion of sources of the offending substances.) But, given the availability of control measures, we are still faced with the difficult questions of where to apply such measures and who should have the responsibility for carrying them out. For questions of outdoor air quality any particular issue is of interest to the entire community in the area affected; as a result there is a clear basis for joint (community or governmental) action. However, many buildings and most homes are under the control of individuals, and so it is they who have a great deal of influence (whether they know it or not) on indoor air quality. Individuals can, in principle, take the responsibility for providing any of the remedies indicated above.

However, members of the general public are not often in a position to select limits on concentrations. Nor can they effectively determine whether pollution in their own environments exceeds agreed-upon standards. And, finally, they are not often able to judge which control technique is appropriate to their circumstances. These difficulties suggest the important role to be played by central authorities, even if the owner or occupant of a building ultimately must implement the solution.

Moreover, there are some functions that can only be accomplished in a practical way by standard-setting authorities. For example, general standards for ventilation usually appear in building codes that draw on standards that are formulated centrally. Similarly, an individual or builder

An air-to-air heat exchanger transfers energy between an exhaust airstream in a building and a stream of air entering from the outdoors. The outgoing airstream is used to condition (heat or cool) the incoming one, thereby avoiding the expenditure of additional energy to do so.

Illustration by John Draves

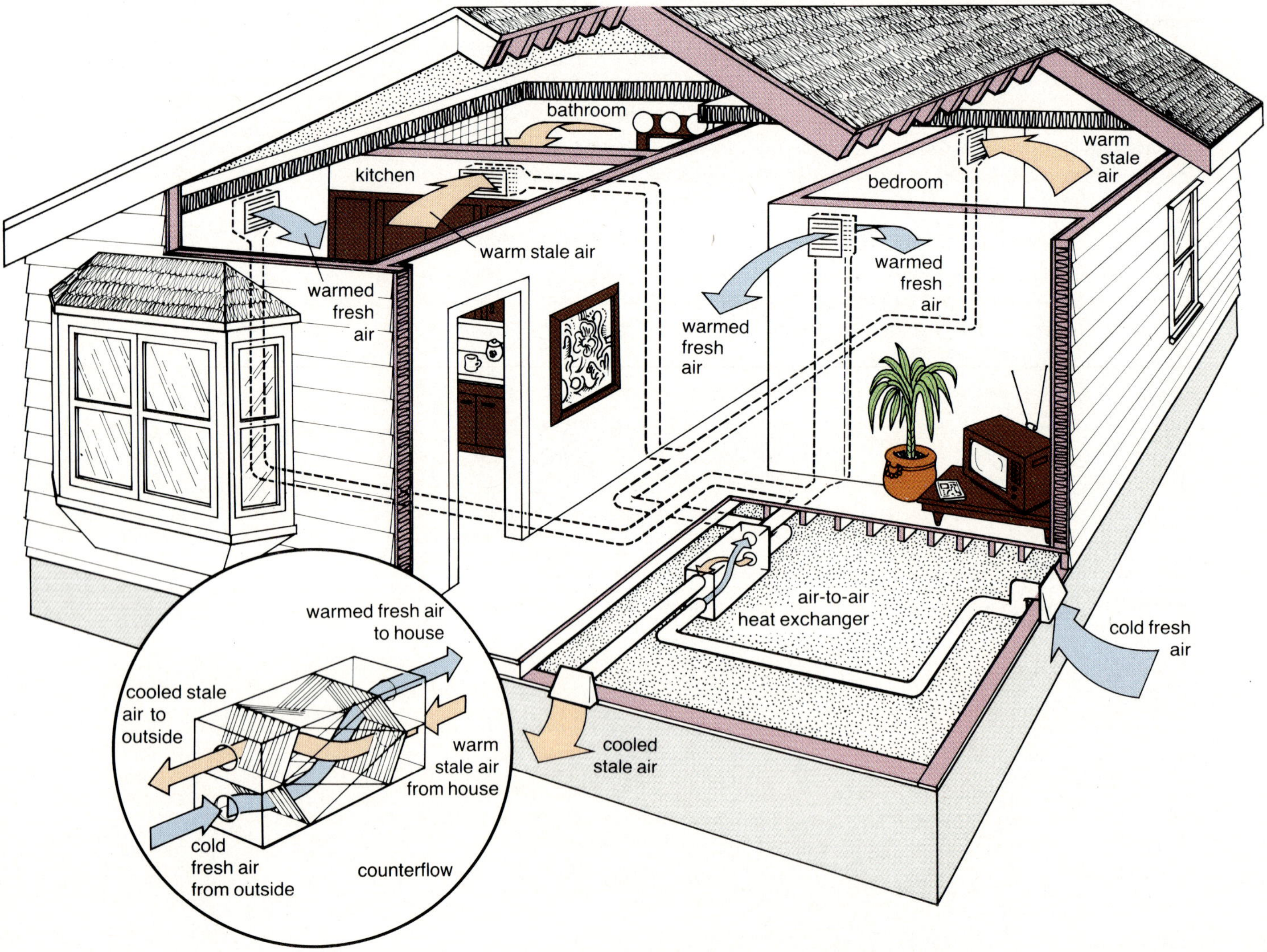

tends to purchase equipment or materials with the expectation that it will meet common standards of health and safety.

The case of radon presents a special challenge because the source of excessive concentrations is usually the ground rather than a manufactured product, and individuals or communities cannot be expected to have the means to judge whether they have a problem or what to do about it. As a result the most effective means of identifying areas or homes with high radon concentrations appears to be systematic surveys or analytical efforts that can best be formulated and carried out through a broad, centrally designed strategy.

The question of indoor air quality thus has a complexity and importance that is comparable to many other environmental issues. It is distinctive in that it occurs in the environments in which people spend most of their time and over which they have most control. People choose to live and work indoors in part because doing so offers them the means to control their environments. But it has become apparent that living indoors requires a reasonable degree of care in building and furnishing homes and other structures.

Assuring the quality of indoor air is not a simple matter. It requires substantial attention to the occurrence and behavior of indoor pollutants, to the effects that these can have on health, and to methods for limiting exposure. And, as just indicated, someone—whether individuals or collective authorities—must decide which exposures are excessive and formulate means for identifying and controlling those that are. We now know enough to characterize the problem in general and to frame broad approaches for control. Whether we have the resolve and resources to reduce excessive concentrations in millions of indoor environments remains to be seen.

Art Pahlke

Many buildings and most homes are under the control of individuals, who, therefore, can significantly influence the air quality within them. Such a simple step as opening a window can increase ventilation and thus lower the level of the pollution indoors.

FOR ADDITIONAL READING

R. R. Miksch, C. D. Hollowell, and H. E. Schmidt, "Trace Organic Contaminants in Office Spaces," pp. 129–37 in *Indoor Air Pollution* (J. Spengler, C. Hollowell, D. Moschandreas, and O. Fanger, eds.), a special issue of *Environment International* (1982).

National Research Council, *Indoor Pollutants* (National Academy Press, 1981).

A. V. Nero, Jr., "Indoor Radiation Exposures from ^{222}Rn and Its Daughters: A View of the Issue," pp. 277–88 in *Indoor Radon* (A. V. Nero, Jr., and W. M. Lowder, eds.), a special issue of *Health Physics* (August 1983).

F. J. Offermann *et al.*, "Control of Respirable Particulates and Radon Progeny with Portable Air Cleaners," Lawrence Berkeley Laboratory report LBL-16659 (February 1984).

J. D. Spengler and K. Sexton, "Indoor Air Pollution: A Public Health Perspective," *Science* (July 1, 1983, pp. 9–17).

Isaac Turiel, *Indoor Air Quality and Human Health* (Stanford University Press, 1985).

THE MANAGEMENT OF LAKES

by Herman S. Forest

A better understanding of lakes and the life they contain is necessary to preserve these precious freshwater resources.

The water of natural freshwater lakes and artificial reservoirs is one of the most vital resources for man and other organisms. Man's thirst for more lake water has never been satisfied, even though he has continued to control, create, and use more. In addition, the quality of the water has become increasingly important, for uses ranging from irrigation to drinking and recreation.

The quality and quantity of the Earth's limited freshwater resources can be managed wisely only through basic scientific studies of the natural ecosystems of which the lakes are a part. A new level of scientific understanding of lakes is being achieved, one that is a result of studies that consider lakes and their surrounding regions as distinct ecosystems.

The most important characteristic of an ecosystem is its self-regulation, its ability to respond to external disturbances so as to maintain itself in essentially the same form. Human intervention and occasional natural catastrophes can, however, damage or destroy this ability. The aim of ecological management is to take advantage of natural self-regulation to preserve ecosystems. For example, by 1965 it had become evident that the greatest freshwater resources of North America, the Great Lakes, were in trouble. Federal, state, and provincial governments in the U.S. and Canada began to work toward a change of view and a change of management policy. The state of New York assembled a scientific advisory group and asked, "How should we manage the Great Lakes?" The scientists replied, "Let nature do the managing as much as possible."

***HERMAN S. FOREST** is Professor of Biology at the State University College of Arts and Science, Geneseo, New York. He has written extensively on the subject of lake ecology.*

(Opposite page) Adeline Haaga—Focal Point

Two worlds: the models of engineers and ecologists

When considering a resource, an engineer tries to determine how much of it can be extracted or modified to serve a particular end. For example, the goal of improved navigation on the Great Lakes was achieved by

building canals to join the lakes and other waters. Ecologists, on the other hand, regard such a resource as the Great Lakes as a limited ecosystem with the parts related. From their point of view the canals brought to the Great Lakes people, industry, waste, and exotic fish that destroyed the natural ecosystem forever.

Engineers have been given the tasks of obtaining drinking water from distant lakes and even of creating the lakes when necessary. Measures to protect water supply sources were generally limited to the water itself and the immediate vicinity of the shore. Drawing water from distant sources or diluting it seemed adequate to preserve or restore its purity.

The next major engineering mission was the building of dams for the generation of electricity. The Tennessee Valley Authority (TVA) began to do this in 1933 by building dams on the Tennessee River and its tributaries and continued for almost 50 years. Along with generating electricity, the TVA projects achieved improved navigation, flood control, and recreational facilities. The problem of urban sewage, however, could not be ignored in the reservoirs formed by the dams, and local governments were aroused to the importance of treating wastewater.

The public awakening to the existence of water pollution brought support for abatement measures. Engineers were given unprecedented opportunities to design collecting systems and wastewater treatment facilities. However, they also had to consider the consequences to the waters that received the treated waste. The amount of human waste overtaxed the capacity of the ecosystem's self-regulating mechanism to render wastes harmless. The variety of recognized pollutants grew: heat, salt, fertilizer minerals, and heavy metals were added to microbes and organic matter. It has been only recently that engineers have begun to work with the data of lake ecology, essentially the flows and storage of matter and energy.

By 1984 it became possible to publish a book that assembles engineering models of lake ecosystems and offers their applications in manage-

Fontana Dam (below) in North Carolina and Lake Chickamauga (below right) in Tennessee are two of the projects of the Tennessee Valley Authority, which built a series of dams to generate electricity, improve navigation, control floods, and provide recreational facilities.

Photos, Tennessee Valley Authority

(Left) Herman S. Forest; (right) William Felger—Grant Heilman

Wastewater treatment plant in Pennsylvania (above) was designed to prevent discharges into lakes such as that entering Lake Ontario at the left. The lake is being discolored and polluted by organic matter, fertilizing minerals, and toxic substances from industrial wastes.

ment. During recent years a different tradition had also been developing in academic surroundings. There, lake research originated in biology, with chemical and physical data added later. From this beginning, "limnology" (lake study) evolved into a field distinct from academic ecology and from practical engineering.

Lake study diverged considerably from the mainstream of ecology partly because of the special nature of the lake ecosystem and the difficulties in collecting data from it. Biotic communities on land replace each other in succession, reaching a climax community marked by a strong self-regulating capacity. This climax ecosystem remains rather stable. For example, a Middle Western prairie or an eastern hardwood forest is much the same for thousands of years, and most disturbances from fire, erosion, or cutting are repaired within a century or sooner. By contrast, lakes fill up, and stability is limited both in capacity and in duration. Plant growth in lakes is not limited by the availability of water but of nutrient minerals, oxygen, or light. It soon becomes evident that some lakes produce more plants and more animals than others.

A distinctive model developed among limnologists. It can be described as the developmental-trophic (nutritional or "richness") model. To describe this model most simply, a lake begins its existence clear and infertile (oligotrophic); it then gradually becomes more fertile (eutrophic), finally filling with vegetation, animal remains, and sediment eroded from the watershed. Watersheds with limestone rock or calcareous soil have been observed to contain more fertile lakes (hard water) than those found in sandstone or granite (soft water) areas. In comparison with the soft water lakes the hard water lakes are less acid, and their mineral nutrients are more available.

Finally, man was recognized as a major agent in fertilizing lakes, and the term cultural eutrophication was coined by Arthur Hasler of Wisconsin in the mid-1940s. Yet even as of the early 1960s the term eutrophication as a key concept in limnology hardly appeared in scientific writings. Then rather suddenly in the middle 1960s eutrophication became the target of water pollution abatement programs. The term and concept were easily grasped and described the problem more precisely than did "pollution." People perceived that it was not enough to kill microbes and reduce oxygen-consuming organic matter. The enemy had become fertilizing minerals, the same nitrogen and phosphorus that is sold in commercial fertilizer. Eutrophication to the aroused citizenry was the "aging" of lakes. "Premature aging," it was believed, must lead to premature death.

The phosphorus controversy

When eutrophication was understood to be the major new problem with lakes, limnologists decided that the solution would be to reduce the input of nutrients. The question then arose as to whether all nutrients should be reduced. This was soon rejected. Some elements are so abundant in nature that input control would be pointless. Of the three major nutrients in commercial fertilizer, one—potassium—was already plentiful in lakes. Either of the other two, nitrogen or phosphorus, may be in short supply depending upon the particular lake. Plant growth requires much

Many factors interact to determine such characteristics of a lake as the composition and distribution of the plants and animals living within it and the rates at which nutrients are recycled. Particular emphasis for many lakes should be placed on the "human impact" factor.

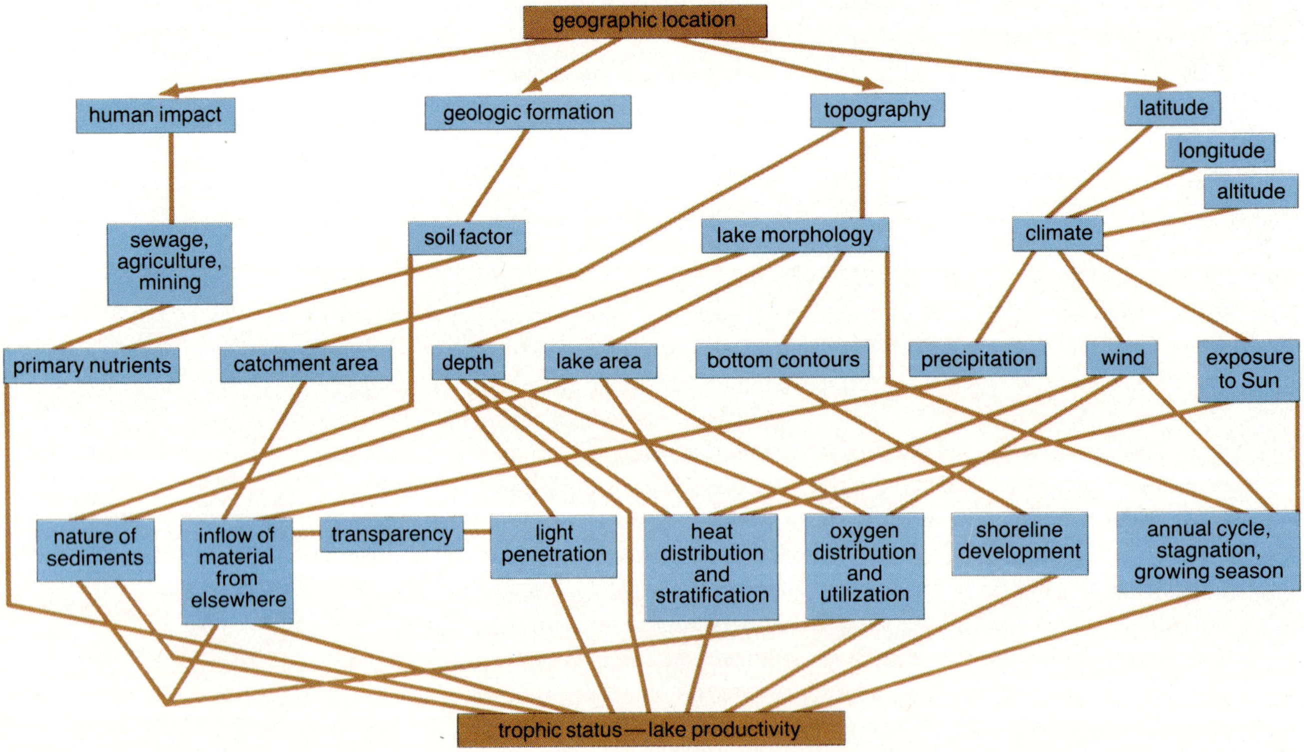

Cladophora, a variety of green algae, rots on a Lake Ontario beach. Nutrient minerals discharged into the lake in municipal wastewater fostered the growth of the algae.

less phosphorus than nitrogen. Consequently, by 1967 lake scientists had selected phosphorus as the most critical element to control.

The decision to concentrate on phosphorus was published in a symposium volume by the American Association for the Advancement of Science. Despite this consensus, however, the controversy continued, as manufacturers using phosphorus looked for other culprits. Silicon briefly became a target because it was known that the number of diatoms, single-celled algae, was limited at times by the available silicon. It was carbon, however, that soon became the rallying banner for the major detergent manufacturers and allied chemical companies. If carbon were the controllable limiting nutrient for plant growth, there would be no point in reducing phosphorus in detergents or removing it from wastewater. Carbon (as carbon dioxide) is plentiful in air, but even in water the supply is sufficient for plant growth in natural water of the temperate zone, particularly in hard waters of limestone areas. In nutrient-saturated waters such as sewage lagoons, however, the supply is soon used.

It was a curious episode. The eutrophication theory had just escaped from its academic cradle and was immediately buffeted about in the

Herman S. Forest

Detergent foam from wastewater discharges collects in a whirlpool below Niagara Falls. Foam persisted in water because it could not be attacked readily by bacteria. When biodegradable detergents were substituted, the foam disappeared, but the substitution alone did not reduce the amount of pollutants.

marketplace. Virtually every accredited limnologist favored phosphorus control, and the major detergent manufacturers opposed it with a massive public relations and lobbying campaign. Detergents had rapidly replaced soaps after World War II, and phosphate compounds were added to them in order to aid the surface-active agent. How much of the compounds was added was secret until independent analyses showed amounts ranging to more than 50% in washing products. In total, phosphorus added to lakes from detergents equaled the amount derived from food and discharged as human waste.

Despite the efforts of the manufacturers phosphorus was eventually chosen as the controllable limiting nutrient. Control strategy and the beginning of its implementation followed in less than five years. Phosphorus was reduced and then virtually eliminated from detergents by federal, state, and provincial governments in the Great Lakes area and elsewhere. By 1975 effective phosphorus-free detergents were generally available. This eliminated half the load of phosphorus that had to be reduced at municipal treatment plants as well as half the discharged load from the private homes that surround many lakes.

The remaining half of the phosphorus load (from human waste) was concentrated in the large urban areas. These regions were top priority for the continuing campaign to reduce phosphorus discharge. Then waste from smaller cities and villages in the lake watersheds was collected and treated. Finally, it was realized that nutrients and other pollutants from the land itself also needed to be reduced for the best practical protection of lakes. The heavy use of fertilizers on land and the change to concentrating farm animals in feedlots vastly increased the runoff pollution in water. Land use controls are much more difficult to implement than the building of sewers and treatment plants, and thus the problem of pollution from land use remains.

In 1984 Hugh F. H. Dobson, of the National Water Institute of Canada at Burlington, Ontario, completed an analysis of 17 years of data on Lake Ontario. His chemical atlas of the lake was summarized as follows:

Trout and salmon are dropped into Lake Ontario, thereby rerouting the trophic chain that once ended with dead alewives and decay bacteria and also generating in the lake a valuable sports fishing industry.

John Goerg—New York State Department of Environmental Conservation

Dan Morrill

Dead alewives litter a Lake Michigan beach in Chicago. Alewives entered the Great Lakes from the Atlantic Ocean through the St. Lawrence Seaway and soon increased greatly in numbers, overcoming many of the native fishes. The later introduction of salmon into the lakes helped control the alewife population.

The lake's recent recovery from eutrophication is a major success story in environmental control. Phosphorus levels in the lake are now near the target [set about 15 years earlier]. Increased nitrogen-to-phosphorus ratios . . . will probably prevent the lake from having noxious blue-green algal types. The governments and the public should be satisfied with their choice of controlling phosphorus loadings, and control should definitely continue while the human population still rises.

Control of fish and toxic chemicals

Repair of the Great Lakes ecosystem was also undertaken at the top of the trophic pyramid. The introduction of salmon and trout into Lake Ontario produced a crop of large tasty fish by rerouting the trophic chain that had once ended with dead alewives and decay bacteria. A valuable sports fishing industry then developed rapidly.

Commercial and sports fishing had previously almost disappeared from the lower Great Lakes. This had occurred because organic matter discharged in spawning streams consumed oxygen necessary for fish to survive. As an example, sawdust dumped into streams destroyed the sturgeon population of lakes. (At one time there were enough sturgeon in U.S. lakes to produce caviar for domestic consumption.) Fertilizing minerals such as phosphorus produced the same effect by encouraging the growth of plants that competed with fish for oxygen.

However, much of the damage to the ecosystem was not initiated by such discharge of matter but was inflicted directly on fish. The lakes were simply overfished until the most-caught species could no longer reproduce adequately.

The ecosystem of the Great Lakes also received a deadly blow from the introduction of exotic fish. The engineers who built the Erie and Welland canals were not expected to consider ecology and did not do so. Consequently, the alewife, a herringlike fish, entered the lakes from the Atlantic Ocean through the canals. It overpopulated the lakes, taking over the place of a number of native fish. The weakened native fish populations were also directly struck by a devastating parasite, the lamprey, which entered the lakes by the same route as the alewife. Both the alewife and the lamprey also migrated or were carelessly introduced to smaller inland lakes.

The fish restocking program was successful for a number of reasons, including lamprey control efforts and the decrease in pollution of spawning streams. In addition, the technology of breeding and raising young fish improved. Yet, in ecological terms the program was not a complete success. The ecosystem was being farmed rather than regulating itself. Natural spawning of the stocked fish had not reached a level sufficient to maintain the large population that had been achieved. Stocking, therefore, must continue indefinitely. More environmental improvement, particularly of tributary streams, would be necessary to achieve a self-sustaining system.

The latest of the lake ecosystem injuries to be identified includes poisons or toxic chemicals. The metal mercury was found in hazardous amounts, and then hazardous organic chemicals such as the insecticide Mirex were detected. In lake ecosystems chemicals are transported far from their source and sometimes are transformed to toxic substances. Poisons that are not eliminated can increase a hundredfold as they travel through the food chain from alga to insect to fish. Even the attempt to destroy bacteria with chlorine was found to produce poisons (trihalomethanes or THM's).

Poisons can be transmitted through the ecosystem to fish and then to humans. For Lake Ontario this hazard was attacked at the source by determining discharge points of the poisons (specific industries or municipalities) and then removing or reducing the discharge. Unfortunately, some of the most toxic chemicals are not decomposed but remain for a long time in the ecosystem. DDT still appears in the fish of some lakes, and Mirex was still being detected some time after its discharge into waters had ceased. The concentration even appears to have increased in certain fish and locations.

Control of discharge and monitoring of the waters cannot remove poisons already present. The human consumer of fish must accept restraints for safety. However, it has been difficult to prohibit the eating of potentially hazardous fish.

Small lakes, bays, and reservoirs

Although there are fundamental differences between large lakes and smaller bodies of water, their problems and the solutions to them are similar. Small bodies of water can be studied more intensively than can a great lake, and comparative studies are meaningful because a large number of such lakes can be analyzed. The small scale also permits the application of a larger number of management measures.

The recent history of Irondequoit Bay, adjacent to Rochester, New York, serves to illustrate much of the recent progress in lake ecology and management. The bay is long and narrow and is joined to Lake Ontario only through a narrow outlet. Except for water level and the passage of fish, Lake Ontario has little influence on the bay; therefore, Irondequoit is more like an inland lake than a bay.

The acid rain problem

One ton of lime per minute is pumped into a lake in Sweden in an effort to neutralize the excess acidity of the lake water caused by acid rain. Although lime reduces acidity, some scientists believe that it seriously disturbs a lake's ecosystem.

Ted Spiegel—Black Star

Acid precipitation has been the focus of an outburst of interest and has been ranked by some as the major environmental problem of the 1980s. Studies directed at tracing atmospheric acid, most often emitted through the smokestacks of heavy industry, have revealed that a significant amount of nutrient minerals and poisons can be transported long distances. The effect is that a lake ecosystem becomes vastly expanded.

Two considerations are appropriate here. First, acid precipitation does not seem to damage hard water lakes, such as those generally found in limestone areas. However, lakes without the buffering action of the carbonates found in limestone are severely damaged. Second, a technological remedy, the liming of acid lakes, was developed in Sweden, where the acid problem was first recognized and studied. Norway, however, is highly skeptical of liming. Lime (limestone) was not a natural part of the soft water ecosystem, and, while liming may reduce acidity, the lime itself can be a seriously disturbing factor. The simple solution is to stop the air pollution at its source and tamper with the lake ecosystem only if it is too damaged to recover on its own.

Rainwater is collected and analyzed for acid content near a lake in the Adirondack Mountains of New York. Many lakes in the Adirondack region have high levels of acidity because of acid rain.

Reprinted with permission of the Electric Power Research Institute

Irondequoit Bay had a natural tendency toward overfertilization because its watershed is very large in comparison to the water body, about 60:1. The typical ratio for a lake or bay in the region would be about 12:1. Its stability was also vulnerable because variety, or diversity, was low for its submersed green plants.

Then, willy-nilly, man piled on insults. The city of Rochester and its rapidly growing suburbs added a waste burden from 100,000 or more people. Commercial sand removal and construction increased erosion and sent sediment to muddy the bay. By 1939 the ecosystem's self-regulating capacity was completely overcome, and stability was lost. By 1970 the water was virtually saturated with nutrients: phosphorus, nitrogen, and others. A number of plant and fish species disappeared, and algae grew so thickly in the open water that light scarcely penetrated through them. Rooted plants grew only to a depth of 1.5 meters (5 feet), in contrast to clearer lakes in nearby upstate New York where they could be found at 6 meters (20 feet).

It was also discovered that salt collected in the bay as a result of deicing of roads. The salt both delayed the circulation of water during the spring and increased the deficiency of oxygen. Lack of oxygen enables the release of phosphorus from sediments to the water, adding to the phosphorus overload. A public information effort that began in the early 1970s succeeded in reducing salt use by more than half, and the bay began to regulate itself again. Many sewers were built in the watershed, and the several small wastewater treatment plants were closed; the wastes were diverted from the bay to a large plant designed to put well-treated effluent into the deeper waters of Lake Ontario.

By 1980 Irondequoit Bay had begun to recover. More species of rooted plants grew in the clearer waters, and trout now traversed the bay into the cleaner Irondequoit Creek. However, pressures for land development on the highly erosive land around the bay increased, both for residential use and to serve the booming sports fishing industry. In 1984 the various political units around the bay agreed to adopt uniform standards for development to protect the bay.

Two important lake management ideas were considered during studies of Irondequoit Bay. These were conducted chiefly by Thomas Bannister of the University of Rochester, but county, state, and federal agencies cooperated actively, or in support. Bannister discovered an unexpectedly great reduction in the phosphorus available for algal growth after the phosphorus input from the watershed was reduced. The effect was suggested in earlier observations at nearby Conesus Lake. A possible explanation is that phosphorus adheres to particles (clays perhaps) that are suspended in the water; also, phosphorus is settled out by particles containing calcium carbonate. In the intricate regulatory machinery of the ecosystem the small particles may make the water a little less clear, but they also reduce the fertility. Man and nature together achieved a 90% reduction in phosphorus, a major management victory.

Other investigators began to consider the cattail marshes extending upstream from the head of the bay along Irondequoit Creek. Marshes

Courtesy, Atlas Copco Management Consulting AB

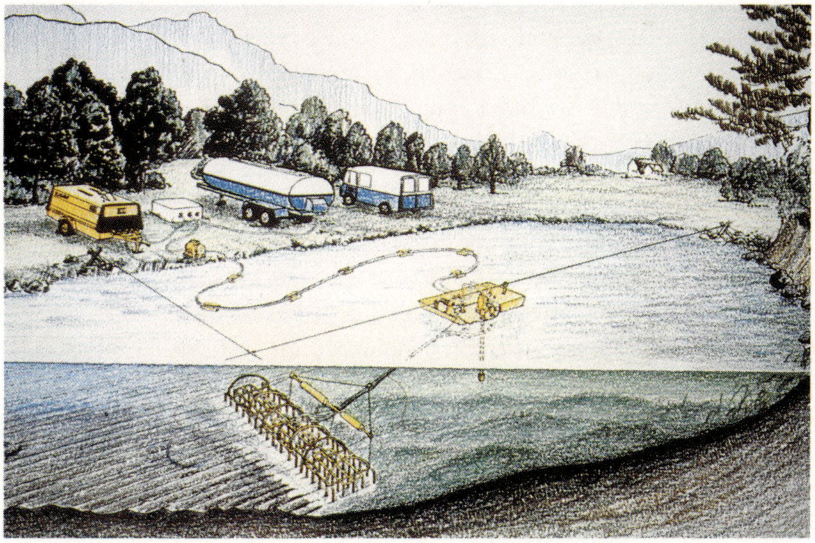

RIPLOX method for increasing the productivity of a lake involves injecting liquid calcium nitrate into the top layer of lake sediment through the pins of a 10-meter (33-foot)-wide harrow that is pulled along the lake bottom by a motorized raft on the surface. The calcium nitrate quickly oxidizes the sediment, which reduces both further oxygen consumption by the sediment and the release into the water of the sediment's accumulated phosphorus. Thus the lake will no longer be depleted of its oxygen content and can support marine life; also, production of algae will be decreased by the reduction of phosphorus.

trap sediment and nutrients, but they also release them when plants decay or during heavy washouts. A management strategy was devised to increase the trapping by means of creating new channels from the main stream. In addition, the cattails could be harvested toward the end of the growing season so as to remove the nutrient-rich leaves.

Other cleanup methods have been suggested, primarily to reduce the return of phosphorus from sediments to water. Fly ash can coat the bottom of a lake and physically seal it to some extent. Alum is used in water treatment to precipitate suspended matter, and artificial aeration can be used to prevent the oxygen deficiency under which phosphorus is changed to soluble forms.

Such technological solutions may be workable, but they are not necessarily cost-effective or without possible harmful side effects. A preferred ecological solution to lake problems is one that leaves as much management as possible to nature.

Historically, natural lakes have been the principal subject of limnology. Although the number of man-made reservoirs has grown rapidly, they were primarily engineering projects for such purposes as flood control and power generation with some limited considerations for public health and recreation. Consequently, they are poorly understood as ecological systems.

More study is needed in order to understand reservoir-watershed relations and the differences among reservoirs. The distinctive features that merit attention include the interactions among a series of reservoirs such as those constructed along most of the Tennessee River system.

A most intriguing feature recognized in reservoirs is their heterogeneity. A reservoir may be highly fertile (eutrophic) at its upper end and infertile (oligotrophic) near its dam. Like many natural lakes, reservoirs may develop areas of oxygen deficiency. In natural lakes these remain in the deepest parts, but in reservoirs such areas may form in the shallow water upstream and then shift into deeper and deeper waters.

Bruce Kimmel of the Oak Ridge (Tennessee) National Laboratory pointed out that the heterogeneity of reservoirs can present an unusual management opportunity. Provided the ecosystem of a reservoir is understood and wisely managed, this heterogeneity can allow apparently incompatible objectives (high water quality, rewarding sports fishery, flood control) to be met.

Simple models and complex lakes

Eutrophication, particularly that caused by man, proved to be an appealing idea for both scientists and the public. It provided an understandable reason for trying to reduce water pollution, a reason that went beyond the older considerations for human health and oxygen for fish. As the support for data gathering grew, it became possible to calculate mass balances for lakes: the total input of water and minerals from the watershed, their total retention time in the lake, and the total chlorophyll of suspended (plankton) algae. This kind of model and the resulting data were particularly appealing to engineers.

By 1975, however, the eutrophication model was being contradicted by field observations. In *Lakes of New York State* I (1978), William Schaffner and Ray Oglesby remarked that, according to the most highly regarded scale of that time, most of the Finger Lakes of New York State were hopelessly eutrophic. But actual observations revealed that most were rather clear and stable. Herman Forest *et al.*, in the same publication, noted that Hemlock and Conesus lakes ranked close to each other on the trophic scale but were obviously very different kinds of lakes. Conesus had many submersed plants, good fishing, and was clear

Homes line the fully developed shore of Lake Conesus in New York State. Although a heavy load of wastes has long been discharged into the lake, it remained remarkably stable for many years until a sewer system was built on its perimeter. Rooted plants in shallow water in the lake (inset) helped maintain the stability by taking up a portion of the available nutrients and thereby preventing algae from becoming dominant.

Photos, Herman S. Forest

in spite receiving a tenfold increase in human waste during the previous 50 years. Hemlock, used as a water supply reservoir, had no cottages around it and a largely forested watershed, but the water was sometimes murky and the fishing poor.

Robert Carlson of Kent State University in Ohio questioned whether the eutrophication model was useful as a guide to lake management and employed a comparison to help illuminate his point. He suggested that obesity in a man is a model for a lake in unhealthy condition. The measures of eutrophication, to extend the metaphor, tell the weight of the man, not whether he is obese or heavily muscled with no fat. Eutrophication theory thus does not adequately discriminate the condition of the ecosystem.

Furthermore, natural lakes as well as reservoirs are heterogeneous. As meticulously demonstrated by Kenton Stewart at the State University of New York at Buffalo, some are riverlike, others pondlike. Within a lake the location of inlet and outlet streams, the flow of lateral streams under ice, the direction of current and wind, and the history of past use all can and do impart individuality.

A particularly revealing lake is Canandaigua, westernmost of the major Finger Lakes. Midlake studies of plankton chlorophyll indicated little change in quantity in recent decades, although the water transparency is not high in comparison to some smaller, shallower, and more fertile lakes. Examination of inshore areas, particularly near stream inlets, showed a very different story. A number of species of submersed plants had disappeared, and the "weedy" milfoil grew rampantly in the most disturbed areas. Furthermore, blue-green algae covered the bottom near septic systems on shore. The pollution sources were residences, food processors, and possibly fertilizer from the vineyards that flank the lake. Most of the shoreline continues to show a limited growth of plants characteristic of clean infertile waters. Consequently, the fertilizing effect of streams is conspicuous: every stream enriched the plant community in species and amount. The affected area closely resembles a much more fertile lake.

Conesus and Hemlock lakes rank about equal in fertility. However, Conesus has more diverse organisms. Available nutrients are divided between algae and rooted plants, and because predator fish keep plankton feeders in check, there are many near-microscopic animals that virtually vacuum the lake clean of small algae and organic debris.

Joseph Shapiro of Minnesota has been a pioneer and advocate of biomanipulation, the adjustment of organisms in an ecological system. Hemlock Lake could be helped with biomanipulation. It is plagued with alewives, possibly introduced by releasing bait fish. Heavy stocking of alewife predators, such as trout and walleye, would suppress the alewife population and give the grazing animals, such as *Daphnia* (small crustaceans), a chance to clean up small algae and debris. Rooted plants could be introduced. Ordinarily, fish managers try to stock lakes with fish that grow rapidly. In this type of biomanipulation, however, the predator fish would not grow rapidly because too many would be stocked; rapid

(Left) Joseph Shapiro, University of Minnesota; (right) Herman S. Forest

Hemlock Lake (right) in New York State, though undeveloped along its shoreline, is unstable. Its riverlike shape exposes it to wind action, and there are few rooted plants to stabilize the bottom. In addition, it is plagued with alewives. Stocking the lake with such alewife predators as trout and walleye would suppress the alewives and allow such small grazing animals as Daphnia *(below) a chance to clean up the algae.*

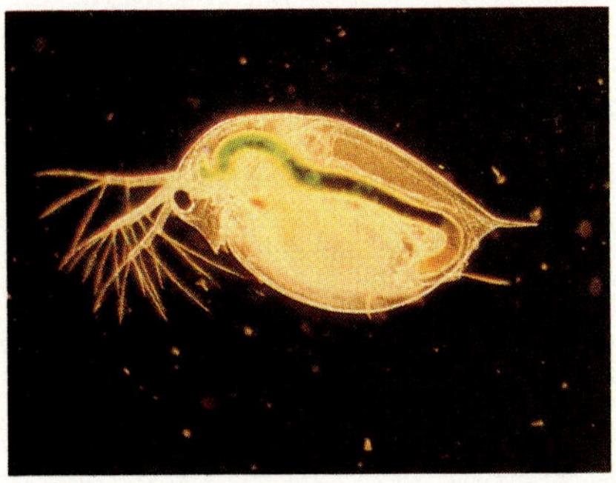

growth is sacrificed to keep them hungry so that they will continue to eat alewives. Thus, immediate fishing satisfaction would be sacrificed.

At Cornell University Ray Oglesby along with his associates and students contributed greatly to the definition and implementation of biomanipulation and the ecosystem approach. Also, Richard Osgood in Minnesota contributed to the rethinking of lakes as ecosystems. He pointed out that nutrient levels should not be ignored in the enthusiasm for biomanipulation. Indeed, no element of the ecosystem should be ignored. The scientific limitation is simply that no one has enough equipment, time, and wisdom to study and understand everything. Models and indexes permit management at the cost of some error and ignorance.

Ecosystem engineering

It is engineers, not ecologists, who usually occupy positions as key technical advisers. Engineers head public works departments, water supply bureaus, electricity generation installations, flood control works, and shore erosion and navigation projects. Moreover, engineering consulting firms designed and supervised the building of the facilities for the major water pollution abatement program of recent years. Engineers have been responsible for sewer systems and waste treatment plants, and have diverted streams to help clean lakes. Ecologists engaged in such projects virtually always have been given a limited and subsidiary role.

It is this situation that may account for the frequent lack of attention to ecosystem structure and function when designing and building water-control projects. When engineers have been pushed to imaginative designing by public pressure, some remarkable achievements have resulted; one of the greatest is the storm control system (mammoth gates instead of permanent dams) to protect the low-lying coast of The Netherlands from the sea. Dams would have destroyed shallow-water ecosystems and the economic resource of some fishermen as well.

One of the major challenges facing lake managers concerns the disposal of wastewater. Partly treated wastewater has been used on land for fertilizing soil, and marine estuaries have been enriched for shellfish production when untreated or treated sewage has been added to them. Wastewater has been treated by both simple and elaborate engineering designs. One of the simplest systems is used in Czechoslovak fish production impoundments, where ducks are raised beside the ponds to fertilize them and waste from swine barns is pumped to the water.

For most lakes, however, limitation rather than addition of nutrients is needed. The Czechoslovak ecologists are now struggling with the fact that the nation's tightly knit fish-farming region cannot absorb more waste without contaminating land and water. The greater distances and lower population densities of North America have only postponed the time when limits to adding nutrients to water there must be considered regionally as well as locally.

In trying to deal with this problem few innovations have been offered and fewer attempted. In 1973 Jacob Verduin of Southern Illinois University received an award for suggesting the creation of wetlands to receive waste from municipalities along the shores of the Great Lakes. They would provide a harvest of plants and fish, thereby recycling the nutrients before they reach the lakes. The proposal was not seriously considered for implementation because wastewater could still be discharged into the deep waters of the Great Lakes, where little biological activity or fertilizer effect is expected. This is an example of North American "dumping space," which does not exist in Czechoslovakia.

Hemlock Lake in New York demonstrates what ecosystem engineering could do. In the past it was managed to achieve the limited goal of supplying plentiful and cheap water (delivered largely by gravity). Demand on the supply rose sharply, and eventually the overfilling and severe drawdown badly damaged the lake ecosystem. Erosion and turbidity increased, valuable submersed plants were destroyed, and the spawning of fish was inhibited. The traditional solution for such a situation is to treat the water after it is withdrawn from the lake. This protects humans from disease but does not help the health of the lake. Turbidity increases the cost of human protection, however, and so a healthy lake and pure water are mutual goals.

The ecosystem model approaches the problem of turbidity from two directions. First, ecosystem engineers ask what can be done for the lake itself to reduce turbidity. In the case of Hemlock traditional erosion control measures can be applied throughout the watershed, particularly on the extremely steep slopes flanking the lake. In addition, an imaginative engineer with ecological awareness could design wetlands by simple diking at the most critical sources of sediment input to the lake and by adding baffles to reduce current action; this would be effective with Hemlock since it is riverine and sediments are apt to be resuspended along the shores. The latter solution meshes with the biomanipulation technique of increasing submersed plants to stabilize the bottom; both wetlands and baffles encourage the growth of such vegetation.

(Top) courtesy, Dutch Ministry of Transport and Public Works; (bottom) Aerocamera, Rotterdam

Massive concrete piers (top), each one weighing almost 18,000 tons, stand in a flooded construction dock on the coast of The Netherlands. When completed, they were among the 66 piers that supported 63 steel gates to form a storm-surge barrier (bottom) across the Eastern Scheldt (Oosterschelde) estuary in The Netherlands. The gates can be left open and are closed only when storm and tide conditions threaten to flood the estuary. This protects the low-lying coast of The Netherlands but, unlike a permanent dam, does not destroy the shallow-water ecosystems along the shore that support valuable fish.

The second direction for the ecosystem engineer concerns the distribution system. There is an alternate water source for the users—Lake Ontario. Indeed, more water is drawn now from the great lake than from the small one. Some reduction of the water taken from Hemlock, carefully planned, would reduce the fluctuation from over 5.5 meters (18 feet) to about 1.2 meters (4 feet). With the high and low levels near the natural elevations the lake would have a chance to recover the biological and physical regulatory features that keep an ecosystem healthy.

Future prospects

To deal effectively with their wide range of problems, it is important for lake managers to develop and adhere to a scientifically derived model of lake ecosystems. Of central importance in this model is the concept of self-regulation—homeostasis.

The property of homeostasis has often been called "stability" in ecological reports, but the concepts of "response" and "adjustment" rather than mere resistance to disturbance should be understood. It is an ecological principle that diversity in an ecosystem increases its homeostatic capacity. Like most theories applied to complex systems it does not always fit the actual situation, but usually it works rather well.

The diversity-stability theory must be invoked to explain why some lakes absorb abuse without marked disruption, while others do not. Indeed, a diverse lake ecosystem is analogous to a diverse climax forest. It is difficult for intruder organisms to get into either one, and, if left alone, each restores itself after a disturbance. Infertile lakes, having little diversity, may by contrast be likened to bare ground—weeds will become established easily. When people also apply fertilizer to infertile lakes, they should expect "weed" growth, not simply rooted plants but also weed algae and weed fish.

The ecosystem model informs the lake manager that there are several potentially stable states or ecosystem structures that can be achieved. The task of lake management is to discern the potential of a lake ecosystem and then either to manipulate it or leave it to nature. Lake management consists both of working with nature (probably still the best manager) and in using cultivation when it is warranted.

Although "lake restoration" has been a stated goal, it is not possible to restore many lakes to their previous healthy condition. Previously established biological and environmental components may have been destroyed and others permanently installed. What may be restored, instead, is a stable state.

Henry Regier of Toronto and his collaborators clearly perceived and stated the kinds of goals for lake ecology and management. When restoration is not possible and not desirable in managerial decision systems, two goals are pursued in practice. The first of these, enhancement, means to "improve" one or more features or simply increase the amount of a component. For example, more waterfowl and/or more salmon may require planting and farming if the natural ecosystem cannot carry the burden of the enhanced feature. The only way to reduce unwanted submersed vegetation in many cases is to keep cutting it as one would cut a lawn. Indeed, the lawn is a farmed system which needs man's subsidy to battle natural regulation and succession.

The second goal, rehabilitation, is more complex and more demanding. The metaphor of human health can be applied here. Although the patient—the lake—may not be restored to an earlier state, it can recover. A new condition that is acceptable or desirable in the value system of society can be attained. Prudent management requires another characteristic of the rehabilitated ecosystem: it must be maintained as much as possible by nature. It is rehabilitation that offers the greatest opportunity and challenge for science and management.

THE SCIENTIST IN LITERATURE

by Gregory Benford

In the past two centuries the image of the scientist in fiction has swung from monster maker to superhero and halfway back again. Nevertheless, literature can scarcely claim to have depicted scientists as they really work and live.

Darwin, Freud, Einstein—these resounding names recall revolutions that have jolted our very ideas of who and where and what we are. Modern science has given us vast new perspectives, and its handmaiden, technology, is transforming the face of our planet.

Yet modern literature reflects all this human activity only dimly. The scientist himself is seldom the central figure in literary works, and it is certainly fair to wonder why.

From the beginning of modern times the humanistic world had trouble welcoming this new figure on the world stage. Before the 20th century, literature often kept its distance from science, as when William Wordsworth respectfully observed a statue

> Of Newton with his prism and silent face,
> The marble index of a mind for ever
> Voyaging through strange seas of Thought, alone.

The key word is alone. Wordsworth saw the scientist as essentially a solitary figure, self-condemned to exile from common humanity. Others railed against what they conceived of as science's narrow, reductionist worldview. William Blake favored the "wise passiveness" of intuition more highly than "analytic industry" and prayed,

> May God us keep
> From single vision and Newton's sleep.

Jonathan Swift's *Gulliver's Travels* showed the scientist as merely ridiculous. John Keats toasted the destruction of the man who had explained the rainbow and so robbed it of its poetry. Matters were hardly helped by Darwin's assertion that, while he had once taken pleasure in Milton and Wordsworth, they later seemed "intolerably silly" and that in rereading Shakespeare he experienced a boredom so intense that it made him feel physically sick.

(Facing page) Portraits of scientists—real, fictional, or both: Victor Frankenstein, as portrayed by Colin Clive (top left); Albert Einstein; James Watson; and Thomas Edison (bottom, left to right).

By courtesy of the Masters and Fellows of Trinity College, Cambridge; photo, the Bridgeman Art Library

The preoccupied expression on the marble sculpture of Newton at Trinity College, Cambridge, moved Wordsworth to write of "a mind forever / Voyaging through strange seas of Thought, alone." Wordsworth's assessment of the self-exiled scientist was at least respectful, if awed. Some of the poet's contemporaries ridiculed or condemned scientists for their narrow, prosaic views of the world.

The scientist and the monster

The coexistence of literature and modern science started off on the wrong foot in part because the scientist inherited the mantle and public image of the medieval alchemists, astrologers, and sorcerers. This identification persists: many people think astrology is a science or use "astrology" and "astronomy" interchangeably. Such associations led to the first major image of the scientist—the obsessive neurotic, with Victor Frankenstein as the emblematic case.

Mary Shelley regarded her "ghost story" as plausible, and the first sentence of her preface states, "The event on which this fiction is founded has been supposed, by Dr. [Erasmus] Darwin, and some of the physiological writers of Germany, as not of impossible occurrence." She portrayed the monster as an anthology of parts from hospitals and graveyards, animated by a stroke of lightning—an image that crops up in many later depictions of scientists. The shambling nightmare "demon," created deliberately by man's ingenuity, is filled with utterly human love for its irresponsible creator. A key realization for the reader is that the monster acts out deeds that its master is unwilling to do himself, even killing his bride. But throughout, the scientist is blind to the fearful power of the "demon," evades responsibility, and even falls into a convenient swoon at crucial moments. He is beset by romantic gloom, sorrow, miseries, and despair—mere luxuries when clearly action is required—and loses the reader's sympathy. Small wonder, since the novel plainly concerns the unacknowledged evil in the scientist's soul.

Universally, readers identify with the monster. They seldom know that the novel was written by a 19-year-old woman in the grip of the Gothic tradition, but they understand Boris Karloff's film portrayal, which distorted the novel considerably but conveyed the ethical implications of the metaphor: the scientist who is hugely powerful but naive about the implications of his work. It is no accident that the name Frankenstein now summons forth the monster, which was unnamed in the narrative, and not the man who made it. The significance of the tale lies in the expression, particularly relevant for us, "We have created a Frankenstein."

Victor Frankenstein has replaced Prometheus in that small company of archetypes—for example, Alice, Don Quixote, Sherlock Holmes, Dracula, Tom Sawyer—who leap from literature into the collective consciousness, slipping free of the details of the original work. Sherlock Holmes, too, has connections to science, since his methods are those of rigorous deduction; his echoes resound in such recent figures as Mr. Spock, the scientific officer of "Star Trek." But unfortunately it is Victor Frankenstein who more than any literary figure brought into common knowledge the worst form of the obsessive neurotic—the mad scientist.

Frankenstein was described as a highly idealistic romantic: "noble," "cultivated," a "celestial spirit." Gradually, in the hands of 19th-century novelists, the romantic aura that surrounded him faded. The image of the scientist became transformed in the cultural consciousness into someone closer to H. G. Wells's vivisectionist Dr. Moreau, who boasted, "You cannot imagine the strange colorless delight of these intellectual

__GREGORY BENFORD,__ Professor of Physics at the University of California at Irvine, is the author of numerous works of fiction, including Timescape *and* Artifact.

From the collection of Sam Moskowitz; photo, Christine E. Haycock, M.D.

The image of scientists as alchemists and sorcerers persisted in fiction and in the public mind long after their medieval forebears had vanished. At left, an illustration from "Old Doctor Rutherford," published in the September 1891 issue of The Ludgate Monthly, *dwells on a favorite fictional occupation of scientists, the search for the "elixir of life."*

desires. The thing before you is no longer an animal, a fellow-creature, but a problem."

This latter image of a scientist has been remarkably persistent. When people use the term mad scientist today, we never wonder for a moment if perhaps they mean that the scientist is merely angry.

The "technohero"

The figure who did the most to offset the mad scientist stereotype, first in the real world and then in fiction, was Thomas Alva Edison. To many he became a world-class metaphor of a new age galvanized by "invention optimism." Edison appeared in such popular fiction as Garrett P. Serviss's *Edison's Conquest of Mars* (1898) and the long series of dime novels featuring Tom Edison, Jr.

Even writers like Jules Verne (who was far more fascinated with the machines that scientists made than by those who made them) and Wells (who prized discovery without paying great attention to the discoverers) temporarily fell prey to the mad scientist distortion. In his early works Wells created a series of near-lunatic scientists, including Moreau, before he realized how this clashed with his faith in science. Thereafter he carefully prepared a different, serenely good image of scientific workers in such novels as *The Food of the Gods and How It Came to Earth* (1904). These later figures were far less striking and had less effect on his readers. Optimism is often bloodless.

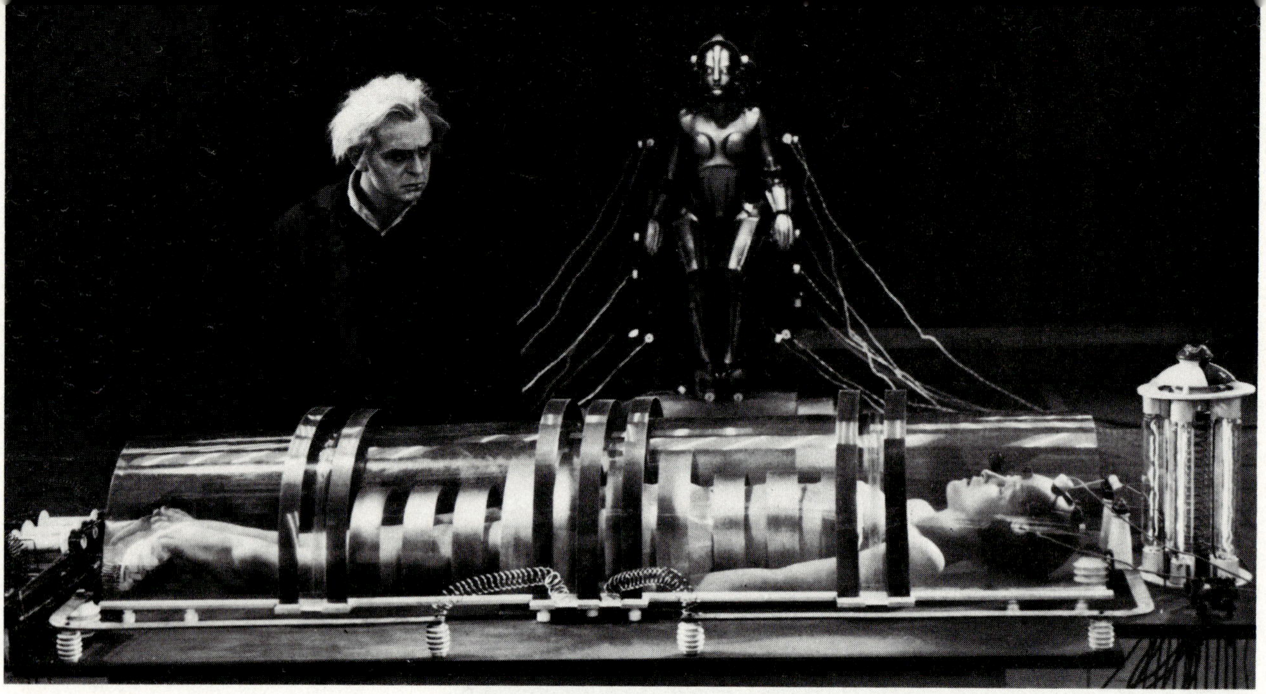

The scientist-romantic embodied in Mary Shelley's Victor Frankenstein (right, as depicted in a 1931 film adaptation) was transformed by later writers into the depraved maker of monsters. In The Island of Doctor Moreau *(1896; below) H. G. Wells describes a scientist who populates a remote island with humanlike creatures surgically reshaped from beasts. The mad Rotwang (above) prepares to transfer the visual and behavioral likeness of a human being to a robot in the 1926 German silent-film classic* Metropolis.

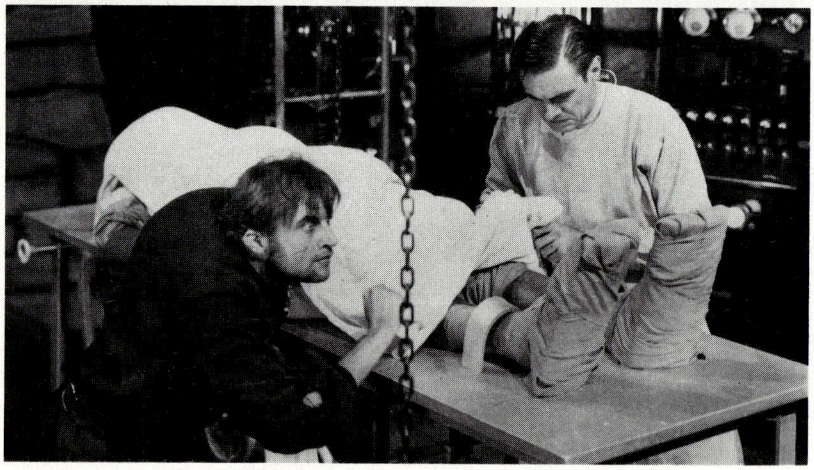

(Top) The Museum of Modern Art/Film Stills Archive, New York; (center) The Bettmann Archive; (bottom) Jean-Loup Charmet, Paris

Although Verne produced the inventor-turned-madman Robur late in his career, most of his literary efforts contributed to the growing countercurrent against the mad scientist. In *Twenty Thousand Leagues Under the Sea* (1870) Captain Nemo is a distant, benevolent overseer trying to prevent war by destroying the weapons with advanced technology. Verne even satirized the less solemn habits of science by showing an arrogant Professor Aronnax who tries to explain the fine points of ichthyology to a harpooner. Aronnax asks the man if he knows how fish are classified; the man of the sea replies simply and sensibly, "They're classified into those you can eat and those you can't," neatly skewering the impracticality of ivory-tower pretentiousness.

Singing hosannas to the technohero reached its zenith in Hugo Gernsback's *Ralph 124C 41+* (1911–12), which depicts a future state in which a technological fix exists for just about everything except, apparently, wooden writing. (The book is notable, though, for its prediction of television and several other inventions.) Ralph's name, we learn at the heavy-handed conclusion, marks him as One To Foresee For 1+

(1+ = the many). Good, problem-solving scientists versus problem-creating villains became a staple of pulp science fiction. In the 1940s Isaac Asimov's *The Foundation Trilogy* carried hero worship of the social scientist to an extreme by chronicling the application of a totally predictive theory that could anticipate the ebb and flow of societies for millennia to come. But again, the genius, Hari Seldon, is a cardboard figure obsessed with his work and unable to see beyond it to broad humanistic concerns.

When someone remarked to Lord Rutherford, the central figure in atomic experiments of the early 20th century, that he seemed to be riding the crest of a wave, he replied, "Well, I made the wave, didn't I?" The self-confidence of that era assumed that there was no limit to human capabilities and that rational progress would overcome the darker side of human desires.

This image of the scientist persists in more recent figures such as Jonas Salk, the man most closely identified with discovery of a polio vaccine, who appeared as a savior to anxious parents in the 1950s. Salk's scrupulously clean white lab smock, conveying an air of certainty, is now common in depictions of the scientist. It invokes a reassuring control over disease and disorder, finding use in everything from television commercials for health products to carefully orchestrated congressional testimony.

Strange loves of the mind

In literature the pendulum swung back soon enough. A scientist who could somehow create a world in which his rigorous but limited equations would work might well be seen as a threat to all.

The inability to translate technical grace into personal adroitness often has tragic overtones. In 1924 Yevgeny Zamyatin sounded this warning note in the influential anti-utopian novel *We*. In his alienating, science-

Both as a prolific inventor and as a fictional character Thomas Edison played a key role in offsetting the mad scientist stereotype. A newspaper illustration (below) from Edison's Conquest of Mars *(1898) shows Edison and the work's author, Garrett P. Serviss, discussing ways to repay the Martians for the invasion of the Earth that H. G. Wells described in his* The War of the Worlds. *In Serviss's story, a serialized sequel to Wells's novel, Edison eventually reaches Mars with a fleet of spaceships financed by the world's governments and staffed by some of the most famous scientists of the 19th century. (Below left)* Tom Edison Jr's Electric Sea Spider *(1892) exemplifies the popular, juvenile-oriented dime novels of the late 19th and early 20th centuries, which recounted the fictional heroics of technologically ingenious young men.*

From the collection of Sam Moskowitz; photos, Christine E. Haycock, M.D.

With godlike bearing, scientists of a future utopia contemplate the cosmos in a scene from a 1936 film version of H. G. Wells's The Shape of Things to Come *(1933). In many such works Wells (inset) expresses his faith in man's ability to live happily in a technologically transformed world.*

The Museum of Modern Art/Film Stills Archive, New York; (inset) The Bettmann Archive/BBC Hulton

saturated, totalitarian future society he carried the rhetoric of science to an extreme when his narrator, D-503, notes, "$L = f(D)$, love is the function of death." The Soviet state saw Zamyatin's work as critical of socialist regimentation, and it is still banned in the U.S.S.R. It greatly influenced such works as Ayn Rand's *Anthem* (1938), George Orwell's *Nineteen Eighty-Four* (1949), and George Lucas's film *THX 1138* (1971). But it also condemned the willy-nilly use of scientific postures and language for emotional and humanistic issues to which they do not apply. This theme of the scientist as cultural imperialist has blended with the mad scientist image to produce many clichéd warnings against overuse

The many contributions of Jules Verne (right) to the image of the scientist as superhero include Captain Nemo, who first appeared in Twenty Thousand Leagues Under the Sea *(1870). In the engraving at the far right, from an early edition of the novel, a bearded Nemo takes Professor Aronnax through the engine room of his submarine, the "Nautilus," a product of high science dedicated to the destruction of war makers and the elimination of oppression.*

(Left) The Bettmann Archive/BBC Hulton; (right) The Granger Collection

From the collection of San Moskowitz; photos, Christine E. Haycock, M.D.

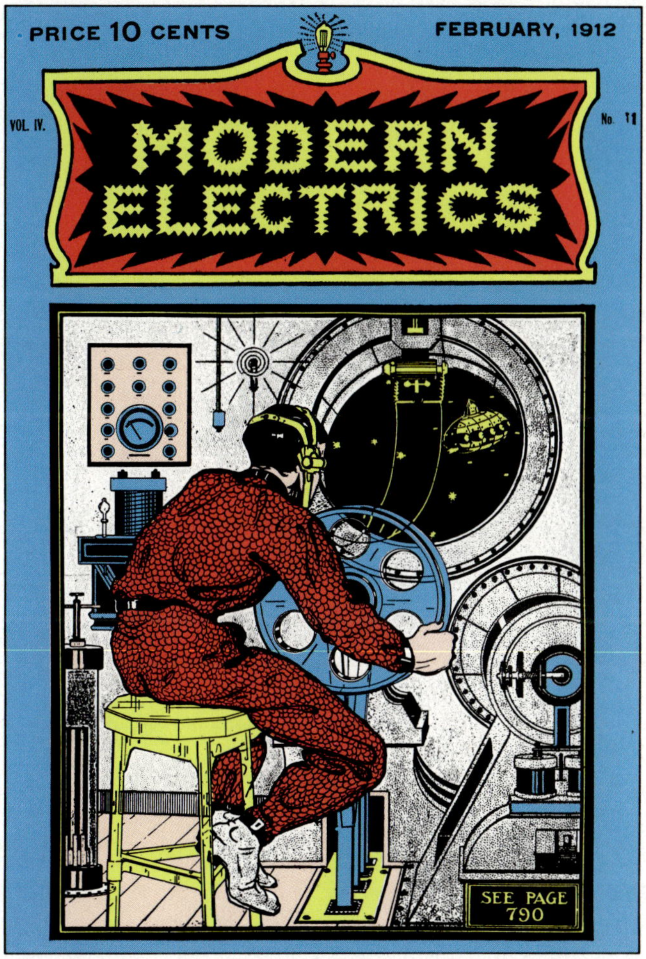

Professor Challenger (above), shown fighting to stay alive in a poisoned atmosphere by breathing bottled oxygen, is Arthur Conan Doyle's fictional contribution to the hero worship of the scientist; the scene is from an episode of The Poison Belt, *serialized in 1913. Scientist, prophet, and adventurer are all apt descriptions for Ralph 124C 41+ (right), Hugo Gernsback's superman of the future.*

of the social sciences to control populations. Here the scientist is seen as out of control himself, wielding power over the nonspecialists, amoral, working for whoever will hire him.

It is just this picture of the inhuman, casually malevolent scientist that yields the updated Frankenstein image of *Dr. Strangelove* (1963). The central scene of Stanley Kubrick's landmark film is Strangelove's long soliloquy, his imagination leaping beyond the coming nuclear destruction to propose more complicated scenarios for human survival while his mechanical arm struggles to strangle him. This is a vivid visualization of the power of a scientist who has lost his moorings of both common sense and morality and is unknowingly suicidal. Again, as with Victor Frankenstein, the scientist knows not what he does.

In the nuclear age the Edison image also evolved. Albert Einstein combined abstract mystery, greatness of perspective, and commanding wisdom. No mere Edisonian technocrat, he revolutionized our worldview. His gray, bushy-haired visage, sloppy clothes, and pacifist sentiments posed a counterfigure to Strangelove. Curiously, there have been few attempts to depict the Einstein figure in fiction; indeed, the theoretical scientist is often seen as benign but fearful of humanity. In Mitchell

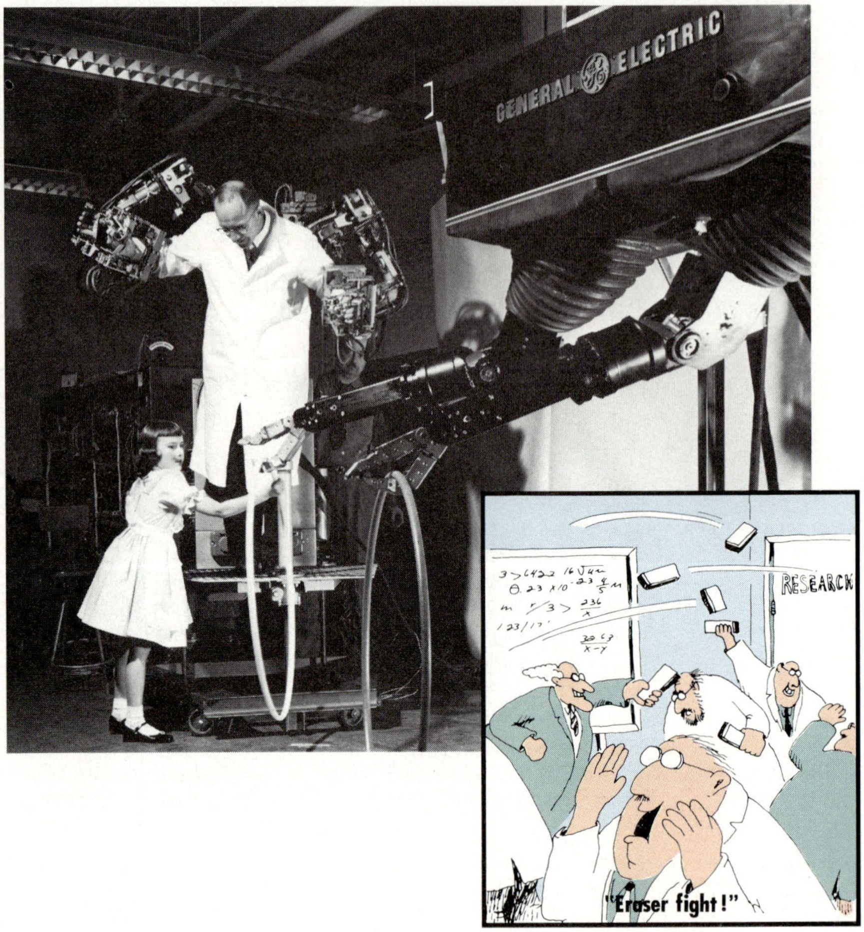

The apparently boundless confidence in science that found expression in early 20th-century fiction has persisted in the visual cliché of the lab coat. Media photos like the one above of Jonas Salk, the vanquisher of a crippling childhood disease, present scientists as white-garbed saviors. The industrial publicity photo at the top right, from the late 1950s, attempts a similar message; ironically it can also be interpreted as an unintentional sequel to Frankenstein, *complete with inhuman monster, looming creator, and innocent victim. At the far right, the lab coats of solemn science clash with their wearers' behavior to generate humor in a modern cartoon.*

(Left) The Bettmann Archive; (right) from the collection of Sam Moskowitz; (inset) drawing by Gary Larson, reprinted by permission of Chronicle Features, San Francisco

Wilson's *Live with Lightning* (1949) a physicist finds refuge from the ferocious world in his work: "At night, long after the place was deserted by everyone, he put down his journal, or his calculations, like a doomed man leaving his cell for the block." Only after he is in bed does "his mind begin to soar again into the still heights of perfect knowledge." Interestingly, the novel's hero is engaged in high-voltage experiments that produce lightning in the laboratory, a curious echo of Frankenstein.

The two modern stereotypes of the scientist thus are Good versus Evil or, cast somewhat more specifically, Einstein versus Strangelove. As Milton found with God and the Devil, evil is easier to characterize and rather more interesting. But the real, working scientist can rightly claim that he has not been characterized at all.

The whole scientist

Oversimplifications are not progress, even if they counter the darker toned misunderstandings of the past. It is striking today how little literature has paid attention to the way scientists actually work and live.

Some attempts at depicting scientists come from works about doctors. Sinclair Lewis's *Arrowsmith* (1925) concerns medical research but does not delve far into medicine as a science. Often, when men trained as physicians, such as W. Somerset Maugham and Arthur Conan Doyle, turned to writing, they did not portray doctors as scientists.

Because scientific ideas can be used to represent conditions in the world, many authors have used them, but in a form that filters out the scientist himself. Lawrence Durrell's *The Alexandria Quartet* (1957–60) concerns the wrenchingly different ways the same events look from different people's viewpoints, and it nods in the direction of relativity and quantum mechanics as metaphors. The hero of Robert Coover's *The Universal Baseball Association, Inc., J. Henry Waugh, Prop.* (1968) constructs an entire baseball league using the theory of chance. The novel is a comic meditation on science's ideas of order and entropy, or increasing disorder, translated into the humanizing perspectives of baseball. In the short story "Entropy" (1960), Thomas Pynchon tries to look at the world from a scientist's view by studying two men in an apartment house. They embody different approaches to the second law of thermodynamics, which states that systems tend naturally toward maximum disorder and inert uniformity. One man conducts a lease-breaking party, whose roar invades the carefully closed space of the other. It is not clear whether the outside city's intruding disorder represents entropic leveling, the second man realizes, in the same way that it applies to "galaxy, engine, human being, culture, whatever." The story reminds us that easy translations of the idea of entropy into everyday life are not simple and that the language of the scientist is special.

Indeed, even scientists have trouble making the transition from work to life. In *Gravity's Rainbow* (1973) Pynchon depicts the mind-set of engineers and scientists building the V-2 rockets of World War II. The scientists have a definite, idealistic way of approaching their problems:

> you use dimensionless coefficients: ratios of this to that—centimeters, grams, seconds neatly all cancelling out above and below. This allows you to use models, arrange an airflow to measure what you're interested in, and scale the results all

In the nuclear age the unkempt, benign visage of Einstein has its antithesis in actor Peter Sellers's film portrayal of the unbalanced, suicidal Dr. Strangelove. Both figures have become modern, albeit inaccurate, stereotypes of the scientist.

(Left) Brown Brothers; (right) photo, The Museum of Modern Art/Film Stills Archive, New York

AP/Wide World

C. P. Snow crafted novels that mirror his long and varied career as a physicist, university administrator, and bureaucrat. Much of the strength of his writing lies in the meticulously detailed lives of his characters, many of whom are scientists.

the way up to reality, without running into too many unknowns, because these coefficients are good for *all* dimensions.

A character then fails to understand the world because in a sense he cannot find a nondimensional coefficient for himself:

the parameters breed like mosquitoes in the bayou, faster than he can knock them off. Hunger, compromise, money, paranoia, memory, comfort, guilt. . . .

An outstanding example of a novelist's dealing with the real lives of career scientists is C. P. Snow's *The Search* (1934) followed by his 11-volume series collectively called "Strangers and Brothers" (1940–70). The latter books concern the academic and administrative career of an Englishman, Lewis Eliot, who begins by studying law and ends in the corridors of power, a path that parallels Snow's life of physicist turned bureaucrat. Their general direction is prefigured by *The Search,* which tells of the tedium and occasional zest of research in crystallography, set in London and Cambridge. Its hero, physicist Arthur Miles, works his way up through a demanding system but does not attain the directorship he covets. Somewhat disappointed, he discovers that an old friend, Sheriff, has taken up one of Miles's abandoned projects against his advice. Sheriff "cooks" the findings and publishes without telling Miles, hoping to win an appointment to a vacant chair at a university before the fakery is discovered. Miles knows precisely where Sheriff falsified the data. What should he do? He writes a letter to the journal that published the paper, but then hesitates. He knows how badly it will hurt his friend. Miles muses,

It was twenty years and more ago, I thought, that night when my scientific passion first broke hot upon me; and, through curiosity, satisfaction, ecstasy, strenuous work, a career, disaster, recovery, partial severance, I had come to this. The passion was over now. I lit a match and put it to the corner of the letter; the flame was steady in the still air, golden and smoky-edged. The passion was over. I had repudiated it, and I should never feel it again.

Miles sees a human truth beyond the standards of science.

After *The Search,* with its nostalgic farewell to the "passion" of research, Snow moved on to academic politics and the study of bureaucratic man, getting further from science as an electrifying process of discovery and more into its growing role as the lever of government power. Thus it mirrors the lives of many scientists, reflecting the sobering fact that the path to higher salary and greater influence leads them away from the primary experience that got them into science in the first place. This bittersweet truth is one of the major facts scientists face. It intersects another fact seldom rendered in literature: research as a group effort.

The romantic ideal of the isolated genius who changes the world's perceptions is, for nearly all scientists, a complete falsification. It may well derive from the old notion of the artist—starving in a garret, inflamed by wild visions of possibility. Nothing is further from the true life of scientists, who in the 20th century work ever more often in teams. The sheer size of giant accelerators and laboratories demands organization of scientific work, much as industrial production depends on the smooth

The real scientist, most likely a team member struggling to advance the frontiers of knowledge by inches through cooperation and hard work, belies the romantic ideal of the isolated scientific genius. At left, the scientists posed in front of a bubble chamber at Brookhaven National Laboratory are a small fraction of the 26 physicists, 14 engineers, 45 technicians, 18 photo scanners, 11 beam crewmen, and 2 programmers who shared in the discovery of the omega minus particle in 1964. Similarly, the discoveries of the intermediate vector bosons in 1983 involved the participation of more than 200 scientists from at least ten countries.

Brookhaven National Laboratory

intermeshing of complicated processes to achieve a result. The people who man such huge research groups are not eccentric geniuses. In the government laboratory at Los Alamos, New Mexico, shown in Thomas McMahon's *Principles of American Nuclear Chemistry: A Novel* (1970), the focus is not on blithe spirits but on steady progress, "milestone" targets, and scheduled results. Maxwell Griffith's *The Gadget Maker* (1954) traces the life of an ordinary engineer who later works on scientific rocket design, and the emphasis is on the hard labor:

> He had learned to spell in spelling bees and not by looking at cute pictures of apples, cats and dogs; he had learned to multiply by memorizing multiplication tables; he had never finger-painted or been encouraged to express himself or to cooperate with the group; he had never spent time in a classroom listening to phonograph records of fairy tales recorded by stars of stage, screen and radio; and he was unaware of the serious damage to his psyche that sitting in a corner and wearing a dunce's cap had done.
>
> . . . alone in an echoing dormitory, he felt a kinship with all the men who have ever toiled over books in lonely rooms. He felt akin to great men, great intellects, for unlike so many old drudges and inexperienced youths he did not believe creative men idled pleasantly and then, when the mystical spirit moved them, whipped out works of merit in short, furious bursts of easy inspiration.

A largely unrecognized asset of science is that ordinary people, following the scientific method, can produce useful work. This is not true of the world of art, where advances come from the towering figures. The team-player scientist is often a rather ordinary-seeming person, far from the lurid excess of the romantic Frankenstein or the antiseptic certainty of Edison. In *Live with Lightning,* as it traces the career of a middling-good physicist through the great era of nuclear physics, an experimentalist says,

The white-coated scientist idea is all very well for drugstore chemists or advertising pictures, but here we use monkey wrenches, screw drivers, lathes and torches. If you have to get down on the floor, you get down on the floor.

But, in that novel, scientists are still not as lowly as engineers. As a senior physicist sardonically remarks,

. . . we service them with information. . . . Put it this way—our attitude towards engineers is about the same kind of tolerant contempt that a $25-a-week repairman in a service station feels for the millionaire whose Rolls he has just fixed.

As scientists see themselves

Though C. P. Snow is the best-known scientist who wrote fiction, there are other telling depictions of scientists by themselves. Leo Szilard, who played a central role in nuclear physics and alerted Einstein to the possibilities of a bomb, wrote parablelike short stories about science's future impact. "The Voice of the Dolphins" is told as a report on the discovery of dolphin intelligence. Its stiff tones do not conceal the scientists' realization that dolphins are of a vastly different, higher order intelligence. Ever loyal to intellect, the scientists then serve as front men in a clandestine cetacean takeover of the world—all to the eventual good of everyone, of course. Szilard usually portrayed scientists as the only figures who truly understood the implications of their own ideas. Nevertheless, he failed to show them as human beings beneath their mandarin masks.

The same is generally true of isolated short stories by such eminent scientists as J. B. S. Haldane, Norbert Wiener, and Julian Huxley—strong, pertinent warnings, delivered as from a pulpit. These reinforced the picture of scientists as disembodied authority figures, and it fell to

James Watson (left), who with Francis Crick (right) unraveled the molecular structure of DNA *in the 1950s, undermined the popular picture of scientists as authority figures in his autobiographical* The Double Helix *(1968). His account of discovery makes clear that scientists bring the same weaknesses and prejudices to their jobs as the rest of humanity.*

Original model of the double helix from J. D. Watson *The Double Helix*, Atheneum, N.Y., p. 206c, 1968 by J. D. Watson

Sidney Harris

Chapter 7
THE STRUCTURE OF THE NUCLEUS OF THE ATOM
"What?" exclaimed Roger, as Karen rolled over on the bed, and rested her warm body against his. "I know some nuclei are spherical and some are ellipsoidal, but where did you find out that some fluctuate in between?"
Karen pursed her lips. "They've been observed with a short-wavelength probe . . ."

James Watson to undermine this view with his autobiographical *The Double Helix* (1968), which traces the discovery of DNA. Some scientists have scathingly called the book fiction because it is highly personalized, but more probably because it blows the cover on a sometimes disagreeable subject: scientific rivalry.

Even before Watson's account, the conflict between Jonas Salk and Albert Sabin over the polio vaccine had moved writers to make battles over right of discovery, stolen secrets, and races for publication the fulcrum for many fiction stories that stress the combative features of research, as in Eleazar Lipsky's *The Scientists* (1959). This kind of conflict is dramatically convenient and often true. Popular belief still has it that scientists, unlike ordinary folk, do not stick blindly to their first principles but rather follow Francis Bacon's maxims, carefully and dispassionately laboring from data to general axioms even if those new findings undermine their most cherished beliefs. Sometimes they do, but we should remember that even the saintly Einstein, who shattered centuries of vaunted preconceptions, refused to accept quantum mechanics because, as he said, "I refuse to believe that God plays dice with the universe." Nobel laureate Max Planck underlined this:

> A new scientific truth does not triumph by convincing its opponents and making them see the light, but rather because its opponents eventually die, and a new generation grows up that is familar with it.

Watson was anticipated by another Cambridge scientist, astronomer Fred Hoyle. His novel *The Black Cloud* (1957) follows astronomers as they study the approach of a huge molecular cloud from interstellar space that threatens to blot out the Sun. They deal analytically with the problem while politicians dither and lurch. They have rival explanations of what is happening, but these are settled by events, whereas the political conflicts churn on endlessly. Through all this turmoil the sardonic

(Left) John Bowden © *Discover* Magazine, Time Inc.; (right) © Magnum Photos; (bottom) The MIT Museum

Scientists who have depicted their own profession in fiction include physicist Gregory Benford (top left), astronomer Fred Hoyle (top right), and mathematician Norbert Wiener (above).

protagonist is an unapologetic elitist who resembles Hoyle himself. One suspects he is a blunt expression of many scientists' private views.

Hoyle's fiction is one of the few attempts to show how scientists would deal with genuinely unexpected new knowledge. A novel that does so in detail, showing how several types of scientists react to the unsettling possibility of communication backward in time, is Gregory Benford's *Timescape* (1980). One entire chapter deals with a theorist's calculations; taking a break, he muses,

> Galileo's blocks gliding across the marble Italian foyers, their slick slide obeying inertia's steady hand—they were cartoons of the world, really. Aristotle had understood in his gut the awful fact that friction ruled, all things groaned to a stop. *That* was the world of man. Only the childlike game of infinite planes and smooth bodies, reality unwrinkled, cast a web of consoling order, infinite trajectories, harmonic life. From that cartoon world it was always necessary to slip back, cloaking exhilarating flights in a respectable, deductive style. But that did not mean, when the papers appeared in their disguise of abstracts and Germanic mannerisms, that you had not been to that other place, the place you seldom spoke of.

This is a central experience of science and among the hardest to convey in literature.

Scientists as writers

Scientists are not likely to become their own bards. Research institutions do not teach the writing art of character development or style. Moreover, people do not become scientists as a part-time job; so there is little chance they would serendipitously drift into writing. The organization of the brain also seems influential: quantitative thinkers are problem-directed and not often verbally skillful. There is a natural self-selection of people into groups according to whether they think of the world in pictures or in words (or even, as Einstein said of himself, in kinesthetics).

Working scientists have little inclination to describe themselves in words and, perhaps most important, they have no time. They pour themselves into their research, and if they have any hobby it tends to be athletic. Sadly, even the scientific journals work against expressive writing.

Journal editors seem to like papers that are impersonal, methodical to the point of dullness, remorselessly literal. Most use the passive voice—"the electron is directed into the chamber and made to resonate"—which, after a few paragraphs, creates a feeling akin to hearing music through cotton wads. Bureaucrats prefer this voice, since it hides what is going on, who does it, and why. Some journals actually require it, suggestive of the increasing connection between government and science.

On top of stylistic barnacles, scientists learn to revere accuracy, to recheck their results constantly, and to be wary of speculation. These are all fine standards for finding facts but often are impediments when trying to get at the supple contours of personal relations, where there is no final "truth," only relative views. Often in scientists' fiction, characterization dies in a blizzard of objective prose, free of bias, couched in value-free sentences. C. P. Snow's depictions often read like the descriptions job interviewers write—an outside picture, tracing probable surface motivations but blind to the true driving force inside.

Often writer-scientists use fiction as a freight hauler for "larger issues" such as philosophy or politics. They neglect character, the very area in which they have the most experience. On one hand the literary-critic fraternity still seems to feel that no one can write about the "human condition" and simultaneously understand the workings of a refrigerator. On the other hand scientists do seldom communicate how they think or what they believe sets them apart.

Of course, seen from the outside, a scientist need not be any more interesting than a bank clerk. At least the reader does not struggle to understand what the clerk is doing. (T. S. Eliot wrote some of his great, bleak verse while canceling vouchers and doing sums, and one can well wonder if that affected his poetry.) The apparent dullness of scientists may be superficial, but several authors have suggested that it is a real limitation. In Raymond F. Jones's short story "Noise Level" (1952), a group of scientists is shown a film of an apparently bona fide working antigravity device, the inventor of which has been killed. Attempting to duplicate it, the group breaks through into a new understanding of fundamental physics. Only later does it emerge that the original film was a fraud, used to exert psychological pressure to break down the viewers' mental set and stir their imagination.

Fiction could be of great help in spanning the chasm separating the rest of humanity from the real, everyday world of the scientist. More should realize, as Nobel laureate P. B. Medawar observes in *The Art of the Soluble* (1967),

> scientists are people of very dissimilar temperaments doing different things in very different ways. Among scientists are collectors, classifiers, and compulsive tidiers-up; many are detectives by temperament and many are explorers; some are artists and others artisans. There are poet-scientists and philosopher-scientists and even a few mystics . . . most people who are in fact scientists could easily have been something else instead.

Perhaps, but they might not have had as much fun. Fiction should convey that fact, too.

"Actually I started out in quantum mechanics, but somewhere along the way I took a wrong turn."

Sidney Harris

SUBJECTIVITY •IN SCIENCE•

by Stephen Jay Gould

Scientists like to believe that they are always as objective as possible in regard to their research, but a look at recent history reveals that their cultural expectations and biases can cause them to become subjective without realizing it.

Thomas Henry Huxley, grieving over the death of a beloved son, wrote of his sustaining faith in science in a famous statement considered ever since as a mark of scientific method at its most admirable:

> Sit down before fact as a little child, be prepared to give up every preconceived notion, follow humbly wherever and to whatever abyss nature leads, or you shall learn nothing. I have only begun to learn content and peace of mind since I have resolved at all risks to do this.

Taken out of the context Huxley intended, this statement has usually been interpreted as arguing against all subjectivity in science, against any intrusion of personality into scientific activity. The arts and humanities may thrive on the interplay of personality and subject, but science must be done in a rigorously objective manner—that is, in the same way by all proper practitioners. This false dichotomy of "two cultures" (science and humanities), two ways of knowing, has led us to stereotype scientists as people who can shed all hope, preference, prejudice, and constraints of culture and therefore approach the natural world as fact-gathering automata who can then process the objective information received according to strict rules of logic and inference that yield certain conclusions. Science then becomes something separate from the rest of human activity, something severe and forbidding, accessible only to a trained elite.

This stereotype is a myth that must be fought, if only to teach us that scientists are like all people, that scientific activity is like all creative human thought, and that science is accessible to all people of reasonable education. We live in a technological world, and science enters our daily lives in a hundred ways (consider the current computer revolution); it therefore behooves us all to understand science and to recognize that it is not an arcane profession accessible only to a trained priesthood, but one expression of a human reason common to all.

A balanced view

A frank recognition of the subjective and human side of science provides an excellent path toward this desired accessibility. Science does seek

OBJECTIVE
TRUTH
INDUCTIVIST MYTH
RELATIVISTIC MYTH
FINAGLING
DISABLING BIAS
FRAUD

STEPHEN JAY GOULD *is Alexander Agassiz Professor of Zoology at Harvard University, Cambridge, Massachusetts, and has written several books, including* Ever Since Darwin *(1977) and* The Mismeasure of Man *(1981).*

truth, and subjective does not mean false—only that human passions and distinctive modes of thought have entered the inquiry. In fact, and for two major reasons, we must frankly acknowledge this human side in order to have a better chance of obtaining truthful answers. First, scientists are human beings necessarily enmeshed in their cultures and strongly influenced by personal hopes and expectations. If we try to deny these influences and claim that we act with pure objectivity, in obedience only to the facts of nature, we make terrible mistakes because we do not honestly recognize our biases. Bias that leads to ignoring or distorting data must be recognized and rigorously eliminated; acknowledgement of the human side of science fosters this activity.

Second, subjective elements do not always drive science toward error; they may suggest fruitful approaches and break impasses in old modes of thought. Long ago, positivist philosophers made a crucial distinction between "context of justification" (how we judge whether or not an argument is true) and "context of discovery" (how arguments arise). Subjective elements must be considered when we assess context of discovery—why one scientist rather than another developed an idea, why a theory arose in a certain time and place, for example. Ideas arise for all sorts of peculiar reasons. Some great theories (that later turned out to be correct) first jelled in their creator's minds during dreams; others (that turned out to be dead wrong) developed inductively from careful and rigorous experiments incorrectly interpreted. For example, Darwin's principle of natural selection arose largely by analogy and creative metaphor (context of discovery)—analogy to observed changes in agricultural crops and livestock by artificial selection practiced by breeders, and metaphor with the laissez-faire economic theory of Adam Smith, in which ordered economies arise as a by-product of individuals struggling with others to maximize profits. (In Darwin's system ordered ecosystems emerge from the struggle among individual organisms for increased success in reproduction.) Yet natural selection, by context of justification, seems to be largely true as a description of how evolution operates in nature. Masses of empirical evidence, gathered later to test Darwin's theory, have continually affirmed it. As Darwin wrote in a famous letter in 1853: "How can anyone not see that all observation must be for or against some view if it is to be of any service."

We must follow the proper and intermediate path, the golden mean, between two extreme and misleading views. One extreme, the inductivist myth cited earlier, holds that science proceeds by clearing the mind of all expectations and then simply recording the facts of nature and building theories from the results. The other extreme, the relativistic myth, holds that science is so embedded in cultural and personal biases that truth can never be discovered and is, in fact, a meaningless concept. Science, in this view, does not progress; its changes merely record a series of fashions, no different from the history of dress. The intermediate, and, in my opinion, correct, view holds that external (or objective) truth exists and that science can find it, but that the routes to correct answers are complex and tortuous. The human or subjective side of science can

Illustrations by John Craig

both aid or retard the search for objective truth. We must distinguish between the helpful aspects (leading to creative insight that can be tested and affirmed in nature) and the regrettably far more common stiflers of progress based upon prejudice and superstition. If we can learn to identify and eliminate these harmful aspects of subjectivism, then we may make the path of scientific advance a bit more straight.

For the best illustration of this negative side of subjectivism, and also of the claim that it can be recognized and eliminated, one should choose a subject where the ratio of empirical data available to social importance of claims advanced is so low that the history of changing ideas proposed by scientists is more a tale of altered social contexts than of accumulating information. The history of views on the measurement of human intelligence provides an excellent example because the social stakes are so high and the objective information so meager.

I have made such a study by reanalyzing, with modern statistical methods, the classical sets of data used to advance claims for inequality of human races, social classes, or national groups based either on anatomical measurements of the brain and skull or on the results of so-called tests of intelligence (*The Mismeasure of Man,* W. W. Norton, 1981). Since scientists are now reasonably confident that the evolutionary and genetic distances between human races are far too small to permit the kind of differences in average intellectual capacity previously accepted by most white male researchers (see final section on the correction of socially conditioned error), this subject reveals particularly good examples of both the negative side of subjectivism and of strategies for recognizing and correcting it. This perennially contentious subject also has much to teach us about the political misuse of science, since pseudoscientific claims for inherent inequality of races serve as such a convenient argument for slavery, colonialism, racism, and exploitation.

The negative side of subjectivism can be classifield into three main divisions: fraud (or outright and conscious fabrication of data), finagling (or unconscious fiddling with genuine data to push them toward preconceived conclusions), and disabling bias (or interpretation of genuine data in peculiar and illogical ways that preserve prior expectations). Of the three, fraud has received most attention, if only because tales of human frailty have a strong appeal. However, it is the least interesting among the three categories because it teaches us nothing about the strong social and personal biases that must influence science and that confute the myth of objectivity. Perpetrators of fraud know exactly what they are doing and are not fooled by illusions of objectivity (indeed, they trade upon the illusion in attempts to avoid detection). But finaglers believe that they are searching for truth; their curious manipulations record the unconscious power of prejudice and dramatically demonstrate the importance of continual vigilance and scrutiny. The following examples, all drawn from well-known studies in the quantitative assessment of intelligence, are also instructive because they all involve the very subject that the myths of science proclaim most invulnerable to disabling prejudice—supposedly objective numbers and measurement.

Fraud

Among the (unfortunately) numerous cases of scientific fraud established during the past few years, none was more spectacular or more unfortunate in its effects than the case of Sir Cyril Burt, dean of British intelligence testers and a leading theorist of hereditarianism. In a long career spent first as head of testing for London's schools and then as professor in Britain's leading chair of psychology, Burt had dominated his field. From his first paper in 1909 to his last (posthumous) effort of 1972, he relentlessly pursued the theme of hereditarianism, arguing that intelligence is a single, fixed, biological entity and that social classes differed in their average endowment, poor people being (in general) stupider. Millions of lives were influenced (and restricted) by educational policies based upon his ideas.

Identical twins separated at birth and reared apart in different social circumstances provide the best potential test for hereditarian claims. Since identical twins form from the same fertilized egg, they share all the same genes. If intelligence is a biological "thing" determined by genetic makeup, then identical twins should display similar intelligence, whatever their upbringing. But if environment strongly influences intellectual performance, then identical twins reared differently might be no more alike than any unrelated pair of children (of same age and sex) reared in similar environments.

Thus, identical twins reared apart have long been recognized as the ideal test case for arguments about the inheritance of intelligence. But such examples are hard to find (identical twins are rare enough; how often is a pair separated at birth and raised differently?). No one had succeeded in locating enough cases for a convincing study until Cyril Burt, in the 1950s, published data on more than 50 pairs. He found a high correlation between twins raised separately and unaware of their sibling's existence—and these results confirmed the hereditarian doctrine that he had advocated for fifty years.

Burt's results went unchallenged for nearly twenty years until Leon Kamin, an American psychologist, examined Burt's figures and found the following impossible inconsistency. Burt had published several papers, each supposedly improving his results by adding more pairs of twins. Yet Kamin found that the measure of similarity in intelligence between twins—the correlation coefficient—remained unchanged to the third decimal place as Burt increased his sample from fewer than twenty to more than fifty pairs. (One might think, naively, that better numerical agreement means more precision and firmer results, but numbers can also be too good to be believed. Statistical studies should yield approximately similar, not identical, results in repeated trials. If someone tells me he flipped a coin 100 times on 50 different occasions and got exactly 50 heads each time, I will be virtually sure that he made up his data.)

Burt's supporters rallied behind his memory and developed improbable rationales to explain Kamin's observation. But the dam broke in 1976 when Oliver Gillie, a medical reporter for the London *Sunday Times*, investigated fully and found an unmistakable pattern of fraud on

FRAUD

a much larger scale than Kamin had proposed. Burt had invented two nonexistent coauthors for his papers; he had written letters to his own journal under several pseudonyms to support his own theories; he had, apparently, invented the twin data entirely, not merely "improved" upon some real information. Finally, a full scholarly biography of Burt, written by a man who began as a supporter and admirer (*Cyril Burt, Psychologist*, by L. S. Hearnshaw, 1979), conclusively established that Burt had perpetrated at least four extensive and significant frauds during the late stages of his career.

One may accept Hearnshaw's argument that Burt became mentally ill and therefore approach him with compassion, but such legitimate personal mitigation does not answer the really troubling question. Burt's frauds were neither difficult to detect nor published by a nobody in a journal read by no one. Burt was Britain's leading statistical psychologist, and his works were widely read, discussed, and implemented as public policy. Yet his fraud was blatant and easy to uncover once someone asked the right question. The acceptance of his phony results for twenty years can only mean that Burt was saying what scientists expected and wanted to hear. Scientists are often insufficiently critical, especially when published results support expected conclusions.

Burt's is not the only case of fraud in the annals of hereditarianism. I discovered a pictorial example in 1979. In 1912 H. H. Goddard published a famous study on the supposed inheritance of immoral behavior and feeblemindedness in a New Jersey family, the Kallikaks. His story became standard textbook material for several generations of students and was an influential weapon in the sometimes successful campaign for institutionalization and forced sterilization of so-called mental defectives. Goddard presented no real evidence for widespread mental deficiency in the Kallikaks beyond the impressions of nonprofessional interviewers. I found that he had enhanced the impression of evil and imbecility by touching up facial features of all photos showing Kallikaks living freely in society where they could reproduce and perpetuate their kind. But all photos of Kallikaks confined in institutions—Goddard's proposed solution for them—are unaltered, and the individuals appear normal and non-sinister. Again, this crude fabrication went undetected for more than sixty years.

Finagling

Despite all the revelations of recent years true and conscious fraud appears to be rare in science—if only because the intellectual penalties of detection (ruined careers) are so severe. Much more common, and therefore in the aggregate more troubling, is unconscious finagling, or manipulation of data to support a preferred hypothesis. Modern canons of statistical and experimental procedure have limited the realm of such finagling considerably, but some extraordinary past examples indicate how remarkably far a basically good scientist can go without his own conscious knowledge. I uncovered the following case in 1978. The finagler published all his raw data and stated all his procedures in detail.

FINAGLING
BLACK
OZ 14 12 10 8 6 4 2
SHOT
WHITE
MUSTARD
SEED

Therefore, it must be assumed that he was unaware of what he was doing. Perpetrators of fraud try to cover their tracks.

Samuel George Morton, a Philadelphia physician, gathered the world's finest collection of human skulls during the 1830s. He used it to establish the relative "worth" of human races according to the size of their brains, and he "found" exactly what he expected and what most successful white males of the time assumed—a hierarchy with whites at the top (average brain size of 87 cubic inches), American Indians in the middle (at 82 cubic inches), and blacks at the bottom (78 cubic inches). Brain size is no guide to intelligence among normal individuals within a species (Anatole France at just over 1,000 cc did about as well as Turgenev with over 2,000), but Morton's results must give us pause because a century of subsequent work has shown quite conclusively that, when appropriate corrections are made for body size and sex, no substantial differences of brain size can be found among human groups. How then did Morton achieve his expected results?

Morton measured brain size by filling each cranium with white mustard seed (he later switched to lead shot, and this is an important part of the story). He then poured the seed out into a graduated cylinder and measured its volume. The method sounds simple and objective, not subject to finagling.

Considering Indians first, Morton's low value arose from the unequal representation of different groups in his sample. Some Indians, Iroquois for example, had brains larger than the white average, but Morton's collection included few Iroquois skulls. However, more than 25% of his skulls came from very short-statured people of the old Inca empire. Short people have relatively small heads, and corrections for body size must be made. But Morton made no such corrections, and the large Inca subsample brought the Indian average well below that of the white.

One might argue that, in those embryonic days of statistics, Morton just had not realized the need for corrections and that the error had nothing to do with his hopes for a low Indian average. Yet Morton baldly states (the basis of my assumption that he could not have been a conscious fraud) that he had eliminated about 25% of his Caucasian sample because a subgroup (Asian Indians) had such small skulls. In other words he made the correction for whites (where he wanted to find a high average), but left the same number of small-skulled people in his Indian sample. If the Asian Indians are put back into the Caucasian sample (or the Incas taken out of the American Indian sample) and a few other errors corrected as well (all made by Morton in a direction favorable to his hypothesis), the white and Indian averages become indistinguishable.

But what about blacks? Morton originally used mustard seed to measure all his skulls. But the seeds were light and uneven in size; they did not pack the skull well, and Morton found that he could not get an accurate and repeatable measure. So he switched to heavy, uniform lead shot and remeasured all his skulls. Thus, he changed from a rough and imprecise measure that could be unconsciously finagled to a precise test that gave consistent results.

Morton's own data for the same skulls, measured by seed and shot, can be compared. Shot always gives a larger volume because the heavy lead spheres fill the skull completely. The difference between white skulls measured by seed and the same skulls measured by shot averages 1.8 cubic inches. But the average difference for black skulls is 5.4 cubic inches. In other words, when Morton used an imprecise method subject to unconscious finagling based on prior expectations, he systematically undermeasured black skulls (relative to white) by more than 3½ cubic inches. It may seem surprising that something so apparently simple and objective as pouring seed into a skull and pouring it out again could be fudged, but the influence of prior expectation can be subtle and pervasive. We do not know how Morton obtained his systematically inaccurate result, but several plausible scenarios come to mind. For example, he might take a large black skull, pour the seed in, shake it gently (thus obtaining minimal settling) and pour it out again. Then he might take a disturbingly small white skull, pour the seed in, shake vigorously, push hard with his thumb on the foramen magnum (entrance hole for the seed), and pour more into the space left by settling. It's easy to do without even realizing what you are doing.

Correction of this difference halves the disparity between white and black averages. The remaining difference arises almost entirely because Morton's black sample happened to contain a higher percentage of female skulls and women's bodies are, on average, smaller than men's. Again, Morton did not make a correction that would have threatened his hypothesis. When the correction is made, the white and black averages become indistinguishable. In other words, Morton's own data, properly corrected, show no differences in skull size among races.

Disabling bias

In this most common category the data themselves are accurately recorded and reported, but interpretations are so skewed by prior expectation that results record personal or social bias rather than nature's message. The work of Paul Broca, famous medical anatomist and leading statistical anthropologist of France in the mid-19th century, provides a series of striking examples. Broca was meticulously careful about data. I have examined his work closely and never found any suggestion that the actual measures he obtained were in any way influenced, as were Morton's, by his expected hypothesis. But Broca was committed to the conventional racist and sexist orderings of whites better than blacks and males better than females, and his interpretations of his own sound numbers are often almost ludicrously forced in the interest of his favored theories. I describe just one example below, again involving the use (or non-use) of corrections.

When in 1861 a colleague argued in print that brain size could not reflect intelligence because Germans had bigger brains than Frenchmen, Broca sternly met this challenge to his nation's honor and superiority. He made—and quite reasonably—every legitimate correction to his colleague's raw data. He found that Frenchmen in the sample had been

DISABLING BIAS
MALE IQ
FEMALE IQ
fem.
male
METRE

significantly older, smaller, and subject to death by degenerative disease than the Germans. (Brain size does decline with age and during prolonged disease, and small men have relatively small brains.) The French average surpassed the German in his corrected data.

Broca then performed an extensive study on brains of men and women. Although the women were also (and to a substantially larger degree) smaller and older on average than the men, Broca in this case made no corrections but simply reported the raw data as evidence for male superiority. In fact, he even explicitly argued that he did not need to bother with corrections because everyone knew that women were less intelligent than men (the notion that he was supposedly trying to demonstrate, not use as an assumed basis for interpretation). Broca wrote in 1861:

> We might ask if the small size of the female brain depends exclusively upon the small size of her body. Tiedemann has proposed this explanation. But we must not forget that women are, on the average, a little less intelligent than men. . . . We are therefore permitted to suppose that the relatively small size of the female brain depends in part upon her physical inferiority and in part upon her intellectual inferiority.

When Léonce Manouvrier, Broca's statistical colleague, made the appropriate corrections a few years later, he found little difference worth mentioning—if anything, a slight advantage for females.

Grounds for optimism

These stories—and they are average tales from large sets of similar cases, not unique and egregious errors—might seem to deliver a pessimistic message about science and the possibility of adequate knowledge about nature. But I read the "bottom line" differently and in an optimistic manner. Science can never be divorced from subjective elements since it is done by passionate human beings, not robots. Subjective elements (part of the context of discovery) can lead to correct conclusions (context of justification). The negative side of subjectivity—distortion of data in the categories designated as fraud, finagling, and disabling bias—does harm science and, as in the cases discussed here, can lead to serious social misuse. But such distortions can be recognized and exposed. Indeed, all cases discussed here have been so corrected. Moreover, by vigilance, scrutiny, and honest recognition of science's social context, we may identify the sources of distortion and work to eliminate them.

All the examples discussed above lie in an area where, as stated previously, the ratio of data to social importance has been so low that conclusions tend to reflect social prejudice. Yet, even in this subject of biological views on human races, one can trace a series of retreats from an original hardline claim for separate and unequal races toward the current recognition that equality of human racial and social groups is a contingent fact of history. Much of this change must be attributed to altering social contexts—to sensitivity awakened by the genocidal use of racism in Nazi Germany, for example. But developing scientific knowledge also made a strong contribution to the modern view. Good science does drive out bad.

Before Darwin, for example, many scientists claimed that human races had been created as separate species: whites on top, blacks on the bottom. The concept of evolution made this hardline version of racism impossible, but many scientists continued to argue that human races had diverged so long ago that they had become substantially different, whites again better than blacks. Recent paleontological discoveries have shown that the division of modern humans into races occurred only with the evolution of our own species, *Homo sapiens,* and not longer than a few hundreds of thousands of years ago at most (and probably a good deal less)—a geological nothing in the fullness of time. Racism has not disappeared (even among scientists), but this old and favorite argument based on time of origin has been dismantled by recent scientific discovery.

As another example, the racist tradition long held that humans evolved in Asia and degenerated in Africa. But discovery of the australopithecines in eastern and southern Africa, starting in 1924 and accelerating ever since, has demonstrated an African home for humanity. The racist line then shifted to an intermediate position: humans evolved from apes in Africa but made the transition to intelligence in Asia in the form of *Homo erectus* (so-called Java or Peking Man), the first species to migrate from Africa. But more recent (and overwhelming) evidence now shows that *Homo erectus* also evolved in Africa. Again, good science drives out bad.

Truth and honesty have their undeniable ethical value. But their practical worth is also inestimable. The world is such a complex and confusing place. Accuracy represents our only hope for coping.

Science Year in Review

Science Year in Review Contents

Anthropology

Human evolution. In October 1984 the National Geographic Society and the National Museums of Kenya jointly announced the discovery of the most complete skeleton of an early human ancestor ever found. The bones were discovered on the western side of Lake Turkana in northern Kenya by an expedition led jointly by Alan Walker of Johns Hopkins University Medical School in Baltimore, Md., and Richard Leakey, director of the National Museums of Kenya in Nairobi. Fossils of human ancestors had been found earlier on the northern and eastern sides of Lake Turkana.

The remains were those of a 12-year-old male *Homo erectus* dated 1.6 million years ago, a very early date for this form of early human. The skeleton lacked only the left arm and hand, the right arm from the elbow down, and most of both feet. It was hoped that the missing bones would be found in the summer of 1985, when excavations at the site would be resumed.

H. erectus is generally regarded by paleontologists as an ancestor of modern humans. Its remains were first found in Java and were later recovered in China near Beijing (Peking), in Europe, and in Africa. It is the first hominid species to have been found with certainty outside of Africa and the tropics, which many believe indicates increased ability to adapt culturally to a wider range of environments.

The stone tools usually found with *H. erectus* are called Acheulian and consist of large, well-made stone hand axes. They contrast sharply with the cruder Oldowan choppers and other tools made by earlier hominid forms and seem to indicate a markedly increased intelligence, also attested by the greater size and complexity of *H. erectus*'s brain. *H. erectus* lived until perhaps 300,000 years ago, gradually evolving into an archaic form of *Homo sapiens* (modern man). Essentially modern forms of humans appear in the fossil record about 40,000 years ago.

The young male *H. erectus* stood about five feet six inches tall, and it is estimated he would have reached six feet if he had lived to maturity. This contradicts the assumption by some scientists that early hominids were considerably smaller than modern humans. There was no indication of disease or any other cause of death, which increased the value of the fossil to scientists. The only other *H. erectus* fossil found with extensive bones below the neck was discovered on the eastern shore of Lake Turkana. This individual had suffered from a serious bone disease in life.

The discovery of a nearly complete skeleton of a juvenile will enable researchers to answer a variety of questions about the anatomy of previously unknown parts of the skeleton, as well as offering valuable insights into the growth and development of early hominids. Before such studies could begin, however, the bones had to be cleaned of their rock matrix. As of early 1985 only the skull and lower jaw had been processed.

Even before detailed studies of the bones began, the discovery reopened discussion of the mechanisms of human evolution. While few juvenile hominid fossils had been found, those that were known were strikingly similar to more modern forms of humans, with their lighter bones and large brain cases in re-

Richard Leakey (left) in Kenya digs out a bone of a 1.6-million-year-old Homo erectus. *Other bones of this most complete skeleton of a human ancestor ever found are inspected by Alan Walker (right).*

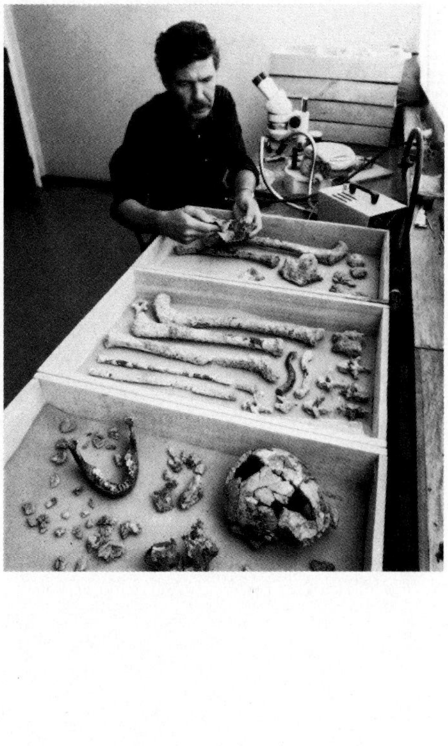

Janice Beran, Department of Physical Education, Iowa State University, Ames

Boys in northern Nigeria play "Bzir Kutli," a traditional children's game in which one player tries to walk on the backs of his or her teammates without falling off. A number of anthropologists during the past year were devoting increased attention to the study of human play.

lation to the rest of the skull. In many ways modern humans are anatomically similar to fetal apes, at least as far as the skull is concerned.

One view of human evolution, sometimes called neoteny, holds that the simplest way for the ape skull to have evolved into the modern human cranium would have been through interruption of the physical maturation of the skull before it was complete, while sexual maturation proceeded. This view of human evolution has never been popular, but the skull of the juvenile *H. erectus* appears, at first glance, to be more like the skull of an archaic *H. sapiens* than that of an adult *H. erectus*. Further analysis may support the proponents of neoteny.

Anthropological method. Two distinguishing characteristics of anthropology as an academic discipline have been its focus on all human cultures and its method of studying those cultures. If culture is defined as a set of learned and shared rules for living, the anthropologist's method for studying cultures is called participant observation—learning the rules of a culture by living in it and sharing the life-style of its members.

Other disciplines have adopted this method only rarely. Professional educators have used what has come to be called the "ethnographic method" of evaluation, but this is just another name for the long-standing practice of having an administrator or supervisor observe classes in order to evaluate the teacher's performance. As the importance of the anthropological perspective for modern-day decision making and the implementation of development projects—in other words, applied anthropology—has grown, however, so has familiarity with anthropological methods. Development agencies have come to depend on anthropologists, using the method of participant observation and other social science techniques, to aid in planning projects and their implementation, especially in third world countries.

For example, the World Bank recently announced a pilot project using participant observers to evaluate two projects aimed at improving the status of urban poor. The bank was interested not only in testing participant observation as a supplement to standard statistical surveys but also in determining whether such evaluations might replace the more expensive survey technique. Preliminary results of

"Anthropologists! Anthropologists!"

Drawing by Gary Larson, reprinted by permission of Chronicle Features, San Francisco

the study were reported to be encouraging, and there was some evidence that participant observation is a better method than the statistical survey for determining the relationship of local politics to project management, assessing cultural and attitudinal factors, and generating hypotheses for further study.

The study of play. Anthropologists continued to broaden the subject matter of the discipline. An example was the Association for the Anthropological Study of Play, founded in 1973 by anthropologists and others interested in applying participant observation and other methods to the study of the ubiquitous human interest in play. This interest culminated in 1984 with the publication of *The Anthropology of Sport: An Introduction* by Kendall Blanchard and Alyce Cheska, both specialists in the field. Anthropologists take what they call a holistic view of culture; they believe that all parts of the human experience are important. Thus, play is seen not as trivial but as a vital aspect of human life, one that literally re-creates the participants.

—Charles W. McNett, Jr.

Archaeology

Major developments in archaeology during the year included the discoveries of a late Ice Age settled community in southern Chile, a Late Bronze Age iron-smelting site in Israel, and a variety of domesticated seed plants around 2,000 years old in eastern North America. New radiocarbon dates from human remains in California shed light on the time of colonization of the New World, and recent magnetic dates from the island of Crete suggest that seismic activity laid the foundation for some Greek and biblical legends.

Promising major developments to come was the announcement early in 1985 by Peruvian and University of Colorado archaeologists that exploration of a "lost city" in the Peruvian Andes would begin during the summer. The site of Gran Pajaten, in San Martín department, was built by an unknown pre-Hispanic people who occupied it between about AD 500 and 1500. Archaeologists who examined it in the summer of 1984 reported that it indicated the presence of an advanced civilization.

Earliest settlement in Chile. The rich and diverse forests of southern Chile supported the inhabitants of a late Pleistocene settlement, the earliest found in the New World. Tom Dillehay and a team of archaeologists, geologists, and other specialists from the Universidad Austral de Chile recovered a remarkable wealth of artifacts and environmental information from the site of Monte Verde, radiocarbon dated to between 12,500 and 13,000 years ago. Log and plank foundations of 12 structures were excavated, revealing a complex of rectangular dwellings made of saplings covered with animal hides. Plant remains, stone tools, food stains, and shallow heating pits were found within each small hut.

Collective cooking hearths, wooden mortars, and grinding stones were discovered in central locations outside the dwellings, indicating a high degree of community organization. About 30 m (98 ft) from the habitation complex the archaeologists discovered a wishbone-shaped structure containing several heating pits, bits of animal hide, burned seeds, and plant remains. The area around the structure contained hearths, firewood piles, stone and wooden tools, and mastodon bones. Activities such as big-game meat dressing, stone tool manufacture, and medicinal practices were apparently carried out in this area.

The ancient Monte Verdeans lived primarily on wild plants, supplemented by several varieties of meat. Nearby bogs and marshes, as well as the surrounding forest, provided most of the food plants, although remains of plants from the Pacific coast were also found. Bones of seven mastodons made up the bulk of animal remains, but the researchers also discovered the shoulder of an ancient camelid and freshwater-mollusk shells.

Eleven flaked stone tools were found at the site. Four were flaked on both of their faces, while the remainder were flaked on only one side. The bifacially flaked hand axes and choppers were found among the mastodon bones near the wishbone structure; the unifacially flaked tools were collected near the dwellings. Other stone tools included spherical smoothed or grooved bola stones and scrapers.

Research at Monte Verde provided a more complete picture than ever before of some of the earliest people in the New World. The period at the end of the Ice Age has long been thought to hold the key to understanding the occupation of North and South America, but little conclusive evidence has been found for the existence of human groups in the hemisphere prior to 11,500 years ago. Not only does Monte Verde strongly suggest that people were well established in South America 1,000 years earlier than previously believed, but the collection from the site also demonstrates the importance of wooden artifacts and plant resources, as well as a broad economic base at this early time.

Redating human skeletal remains. While work at Monte Verde was establishing a human presence in the New World before the end of the Pleistocene, new radiocarbon dating techniques showed that human remains previously thought to date to the early and middle Pleistocene are of a considerably younger age. On the basis of chemical changes that take place in amino acids after the death of an organism (racemization), amino acids from skeletons in California were dated to 70,000 and 50,000 years ago by J. L. Bada and his colleagues at the Scripps Institution. Additional dates around 23,000 and 17,000 years ago were obtained on similar material by researchers from the University of California at Los Angeles. These dates were the subject of considerable controversy during the 1970s.

The recent development of accelerator mass spectrometric (AMS) radiocarbon dating allowed Bada and colleagues at Scripps and Oxford University to redate the original amino acids from the skeletal material. This technique requires only a few milligrams of sample for dating, as opposed to the gram quantities required by conventional methods.

One of the problems with the original acid racemization dates was their calibration on a radiocarbon date of 17,150 years from a skeleton found in Laguna Beach, Calif. A new date of 5,100 years ago was obtained from amino acids of this same skeleton, using the AMS method. Recalibration with this date provided new racemization dates of around 14,000 years on two skeletons (Sunnyvale and Del Mar), previously dated older than 50,000 years. AMS dating of these skeletons suggests even younger ages, around 5,500 to 6,500 years. Recalibrated racemization dates on the Los Angeles Man skeleton, previously dated older than 23,000 years, suggest that 8,000 years is more accurate. These results indicate that Paleo-Indian skeletons from California are of Holocene, not middle or late Pleistocene ages.

Archaeomagnetic dating. A relatively new method of dating archaeological deposits was used to check the association of volcanic activity in the eastern Mediterranean with certain ancient legends. Early excavations in the area led to speculation that the highly advanced Minoan civilization formed the basis for Homeric legends and Plato's description of Atlantis. Later excavations suggested that the explosive eruption of Thera, on the island of Santorini, 120 km (75 mi) north of Crete, created widespread devastation and the collapse of Minoan culture. Aspects of this event were consistent with the Atlantis legend and could possibly account for the plagues and other events associated with Moses and the Exodus of the Israelites from Egypt. The problem was that no dating method was sufficiently accurate to determine the exact date of the eruption or of the destruction levels on Crete. Without evidence that the events corresponded in time, such suggestions remain highly speculative.

Various dates between 1700 and 1200 BC have been determined for the end of Minoan civilization, and the Santorini eruption has been dated to 1626 BC and 1390 BC. The Exodus is believed to have occurred 480 years before Solomon's fourth year, or around 1446 BC, although 1250 BC has also been suggested. In an attempt to clarify the situation, archaeomagnetic dating of fired materials on Crete and Santorini was reported by W. S. Downey of the University of Naples, Italy, and D. H. Tarling of the University of Newcastle upon Tyne, England.

Archaeomagnetism is based on the fact that fired materials retain a magnetic reference to the direction and intensity of the geomagnetic field at the time of their cooling. Materials heated to a certain temperature "lock" onto and keep a record of the Earth's magnetic field as they cool. Since the direction and intensity of the field change by increments, relative time of firing can be established through comparisons with previous geomagnetic fields. In areas where changes in the field have been accurately radiocarbon dated, the absolute date of firing can be established. However, because few good records have been established for the Minoan period, only relative dates could be determined. This posed no great problem because synchronicity of the eruption and destruction events was being tested, not necessarily the absolute time of the events. Materials magnetized at the same time in the same region usually have identical directions and nearly identical intensities; archaeomagnetic samples from Crete should have records similar to those from Santorini if the events were related in time.

Mud bricks burned during the destruction of Minoan sites on Crete were sampled, providing a data collection of nearly 300 samples from nine sites. Samples from Santorini were taken from the Mi-

Wishbone-shaped foundation 3.9 meters long and 3 meters wide was discovered at Monte Verde in Chile. Some 12,500 to 13,000 years old, it probably served as a site for dressing meat and for medicinal practices.

Tom D. Dillehay, University of Kentucky, Lexington

Photos, (left) George E. Stuart, (right) Wilbur E. Garrett—© 1984 National Geographic Society

Mayan burial chamber in northeastern Guatemala (right) contains wall paintings and ceramic pots. A twist-lid jug (above) is the most valuable find in the tomb, which has been dated to between AD *450 and 500.*

noan city of Akrotiri, as well as pumice and surge deposits from several other locations on the island. The Cretan archaeological samples and the non-archaeological samples from Santorini proved to be highly stable and allowed precise directional determinations. Archaeological samples from Akrotiri were not as stable, and directions were not precise.

Identical directions and intensities were found on the sites from eastern Crete. Directions and intensities from the central portion of the island were significantly different from those of the east but were mutually consistent. This was unexpected because it had always been assumed that the final destruction levels on Crete were of the same age. Even more surprising were results showing that the directions and intensities found on central Crete were nearly identical to those from pumice deposits on Santorini, while those from eastern Crete were similar to tephra ejected during the later explosive eruption. A time gap of 10 to 30 years existed between the burning destruction of sites in central Crete and those on the eastern portion of the island.

The island of Santorini would have been uninhabitable after the deep ashfall of the initial eruption, but only about five centimeters (two inches) of ash are believed to have fallen on Crete. Downey and Tarling concluded that destruction on central Crete was almost certainly caused by earthquakes associated with the first eruption. The city of Knossos was the only central Cretan site to be reoccupied after this event. Some 10 to 30 years later, percussive air waves, sea waves, and earthquakes accompanied the violent explosion of Thera, causing widespread destruction on eastern Crete and the collapse of Minoan civilization. Dense ash clouds from the eruption may have destroyed crops as far away as Egypt, and severe temblors probably were felt throughout the region. Because no specific dates may be assigned to these events, their bearing on the Atlantis and Exodus legends remains unclear. However, Plato's descriptions of the destruction of Atlantis are consistent with an explosive eruption.

Late Bronze Age iron smelting. Another exciting discovery was reported by Harold Liebowitz and Robert Folk of the University of Texas at Austin. Several years of excavations at Tel Yin'am, Israel, shed new light on the origins of iron smelting in the Near East. Tradition gives the Hittites a monopoly on ironworking in the region during the Late Bronze Age, although there is little archaeological evidence to support such a claim. For example, bronze weapons far outnumber iron ones in Hittite sites in Asia Minor, and such iron as has been found is inferior to that from sites farther south. In addition, a possible iron-smelting installation dating to the 15th century BC has been found in the Bekaa Valley of Lebanon, and another appears to have been in operation during the 13th century BC at Tel Yin'am, possibly 50 years before the southward migration of Hittites into Palestine.

The ironworking installation at Tel Yin'am is located in the northwest corner of a large public structure built early in the 13th century BC. The northwest storeroom of the structure was reused as an iron smelter around the end of the century. The walls were a superstructure of mud brick built on a basalt slab foundation; the floor was cobblestone. About 1.5 m (5 ft) of industrial debris, consisting of poorly preserved semispherical smelting furnaces and vitrified slag deposits, covered the floor. The slag consisted of 8–9% iron oxide, layers of plant remains consumed as fuel, chunks of red ochre, and pottery. The absence of copper in the slag and the presence of ochre-smeared, refired ceramic vessels, hematite ore, smelting furnaces, and color changes indicating a reducing atmosphere all support an iron-smelting interpretation.

The authors do not believe the smelter was successful. Locally available ore is not sufficiently rich in iron to be worthwhile. The inhabitants of the site would have been unable to produce quantities of metallic iron from this ore, and so the smelter was used only a few times. Soon after the installation was abandoned, its parent structure was destroyed.

Early domesticates in eastern North America. About 2,000 years ago garden plots in eastern North America contained four species of domesticated plants: the bottle gourd, squash, marsh elder, and sunflower. Tobacco, maygrass, knotweed, little barley, and goosefoot were thought to have been important species, but they had not been morphologically altered to the same extent as domesticated species. Recently, however, analysis of floral remains from Russell Cave, Alabama, showed that goosefoot (*Chenopodium*) was indeed present as a domesticate at least 2,000 years ago and may be added to the list.

More than 50,000 carbonized goosefoot fruits were recovered from an uncataloged basket from Russell Cave collections in the National Museum of Natural History, Washington, D.C. The fruits were a part of a mixed botanical assemblage containing wood, leaf, grass stem, hickory nut, and basket fragments. They represent the earliest known evidence for the deliberate storage of goosefoot fruit and seeds in the Eastern Woodlands. Deliberate storage and subsequent spring planting of seed stocks mark the beginning of domestication.

The process of domestication produces changes in the seed crop that constitute the adaptive response to the artificial environment created by human intervention in the plant's life cycle. These changes include the loss of natural shatter mechanisms, uniform maturation of fruit, fruit ripening at the ends of stalks, and enlarged food reserves surrounding the seed. They result in more nutritious and more easily gathered food resources for human beings.

The Russell Cave assemblage did not exhibit many of these changes because of its poor state of preservation; evidence for domestication was limited to indications of a reduction in dormancy and increased seedling vitality. Reduction in dormancy was shown by a thin outer seed coat approximately one quarter that of wild species. Increased seedling vitality was exhibited in increased food supplies within the seed, reflecting selective pressure for seedling viability. These goosefoot seeds would have sprouted earlier and would have had internal reserves to spur growth.

Whether this early domesticate was developed in the Eastern Woodlands or introduced from Mexico is not known. Domesticated *Chenopodium* species existed in both Mexico and South America prior to 2,000 years ago, but Russell Cave contains the earliest evidence for domestication in North America.

See also Feature Article: DIVING FOR THE ANCIENT PAST.

—James D. Wilde

Architecture and civil engineering

Architecture. The design of new buildings during the past year continued to depart from the slab shapes and glass boxes that had dominated architecture

PPG Place in Pittsburgh, Pennsylvania, consists of a 40-story tower, one 14-story office building, and four 6-story buildings grouped around a tiled plaza. The walls of the buildings are of bluish-gray reflecting glass.

Courtesy, John Burgee Architects with Philip Johnson

© R. Greg Hursley, Inc.; courtesy, E. Fay Jones Architect

Residence in Hogeye, Arkansas, was designed to blend with its site in the Ozark Mountains. Building materials include red cedar, to match the trees, and local fieldstone.

from the 1950s to the late 1970s. Throughout the world buildings began to reflect the character of the neighborhood, city, region, or country in which they were built.

The desire for architecture to express the tastes and styles of ordinary people also was becoming apparent. "For far too long, it seems to me, some planners and architects have consistently ignored the feelings and wishes of the mass of ordinary people in this country," said Charles, prince of Wales, at the 150th anniversary celebration of the Royal Institute of British Architects. "Consequently, a large number of us have developed a feeling that architects tend to design houses for the approval of fellow architects and critics, not for the tenants."

Prince Charles went on to say that "community architecture," or buildings that are designed to include the needs and desires of the people using them, has shown ordinary people that their views are worth having, and "that architects and planners do not necessarily have the monopoly of knowing best about taste, style, and planning." He added that there is a growing number of architects who will listen to ordinary people.

Although far from ordinary, the needs and desires of the people of Pittsburgh were definitely considered when architects John Burgee and Philip Johnson of New York City designed PPG Place in downtown Pittsburgh. The home of PPG Industries, a leading manufacturer of glass, chemicals, paint, and fiberglass, PPG Place is a sparkling castlelike complex of six buildings—one 40-story tower, one 14-story office building, and four 6-story office buildings—arranged around a tiled plaza. The buildings' walls consist of bluish-gray reflecting glass pleated with rectangular and triangular bays reaching from the ground to above the roofs. More than 200 towers and turrets ending in spires or pyramids cap the complex. The result is a gothic fantasyland of reflected arches and spires blending with the small shops and boxlike high rises that surround it.

Despite the size of the complex it is not threatening. The four 6-story office buildings are scaled to blend with the older existing buildings and point up-

Courtesy, McIntyre Partnership

Sea house lies on a cliff overlooking the coast of Australia near Melbourne. The architect situated the house in a pocket of still air formed by the wind as it blew up the cliff so that squalls from the ocean would not tear at the house.

ward to the two towers. The small shops in adjoining Market Square and the surrounding flattop high rises seem comfortably woven around the complex.

Office towers and urban complexes were not the only buildings that were being designed to complement their environment. For example, architect Fay Jones of Fayetteville, Ark., listened when his client said that he wanted a house designed for the hills, one that would blend in with the hickories, cedars, and oaks of the Ozarks.

To do this Jones used western red cedar framing and cedar shake roofing to blend with the color of the tree limbs. He used local fieldstone for the basement. The result was that the fieldstone base of the house resembled a natural rock outcropping of the land. The house, thus, seemed to spring from its hillside habitat.

The owner also wanted the house to be energy efficient and inexpensive. Jones, therefore, kept the design simple. The centerpiece of the living area is a woodburning stove and two floor-to-roof clay tile flues, which distribute heat more efficiently than masonry. The house is cooled by breezes entering the big hayloft window and by two ceiling fans over the living space. The fans gently push the warm air down from the ceiling during the winter.

Another architectural gem that seemed as though it grew right out of the land is the sea house of Australian architect Peter McIntyre. It hangs on the shore of Port Phillip Bay near Melbourne on the southern coast of Australia, looking as though it was formed by driftwood hurled by a huge wave against the face of Mornington cliff.

At first glance the design of the sea house appears delightfully simple. Yet, in reality it is complex. The basic house is diamond-shaped with a pitched roof. An octagonal structure with gables is added at each end. One contains the entrance and the veranda and outer living pod. Exposed interior timbers are assembled with mathematical precision.

The wind was also an integral design consideration. McIntyre positioned his house within a pocket of still air formed by the wind as it blew up the cliff. Squalls from the bay do not tear at the house, yet wooden screens on the outside of the doors and windows slide back to open the interior to the summer breezes.

The "community" approach to architecture was successfully demonstrated by the Provincial Electric Co. (PEN) telephone company building in Alkmaar, Neth. The 24,000-sq m (78,700-sq ft) complex of low-rise brick buildings with tile roofs was carefully scaled to residential proportions, and the buildings themselves were pleasingly woven among wooden walkways, landscaping, and water.

A major success of the project was the design process itself. A committee of PEN employees decided on some specific requirements for the building: open

and closed offices with movable partitions; roof gardens accessible from every floor; 60% better energy efficiency than comparable buildings; exposed service equipment to emphasize the company's technical orientation; and the use of natural construction materials throughout the structure.

Architect A. Bonnema used a low-cost system of prefabricated concrete modules on a 7.2 by 7.2-m (23.6 by 23.6-ft) grid to create strong visual variety. The result of this grid system was that the offices were designed in a number of delightfully odd shapes that followed the buildings' rooflines. The offices are warm and well lighted, and most have two exterior walls.

Civil engineering. Despite the many problems and high costs that the approximately 40 nuclear power plants under construction in the U.S. were experiencing, the St. Lucie II 802-Mw nuclear plant of the Florida Power and Light Co. was completed on time within the budget. From construction permit to operating license St. Lucie II was brought on-line in 72 months at a cost of $1.4 billion.

Begun in 1977 on Florida's Atlantic coast between Palm Beach and the John F. Kennedy Space Center, St. Lucie II stands alone as a success among a host of U.S. nuclear power projects that have been mired in financial struggles and construction delays. For those plants begun in the mid-to-late 1970s the completion schedules were stretching up to six years beyond initial estimates.

PEN telephone company complex in Alkmaar, The Netherlands (above and below), consists of low-rise brick buildings woven among wooden walkways, landscaping, and water. PEN employees made some of the decisions regarding the design of the complex.

Photos, Jan and Fridtjof Versnel

Even before the first concrete was poured, 75% of the engineering design for St. Lucie II was accomplished. The site was developed, the foundation was prepared, and most of the labor disputes were settled. A detailed management plan had been agreed upon by a tightly knit and experienced project team. The project, Florida Power and Light's fourth reactor, was constructed by the same firm, Ebasco Services, Inc., of New York City, that had built the earlier St. Lucie I.

The project team had to overcome the usual assortment of problems and delays—a hurricane, a tipped crane, problems with equipment vendors and contractors, and Nuclear Regulatory Commission (NRC) safety regulations. The biggest problem, which was an attitude that things could not be done on time, was overcome by team visits to each of the major participants to explain that Florida Power and Light was serious about meeting schedules and about penalties for not meeting deadlines.

Some of the toughest skeptics were at the NRC. Normally, the NRC's final safety analysis reviews drag on for more than 30 months. But after more than 40 face-to-face meetings in Washington, D.C., involving the company team and the NRC representatives, the commission gave St. Lucie II its regulatory sanction in nine months.

Another remarkable project recently completed was the Luling-Destrehan Bridge (renamed the Hale Boggs Bridge in July 1984) over the Mississippi River just above New Orleans. Chosen the outstanding civil engineering achievement of 1984 by the American Society of Civil Engineers, it was the first high-level, long-span, cable-stayed bridge in the U.S.

Built of high-strength, corrosion-resistant steel, the bridge was designed to survive Gulf coast hurricanes. A-frame towers soar 85 m (280 ft) above the roadway, which is 40 m (133 ft) above the navigation channel. Each tower supports 12 stays of either two or four cables, depending on position and load. A permanent socket with a conical cavity houses the end of each cable, and the void is filled with small steel balls and epoxy resin. This type of anchor has fatigue strength far greater than that of the conventional zinc sockets used in suspension bridges. Joint venturers Frankland & Lienhard/Modjeski & Masters were the designing engineers.

Location and cost make the water-distribution system project in Barrow, Alaska, an engineering curiosity. Barrow is the northernmost town in the U.S., an oil-rich but barren little community where raw sewage is left in plastic containers for collection and where melted snow is used for drinking water. Barrow's leaders decided to build an underground system that would hold water and sewage pipes, and electric and telephone lines, at an estimated cost of $349 million.

Besides 188 km (117 mi) of polyethylene pipe and 12 km (8 mi) of installed utility sections, the project includes a system to cool treated drinking water so that it does not melt the permafrost. Because of the fear of melting the permafrost, which would cause ground subsidence, all construction is done during the winter months. Crews often worked for only half-hour periods because temperatures frequently drop to −52° C (−60° F). Artifical lighting is provided for the 64 days in winter when the Sun does not rise above the Arctic Circle. If the work

St. Lucie II nuclear power plant of the Florida Power and Light Co., located on Florida's Atlantic coast, is notable among nuclear plants because it was completed on time and within the budget.

Courtesy, Ebasco Services Incorporated; photo, Rick Tango

Louisiana Department of Transportation & Development

Luling-Destrehan Bridge (later renamed the Hale Boggs Bridge) over the Mississippi River in Louisiana was named the outstanding civil engineering achievement of 1984 by the American Society of Civil Engineers.

were done in warm weather, the permafrost would thaw and turn the work sites into quagmires.

Trenches for the frame of the utility sections were blasted out of the frozen ground with dynamite after initial slices had been made with a giant chain saw. A 5-m (16-ft) cadmium-alloy blade was used to make two 4.2-m (14-ft)-deep cuts on both sides of a 2.1-m (7-ft)-wide trench.

Singapore began work on a $300 million computerized water-supply management system to compensate for a shortage of drinking water. The small island-nation of 2.4 million people imported much of its drinking water from nearby Malaysia. The new system was designed to collect urban runoff in ten ponds spread out over the eastern part of the nation's main island and then to pump the water into two large reservoirs that were under construction. The reservoirs in turn would supply a new water treatment plant. The computers would allow operators at the treatment plant to control the flow of water from the ponds to the reservoirs.

Singapore's only source of raw water is its approximately 250 cm (100 in) of annual rainfall that descends in furious tropical storms. Much of the runoff from these storms is lost. "We needed a collection and pumping system that could respond quickly to a large volume of water moving from the ponds to the reservoirs," said Brendan Harley of Camp Dresser & McKee of Boston, designer of the system.

The water distribution operation was to be performed from Badok, one of the two reservoirs. Sensors in the reservoir dam would read both the tide and reservoir levels and then signal a microprocessor in the drainpipe to open or close the valve. Operators at the central station would monitor the process by computer. According to Camp Dresser & McKee, when the system is complete, probably in 1985, Singapore may well have the most sophisticated water-management system in the world.

See also Feature Article: THE DEVELOPMENT OF SKYSCRAPERS: THE CHANGING SKYLINE.

—John Davis

Astronomy

During 1984 ground-based observations helped to interpret data previously returned from satellites by supplying optical and radio data on objects detected in other portions of the spectrum. Observing techniques were continuously improved, leading to better resolution of many objects and the detection of fainter light sources.

Instrumentation. The 15-m National New Technology Telescope came one step closer to existence when an advisory panel to the National Optical Astronomy Observatories chose a multimirror instrument concept based on four 7.5-m mirrors. The multiple mirror telescope, in operation since 1979 on Mt. Hopkins in Arizona, has demonstrated the feasibility of combining the images of several mirrors mounted in a common structure to provide the light-gathering power and image quality of a single large instrument. The major hurdle remaining to be overcome is the construction of the 7.5-m component mirrors, which must combine light weight, high rigidity, and low thermal inertia to stop image deterioration due to thermal stress. In addition the mirrors should have short focal lengths, equal to their diameters, in order to reduce the size of the telescope and its housing.

Roger Angel at the University of Arizona devised a novel technique to achieve the desired mirror characteristics. The mirrors were to be strongly honeycombed in order to reduce the required mass of glass, thus yielding good thermal properties but also achieving high rigidity. Furthermore, to reduce the great amount of glass that would have to be ground away in a conventional mirror blank to form the short focal length desired, Angel proposed that the mirrors be cast in a huge, rotating oven. The rotating molten glass would automatically assume the parabolic surface needed for an astronomical mirror. The Arizona group demonstrated the technique by making a 1.8-m prototype, and by 1985 they were in the process of constructing the large furnace needed for the 7.5-m mirrors. Plans called for the casting, grinding, and polishing of a 3.5-m mirror as a first step toward making the 7.5-m mirrors.

A particularly exciting event took place in April 1984 when repairs were made to the Solar Maximum Mission satellite by astronauts aboard the U.S. space shuttle "Challenger." This satellite, which had supplied much data on the Sun, once again became fully functional. The ability to repair scientific satellites in space, as demonstrated by the shuttle crew, promised to be of great importance to the future of space observations.

Mirror blank measuring 1.8 meters in diameter is honeycombed in order to reduce the required mass of glass. The mirror is a prototype for those that will comprise the National New Technology Telescope.

Roger Angel, University of Arizona, Tucson

New instrumental techniques utilizing charge-coupled devices (CCD's) were used to overcome the degradation of images caused by the Earth's atmosphere. Astronomers refer to this distortion of telescopic images as "seeing." It has long been known to visual observers that even during times of bad to moderate seeing, moments of particularly good clarity can occur. Unfortunately, during long exposures on faint objects these good moments are submerged in the badly formed images. CCD's are particularly useful in adding and retaining images selected during short intervals of exceptional clarity because of their high sensitivity, excellent linearity, and good integration capability.

One method of overcoming the seeing problem was described by Laird Thompson of the University of Hawaii, where an instrument called ISIS (Image Stabilizing Instrument System) was in use. The correction of image motion begins by detecting the motion of the image of the guide star near the object of interest. Since image motions of objects over a small region of the sky are correlated, the measured movement of the guide star can be used to control a mirror in the light path of the object being studied to deflect the image to a fixed position. In addition, ISIS has a fast-acting shutter that closes whenever the image becomes fuzzy due to an increase in size. The computer-controlled mirror can remove image motion greater than 0.2 arc second at frequencies up to 50 Hz. The shutter can defeat image blowup by closing as many as 20 times per second. The stabilized image is detected by a CCD. In a 20-minute exposure of the strong radio source galaxy, Cygnus A, made using the 88-in University of Hawaii telescope on Mauna Kea, the shutter remained open during 80% of the exposure and the mirror held the seeing motions steady enough to provide field star images resolved to 0.65 arc second. The image of Cygnus A, which, although bright at radio wavelengths, is only a 17th-magnitude object at visual wavelengths, revealed much clearer detail than ever obtained before. This high-resolution observation of Cygnus A confirmed earlier suggestions that the strong radio activity of this object is the result of the collision of two galaxies. Even though ISIS performed well in its first telescopic use, Thompson expected even better results as its operation was improved.

Solar system. One of the surprising finds among the many meteorites recovered in Antarctica was a meteorite so characteristic of lunar rock materials that it almost certainly originated on the Moon. A second lunar meteorite was found among the collec-

Laird A. Thompson, Institute for Astronomy, University of Hawaii

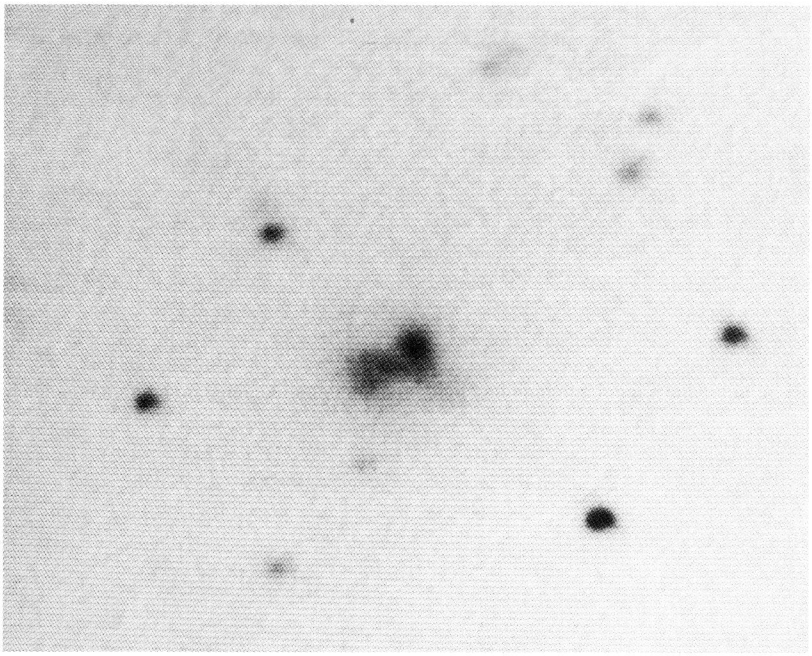

Optical image of the strong radio source galaxy Cygnus A (center) was obtained by using the Image Stabilizing Instrument System, which employs a fast-acting shutter that closes whenever an image becomes fuzzy due to an increase in size. In this 20-minute exposure Cygnus A was revealed more clearly in the optical wavelengths than it ever had been before.

tion held by the National Institute of Polar Research in Tokyo. Keizo Yanai and Hideyasu Kojima of the institute reported their find during a symposium on Antarctic meteorites held in March 1984. The small rock, only 25 g in weight, has both the appearance and chemical composition characteristic of lunar highland materials. The Yamato meteorite, as it is called, is sufficiently different from the earlier one discovered by U.S. scientists that it probably originated at a different impact-ejection site.

The frequency of meteorite falls on the Earth was studied by Ian Halliday, Alan T. Blackwell, and Arthur A. Griffin of the Herzberg Institute of Astrophysics in Canada. By using a network of 60 cameras to photograph meteors that are bright enough to result in meteorites and observing for a nine-year period, they were able to derive a good measure of the impact rates of meteorites. All told, they detected 43 events that involved meteorite masses between 0.1 and 12 kg. Correcting for cloudy conditions during the observing period, they concluded that the annual rate of meteorite falls is eight per million sq km for masses of 1 kg or more. Extrapolating this to the entire surface of the Earth gives a number in excess of 4,000 per year. If meteorites as light as 0.1 kg are included, this number rises to 19,000. Since only a dozen or so meteorites are recovered each year outside of Antarctica, it is obvious that most meteorites that strike the Earth are not discovered.

On occasion, abnormally high meteorite impact rates have occurred in the past. Kenneth Brecher of Boston University believes that he has identified a collection of bodies orbiting the Sun that give rise to some of these high incidence rates. He noted that in June 1178, the monk Gervase of Canterbury in England reported that he saw an impact on the Moon that caused the upper horn of the crescent Moon to appear to split in two and flame to shoot from it. A very young impact crater named for Giordano Bruno is located exactly where Gervase reported seeing the event that probably formed the crater. Then in June 1908 the great Tunguska meteorite impacted on the Earth. In addition, seismometers left on the Moon by the Apollo astronauts recorded in June 1975 an abnormally high meteorite impact rate of 10 to 15 per day rather than the usual one per day. Brecher concluded that these events were all connected and arose from material left over from a fragment of Comet Enke, which he thinks broke off about 1,500 years ago. The Earth intersects the orbit of these bodies every June but fortunately does not encounter the main swarm at every intersection. Brecher named the concentration of particles the "Canterbury Swarm" in honor of the early sighting by Gervase. The next time that the swarm itself should intersect the Earth-Moon system will be in June 2042.

Ever since the Italian astronomer Giuseppe Piazzi discovered the first and largest asteroid, Ceres, in 1801, the number of known asteroids has been increasing. In February 1984, the 3,000th asteroid was assigned a number by the Minor Planet Center in Cambridge, Mass. This asteroid had been discovered in 1961 by S. J. Bus using the 1.2-m Schmidt telescope at Siding Springs in Australia. To be assigned a number, an asteroid must have a well-determined orbit, good enough to ensure that the asteroid will not be lost. As of 1985 there were another 6,000 or so asteroids for which orbits were not yet well enough determined to be assigned a number. Brian Marsden, director of the Minor Planet Center, concluded that the discovery of asteroids down to the 16th magnitude in brightness was essentially complete and that to keep up the rate of discovery of recent years would require a sustained new observing program to find and follow up on the much fainter remaining asteroids.

Knowledge of the cloud-shrouded surface of the planet Venus was markedly increased by data returned by the Soviet satellites Venera 15 and 16, which were orbiting Venus with radar scanners. The Venera radars can resolve objects as small as 1.3 km in diameter compared with 25 km for earlier U.S. satellites and 2 km for Earth-bound radars. This increased resolution allowed the Soviet researchers to distinguish impact craters on the planet's surface from volcanic craters. By comparing the number of impact craters seen with that expected from the rate at which impacts should occur on Venus, they concluded that the present surface is quite young, only about 1 billion years out of the 4.5 billion years that Venus has existed. Thus the Venusian surface has been completely reworked in comparatively recent times. The method of reworking is not yet clear, however. The new data do not favor the plate tectonic processes currently taking place on the Earth. Possibly the predominant mechanism for change on Venus involves volcanic activity not unlike that which is believed to have occurred on the young Earth.

Continued examination of the data returned by Pioneer 11, Voyager 1, and Voyager 2 on the rings of Saturn led Kevin D. Pang, Charles C. Voge, Jack W. Rhoads, and Joseph M. Ajello of the Jet Propulsion Laboratory (JPL) to propose a new explanation for the formation and structure of Ring E. Rings about planets are generally thought to have been formed by collisions between satellites, the tidal disruption of a body, or material left over from the planet's formation. The JPL investigators noted that Ring E is brightest in the vicinity of the small satellite Enceladus. Furthermore, analysis of the optical properties of the ring particles show them to be tiny ice spheres about two micrometers in diameter. (One micrometer is one-millionth of a meter.) Thus the scientists concluded that the particles arose from volcanic eruptions on Enceladus that were induced by tidal heating of the satellite during its eccentric orbit about Saturn.

Stars. Continued study of the data from the now defunct Infrared Astronomical Satellite (*see* Feature Article: THE COSMOS ACCORDING TO IRAS) revealed dust rings around the stars Fomalhaut, Epsilon Eradini, and Beta Pictoris in addition to those around Vega that had been reported earlier. Following up with Earth-based observations, Bradford A. Smith of the University of Arizona and Richard J. Terrile of the JPL detected the ring about Beta Pictoris using the 2.5-m duPont telescope in Chile. They used a CCD to image the circumstellar ring, but in order to detect it they had to block the strong radiation of the parent star with an occulting disk. Even so, the ring was still several times fainter than the scattered light from the star, and only the use of the CCD allowed it to be extracted from the background light. The ring was revealed in much better detail than by IRAS and was seen to extend over 400 astronomical units from the star in a nearly edge-on aspect. (One astronomical unit is the mean distance of the Earth from the Sun, about 150 million km or 93 million mi.) Smith and Terrile constructed a mathematical model to fit their observations, leading them to the conjecture that the ring might be a planetary system undergoing formation or even material left over from a recently formed planetary system. Since IRAS found more than 40 stars with excess infrared light, which is characteristic of circumstellar rings, there appears to be a future in studying these phenomena.

During the mid-1960s George Van Biesbroeck of the University of Chicago catalogued a number of faint red stars that were relatively close to the Sun. Work done at the U.S. Naval Observatory, announced in 1983, indicated that two of his stars, Van Biesbroeck 8 and 10, had small motions suggestive of unseen companions. Donald W. McCarthy and Frank J. Low at the University of Arizona and Ronald G. Probst of the U.S. National Optical Astronomy Observatories observed the stars in the infrared with the 4-m telescope on Kitt Peak. Using the technique of speckle interferometry to overcome image distortion by the atmosphere, they found the companion to Van Biesbroeck 8. It was a large ball of hydrogen and helium, glowing with a temperature of 1,400 K, with a mass several dozen times that of Jupiter. (On the Kelvin scale zero is −273.15° C.) It appeared to be a "brown dwarf," a body too large to collapse to a planet and too small to begin hydrogen fusion and become a normal star. Brown dwarfs were only a theoretical conjecture until this observation. No companion was seen with Van Biesbroeck 10, implying that the unseen body causing its motion might in fact be a Jupiter-sized planet, too faint to be seen by this technique.

Albert D. Grauer of the University of Arkansas and Howard E. Bond of Louisiana State University discovered the first central star in a planetary nebula to show oscillations. The star illuminates the nebula designated Kohoutek 1-16. Because the period of the oscillations, detected by brightness changes in the star, should change as the star cools and shrinks, this object will provide the first direct observational test of the theory of the evolution of the hot, dense remnant stars in planetary nebulas.

Galactic astronomy. The most distant star yet known to belong to our Galaxy was discovered accidentally by Bruce Margon at the University of Washington, Marc Aaronson and James Liebert at the University of Arizona, and David Monet at Mt. Wilson Observatory in California while searching for optical counterparts of X-ray sources discovered by NASA's satellite "Einstein Observatory." The resolution and, therefore, the positional accuracy of the X-ray observations required that spectra of all visible objects within the error range of the X-ray source positions be obtained. For one source in the constellation of Libra the researchers obtained the spectrum of a faint red star, which, while not the X-ray object itself, turned out to be a carbon star. Stars of this kind are intrinsically bright, and so, since this one appears so dim, it must be at a distance of nearly 400,000 light-years. (One light-year equals 9,460,500,000,000 km or 5,878,000,000,000 mi.) Because the direction of Libra is away from the plane of the Galaxy, little obscuration by interstellar dust is expected, and the faintness of the star is attributed to its great distance alone. The star has a radial velocity (speed in the line of sight) of only 75 km per second, just what might be expected for a distant star loosely bound to the Galaxy. If more such objects can be discovered, a statistical analysis of their radial velocities can be used to find the mass of our Galaxy.

Studies of the Horsehead Nebula in optical and near-infrared wavelengths by Bo Reipurth of the Copenhagen University Observatory and Patrice Bouchet of the European Southern Observatory, using the 3.6-m telescope of Bouchet's observatory in Chile, led the two investigators to propose that the nebula may be on its way to becoming a Bok globule. These are small, dense clouds of interstellar matter that are thought to collapse to form a star. According to Reipurth and Bouchet, the bright nearby star Sigma Orionis is stripping away most of the matter from the nebula by means of its intense radiation. They propose that the remaining concentrated material will become a Bok globule.

Gamma rays, the most energetic form of radiation detected by satellite observations, were first believed to be generated near exotic objects such as black holes or neutron stars. An Anglo-Australian observing team, led by P. A. Riley of the University of Durham in England and by Richard Manchester of the Commonwealth Scientific and Industrial Research Organization in Australia, correlated the positions of molecular clouds with Southern Hemisphere gamma-ray sources found by the European COS-B satellite. The gamma rays are generated by cosmic rays interacting with gas in the molecular clouds. These findings supported a similar correlation of gamma-ray sources with molecular clouds in the Northern Hemisphere; they also strengthened the idea that most gamma rays originate from cosmic-ray interactions.

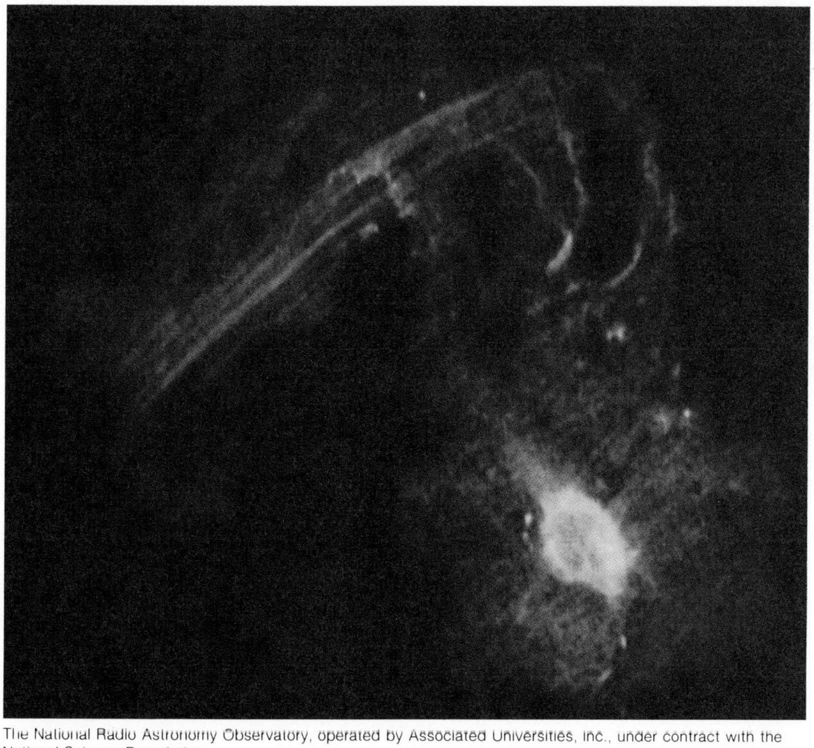

Sagittarius A at the center of the Milky Way Galaxy is revealed by the Very Large Array radio telescope in New Mexico. Among the most interesting findings at this powerful radio source is an arc of radio filaments (top of picture) that are perpendicular to the galactic plane, parallel to one another, and regular and unbroken in their appearance.

The National Radio Astronomy Observatory, operated by Associated Universities, Inc., under contract with the National Science Foundation

Image obtained at Palomar Observatory by Matthew A. Malkan, University of California, Los Angeles

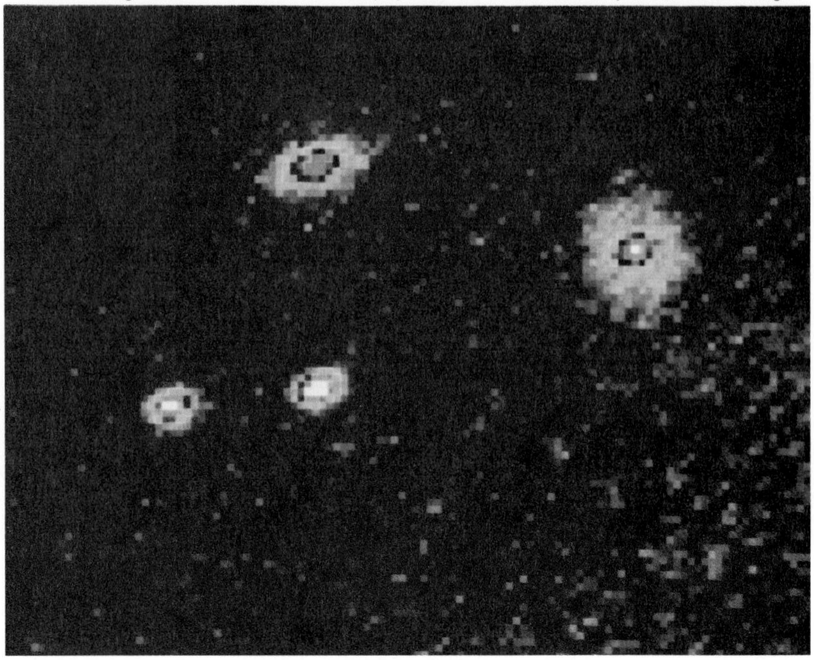

Quasar (right), imaged by an electronic camera on the 1.5-meter telescope at Palomar Observatory, appears as a bright center surrounded by faint light that is characteristic of spiral galaxies.

Farhad Yusef-Zadeh and Don R. Chance of Columbia University and Mark R. Morris of the University of California at Los Angeles used the Very Large Array (VLA) in New Mexico to observe Sagittarius A at the center of the Milky Way Galaxy in 20-cm radio waves. They found an arc of continuous filaments that seem to lie along magnetic field lines associated with the galactic center. The magnetic field appears to be a dipole field similar to fields associated with the Earth, the Sun, and some planets, but very much stronger. The arcs were overlooked in earlier work by others because radio interferometer measurements of the region were interpreted as arising from discrete sources excited by hot young stars near the galactic center. But when the researchers used two different spacings of the component telescopes of the VLA, the continuity of the arc became obvious.

Extragalactic astronomy. Phillip K. Kronberg of the University of Toronto and Richard A. Bramek at the National Radio Astronomy Observatory, Socorro, N.M., for several years used the VLA to observe the galaxy Messier 82. M82 is a close active galaxy, that is, one that shows an abnormally high energy output from its nucleus. Observing at the six-centimeter wavelength, they found that M82 contains about 40 bright, discrete radio sources that cannot be identified with optical sources because of the strong obscuration in the galaxy. The sources are all decaying in brightness, much like supernova remnants observed in our own Galaxy. Assuming that the sources result from supernovas associated with a burst of star formation in the galaxy nucleus, Kronberg and Bramek determined that the energy released would be enough to account for the excess infrared radiation of M82, about equal to ten billion Suns. They planned to continue their observations in the hope of finding new supernovas as they form, at an expected rate of one every several years.

Among the more mysterious objects found by IRAS were the blank-field sources, regions of strong infrared emission that could not be identified with optical counterparts on the Palomar Sky Survey photographs. Marc Aaronson of the University of Arizona and Edward W. Olszewski of Canada's Dominion Astrophysical Observatory used the four-meter telescope at Cerro Tololo, Chile, to identify four of these sources as galaxies with infrared luminosities from 35 to 250 times their blue-light luminosities. The most extreme source, designated 0422+009, has an infrared luminosity more than 30 times that of M82.

An examination of 24 quasars that emit X-rays provided additional evidence that quasars are superluminous nuclei of spiral galaxies. Red images taken with an electronic camera on the 1.5-m telescope on Palomar Mountain in California revealed that all 15 of the quasars with redshifts of 0.4 or less are surrounded by faint light that is characteristic of spiral galaxies. The quasars with larger redshifts were unresolved, their surrounding matter presumably below the limit of detection because of their greater distances.

—W. M. Protheroe

See also Feature Articles: THE INFLATIONARY UNIVERSE; PERIODIC COMET SHOWERS: THE MAINSPRING OF EVOLUTION?

Chemistry

Major accomplishments of the past year in chemistry include the isolation or synthesis of antibiotics, pheromones, toxins, repellents, and other biologically important substances. Investigators improved their understanding of the way electrons are transferred within molecules during certain types of reactions and contributed to speculations about the ultimate fate of the universe. Among other highlights were reports on an artificial-gill material based on the active principle of hemoglobin, polymer fibers many times stronger than steel, and the synthesis of artificial jade.

Inorganic chemistry

Inorganic chemistry continued to move ahead on multiple fronts, making significant fundamental contributions in the fields of main-group-element (*sp*-element) chemistry, transition-metal (*d*-element) chemistry, and lanthanide- and actinide-metal (*f*-element) chemistry. The *s, p, d,* and *f* notations refer to the valence electrons that participate in bonding when these elements form compounds and account for the positions of the elements in the periodic table. In at least three specific areas of research, bioinorganic chemistry, organometallic chemistry, and solid-state chemistry, inorganic chemists reported major breakthroughs.

Long-range electron transfer in proteins. In the 1950s Henry Taube, then at the University of Chicago, showed that electron transfer reactions take place by either an inner-sphere or an outer-sphere mechanism (see *1985 Yearbook of Science and the Future* Year in Review: CHEMISTRY: *Inorganic chemistry*). Furthermore, it was demonstrated that electron transfer from chromium(II) (a chromium atom deficient two electrons, Cr^{2+}) to cobalt(III) (Co^{3+}) in the compound (1) can occur over some distance provided that the bridging group contains multiple bonds, which allow for electron flow. If the bridge is saturated ($-CH_2-CH_2-$), then it insulates electron flow. In spite of research on well-defined inorganic metal complexes of this type, it was long suspected that, in some of the more complicated natural metalloproteins, biological electron transfer reactions take place over considerable distances even though the polypeptides between the two sites are saturated molecules.

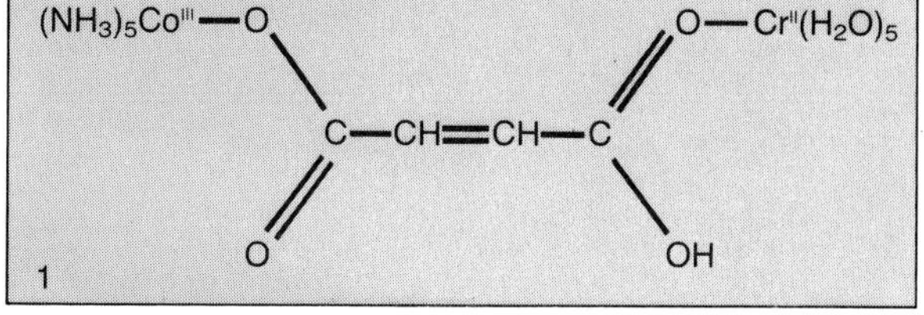

1

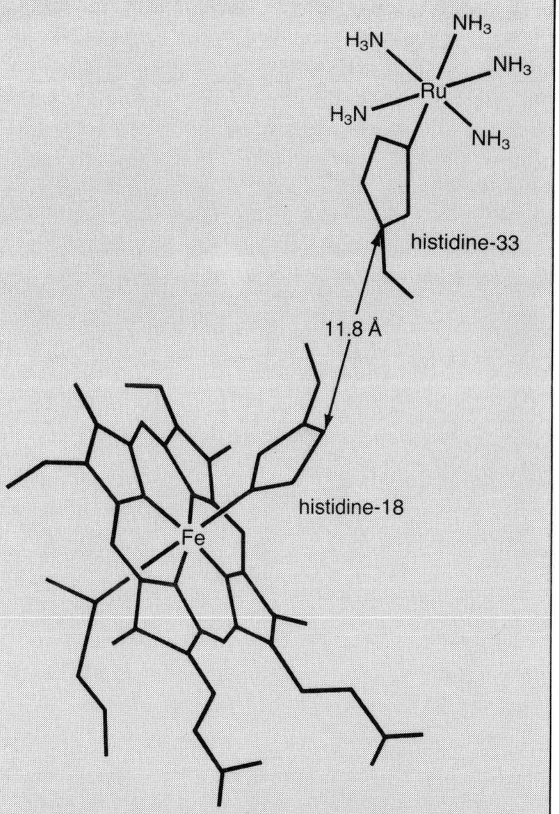

2 structural model for selected parts of $Ru(NH_3)_5$ (histidine-33) -ferricytochrome *c*

Recent independent studies in three different laboratories made the most important observation that electron transfer can occur at biologically significant rates across distances of 10–25 angstroms, depending on the biological system used (one angstrom, Å, is 10^{-8} cm). The three studies were done in the laboratories of Harry B. Gray at the California Institute of Technology, Stephan S. Isied at Rutgers University, New Brunswick, N.J., and Brian M. Hoffman at Northwestern University, Evanston, Ill. The research groups of Gray and Isied used similar approaches to study electron transfer in cytochrome *c*, whereas Hoffman's group used a different approach to investigate electron transfer in hemoglobin.

Every oxygen-breathing organism requires cytochromes, coordination compounds of iron (Fe) that are essential for the breakdown and metabolism of foods and storage of the energy released by this metabolism. Cytochrome *c* is a complicated globular protein comprising a chain of 104 amino acids wrapped around a heme group, a ring-shaped porphyrin molecule with an iron atom at its center.

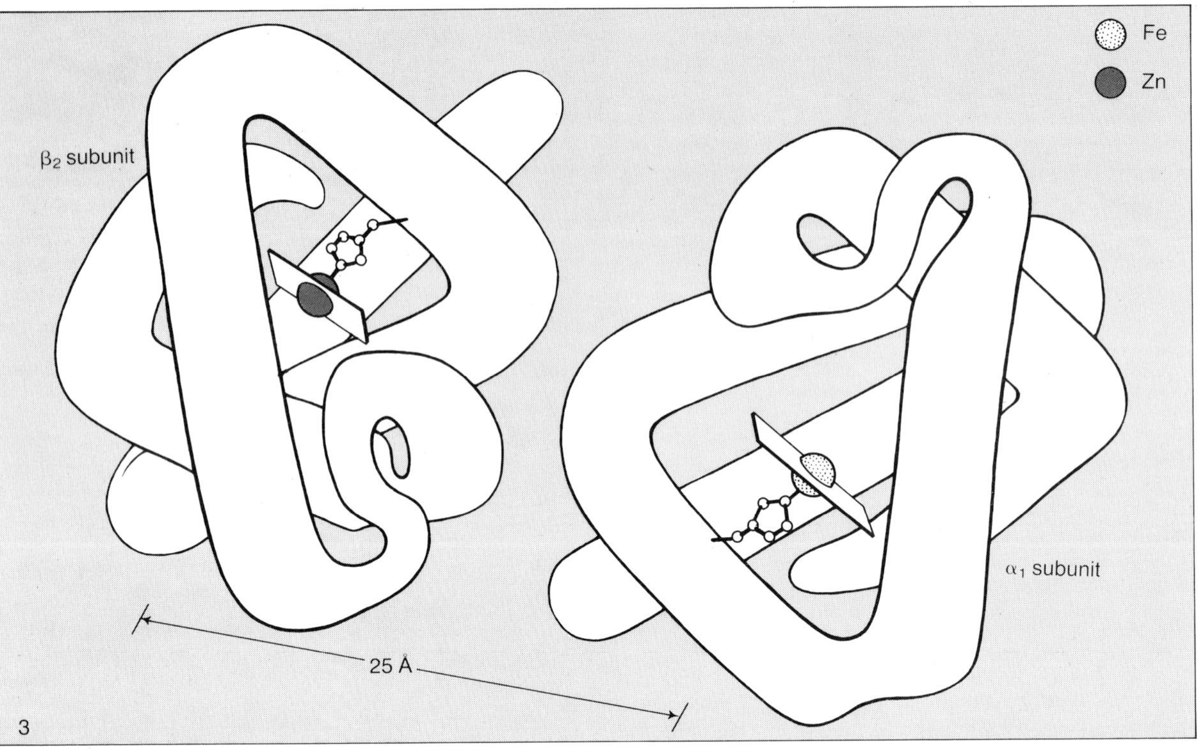

3

Fortunately, the structure has only one exposed imidazole group, on the particular amino acid histidine-33, capable of binding with $[-Ru(NH_3)_5]^{2+}$. Once bound to the imidazole, the ruthenium(II) is about 12 Å distant from Fe(III) (see 2). The reaction studied is the transfer of an electron from Ru(II) to the Fe(III)-porphyrin group to yield Ru(III) and Fe(II). The rate of electron transfer, approximately 40 per second, was found to be virtually independent of temperature. This suggests that the energy barrier for reaction is very small, although the two metals are bridged together through a saturated polypeptide at a distance of approximately 12 Å.

Hoffman's research group worked on a strikingly different system from cytochrome *c*, an iron-zinc hybrid hemoglobin in which two of the molecule's four iron-containing heme groups are replaced with zinc-porphyrin groups. They also used a different experimental approach but obtained similar results: electron transfer is rapid between the widely separated Fe-Zn sites. Hemoglobin consists of two pairs of identical protein subunits, designated α_1, α_2, β_1, and β_2, each containing a heme group and all held together in a complicated specific geometry. The closest metal centers are between α_1 and β_2; in the hybrid hemoglobin these centers—atoms of iron and zinc, respectively—are separated about 25 Å (see 3). The electron transfer reaction was initiated by flash of light to form an excited zinc-porphyrin state in the β_2 subunit, which reduced the Fe(III) in the α_1 subunit to Fe(II). The rate of electron transfer is approximately 100 per second.

This elegant discovery of long-range electron transfer will open up yet another important area of research in bioinorganic chemistry. One immediate major unanswered question is the mechanism of this process. Does the electron tunnel directly from one metal ion to the other? Or does it some-

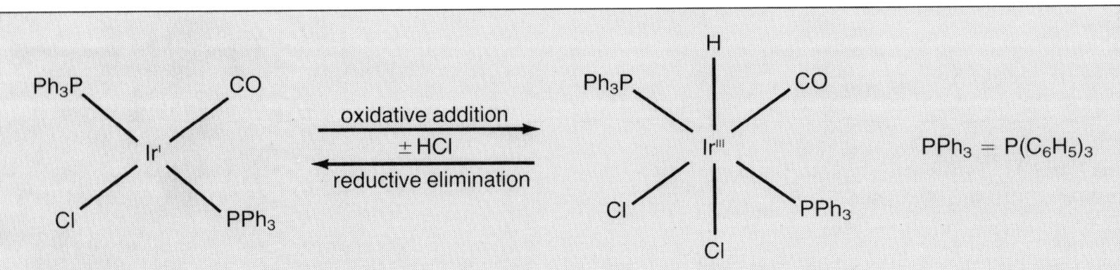

4

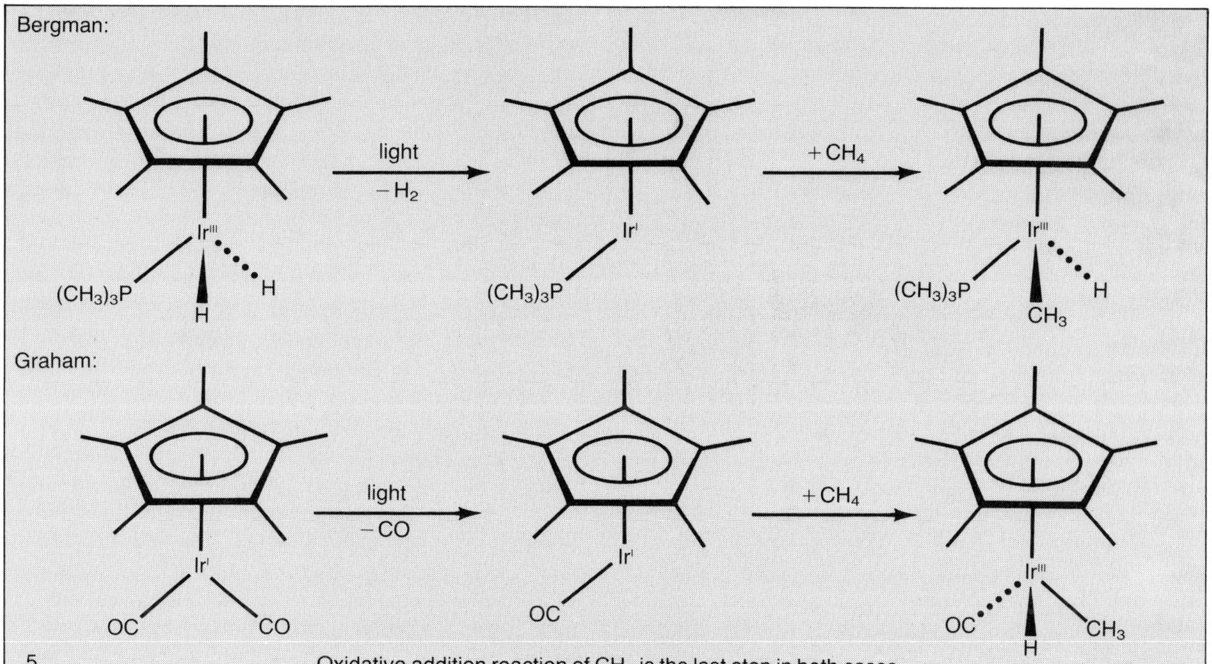

5 Oxidative addition reaction of CH_4 is the last step in both cases.

how move along the polypeptide backbone, utilizing molecular orbitals along the way? These and other questions are certain to be addressed, for electron transfer processes are among the most fundamental processes of life.

Alkane C—H bond activation. In the mid-1960s Lauri Vaska at Clarkson College of Technology, Potsdam, N.Y., worked with an iridium (Ir) compound now often called Vaska's compound, $Ir(PPh_3)_2(CO)Cl$, with which he did some of the seminal research in the chemistry of transition-metal organometallics. For example, he was the first to report on the oxidative addition and reductive elimination of these compounds (see 4). The forward reaction is termed oxidative addition because Ir(I) is oxidized and because HCl is added, resulting in six groups attached to Ir(III). This process is reversed in the reductive-elimination reaction.

These reactions are important in organometallic chemistry because a large variety of compounds can be added and eliminated by means of such metal centers. This technique then provides a route for activating, or increasing the reactivity of, certain organic compounds and for the homogeneous catalysis of their reactions.

Alkanes, $CH_3-(CH_2)_n-CH_3$ and CH_4, are a class of compounds that had resisted all attempts at activating the C—H bond by the method of oxidative addition on a metal complex. Four independent research groups recently reported remarkable success in having achieved C—H bond activation of alkanes. Two groups, those of Robert G. Bergman at the University of California at Berkeley and William A. G. Graham at the University of Alberta, did this by using the oxidative-addition reaction on a "hot" (*i.e.*, very reactive) metal center, while the groups of Patricia L. Watson of Du Pont and Tobin J. Marks at Northwestern University succeeded largely by the more usual method of serendipity.

Petroleum gases and oils constitute the world's resource of these hydrocarbons, which are largely burned to power automobiles and heat or cool homes. Saturated hydrocarbons are stable and are generally chemically unreactive. It is important that the C—H bond in these compounds be activated for selective chemical reactions at reasonable conditions of temperature and pressure. The simplest example of these compounds, methane (CH_4), is stable; its valence electron requirements are satisfied. Methane has no tendency to react either as an electrophile (a seeker of electrons) or a nucleophile (a seeker of protons in nuclei). The only common reaction of methane is its burning in oxygen, and chemists have not been very successful in readily converting methane into other useful compounds.

With the recent success of activating C—H bonds in alkanes, however, there is reason to be optimistic that methane and other hydrocarbons will yield to homogeneous selective chemical reactions. The oxidative addition approaches used by Bergman and by Graham for activation of CH_4 are illustrated in (5). The other reports of C—H bond activation in methane appear not to involve oxidative-addition reactions and make use of lanthanide and actinide

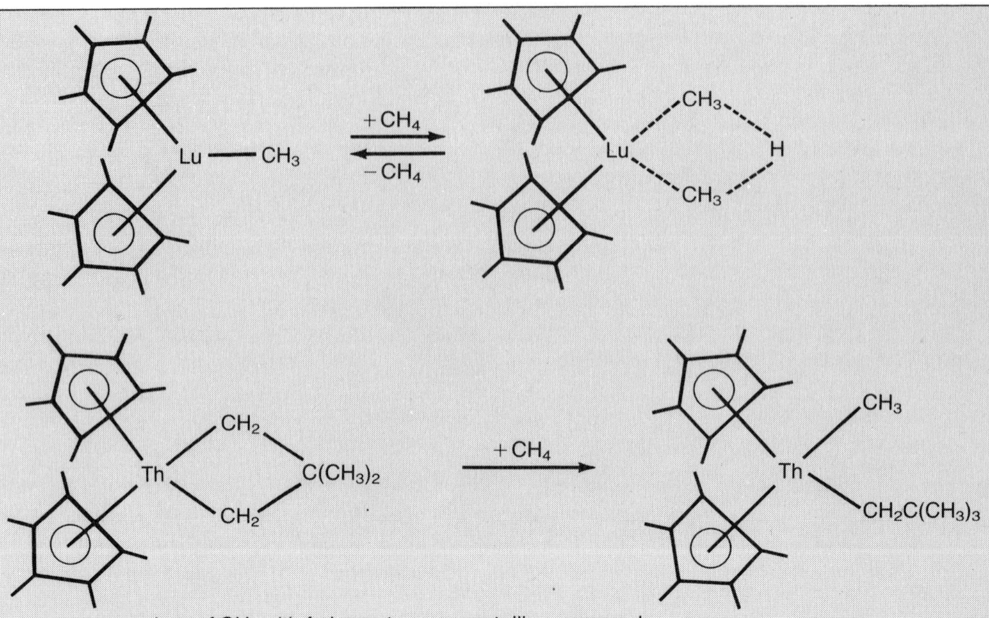

6 reactions of CH_4 with *f*-element organometallic compounds

(*f*-element) organometallic compounds (6). Now that the initial goal of C—H bond activation of saturated hydrocarbons finally has been realized, research is needed to find good homogeneous catalytic reactions that can convert the alkyl-metal intermediates into desired useful organic products.

Polymer solid electrolytes. In 1884 Svante Arrhenius, defending his doctoral thesis at the University of Uppsala, Sweden, proposed most clearly that sodium chloride (NaCl) and related compounds are dissociated into ions (*e.g.*, Na^+ and Cl^-) in water solution. He said further that these ions are responsible for the electrical conductivity of their solutions. Although his faculty mentors ridiculed him for his ideas, the modern theory of electrolytes embodies most of Arrhenius's principal postulates.

A century after the Arrhenius concept of electrical conductivity of salts in aqueous solution, inorganic chemists are finally succeeding in devising solid solutions of salts that behave efficiently as ionic con-

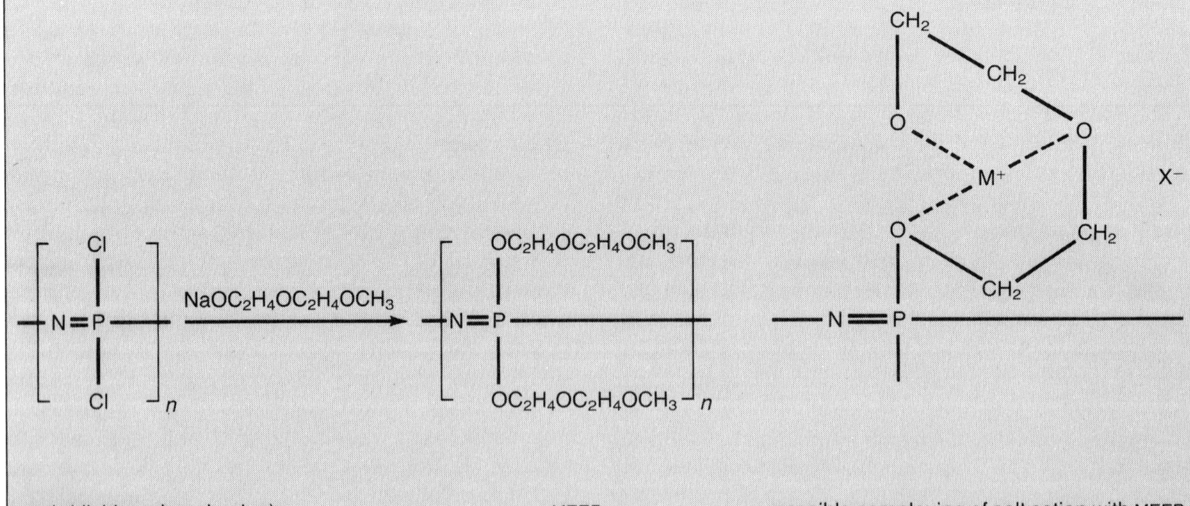

poly(dichlorophosphazine) MEEP possible complexing of salt cation with MEEP

7

ductors. Recently a promising material of this type emerged from the efforts of Duward F. Shriver and co-workers at Northwestern and Harry R. Allcock and co-workers at Pennsylvania State University.

Inorganic ion-conducting solids have attracted attention both for fundamental scientific reasons and for their important potential commercial use in high-energy-density batteries. Researchers have designed radically new batteries that operate at temperatures of about 100° C (212° F) and are based on solvent-free thin-film polymer electrolytes. These electrolyte materials contain ionic salts in host ethylene oxide polymers, $\leftarrow OC_2H_4OC_2H_4O \rightarrow_n$. It is known that ion transport in such polymer-salt electrolytes implicates a liquidlike mechanism, which suggests the use of polymers having a low glass (liquidlike) transition temperature. Shriver decided to explore the use of polyphosphazenes because he knew from the work of Allcock that these polymers have low glass transition temperatures.

The ethylene oxide polymer is believed to work, in part, because its oxo (oxygen-containing) groups complex the ionic salt and bring it into solid solution. This necessary feature was incorporated into the polyphosphazene polymer (given the acronym MEEP), while still maintaining its desirable low glass transition temperature (see 7). The silver salt $Ag^+(SO_3CF_3)^-$ was incorporated into MEEP, and the material was found to be most conductive at about 0.2 silver ion per polymer repeat unit. The new polyphosphazene electrolyte represents a significant advance in inorganic ion-conducting solids and is a good candidate for a thin-film battery that operates at or close to room temperature.

—Fred Basolo

Organic chemistry

Special challenges exist in the synthesis and study of very large molecules as well as small molecules of novel structure. One aspect of large molecule synthesis was the subject of the 1984 Nobel Prize for Chemistry, while several notable advances were made in the synthesis of small molecules. Other highlights of the past year include construction of enzyme models, use of enzymes to facilitate organic synthesis, detailed study of the three-dimensional course of enzymatic reactions, and identification of unusual natural products.

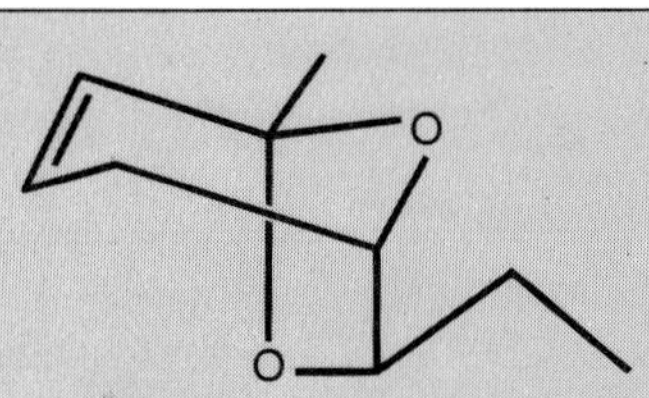

1 pheromone of the mouse

Natural product synthesis. R. Bruce Merrifield of Rockefeller University in New York City (*see* SCIENTISTS OF THE YEAR) was awarded the 1984 Nobel Prize for Chemistry for developing a method for the synthesis of peptides, short proteins made of chains of amino acids, using a polymeric support. The procedure, representing a completely new approach to organic synthesis, has brought about a revolution in peptide and protein chemistry, opening the door to the synthesis of thousands of different peptides as well as other biopolymers, particularly, short pieces of DNA, which are important in genetic research.

The essential feature of Merrifield's method involves covalent bonding of the growing peptide chain to an insoluble solid support, such as a polystyrene resin, during the course of the chain assembly. Since the anchored products are part of the solid phase, by-products and unreacted starting materials can be easily eliminated after each synthetic step by filtration and washing of the polymer. Only when the entire synthesis is complete is the peptide removed from the support. The procedure is fast and efficient; the entire process can even be conducted automatically. Since the work of Merrifield, polymer-bound reagents have found wide use in synthesis.

Elias J. Corey of Harvard University received the American Chemical Society's 1984 Willard Gibbs Medal for his many contributions to the field of organic synthesis. During the past year Corey and co-workers completed syntheses of hybridalactone and (+)-clavulone, biologically active marine compounds from the red alga *Laurencia hybrida* and the coral *Clavularia viridis,* respectively, and bongkrekic acid, a bacterial toxin.

A fascinating compound was isolated from the urine of adult male mice and subsequently synthesized by University of Indiana chemists Donald P. Wiesler, Frank J. Schwende, Marvin Carmack, and Milos Novotny. The compound, *exo*-dehydrobrevicomin (1), is said to be a pheromone (chemical stimulant of behavior) or pheromone adjuvant and in combination with another uniquely male mouse compound, 2-*sec*-butyl-4,5-dihydrothiazole, is an aggression-promoting principle of the adult male mouse. Remarkably, an analogue of (1), lacking the carbon-carbon double bond, had been previously identified as a western bark beetle pheromone.

An intriguing group of natural products that repel sharks was found in secretions of the southern Pacific sole, *Pardachirus pavoninus.* Masahiro Sakaitanai, Kazuo Tachibana, and Koji Nakanishi of the Suntory Institute for Bioorganic Research, Osaka, Japan, identified six related steroid compounds (pavoninin-1 to -6) and a protein as the active toxins, which function as repellents and antifeedants.

Robin N. Boar, Lorraine A. Couchman, Alan J. Jacques, and M. John Perkins of Chelsea College in London, isolated a bicyclic (two-ringed) triterpenoid of unusual structure from gum mastic, the abundantly available resin obtained from the Mediterranean shrub *Pistacia lentiscus* and widely used as a varnish for paintings. Since gum mastic also contains a range of normal four- and five-ringed triterpenoids, it was concluded that the bicyclic compound represents a rare example of a trapped intermediate in the cyclization (formation of ring-containing molecules) of squalene 2,3-epoxide in a system that has the enzymes needed for more complete cyclization.

The triply unsaturated sulfur compound $CH_2{=}CHCH_2S(O)CH_2CH{=}CHSSCH_2CH{=}CH_2$, termed ajoene, which prevents blood platelets from aggregating as part of the normal clotting process, was isolated from garlic preparations by Eric Block and Saleem Ahmad of the State University of New York at Albany, Mahendra K. Jain and Roger Crecely of the University of Delaware, and Rafael Apitz-Castro and Maria R. Cruz of the Venezuelan Institute of Scientific Investigations, Caracas, Venezuela.

Further evidence that most widely used organic reactions in the laboratory have their natural counterparts was provided by Glenn A. Berchtold of the Massachusetts Institute of Technology and Jeremy R. Knowles of Harvard and co-workers, who found that prephenate, the biochemical precursor to the amino acids tyrosine and phenylalanine, is produced enzymatically by concerted rearrangement of chorismate (2) with the same transition-state geometry (called the chair form) as the Claisen rearrangement, a well-known organic reaction.

There has been a growing tendency to incorporate natural biological transformations into organic syntheses; *e.g.,* to use either isolated enzymes or intact microorganisms to perform specific chemical reactions. The advantage lies in the high degree of stereochemical or regiochemical control (control of the orientation and positions of the atoms in a molecule) that can be achieved. Thus, Dieter Seebach and co-workers of the Swiss Federal Institute of Technology in Zürich used the bacterium *Thermoanaerobium brockii* to reduce beta-oxoesters and found that they could get as much as a fivefold preference for one of the two possible forms of the resulting beta-hydoxyesters and that the stereochemical preference is in some cases opposite to that found when the ester is reduced with baker's yeast, another microorganism used in organic reactions.

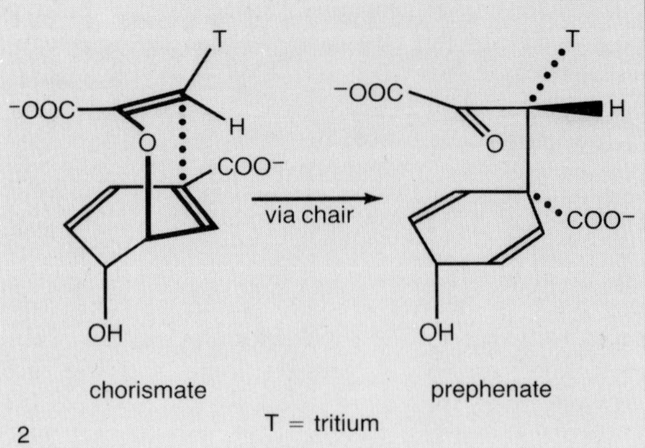

2

In another example Saul Wolfe and his colleagues at Queen's University in Kingston, Ontario, and Jack Baldwin and co-workers at Oxford University found that peptides can be cyclized to either penicillin or cephalosporin antibiotics using enzyme isolates. Wolfe developed a continuous process for these syntheses in which cyclase, epimerase, and expandase enzymes are immobilized through bonding to a resin support in a tube as a solution of the peptide and necessary cofactors are continuously pumped through. This procedure should allow chemists to prepare previously unknown antibiotics by starting with chemically altered peptides.

Nonnatural product chemistry. Numerous synthetic ionophores, counterparts of receptor sites of

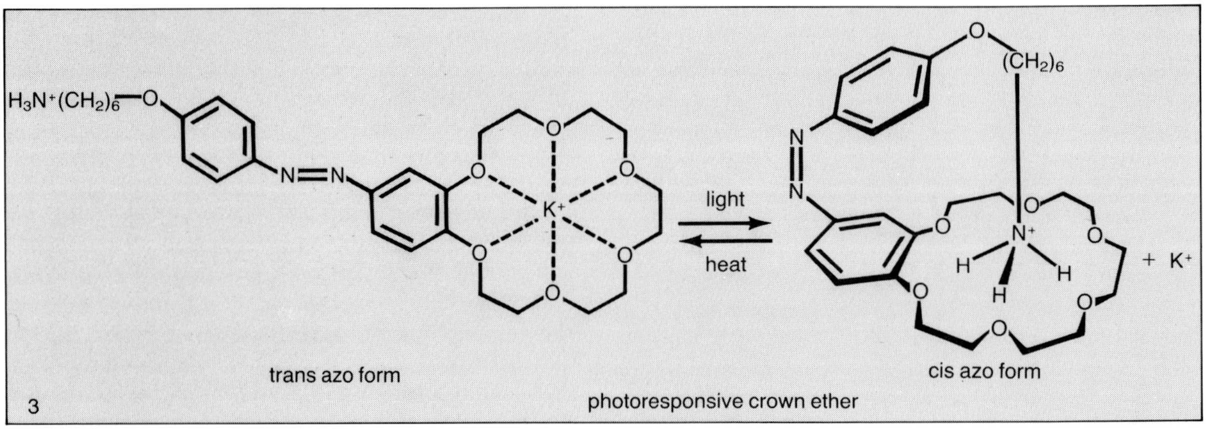

3

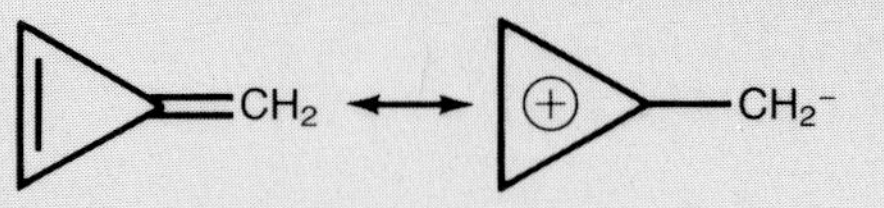

methylenecyclopropene

4

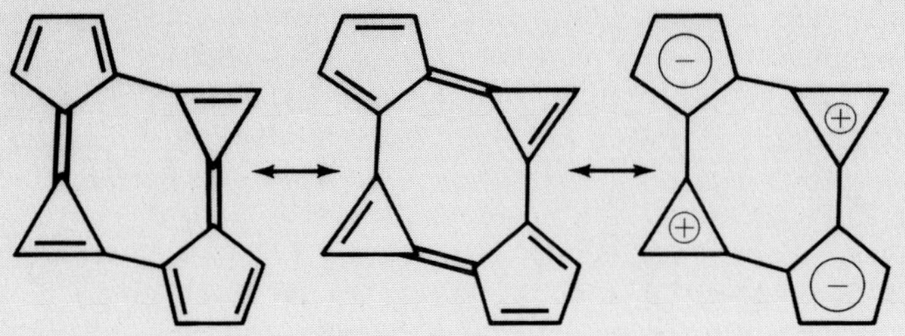

bicalicene

5

enzymes and polyether antibiotics, were prepared by Donald Cram of the University of California at Los Angeles, winner of the 1984 Roger Adams Award in Organic Chemistry. Typically these compounds, called hosts, have a ring-shaped structure containing an array of electron-rich oxygen atoms, which bind positively charged metal and ammonium ions, called guests. Ion transport by ionophores requires that these molecules switch from binding to nonbinding forms. A synthetic model for this process was provided by a photoresponsive crown ether (3) containing an azo (N=N) grouping, synthesized by Seiji Shinkai, Midori Ishihara, Kaori Ueda, and Osamu Manabe of Nagasaki University in Japan. Photolysis (light-stimulated chemical reaction) converts the trans azo form, an effective host for potassium ion, into the cis azo form in which the ammonium-ion "tail" can compete for the binding site.

A number of small molecules made the news in the past year. On a tragic note the loss of more than 2,500 lives in Bhopal, India, resulted from leakage of the toxic molecule methyl isocyanate, $CH_3N{=}C{=}O$, from a storage tank at a pesticide plant. In terms of human lives and suffering this represented the worst chemical catastrophe involving the chemical industry ever to occur. (*See* ENVIRONMENT.)

Chemists Robert W. Murray and Ramasubbu Jeyaraman of the University of Missouri at St. Louis obtained strong evidence for the transfer of an oxygen atom from the three-membered ring compound dimethyldioxirane, $(CH_3)_2\overline{C{-}O{-}O}$, to polycyclic (multiple-ringed) aromatic hydrocarbons, a useful model for an environmental process that may produce carcinogens in polluted air. Orville Chapman and Thomas C. Hess of the University of California at Los Angeles presented the first spectroscopic evidence for a ring-opened isomer of a dioxirane, a carbonyl oxide, $R_2C{=}O{-}O$. This particular carbonyl oxide was formed from the reaction of a carbene R_2C: (specifically cyclopentadienylidene) with oxygen at 10 K (−442° F). West German chemists Waldemar Adam and Wilhelm J. Baader of the University of Würzburg reported the first synthesis of the related four-membered oxygen-containing ring compound 1,2-dioxetane, $\overline{CH_2CH_2{-}O{-}O}$, which easily decomposes generating chemiluminescence (chemically produced light) and formaldehyde.

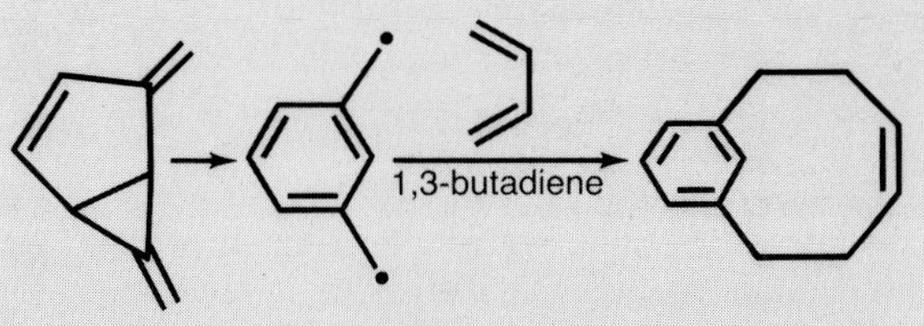

m-quinodimethane

6 7 8

Kosta Steliou and Yves Gareau of the University of Montreal and David N. Harpp of McGill University in Montreal succeeded in forming the sulfur counterpart of oxygen, S_2, and adding it to simple dienes (sulfur normally exists as S_8). On a related note O. P. Strausz, E. M. Lown, and Martina Green of the University of Alberta presented evidence for the reaction of atomic sulfur (S) with dimethyl sulfide $(CH_3)_2S$, giving dimethyl thiosulfide, $(CH_3)_2S{=}S$, a rare example of an organic molecule with a sulfur-sulfur double bond. Another organic sulfur compound with unusual bonding described in the past year is $F_3C{-}C{\equiv}SF_3$. This first example of an organosulfur compound with a triple carbon-sulfur bond was described by Brigitte Potter and Konrad Seppelt of the Free University of Berlin in West Germany.

Several molecules with unusual bonding were reported in recent months. Methylenecyclopropene (4), the simplest representative of a class of cross-conjugated compounds known as fulvenes, was independently synthesized by chemists W. E. Billups, Long-Jin Lin, and Edward W. Casserly of Rice University, Houston, Texas, and Stuart Staley and Timothy D. Norden of the University of Nebraska at Lincoln. Calculations suggested that the molecule should be very polar and highly strained; spectroscopy re-

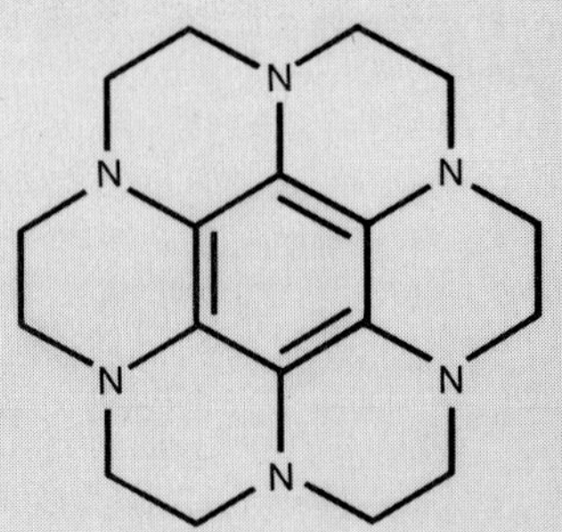

hexaazaoctadecahydrocoronene

9

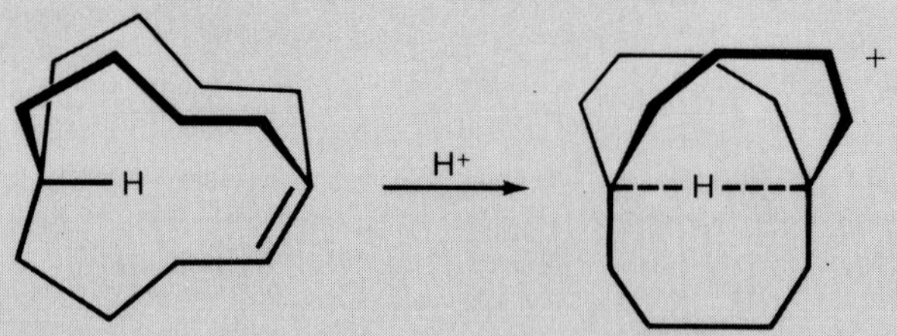

10

inside-outside molecules

vealed that there is indeed significant charge separation and that the compound decomposes in solution above −75° C (−103° F). A multiring compound incorporating (4) in its structure, cyclic bicalicene (5), was prepared by Zen-ichi Yoshida and colleagues at Kyoto University in Japan; it also displays charge separation. Yale University chemists Jerome A. Berson and Joshua L. Goodman found that unstable methylenecyclopropene (6) readily undergoes an electronic reorganization to *m*-quinodimethane (7), which exists not with charge separation as in (4) and (5) but rather as a neutral triplet diradical (two electrons, shown in the structure as dots, remain uninvolved in bonding with parallel spins). If (7) is made in the presence of 1,3-butadiene, the ultimate product is metacyclophane (8).

In connection with a search for ferromagnetic (strongly magnetic) organic salts Ronald Breslow, Przemyslaw Maslak, and John S. Thomaides of Columbia University, New York City, prepared hexaazaoctadecahydrocoronene, or HOC (9), and found that it undergoes sequential loss of electrons together with a series of color changes giving the mono-, di-, tri-, and tetra-cations (HOC^{+}, HOC^{2+}, HOC^{3+}, and HOC^{4+}, respectively). The latter two ions represent the first reported benzene tri-cations and tetra-cations.

The curious inside-outside molecule *in*-bicyclo-[4.4.4]-1-tetradecene was prepared by John E. McMurry and Carl N. Hodge of Cornell University, Ithaca, N.Y. On treatment with acid it gives a novel structure possessing an inside hydrogen bearing a partial negative charge and symmetrically placed between two positively charged bridgehead carbon atoms (10).

—Eric Block

Physical chemistry

During the past year physical chemists shed some light on one of science's biggest mysteries, the ultimate fate of the universe. They also devised a new way to make high-resolution spectra of charged molecules and demonstrated a relationship between the exothermicity of a reaction and its speed that is not described by classical theory.

The end of the universe. Chemical elements in nature arise from three major sources: from processes associated with big bang, the explosive birth of the universe; from synthesis in stars; and from formation by the interactions of galactic cosmic rays with the interstellar medium. According to nuclear chemist Victor E. Viola of Indiana University, Bloomington, the probability of the formation of certain isotopes differs among these sources. These differences provide important information about the total mass of the universe, and a good calculation of this mass in turn can answer unequivocally how the universe will end. If the total mass of the universe is high enough, its gravity will overcome the effects of the big bang explosion. This "closed universe" will eventually stop its present expansion and collapse into a superdense mass. If the total mass does not provide enough gravity, the universe, an "open" one, will continue to expand forever.

In order to get a good estimate of the universe's mass, Viola looked at the occurrence of the isotopes lithium-6 and lithium-7 (^{6}Li and ^{7}Li). Lithium is the lightest metal and the third element of the periodic table. When lithium is made by the action of cosmic rays today, twice as much of the heavier isotope, ^{7}Li, is formed as the light, ^{6}Li. Viola noted, however, that the current ratio of the isotopes is not two-to-one, but twelve-to-one. This excess ^{7}Li can only be explained by assuming that it formed during the big bang. The production of the heavier isotope was overwhelmingly favored at that time and accounts for the current excess.

From equations for the two processes of formation, the current abundances of the isotopes, and the estimated age of the universe, the mass at the time of the big bang can be calculated. That mass should be the same as the current mass. From such calculations Viola found that the total mass of the universe is too low to stop the current expansion;

i.e., the universe is open. In order to collapse, the universe would require about ten times its estimated total mass. Viola's calculations agree well with those based on the abundance of deuterium, a heavy isotope of hydrogen assumed to have been formed almost exclusively during the big bang, as well as with total mass estimates that are based on adding up the masses of all observable matter.

Although this result supports the concept of an open universe, the question is by no means settled. At least two possible developments could add enough mass to current estimates to demonstrate a closed universe. If neutrinos can be shown to have even a tiny mass (currently they are thought to have none) or if a large amount of "dark matter" (matter not easily observed) is detected, then the universe may well be destined for a "big crunch" at the end of time.

A new spectroscopy. Charged molecules, or ions, such as H_3O^+ (the hydrated form of the hydrogen ion H^+ in water solution), have a chemistry that is fundamentally different from that of neutral molecules. Among other things they tend to form strong complexes with other molecules, and they have different intramolecular properties. Understanding their chemistry is necessary for the comprehension of many important chemical and biological processes, but until recently it had been difficult to get much information. In many cases the ionic forms coexist among far greater quantities of neutral molecules, so their spectroscopic signals are typically completely overwhelmed.

Recently researchers at the University of California, Berkeley, taking advantage of Doppler effects, devised a method that permits making high-resolution spectra of charged molecules. The technique, called velocity-modulation laser-absorption spectroscopy, was developed by Richard J. Saykally, Christopher S. Gudeman, Jürgen Pfaff, Eckhard Schäfer, Marianne H. Begemann, and Cecilia C. Martner.

The idea for the technique came from earlier experiments in the microwave absorption spectroscopy of ions that are generated in a direct-current (DC) electric discharge. R. Claude Woods, with whom Saykally worked as a graduate student at the University of Wisconsin, Madison, suspected that a consistent discrepancy in some experimental data was being caused by the motion of the ions in the DC field. This discrepancy looked like a Doppler shift in frequency: under microwave bombardment the ions did not absorb at precisely the expected frequencies but at values slightly shifted along the electromagnetic spectrum, as would be the case if the ions and the microwave source were in relative motion. By moving their detection devices and changing the direction of their microwave irradiation, Woods and his co-workers were able to shift the frequencies in the opposite direction, confirming the cause of the observed discrepancy.

Later in his own laboratory Saykally changed the radiation source from a microwave generator to a tunable infrared laser and used an alternating-current discharge to generate ions instead of direct current. In the alternating field all the charged molecules move back and forth together, while neutral molecules do not. As a result the absorption frequencies of the ions are constantly being shifted up and down the spectrum and thus can be easily picked out from those that do not change.

Detailed spectra already have been obtained for a number of charged molecules, including H_3O^+, HCO^+, HNN^+, NH_4^+, H_2F^+ and H_3^+. Such information is essential to understanding the chemistry of solutions. Problems the new method might help solve include the way metal ions are solvated, the structure of nonclassical cations, and the way ions interact with neutral molecules. Application of the technique has been expanded to visible-light and far-infrared spectroscopy.

Exothermicity and reaction speed. In general, the more heat released by a reaction, the faster the reaction progresses. Explosions tend to release a lot of heat, *i.e.*, to be exothermic, and they proceed very rapidly. Exothermic reactions are a familiar and useful part of daily life. For example, the burning of natural gas, $CH_4 + 2O_2 \rightarrow CO_2 + 2H_2O$, is a common source of home heat. Recently this classical view—the more exothermic, the greater the rate of reaction—was challenged by new data from experiments conducted by John R. Miller of Argonne National Laboratory in Illinois and Gerhard L. Closs and Lidia T. Calcaterra of the University of Chicago. They showed that, for some electron transfer reactions, greater exothermicity can lead to slower rates of reaction.

Detailed infrared spectrum of the hydronium ion (H_3O^+) was obtained with a newly devised technique that allows the spectroscopic signals of ions to be separated from the typically overwhelming signals of neutral molecules.

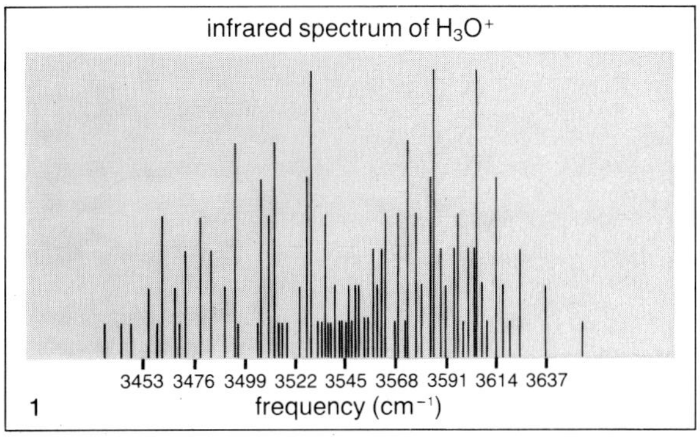

Adapted from information obtained from the University of California, Berkeley

Adapted from "Intramolecular Long-Distance Transfer in Radical Anions. The Effects of Free Energy and Solvent on the Reactions Rates," J. R. Miller, L. T. Calcaterra, and G. L. Closs, *Journal of the American Chemical Society*, vol. 106, no. 10, pp. 3047–3049, 1984, © 1984 American Chemical Society

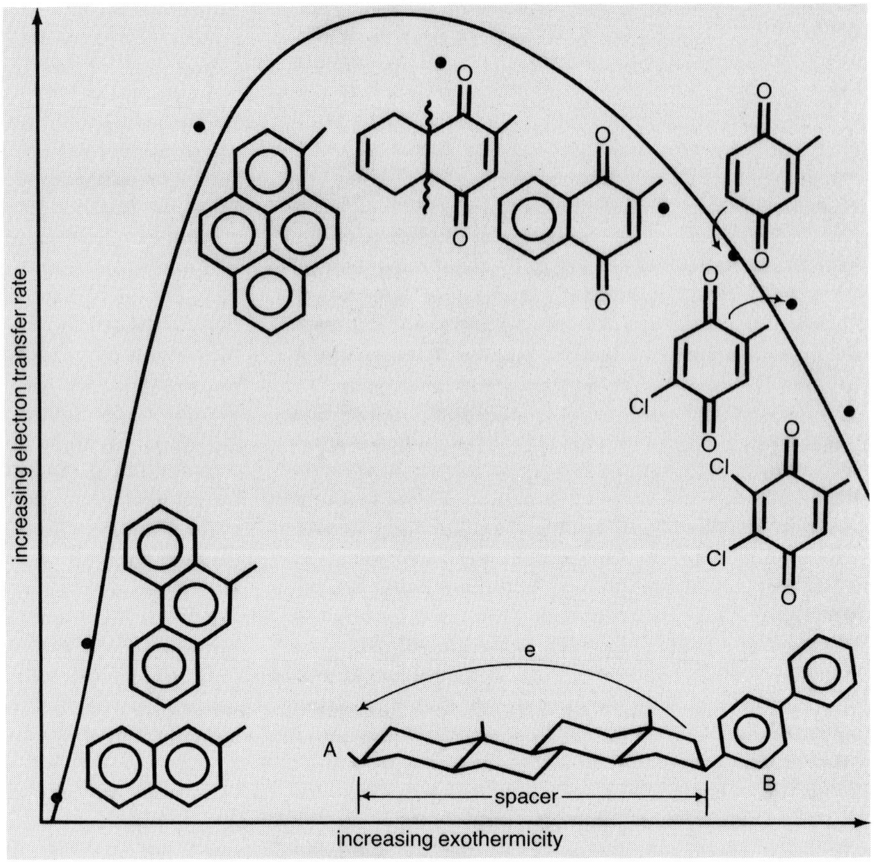

Electron transfer rate constants determined by John Miller, Gerhard Closs, and Lidia Calcaterra for each of eight variants of their A-spacer-B molecule are plotted against the exothermicity of the transfer. The general structure of the molecule is shown just above the base of the graph. The B group remained the same for each determination, while the A groups were varied; the structure of each A group appears next to the data point derived from the molecule containing that group. Electron (e^-) transfer took place from B to A. When a curve is drawn through the data points, the rate of the transfer is seen first to increase with increasing exothermicity and then to fall.

The new finding is not entirely unexpected. In the mid-1950s Rudolph A. Marcus of the California Institute of Technology postulated exceptions to the classical view—so-called inverted regions for electron transfer reactions. His idea was expanded and supported by theorists but seemed to be contradicted by almost all experiments carried out in solutions.

In 1984 Miller, Closs, and Calcaterra reported that they had the first unequivocal proof that an inverted region exists. The effect is subtle and difficult to demonstrate, but they succeeded by looking at the movement of electrons through some specially designed and synthesized molecules.

The molecules have the basic structure of A-spacer-B, in which A and B are molecular groups with different affinities for electrons (A greater than B) and the spacer is a stiff, relatively inert molecular group that is there primarily to keep A and B at a fixed distance apart, about ten angstroms (one angstrom is 10^{-8} cm). The B group remained the same molecule throughout the experiments, while different A groups having a range of electron affinities were tested.

The compounds were dissolved in an organic solvent, methyltetrahydrofuran, and highly energetic electrons from a particle accelerator were pulsed into the liquid. The extra electrons were held by the solution and then captured by group A or B on the test molecules. At first, A and B had equal chances of capturing electrons, but since A has the higher affinity, eventually electrons captured by B were transferred to A, an exothermic reaction.

A complicating factor was the occurrence of intermolecular electron transfers, in which electrons hop from one molecule to another rather than from one molecular group to another on the same molecule. The scientists were able to measure this effect and factor it out by observing transfers in solutions containing A-spacer and B-spacer molecules. In these solutions only intermolecular transfers could take place.

When the electron transfer rate constants were determined for each A-spacer-B molecule and then plotted together on a graph against the exothermicity of the transfer (directly dependent on the electron affinity of the A group), an arching curve appeared, indicating that the rate of the transfer first increased with increasing exothermicity and then fell. This curve almost exactly matched curves predicted by theory. According to the investigators the results provide the first unambiguous evidence that under certain circumstances the rates of reactions can decrease with greater exothermicity. They credit their success to eliminating the inherent problems of intermolecular reactions, which are affected by diffusion rates.

—Peter J. Andrews

Applied chemistry

In recent months research in applied chemistry led to developments in herbicides, polymers and fibers, right- and left-handed compounds, artificial gills, bitter and salty substances, and artificial jade.

Herbicides. The U.S. Department of Agriculture (USDA) estimated that American farmers used 196 million kilograms (432 million pounds) of conventional weed killers in 1984, many of which have not been thoroughly tested for their ability to cause cancer or birth defects. As part of the search for safer herbicides and pesticides plant biochemist Constantin A. Rebeiz and horticultural scientist Herbert J. Hopen of the University of Illinois, Urbana, investigated a photosensitive herbicide whose action is triggered by sunlight. The compound, which occurs naturally in all plants and animal cells, is an amino acid called δ-aminolevulinic acid ($NH_2CH_2COCH_2CH_2COOH$) and abbreviated ALA. An intermediate in chlorophyll biosynthesis, it kills weeds but spares crops and does not harm humans or other animals.

ALA is sprayed at dusk, and during the night it is absorbed by the plants and converted into certain light-sensitive compounds called tetrapyrroles, which are building blocks of chlorophyll. At sunrise the excess concentrations of tetrapyrroles act as photosensitizers, converting oxygen to a highly reactive form, which oxidizes phospholipids of membranes within plant cells in free-radical chain reactions. The damaged membranes become leaky, and the plant loses sap, wilts, and dies within one to four hours. Afterward the lethal substances quickly break down in the soil, in contrast to conventional herbicides, which leave behind long-lasting, dangerous toxins. Susceptibility to the photosensitizing reactions of excess tetrapyrroles is greater for dicotyledons (a group that includes most weeds in cultivated fields) than for monocotyledons (*e.g.*, wheat, barley, corn, and other grains), which have a somewhat different metabolic system. ALA can also be used safely on some dicotyledon crops, such as soybeans and kidney beans, if it is sprayed before the plants have put out their first true leaves.

The compound is already commercially available but is very expensive ($6 to $7 per gram of the hydrochloride of the acid), but the Urbana scientists obtained good results by spraying it at only 75 grams per acre in conjunction with such activators as α,α-dipyridyl. Its success as a herbicide will depend on current efforts to find cheaper ways of making it.

Polymers and fibers. Fibers of polyethylene ten times stronger than steel fiber, twice as strong as carbon fiber, and exceeding even the strength of aramid aromatic polyamide fibers were made by Piet Lamstra and co-workers at DSM in The Netherlands. Its strength is due to a high degree of crystallinity along the polymer's chain in conjunction with its carbon-carbon covalent bonds. The polymer, a high-density polyethylene with few, if any, pendant side chains, has a molecular weight four times that of conventional high-density polyethylene. In a so-called gel-spinning technique, details of which have not been made public, macromolecules of polyethylene in solution are "disentangled" to the desired degree and the solution transformed to a gel, within which the macromolecules can be easily oriented. The molecules must be perfectly parallel and have a precise orientation to effect the proper degree of crystallized regions along the chain lengths. Fibers are spun from solution and drawn in the disentangled gel state. DSM foresees two broad markets for the superstrong fibers in addition to their potential use in making rope: reinforced, lightweight plastic composites for helmets, sporting goods, automotive and industrial parts; and bulletproof vests

University of Illinois, Urbana-Champaign

Cucumber seedlings on the right show the devastating effects of exposure to morning sunlight after application of δ-aminolevulinic acid (ALA) the previous evening, while seedlings on the left, which did not receive ALA, continue to thrive. As a potential new herbicide ALA shows the greatest activity against such dicot weed species as lamb's quarter, mustard, and common purslane and the least against monocot cereal crops like wheat, oats, and corn.

and other high-strength fabrics. The fibers melt at about 150° C (300° F), probably ruling out their use in fire-resistant applications; by contrast, carbon fibers melt above 300° C (570° F). Studies are in the development stage, with the possibility of commercialization some two years away.

Another new material from DSM is a polyamide nylon 46, tradenamed Stanyl. Henk Jacobs, DSM's Chemical Products Division project manager, who developed the new nylon, also devised a new synthesis for one of its starting materials, an expensive compound called 1,4-diaminobutane. Stanyl has characteristics (high melting point, toughness, and dimensional stability at high temperatures) superior to conventional nylons (particularly nylon 6 and nylon 66, which have had no serious competition in half a century) and many engineering plastics.

Temperature-adaptable fibers having markedly enhanced thermal-storage characteristics in the range of 5°–100° C (41°–212° F) were being developed by Tyrone L. Vigo, director of the USDA's Textiles and Clothing Laboratory at Knoxville, Tenn., and Cynthia M. Frost, who were chemically treating textile fibers to increase their ability to store and release heat. Their research has many possible applications; *e.g.*, protective clothing for high- or low-temperature extremes, building insulation, and protection of plants, animals, and food from temperature fluctuations.

Able to extract oxygen from seawater, Hemosponge is a polyurethane foam containing heme, the active component of hemoglobin. The new material may find use in supplying oxygen to underwater combustion engines and to submarine crews.

Scott D. Taylor, Duke University Marine Lab

The USDA researchers were able to increase the heat content of rayon, cotton, and polypropylene fibers in a given temperature range to 3.5–4, 2.5, and 2 times their usual values, respectively, by treating them with aqueous solutions of "plastic crystals," compounds that undergo large changes in heat content not because of phase changes (melting or crystallization) but because of transitions from one solid form to another at temperatures well below their melting points. Two plastic-crystal compounds were found to be particularly effective, 2,2-dimethyl-1,3-propanediol (DMP) and 2-hydroxymethyl-2-methyl-1,3-propanediol (HMP). In a hot environment fabrics made with the treated fibers absorb heat as the plastic crystals change form. Later, when the surroundings cool, the crystals return to their original form, releasing the stored heat.

Left- and right-handed compounds. Certain prescription drugs, pesticides, food additives, organic compounds, and other industrial products may become more potent (with fewer side effects) as well as easier and cheaper to produce because of a discovery by Nobel laureate Herbert C. Brown and Bakthan Singaram of Purdue University, West Lafayette, Ind. Many chemical substances can exist in left- and right-handed forms as mirror images (optical isomers), which can be distinguished by the fact that their separate solutions rotate polarized light in opposite directions. Living organisms usually can metabolize only one of such chemical "twins." For example, the left-handed form of the sugar glucose, although as sweet as the common right-handed variety, passes through the human body untouched, thus making it noncaloric. Traditionally scientists have had to use whichever form of a chemical they found in nature. Reactions in the laboratory usually produce a mixture of both isomers, and their separation is usually a tricky and tedious task.

Brown and Singaram started with α-pinene, a substance produced naturally in great quantities in pine trees and largely responsible for their pungent smell. Although individual trees generate only one or the other of the two isomeric forms, commercial methods generally process the material as a mixture of both forms. The Purdue chemists found that when an α-pinene mixture rich in one form is dissolved in diethyl ether and treated with borane (a compound of the elements boron and hydrogen) to produce diisopinocampheylborane, the product containing the more abundant form of α-pinene crystallizes out of

Photos, General Electric Research and Development Center, Schenectady, N.Y.

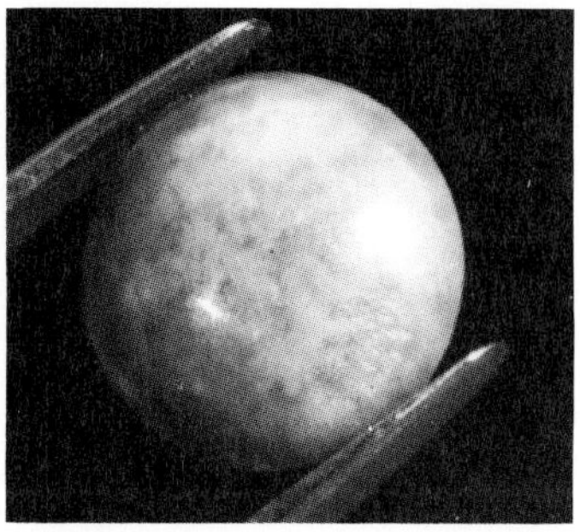

At left, GE scientists James Fleischer (far left) and Robert DeVries load a small furnace containing oxides of aluminum, sodium, and silicon into a high-pressure diamond-making press. Their technique produced samples of gem-quality jadeite (above).

solution, leaving the less abundant form still dissolved. Several more steps convert the crystalline product, now 100% optically pure, into a stock solution that can react with olefins (organic compounds containing double bonds between carbon atoms) to yield a product that is also 100% optically pure. Further treatment displaces the α-pinene so that it can be reused and leaves an optically active alkylborane that can be readily converted into many other optically active organic compounds. With an appropriate olefin theoretically almost any optically active compound can be synthesized with 100% optical purity. The method, which according to the discoverers is "going to open up a big, new field of chemistry," is especially attractive because it can be carried out easily on a large scale and because α-pinene is readily available commercially.

Artificial gills. An artificial gill that filters oxygen out of seawater was developed by Joseph and Celia Bonaventura, a husband and wife team of biochemists who codirect the Duke University Marine Biomedical Laboratory at Beaufort, N.C. Called the Hemosponge, the material is a polyurethane foam in which heme, an oxygen-binding iron complex that is the active component of hemoglobin, is immobilized. The Hemosponge is currently in the early stages of commercial development following the sale of its patent for $1 million to Aquanautics Corp., a San Francisco marine technology firm, which also paid $300,000 to Duke University for continuing research. The company recruited Makai Ocean Engineers of Oahu, Hawaii, to develop an efficient underwater fuel-burning power system that uses oxygen collected by the Hemosponge. According to Aquanautics Corp., a full-scale 15-kilowatt engine should be forthcoming within the next few years.

The Bonaventuras are currently attempting to combine their oxygen-extracting technology with a submersible engine to create an underwater vehicle as practical as an automobile. The greatest obstacle to a practical underwater engine has been the lack of a readily available, abundant oxygen source. A typical internal combustion engine consumes almost eight times as much oxygen as fuel, and transporting the necessary bulky oxygen supplies is not feasible. The Hemosponge solves this problem by transforming the sea itself into a virtually infinite on-site oxygen supply. When seawater is pumped through the Hemosponge, the heme binds the dissolved oxygen molecules. To release the oxygen the sponge is subjected to a vacuum or a weak charge of electricity.

The Hemosponge is also being studied for use in providing oxygen to submarines and to seafloor communities of as many as 150 persons. In the form of a rechargeable oxygen supply it could be incorporated into medical equipment like oxygen tents and masks and into air supplies for welders and miners. In its oxygen-depleted state it could be used in packaging to protect air-sensitive products from oxygen exposure and to increase the shelf life of foods and beverages. Injecting the material with sulfur or other active ingredients could allow it to neutralize toxic fumes from mercury spills or to treat effluents from nuclear reactors.

Bitter and salty substances. Chemists at Atomergic Chemetals Corp. in Plainview, N.Y., discovered the bitterest substance known to date. The white crystalline powder, called denatonium saccharide, is 3,000 times more bitter than quinine and retains its bitter taste at a dilution of one part in 100,000,000. The compound could be added to poisonous household products to prevent accidental ingestion by children and is being investigated as a repellent for sharks, beavers, woodpeckers, and water rats.

In the course of analyzing the bitter fermentation products (peptides) of casein, the chief protein in milk, food chemists Hideo Okai and co-workers at Hiroshima University's Department of Fermentation Technology in Japan created a substance that tastes like salt but contains no sodium, the element in table salt that is believed to raise the blood pressure of susceptible persons. Although not yet tested for safety, the salty substance, which is made from amino acids (the naturally occurring building blocks of proteins), shows promise as a seasoning agent for persons trying to lower their sodium intake. Surprised at finding a salty peptide, the Hiroshima chemists synthesized similarly constructed peptides and found two more, each consisting of two amino acids; on a weight basis one peptide was as salty as sodium chloride and the other 1½ times as salty. The saltier one contains lysine and taurine, while the equally salty one, L-ornithyltaurine, which is more commercially promising because it can be substituted quantity-for-quantity for salt, contains ornithine and taurine. The latter is not likely to be a perfect substitute for salt, however, because it will probably show the same kind of instability in liquids that afflicts aspartame, the newly marketed artificial sweetener, which is also a peptide.

Synthetic jade. One of the last remaining gemstones to be made in the laboratory, jadeite, was added to the list of those previously synthesized (*e.g.*, sapphire, ruby, emerald, opal, turquoise, lapis lazuli, and diamond). Although other laboratories had produced sand-sized grains of jadeite, inorganic chemist Robert C. DeVries and high-pressure research specialist James F. Fleischer of the General Electric Research and Development Center, Schenectady, N.Y., were the first to synthesize the material in sufficient size and quality for polishing into gems.

They mixed powdered oxides of sodium, aluminum, and silicon and heated them in a furnace to about 1,480° C (2,700° F), withdrew the resulting molten liquid, and cooled it to a glassy solid. They crushed the glass and "refined" it at the same temperature in a thumb-sized furnace inside a diamond-making press where it was subjected to pressures as high as 440,000 pounds per square inch. In this way they made dozens of samples of white jadeite that were cylindrical in shape and a fourth of an inch thick by a half-inch long. Other colors were made by adding small amounts of certain minerals to the starting mix; *e.g.*, chromium for green, larger amounts of chromium for black, or manganese for lavender.

—George B. Kauffman

See also Feature Articles: BATTERIES: A TECHNOLOGY RECHARGED; PENDULUMS IN THE TEST TUBE.

Defense research

For the first time since the Apollo Moon-landing program of the 1960s, the resources of the U.S. aerospace research community are being marshaled in a major effort to establish American primacy in space. This time the Department of Defense—not the National Aeronautics and Space Administration—is in charge, and national security—not merely national prestige—is the name of the game.

The effort springs mainly from Pres. Ronald Reagan's call, in his so-called Star Wars speech of March 23, 1983, for the exploratory development of nonnuclear defensive weapons that could make the U.S. secure against nuclear attack. Those weapons, including lasers, particle beams, and hypervelocity guns, would necessarily operate in or through space. Although their technologies had been pursued for many years, Reagan was the first president to order the Department of Defense to accelerate their development and to concentrate it under centralized Pentagon management.

In 1984 the program for carrying out the task was given the name Strategic Defense Initiative (SDI). Encompassing the formerly disparate space-weaponry research programs of the military services and the Defense Advanced Research Projects Agency (DARPA), SDI was made the responsibility of a newly created, top-priority Pentagon office called the Strategic Defense Initiative Organization (SDIO). It is headed by Air Force Lieut. Gen. James A. Abrahamson, formerly NASA's associate administrator in charge of the space shuttle program, who reports directly to the secretary of defense. The administration plans to spend about $26 billion on SDI research and technology programs over a five-year period ending in 1989–90. SDI was funded at a level of $1.4 billion in fiscal 1985 (ending on Sept. 30, 1985), and Congress was asked to appropriate $3.8 billion for it in fiscal 1986.

The SDI program is the most striking example of the Pentagon's drive to make space what its officials have called "the fourth combat medium," along with land, sea, and air. There are, however, several other examples such as the Air Force's antisatellite (ASAT) program and its budding program to develop a manned spacecraft for waging space-to-space and space-to-ground combat. Moreover, the armed services are reorganizing for future space missions. The Air Force created its Space Command in 1982, and the Navy followed suit a year later. In 1984 President Reagan approved the formation of a unified space command that would embody the operational space elements of all the services. Many military officers are convinced that a new Space Force will be needed—and created—as an autonomous branch of the armed forces before the turn of the century.

U.S. Air Force

U.S. air-launched antisatellite weapon hangs from the belly of an F-15 fighter aircraft during an early test flight. Once launched and above the upper atmosphere, the two-stage rocket releases a cylindrical device that homes in on the heat of its target satellite in low Earth orbit.

Pros and cons. For the present the spotlight is on the SDI program. In ordering it into existence Reagan was motivated by the Soviet threat to U.S. land-based intercontinental ballistic missiles (ICBM's), a threat that has become much graver in recent years as a result of Soviet deployment of highly accurate, very powerful ICBM's, with even better ones in the offing. The unprecedented menace of these missiles was the reason the U.S. agonized so long over a basing strategy for its MX ICBM's and, indeed, over building them at all. The Soviet missile threat forged a consensus among U.S. policymakers that there no longer is any certain way to deploy fixed-base ICBM's with impunity, short of defending them with antiballistic missiles (ABM's). This realization, in turn, raised grave doubts about the mutual assured destruction (MAD) policy of strategic deterrence—a policy grounded in the premise that the best defense against ballistic missiles is an overwhelming and secure offensive ballistic missile force—on which the U.S. has bet its survival in the nuclear age.

Such was the context of Reagan's Star Wars proposal, which presaged the most radical change in U.S. strategic policy since World War II. Reagan proposed countering "the awesome Soviet missile threat," as he put it, by gaining the ability to "intercept and destroy strategic ballistic missiles before they reached our own soil or that of our allies" and thus to "save lives [rather] than to avenge them" in a nuclear exchange. His approach has much political and humanistic appeal and has attracted strong support in the U.S. defense policy and defense research communities. But it also has drawbacks and detractors. Some critics such as Harold Brown, a former secretary of defense as well as a nuclear physicist, believe that directed-energy and kinetic-energy weapons being researched in the SDI program indeed could indeed become effective defenders against hostile satellites. But they doubt that SDI could possibly culminate in a sufficiently secure, leakproof defense against the myriad land- and sea-launched ballistic missiles and decoys with which such a defense would have to cope.

Many U.S. military officers are concerned that a national commitment to the headlong development, production, testing, and deployment of SDI weaponry would siphon enormous sums of money from other weapons that the services more urgently need. For example, the fear in some Air Force circles is that ICBM's, bombers, and cruise missiles may go begging for political support and funding if Pentagon policies and budgets become too heavily freighted in favor of defensive strategic weapons. This concern is heightened at a time of increasing political and economic pressure to restrain military spending as a means of reducing the troublesome national debt.

The possiblity of overemphasis on nonnuclear defensive weapons worries America's European allies as well. They generally support the SDI technology program but are far from enthusiastic about it, fearing a diminution of NATO's long-standing reliance on nuclear forces to deter war in Europe. France in particular was outspokenly critical of SDI for this reason in 1984. Finally SDI critics worldwide worry about its possibly adverse effect on arms control, specifically its potential for subverting the U.S.-Soviet antiballistic missile (ABM) treaty of 1972, for expanding the arms race into space (a concern also expressed about the U.S. ASAT program), and for destabilizing the so-called nuclear balance between the two superpowers.

The Reagan administration has comebacks to all such qualms. Its officials point out that the SDI program was set up to determine the feasibility and practicality (including costs) of a space-oriented strategic defensive system, not to foreordain its deployment. Moreover, they claim, SDI critics have underrated its technologies, have overstated its costs and complexities, and are premature in their misgivings about its negating arms control. They note that deployment of SDI weapons is as yet conjectural and that there are no treaty restrictions on their research.

SDI champions in the administration also contend that the Soviets have come very close to breaching (or indeed may have breached) the ABM treaty by having taken liberties in their deployment of ABM radars and ABM command, control, and communications networks, all of which could be harbingers of a future "breakout" of ABM missiles suddenly deployed in the Soviet Union. Furthermore, they claim, the Soviets have long been at work on directed-energy weapons that could be used against U.S. ballistic missiles and satellites and already possess an operational ASAT system of hunter-killer satellites capable of attacking U.S. satellites in low orbit.

To placate critics in the Pentagon itself, the administration's SDI planners deny any intention of abruptly forsaking U.S. offensive strategic weapons in the afterglow of a successful SDI; rather, the administration's goal is a gradual shift away from sole reliance on such weapons toward primary, although perhaps not total, reliance on defensive weapons to deter nuclear war.

Star Wars weaponry. The SDI program is divided into five research and technology-exploration elements: directed-energy weapons; kinetic-energy weapons; surveillance, acquisition, and tracking; systems analysis and battle management; and support. It is aimed at fashioning a "layered" defense that would intercept ballistic missiles and their warheads (reentry vehicles) in all four phases of their flight. The first of these is the boost phase, in which the first-stage and second-stage rocket engines burn and give off intense heat easily detected by infrared sensors from space. In the second, or post-boost, phase the "bus" containing the warheads (as many as 14 on some Soviet ICBM's) separates from the rocket and deploys its warheads and "penetration aids" such as decoy warheads and chaff. In the third, or midcourse, phase the warheads and decoys arc through space on ballistic trajectories. In the fourth, or terminal, phase they reenter the atmosphere.

For the SDI planners the trick is to design space weapons capable of intercepting as many missiles as possible before they disgorge their warheads. As of mid-1985 directed-energy weapons were regarded as the best bets for this task. In that category are four basic SDI concepts: space-based lasers; ground-based lasers whose beams would be reflected to targets from space-based mirrors; space-based, nuclear-generated X-ray weapons; and space-based particle beams.

High-energy lasers would reach targets at the speed of light and burn through them with thermal energy. The SDI program is developing chemical lasers of long wavelengths, on the infrared side of the light spectrum, and excimer and free-electron lasers, both of short wavelengths, on the ultraviolet side. All have advantages and disadvantages. For example, infrared chemical lasers, their beams formed by the chemical reaction of such elements as hydrogen and fluorine, require relatively little stimulative energy but need relatively large beam-focusing mirrors. Because of their shorter wavelengths, excimer lasers and free-electron lasers require smaller mirrors. The shorter the wavelength of a laser, the less mirror surface it needs for precise reflection to its target. But those mirrors must be of exquisite quality to keep the reflected beams in full focus. The U.S. optics industry is said to be capable of building such mirrors, but deploying them or assembling them undamaged in space may be another matter. One nick would make them severely flawed for their task.

Power supply in space could also be a problem. Free-electron lasers operate by passing a beam of accelerated electrons through an array of magnets. As the electrons "wiggle" through the precisely tailored magnetic field, they emit light that can be harnessed into a laser beam. Excimer lasers use a powerful electrical surge to create exotic, short-lived molecules like xenon fluoride, which emit coherent light as they decay. Free-electron and excimer lasers, respectively, require bulky electromagnetic and electrical power sources that may defy space deployment.

Adaptive optics may make it possible to forgo such placement in space by overcoming the problem of laser beams dissipating as they pass through the atmosphere. Laser systems could be positioned on land, most likely on mountaintops where the air is thin, and their beams directed at mirrors in space that would steer them to targets halfway around the world. A sensor stationed in space would monitor atmospheric conditions and continuously transmit atmospheric calibration data to a computer that is part of the laser system. That computer would constantly adjust a "deformable mirror" in the system. The warped mirror would distort the laser beam to compensate for the atmospheric dissipation that the beam would subsequently undergo. As the beam penetrated and interacted with the atmosphere, it would reacquire its coherency and other essential characteristics.

The X-ray laser is made to work by pumping laser rods with X-rays released by a small nuclear explosion. The powerful burst of radiation stimulates each rod to emit an intense, highly directional X-ray pulse of its own. This device would destroy missiles not with heat but with penetrating radiation that disrupts the missiles' innards. Highly classified, this part of the SDI program may be the most ambitious, longest-term project in the weaponry segment of the program. One way to deploy such X-ray weapons is to mount them atop land-based ICBM boosters, which would launch them into space as satellite battle stations on warning of a Soviet attack.

Particle beams composed of light neutral atoms such as hydrogen would head for their targets at only about one-fifth the speed of lasers. They would destroy missiles by disrupting their electronics and other components. Particle-beam weapons require linear accelerators for power generation and thus are intrinsically unwieldy. Nevertheless, SDI planners believe that such accelerators can be made small enough for deployment in space-based, particle-beam battle stations.

Enthusiasm for kinetic-energy weapons—basically externally propelled or self-propelled projectiles—has been growing in SDI circles. Although slower en route to targets, they have advantages over directed-energy weapons. They do not need sophisticated optics and are not susceptible to atmospheric or magnetic distortion. Moreover, they can carry their own seekers and microcomputers for midcourse and terminal guidance to targets.

Kinetic-energy weapons, which destroy by impact, can be chemically or electromagnetically propelled. One example of the former is the air-launched miniature vehicle (ALMV), a cylindrical device that the Air Force has begun testing as an ASAT weapon. It is launched on a two-stage rocket taken aloft by an F-15 fighter aircraft. Having cleared the upper atmosphere and separated from its booster, the ALMV is steered to the vicinity of its target in space by an onboard microcomputer reacting to signals from navigation satellites. Then, powered by small thrusters positioned around its circumference, it homes on the heat of its target satellite and strikes it.

Another example of a kinetic-energy weapon is the projectile that the Army developed in its Homing Overlay Experiment (HOE) program for defense against ballistic missiles. In mid-1984 such a projectile scored a spectacular success after launch atop a Minuteman I missile booster from Meck Island in the central Pacific. Once in space it separated from the booster, and its longwave infrared sensor and guidance computer locked it onto a dummy ICBM warhead that the Air Force had launched from Vandenberg Air Force Base, Calif. The HOE projectile sped toward its prey at a speed of more than 24,000 km/hr (15,000 mph) and intercepted it in its midcourse phase, more than 160 km (100 mi) high and 6,760 km (4,200 mi) downrange from Vandenberg. Just prior to impact the projectile unfurled a 4.6-m (15-ft) metal net studded with weights to enhance its chances of snaring and smashing its target. As one SDI official described the result, it was tantamount to "hitting a bullet with a bullet," and it gave the SDI program a very big lift.

Projectiles launched by means of electromagnetic fields have caught the fancy of SDI planners. Called hypervelocity railguns, such weapons are much swifter than those that derive their thrust from fuel-burning missiles. In concert with aerospace companies, the Air Force Space Technology Center at Kirtland Air Force Base, N.M., is designing such a gun to shoot rounds from an orbiting platform or from the ground. One problem may be the heavy buffeting that the guidance electronics of those rounds would take during their tremendous post-launch acceleration.

C^3I. Developing kinetic-energy and directed-energy weapons is perhaps the easiest task confronting the SDI researchers. Their toughest problem—one that is often slighted in public discussions and writings—may be battle management, the segment of the SDI program that deals with command, control, communications, and intelligence (dubbed C^3I). Without it the weapons would be worthless.

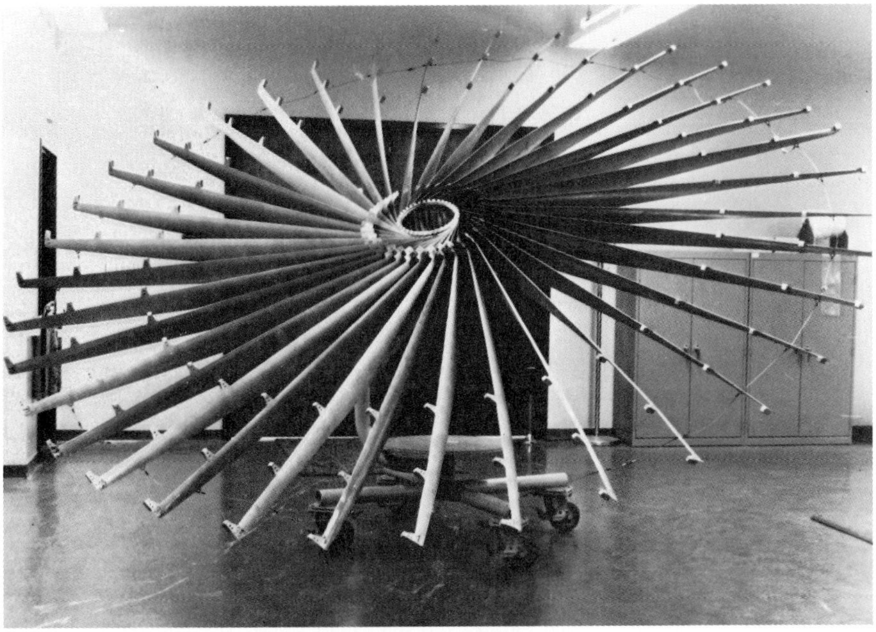

U.S. Department of Defense

Nonexplosive warhead of an antiballistic-missile projectile developed for the U.S. Army's Homing Overlay Experiment (HOE) program comprises a radial array of ribs studded with steel weights. Wound around the neck of the HOE projectile during flight, the ribs unfurl just seconds before impact with an incoming ICBM. The device was successfully tested in flight against a dummy ICBM warhead in 1984.

A multi-tiered ballistic-missile defense system that must react in a flash and without problems to an attack involving thousands of missiles would require flawless, unjammable, high-capacity communications and automatic signal and data processing under extremely stressful conditions both in space and in links between space and ground. For this reason development of C^3I technologies for boost-phase and midcourse-phase target surveillance, acquisition, tracking, and kill assessment (called SATKA) has received the lion's share of attention and funding in the SDI program. SDI officials acknowledge that the required SATKA technologies are presently embryonic at best. Included among them are millimeter-wave and laser space communications; multinode space, air, and terrestrial communications networks; and completely reliable interactions between human operators and space-based and ground-based weapons. Implicit in all this are fault-free or highly fault-tolerant microcomputers and software of unprecedented sophistication, embodying exceptionally "smart" algorithms, or problem-solving procedures. Designing and building such a system would be difficult enough; testing it prior to deployment, or even afterward, may indeed be impossible.

In conjunction with SDI the Pentagon is assembling a program called Strategic Defense Architecture (SDA) 2000. Its ambitious goal is to integrate space defense, ballistic missile defense, and air defense by the 21st century. Such integration would be a function of C^3I and would demand a great deal both of it and from it. In other words, C^3I is being called upon not only to ensure that the U.S. can deter and detect a nuclear strike, and retaliate effectively with nuclear weapons, but also to become the central nervous system of an all-embracing and highly demanding nonnuclear defensive strategic posture as well—a very tall order.

Offensive operations. Space is the key environment for U.S. military operations. It harbors the communications, early-warning, surveillance, photoreconnaissance, navigation, and weather satellites on which U.S. strategic and tactical forces by now almost utterly depend. Defending those satellites against Soviet hunter-killer satellites or lasers is the purpose of the U.S. Air Force's ASAT program. That program, described in the *U.S. Air Force Space Plan* promulgated in 1983, lays the doctrinal groundwork for the Air Force to make space a combat arena. In that respect the plan also discusses "force application," foreshadowing the development of manned spacecraft capable of offensive as well as defensive operations in and from space.

Such machines would be a far cry from the space shuttle, which is launched vertically on boosters and is intended mostly for such benign missions as deploying and repairing satellites. One variant of a combat spacecraft is the transatmospheric vehicle (TAV) being designed by the Air Force and five aerospace companies: Boeing, General Dynamics, Lockheed, McDonnell Douglas, and Rockwell. It is shaping up as a revolutionary, multimission machine capable of taking off from ordinary airfields under its own power, flying into space, and then reentering the atmosphere to carry out strategic or even tactical missions anywhere in the world. The transatmospheric vehicle and other Air Force "advanced aerospace vehicle" concepts and technologies are now being pursued under a newly consolidated military aerospace plane (MAP) program at the Air Force Systems Command.

President Reagan's message to the U.S. military to begin preparing to take the "high ground" in space actually predates his origination of the SDI program in 1983. On July 4, 1982, at the landing of the fourth flight of the space shuttle, Reagan declared what none of his predecessors ever had, that the most important goal of the U.S. space program is to strengthen national security, and he served notice in effect that, unless the U.S. and the Soviet Union can come to terms, the arms race would indeed be extended to space.

—James W. Canan

Earth sciences

Increased international cooperation was a notable feature of the past year in the Earth sciences. The United States, Soviet Union, and many other countries participated in experiments together and also exchanged information gained in their own studies. Geologists continued to analyze catastrophic events of the past in an effort to determine their role in influencing changes of the Earth's surface and among its plants and animals. Hydrologists focused on the widespread and severe drought in Africa and performed follow-up research on the El Niño event of 1982–83.

Atmospheric sciences

During the past year four areas of the atmospheric sciences achieved notable advances. These areas were concerned with the development of specialized monitoring capabilities, the planning and performance of observational field programs, the implementation of new forecasting techniques, and the investigation of inadvertent climatic change.

Monitoring the weather. As in previous years there was continued development of capabilities to monitor the atmosphere. These observational systems included surface-based as well as Earth-orbiting instruments. On Oct. 5, 1984, the Earth Radiation

Budget Satellite of the U.S. National Aeronautics and Space Administration (NASA) was launched by the space shuttle "Challenger." Designed to monitor the Sun's radiation reflected back from the Earth and also the radiation emitted by the Earth, this satellite became a vital component in the system of monitoring climatic change. The increasing frequency of shuttle launches was expected to enhance this capability in the coming years.

Surface-based measurement systems were taking advantage of the Earth-orbiting satellites. Wind, temperature, humidity, and other meteorological and hydrological variables could be sampled in remote locations, after which the data would be transmitted to satellites and then telemetered back to ground stations on the Earth. This could be accomplished in real time so that weather changes near the ground in isolated locations could be closely monitored. The size and shape of the surface-based systems varied considerably, but in general, apart from the small antenna that was necessary, their size was no larger than an average suitcase, with a weight of no more than 50 kg (110 lb).

The monitoring of thunderstorms can also be achieved by using ground-based observational systems called lightning locators. Until recently satellite and radar observations were the chief tools used to track thunderstorm activity. Lightning locators represent a new technique to monitor lightning. At least two types were developed to detect cloud-to-ground lightning strikes. Both record time, location, flash polarity, and stroke count of the strikes. Lightning locator systems were being installed in order to provide continuous monitoring of thunderstorms in many parts of the world.

The value of other types of remote sensing tools continued to be demonstrated. The pulsed carbon dioxide infrared Doppler lidar, as applied by the U.S. National Oceanic and Atmospheric Administration (NOAA) Wave Propagation Laboratory, was being used to monitor a wide range of atmospheric properties, including the thickness of volcanic dust layers in the stratosphere, the prevalence of cirrus clouds, and the spatial and temporal variations in the wind near the ground over distances of a few kilometers. This information was obtained by monitoring the backscatter from coherent electromagnetic radiation having a wavelength of 10.6 micrometers. This is one of the primary tools used to continue to observe the quantity and spatial distribution of residual aerosols in the stratosphere remaining from the El Chichón volcanic eruption of 1982.

The launching of lidar systems on future space shuttle flights was under consideration. Referred to as wind sensing coherent infrared radars, these systems were intended to provide a global mapping of winds as a function of height for use as input to global

NASA

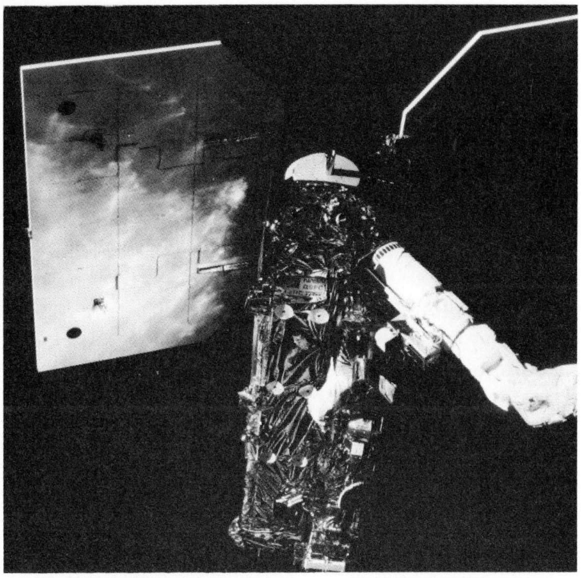

Earth Radiation Budget Satellite of NASA, shown here just before its release into orbit, monitors the Sun's radiation reflected back from the Earth and also the radiation emitted by the Earth.

circulation numerical models. The information contained in the laser radiation that was backscattered from atmospheric aerosols such as water droplets and dust particles is used to determine the winds.

The Doppler radar demonstrated its versatility by observing a wide range of meteorological phenomena, including wind shears dangerous to aircraft, tornadoes, and the wind flow in the vicinity of weather fronts. Employing electromagnetic wavelengths ranging from three centimeters to ten centimeters, scientists used the backscattering of the radar beam from water droplets and aerosols to estimate velocity in the direction of the beam. By using two radars with the same region of coverage (referred to as dual Dopplers), they could obtain a complete description of the three-dimensional wind velocity. These radars can be flown on an aircraft, such as those of the U.S. National Center for Atmospheric Research (NCAR) Research Aviation Facility, or moved around in trailers. In several countries, including the United States and Sweden, Doppler radars were being implemented as part of routine weather forecasting.

Observation programs. In order to enhance the understanding of atmospheric processes a wide range of measurement experiments was performed during the year, and additional research was planned for the future. The Genesis of Atlantic Lows Experiment, planned for January to March 1986, is designed to improve knowledge concerning the mechanisms of winter storm development along the east coast of the United States. During the winter such storms often develop considerable intensity, producing high winds and rain along the coast from Cape Hatteras, N.C., northward and heavy snow inland.

In the summer of 1984 scientists from NOAA and Colorado State University participated in a field experiment to monitor large organized areas of thunderstorms, referred to as mesoscale convective systems, which form east of the Rocky Mountains at night. Flash floods and other severe weather are often associated with these systems, which can cover areas as large as 100,000 sq km (40,000 sq mi). The program was designed in anticipation of a much more ambitious project, known as STORM-Central (Storm-scale Operational and Research Meteorology program in the central United States), which is being planned for the central Great Plains in the late 1980s.

Lightning associated with thunderstorms was the focus of a field program in the late summer of 1984 centered at the New Mexico Institute of Mining and Technology's Irving Langmuir Laboratory. In addition to such observation equipment as aircraft, radars, and balloons, small rockets trailing grounded wires were launched to trigger lightning when conditions were right.

Safe aircraft landings and takeoffs were the focus of the Joint Airport Weather Studies program in 1984. Rapidly sinking air near the ground associated with evaporating precipitation from cumulus clouds, referred to as downbursts, and resultant large wind shears can catastrophically influence aircraft operations through the loss of lift. In Denver, Colo., during the summer of 1984 forecasts of suspected downbursts were transmitted to pilots. A crash almost occurred in Denver during the late spring of 1984, when a commercial pilot ignored the guidance of the forecasters concerning the presence of a downburst in the vicinity of the airport.

Improving the understanding of the fundamental processes by which trace atmospheric constituents are transferred between the troposphere and the stratosphere was the goal of NASA's stratospheric-tropospheric exchange project. During five major missions scheduled to take place from 1984 to 1987, high-altitude aircraft are to measure a range of atmospheric components including ozone, carbon monoxide, water vapor, liquid water, ice crystals, and aerosols. Knowledge of aerosol exchange from the troposphere into the stratosphere is critically important to climate. In contrast to the troposphere, where materials are usually flushed out by precipitation within a few days or weeks, aerosols have a much larger residence time in the stratosphere. An increase in concentration at that height can act to cause cooling at the Earth's surface by reflecting more sunlight back into space.

An international program to determine the effect of clouds on the atmospheric circulation of the Earth was the format of the five-year Project FIRE (First International Satellite Cloud Climatology Program Regional Experiment), begun in late 1984. Separate missions to study cirrus clouds over the midcontinental U.S. and marine stratocumulus clouds off the southwest coast of California were scheduled to be performed in both 1986 and 1988. Knowledge of cloud properties and how they influence the radiation budget of the Earth is essential if an adequate understanding of climate and climate change is to be achieved.

The Tropical Ocean and Global Atmosphere (TOGA) research program is a ten-year international project that began in 1985 as a joint project of the Joint Scientific Committee of the World Climate Research Program and the Committee on Climatic Changes and Ocean. It was designed to investigate the ways in which oceanographic and atmospheric changes in the tropics influence climate worldwide and includes a joint air-sea interaction study in the western Pacific planned by China and the U.S. The project involves two consecutive five-year expeditions, for which China is to provide research vessels and the U.S. instruments and equipment. As many as 36 nations are expected to participate in TOGA.

Forecasting techniques. Improved prediction of weather and climate requires better forecasting techniques. In 1984 significant advances were obtained in both very short-range predictions and in forecasts of significant long-term climate. Over the longer time frame J. J. O'Brien of Florida State University introduced a procedure that permits the prediction of the El Niño four to eight months before it occurs. (The term El Niño refers to a major warming of the water along the South American Pacific coast associated with the reduction or elimination of large-scale wind flow parallel to the coast. Climatologists have concluded that this warming causes major weather disruptions from normal weather patterns throughout the world.) In 1983 the El Niño was blamed for such climatic anomalies as record-setting rains in Peru and Ecuador as well as drought in Australia, Indonesia, Africa, India, and Spain. Using a statistical model, O'Brien forecasts the El Niño by certain early distinctive anomalies in the surface wind stress over the Pacific.

On the shorter time scale NOAA continued to expand its capabilities to forecast hurricane storm surges along the Atlantic and Gulf coasts of the U.S. Potential storm surge impacts on Buzzards Bay in Massachusetts, Narragansett Bay in Rhode Island, Delaware Bay in Delaware and New Jersey, and Matagorda Bay and Lower Laguna Bay on the central and south Texas coast were completed in 1984. Storm surge, the coastal flooding that occurs before the arrival of a tropical cyclone, is the leading cause of deaths in such storms throughout the world.

A new procedure to alert the public to health risks during heat waves was introduced in 1984. A heat index defined by NOAA measured the contribution made by high humidity in association with abnormally high temperatures in reducing the body's ability to cool itself. This index can be considered the "apparent temperature" that the average person feels for various combinations of high temperature and humidity. With an actual air temperature of 38° C (100° F) and a relative humidity of 50%, for example, the apparent temperature would be 49° C (120° F). Sunstrokes and heat exhaustion are likely when the heat index reaches that level.

During the 1984 Olympic Games, held in Los Angeles, this index and a wide range of other weather information were prepared for use by Olympic officials and participants. The most sophisticated weather monitoring and forecasting system ever assembled in support of the Olympics was made available during the events. Solar-powered, automated weather stations were located at each of the outdoor game sites. An Olympic Weather Center was established at the National Weather Service Forecast Office in Los Angeles, with a special weather facility at Long Beach Harbor to support the yachting races. Meteorologists from the Canadian and French national meteorological services participated in preparing the Olympic event forecasts.

The capability of meteorologists to analyze current weather conditions and to provide very short-range forecasts continued to improve in 1984. This scale of weather description and short-period forecasting is referred to as nowcasting. In September 1984 the International Association of Meteorology and Atmospheric Physics, the World Meteorology Organization, the European Space Agency, and the Swedish Meteorological and Hydrological Institute sponsored the second International Symposium on Nowcasting in Norrköping, Sweden. Major advances in nowcasting were reported at that symposium including the United Kingdom Meteorological Office Mesoscale Forecasting System. This system included the first operational prediction model in the world capable of representing the effect of local hills, islands, and different land use patterns on local weather. These small-scale variations are often critical in determining weather conditions observed at a specific location. The French national meteorological service was developing a similar model.

Inadvertent weather modification. A major finding released during the past year by the Illinois State Weather Survey was that frequencies of thunderstorms over areas that are as large as the North American continent show major long-term trends. From 1901 to 1945 there was a general increase of about 15% followed by a general 10% decrease from 1945 to 1980.

Concern continued with respect to the potential impact on climate of the increase of carbon dioxide in the atmosphere. Additions of carbon dioxide to the atmosphere as a result of the combustion of fossil fuel can retard radiational cooling to space, thereby causing a net warming at the Earth's surface. Unless mitigated by other results of human activities, such as reduced sunlight at the ground due to additions of aerosols to the upper atmosphere, this warming could result in major changes in climate patterns. In 1984, as part of the continued study of this phenomenon, NOAA used aircraft to estimate how much carbon dioxide is transferred from the atmosphere into the North Atlantic during winter storms when the cold ocean waters are most efficient in absorbing carbon dioxide.

Acid deposition research also continued to receive considerable attention during 1984. Such deposition, which can fall either in precipitation (acid rain) or be deposited in dry air, has been attributed both to industrial and automotive emissions. NCAR and other groups were attempting to develop models that would identify the source of the acid and the physical and chemical mechanisms by which it is transported to other locations. Despite the lack of complete understanding of the acid deposition problem, specific political measures were enacted. In 1984 the state of New York passed a law that would reduce its sulfur dioxide and nitrogen oxide emissions, the main precursors of acid deposition, by as much as 30% by 1991. In December 1984 seven northeastern states filed a lawsuit against the U.S. Environmental Protection Agency, charging it with failure to curb acid rain.

Heightened interest and concern continued in 1984 regarding the climatic impact of a major nuclear exchange between the superpowers. It was postulated that the enormous volumes of soot and smoke that would be ejected into the upper troposphere and stratosphere by surface thermonuclear releases and by the huge fires following such an exchange would darken the sky for an extended period. Sunlight reaching the ground would be reduced, possibly resulting in subfreezing temperatures at the surface throughout a substantial portion of the world, even during the summer. Considerable uncertainty remained as to the climatic impact of such a catastrophic event, however, and intensive research was expected to continue into 1985 and beyond.

—Roger A. Pielke

Geological sciences

Among the major developments in the geological sciences was the use of seismic waves generated by earthquakes to study processes taking place in the Earth's interior. An earthquake that took place in

California justified a prediction made in 1980. Paleobotanists analyzed fossil pollen to infer climatic conditions during the geologically recent past.

Geology and geochemistry. It has been the fashion in recent years to refer to almost any scientific change as a "revolution." But if that term is reserved for fundamental changes in the theoretical structure of a science, then one would expect, on the basis of past experience, that such revolutions would occur very infrequently. From this point of view there were no revolutions in geology during the past year. There were, on the other hand, scientific developments which, while falling short of revolutionary significance, nonetheless dramatically altered the direction of research.

Catastrophes. A recent development in geology promised to shape the character of research, at least in the near future. This development was a renewed interest in catastrophes, unusual events of great magnitudes as agents of geologic change.

Accounts of cataclysmic events, such as floods and earthquakes, have played a significant role in literature. One such account that has intrigued both scholars and laymen since antiquity is that of the destruction of the continent Atlantis contained in Plato's *Timaeus* and *Critias*. In a special issue of *Earth-Science Reviews* published in 1984 Zdenek Kukal of the Central Geological Survey of Czechoslovakia concluded, on the basis of an exhaustive review of geological and archaeological evidence, that there was no Atlantis and, therefore, no destruction of Atlantis. He argued that Plato used Atlantis and its destruction allegorically in dialogues devoted primarily to moral and political questions.

The current interest in catastrophic events appears to date from 1980, when Luis Alvarez, Walter Alvarez, Frank Asaro, and Helen Michel announced their hypothesis, based largely on geochemical data, that an asteroid of such magnitude as to cause the mass extinction of plants and animals had struck the earth about 65 million years ago. The publication of this hypothesis triggered a widespread renewal of interest in unusual events of great magnitude as significant factors in the history of the Earth and its life. Among the most intriguing suggestions was that by David Raup and John Sepkoski that there is a discernible periodicity to episodes of mass extinction. Walter Alvarez and R. A. Muller reported that a study of impact craters supported the view that the periodicity of large impacts on the Earth's surface was closely correlated with the periodicity of episodes of extinction. (For additional information on this subject *see* Feature Article: PERIODIC COMET SHOWERS: THE MAINSPRING OF EVOLUTION?)

During the 1920s, at a time when geological explanations invoking cataclysmic events were not at all in vogue, J. Harlan Bretz suggested that the unique erosional features of western Washington and the Columbia River drainage system could be explained by the sudden emptying of a late ice-age lake that had covered much of what is now northwestern Montana. Glacial Lake Missoula, as it came to be called, was drained during a short period of time, according to Bretz, when a dam formed by a lobe of glacial ice failed during a period of melting. The erosion caused by the resulting flood removed many of the preflood drainage features of western Washington, resulting in the region described as the "Scablands." So great was the volume of water from Lake Missoula that it flooded the Columbia River gorge to a depth of hundreds of feet. More recent studies, including those of Bretz himself, suggested that there was more than a single flood. In a paper published in the *Journal of Geology* in 1980 R. B. Waitt estimated that there

Scale model of a coal mine is loaded into one end of the 7.6-meter (25-foot) arm of a centrifuge. The arm is then rotated at speeds as great as 155 revolutions per minute. Such a test simulates the high pressures on an actual coal mine caused by gravity and overlying earth.

Sandia National Laboratories

were probably about 40 floods emanating from Lake Missoula during the last glaciation. This estimate was consistent with findings of Brian F. Atwater, who reported that there had been 15 floods from Lake Missoula during just a part of the last period of glaciation.

Sherwood B. Idso, writing in the July 1984 issue of *Geotimes,* suggested that at least one long-predicted catastrophe may not occur. For many years scientists have warned that increasing concentrations of carbon dioxide (CO_2) in the atmosphere that have resulted from the burning of fossil fuels might result in a drastic change in the Earth's climate. It has been supposed that while increased concentrations of CO_2 in the atmosphere will have little effect upon the amount of solar radiation that reaches the Earth's surface, it will significantly decrease the amount of thermal radiation transmitted back into space and thereby cause an increased heating of the atmosphere. Idso claimed that reports by the U.S. National Academy of Science and by the U.S. Environmental Protection Agency on this "greenhouse effect" were deficient in two ways. First they failed to address objections that had been raised against their predictions, and second they failed to acknowledge the possibility that an increase in the temperature of the atmosphere might actually have beneficial consequences. Idso cited experimental studies which indicated that a doubling of the carbon dioxide concentration in the atmosphere would result in increased crop yields of up to one-third and a decrease in water loss by plant evaporation of one-third, which would result in a greater efficiency of water use by plants.

Paleobotany. A foundation for an eventual test of the hypothesis that increasing concentrations of CO_2 in the atmosphere will lead to dramatic changes in climate was being laid by the cooperative Holocene Mapping Project involving scientists from Brown University, University of Wisconsin, University of Minnesota, Oxford University, and the Lamont-Doherty Geological Observatory. The objective of COMAP was to map the global climate at 3,000-year intervals during the past 18,000 years. The principal means of inferring climates of the geologically recent past is provided by pollen preserved in sediments of the Holocene Epoch (the last 10,000 years). It is assumed that plants in the past had climatic requirements similar to those of plants of the same species in the present. Studies by members of the consortium have already cast doubt upon the existence of the altithermal period of 6,000 years ago, when temperatures were supposed to have been appreciably higher than they are today. Studies of the distribution of trees based upon an analysis of fossil pollen suggest that southeastern North America was no warmer 6,000 years ago than it is today.

The assumption that plants in the past lived under much the same climate as plants of similar form in the present serves as a foundation for inferences about climatic conditions in the recent geologic past, but for remoter periods in the Earth's history, when plant life was markedly different, that assumption must be employed with caution. Bruce H. Tiffney and Leo J. Hickey of the Peabody Museum of Natural History at Yale University noted that in studies of ancient climates paleobotanists were augmenting their systematic research on ancient plants with information from other disciplines. For example, of increasing importance in the study of plant distribution and its relationship to ancient climates was the knowledge of the relative positions of the continents during geologic time, knowledge that was gained largely from recent advances in geophysics.

A recent paleobotanical study, while not primarily concerned with climate, also illustrated the interdisciplinary character of contemporary paleobotany. Geologists had for a long time been intrigued by the upright, petrified tree stumps found buried in sediments of the Eocene age (50 million–60 million years before the present) on Specimen Ridge in the northern part of Yellowstone National Park. As long ago as 1879 W. H. Holmes concluded that, judging from their upright position and their relationship to the surrounding sediments, the stumps had been buried in the place where they had grown. In 1980, however, W. J. Fritz suggested that the stumps had in fact been transported by running water and then deposited in their upright position. Later that year Fritz supported his argument by pointing to stumps that had been transported into place on Mt. St. Helens in the aftermath of recent volcanic eruptions. But the controversy over the origin of the stumps on Specimen Ridge promised to continue. In a paper published in *Geology* in 1984 Richard F. Yuretich of the University of Massachusetts concluded that, although there is abundant evidence of fluvial processes similar to those present at Mt. St. Helens, the evidence still pointed to the burial in place of the stumps on Specimen Ridge.

International cooperation. During the past year international cooperation in geological science continued without interruption. Dallas Peck, director of the United States Geological Survey, led a delegation of 343 U.S. geologists in attendance at the 27th International Geological Congress held in Moscow during August. Howard R. Gould, chairman of the U.S. National Committee on Geology, thanked the organizing committee of the congress for providing the opportunity for an exchange of geological knowledge and announced that the council of the 27th Congress had accepted an invitation to hold the next congress, which was scheduled for 1989, in Washington, D.C.

Tass/Sovfoto

At the 27th International Geological Congress, held in Moscow in August 1984, the world's geologists demonstrated increasing agreement on questions of theory.

The International Geological Congress demonstrated not only a cordial relationship among geologists but evidence of increasing theoretical agreement as well. The theoretical differences that separated Soviet and Western geologists had never been so great as those which separated Soviet and Western biologists, but the Soviet geological community had been cautious in its acceptance of the theories of continental drift and plate tectonics. During the past decade, however, Soviet geologists began to invoke plate tectonics in their explanation of crustal motion. In an article on the state of Soviet geology published in 1984 in *Episodes* Yevgeny A. Kozlovsky and Alexandr L. Yanshin, after noting that plate tectonics serves to describe some tectonic phenomena, went on to assert, "There is, for example, little doubt about the existence of independent vertical movement in the crust. These are caused by phase transitions in the Moho and Conrad discontinuities, by the rise to the base of the lithosphere of hot low-density mantle material, by isostatic compensation and by other factors." By contrast, few U.S. geologists continued to point to phase changes in the mantle as a major source of crustal motion.

Reflecting their intense scientific interest in the deep crust and mantle and their desire to explore more deeply for petroleum, the Soviet scientists undertook an ambitious program of deep drilling. They reported during the year that the first of these drill holes had reached a depth of 12 km (7.4 mi). Other deep boreholes were planned. It appeared, however, that because of technical problems and great expense drilling much beyond that depth would be limited. For that reason seismic reflection would remain the principal means of probing the Earth's deep crust.

In a quite different application of seismology 34 countries, including the United States and the Soviet Union, participated in an experiment during 1984 involving the international exchange of seismological information. The three-month-long experiment, sponsored by the Conference on Disarmament in Geneva, was organized in the hope that an exchange of detailed seismological information would improve the means of detecting nuclear explosions and thereby enhance the prospects for verifiable arms treaties. (For additional information on seismology, see *Geophysics*, below.)

Energy. The problem of diminishing supplies of fossil fuels, which commanded much of the attention of the geological community during the 1970s and early 1980s, had by no means been solved by 1985. Its urgency had diminished, however, largely because of declining prices for petroleum on the world market, which had resulted in part from energy conservation efforts by consuming nations. Petroleum exploration continued, especially in those countries hoping to achieve some degree of self-sufficiency in energy supplies, but it did so at a diminished rate. As the readily available sources of petroleum were exploited, the search inevitably became more difficult and expensive.

Much of the remaining undiscovered crude oil within the jurisdiction of the U.S., for example, is believed to occur in offshore areas, and in the Alaskan Arctic. The high cost of exploration in the Arctic was dramatically illustrated by the fact that the average drilling cost for a well in the Beaufort Sea is \$40 million–\$50 million as compared with only \$1.5 million–\$3 million in the Gulf of Mexico. Conventional offshore drilling techniques are unsuitable in Arctic environments, where the principal hazard is ice. Several solutions to the problem were suggested by the experience of the Canadians, who have engaged in Arctic offshore drilling for several decades. One such solution is to artificially thicken the ice pad around the drilling platform during the winter in order to guard against ice motion when such motion is not great. A more dramatic solution, and a very expensive one, is to construct man-made islands from material dredged from the sea bottom. The Canadians succeeded in building such islands to depths of 19 m (63 ft) off the Mackenzie Delta.

The search for alternative sources of energy diminished sharply during the past few years as a result of the easing of the situation in regard to petroleum, but it did not come to a complete halt. Marianne Guffanti of the United States Geological Survey reported that between the end of 1982 and the end of 1983 the geothermal electrical generating capacity increased worldwide from 2,857 Mw to 3,397 Mw. The U.S. continued to lead in the generation of electricity from geothermal energy, and the Philippines, Indonesia, and Kenya all added substantially to their geothermal generating capacity.

The problem of the safe and economical disposal of radioactive waste continued to plague the many countries that planned to increase their use of nuclear energy as a source of power. In an article published in the June issue of *Episodes* James Harrison reviewed the national and international efforts to cope with this problem. The search for disposal sites, not surprisingly, reflected the geological environments of the countries involved. Canada and Sweden, for example, were engaged in research and exploration to assess the suitability of crystalline rocks as disposal sites. On the other hand, in Belgium and The Netherlands, where crystalline rocks are virtually absent, the possibility of storing nuclear waste in salt domes and layers of clay was being tested.

—David B. Kitts

Geophysics. The dynamic processes responsible for, among other things, the distribution of the Earth's oceans and continents and the formation of mountain ranges, ocean trenches, and oceanic islands were studied during the past year in cooperative efforts on the part of scientists from many countries. Two developments were aiding those studies: the acquisition of much high-quality digital data from global networks of seismometers, and the use of sophisticated computer processing methods to obtain information about earthquake sources and seismic velocities and flow patterns in the Earth.

Natural hazards. As usual there were during the year numerous reminders of the Earth's dynamism, as manifested by volcanoes and earthquakes. Fortunately, few lives were lost in those events. Some 59 people were killed in 14 earthquakes in 1984, but no one earthquake was truly devastating. There were fewer than normal large earthquakes. Although on the average about 18 earthquakes larger than magnitude 7 on the Richter scale occur each year, only 6 such events took place in 1984. The low loss of life cannot necessarily be attributed to the abnormally low seismicity, however, for many of the earthquake-caused deaths throughout history can be attributed to the collapse of poorly built dwellings during moderate earthquakes.

One of the moderate earthquakes that struck during the year was interesting not because of the damage it caused (although the damage was estimated to be $7.5 million) but because of its importance in the effort to predict earthquakes. The earthquake had a magnitude of 6.1 and occurred on April 24, 1984, along the Calaveras Fault near Morgan Hill, Calif. The dense seismometer network and repeated geodetic surveys in the area gave no indication of the impending quake, suggesting that short-term premonitory signals may be difficult to observe. On the bright side the earthquake occurred in a segment of the Calaveras Fault along which William Bakun of the U.S. Geological Survey (USGS) had forecast just such an event in a paper published in 1980. His forecast was based on the seismic gap hypothesis, which holds that future earthquakes along an active fault are likely to occur in those segments of the fault that have been quiet in comparison with surrounding regions of the fault. Bakun noted not only the dearth of seismic activity in that area since 1911 but also the relative linearity of this segment of the fault, bounded by bends and offsets of the Calaveras Fault at either end. The lack of seismic activity indicated the place, and the length of fault without geometric complexities suggested the magnitude of the earthquake.

The essential missing ingredient that would have made Bakun's prediction a useful one was the time of the event. Farther southeast, along the San Andreas Fault near Parkfield, Calif., earthquakes have occurred almost periodically about every 22 years (1857, 1881, 1901, 1922, 1934, and 1966). Thus a forecast of the time for future quakes can be made in that area. Bakun and his colleague Allan Lindh, also at the USGS, estimated that the next Parkfield earthquake should occur within a ten-year period centered on 1988. Although such a time estimate is of no help in short-term preparation for an earthquake (such as evacuation of people and mobilizing emergency equipment and personnel), it does provide an excellent reason for intensely monitoring the area in the hope of detecting precursory geophysical signals. That area of the San Andreas Fault is probably the best understood and certainly the most well-monitored fault in the world.

Volcanic activity during 1984 differed little from that in recent years, with eruptions in more than 30 volcanoes distributed in more than 20 countries. Several eruptions were interesting. The world's southernmost volcano, Mt. Erebus in Antarctica, produced the strongest explosive eruptions since its discovery in 1841, and on Luzon Island in the Philippines eruptions and mudflows from Mayon Volcano forced the evacuation of 73,000 people.

Although volcanic activity is usually thought of in terms of spectacular eruptions of volcanic mountains, much of the transfer of molten rock from the depths of the Earth to the surface takes place unobserved, under the cover of the world's oceans. In a continual process of rebirth and destruction of the skin of the Earth, molten rock oozes out of fissures along the boundaries between the huge tectonic plates that comprise the Earth's crust. This magma cools and thus adds to the Earth's surface. A steady state seems to exist, however, for other portions of the plates are being consumed at subduction zones.

Occasionally the submarine volcanic processes are more dramatic than a mere oozing of magma into the seafloor. In 1984 an explosive eruption destroyed Home Reef in the Tonga Islands, and a new island was formed by the subsequent volcanic activity. The eruption produced rafts of lighter-than-water, air-filled rock called pumice. These rafts were so extensive that in Fiji they forced ships back to port. Another underwater event during the last year was worthy of note, if only to emphasize the peculiar ways in which the Earth's dynamic processes can affect its residents. An apparent volcanic explosion beneath Lake Manoum in Cameroon produced shock waves and a five-meter-high water wave, as well as a white foglike poisonous cloud that killed 37 people.

Seismic tomography. Although earthquakes are a result of dynamic thermally driven processes in the Earth, they can be used to investigate those processes as well. Earthquakes radiate seismic waves, some of which travel through the Earth's interior (body waves) and some of which spread out along the surface of the Earth (surface waves). The transit time of these waves from the source to a seismographic station is a function of the density and elastic properties of the rocks through which the waves pass. These properties, in turn, are controlled by the mineralogy of the rocks, the pressure, and the temperature.

Temperature is particularly interesting because variations in the temperature can be clues to such dynamic processes in the Earth as convection currents and plate movements. For example, in a region of upwelling of hot materials beneath a spreading center (usually along mid-ocean ridges), the seismic velocity should be high relative to surrounding areas. The challenge, then, is to use the transit times of earthquake waves to infer the distribution of seismic velocity in the interior of the Earth. Because the variations of seismic velocity are generally much greater with depth than with horizontal position at a given depth, much of seismology has been concerned with deriving the globally averaged depth-dependence of seismic velocity. These investigations led to division of the Earth into crust, upper mantle, lower mantle, outer core, and inner core.

It is the horizontal variations in velocity, however, that convey information about dynamic processes. Defining these lateral variations has captured the attention of seismologists around the world. Not until recently was such research possible, at least on a global scale. By 1985, however, decades of careful measurements of seismic travel times had provided an extensive data base, and the increased power of computers and the development of sophisticated processing techniques made it possible to utilize the data effectively. The processing scheme bears a close resemblance to computer-aided tomography (CAT) in medicine, in which the variations in absorption of X-rays are used to construct a three-dimensional "map" of the body's internal organs. As used in seismology the technique employs the multitude of paths traveled by seismic waves between the earthquake sources and the seismographic stations comprising the data base. For any one ray, deviations of the observed transit time from the time predicted on the basis of a globally averaged reference model of the Earth's seismic velocity reflect variations from the average model along the whole ray path. The transit time deviations for two ray paths that intersect, however, share some information in common about the seismic velocity at the point of intersection. With enough crisscrossing paths the local perturbations in seismic velocity with respect to the reference model can be extracted.

Development and application of seismic tomography were vigorously pursued during the last several years, primarily by groups at Harvard University, the California Institute of Technology, and the University of Utrecht in The Netherlands. Several studies used more than 500,000 rays to map the velocity variations in the Earth. Many of the features, such as slow velocities under the mid-ocean ridge system (indicative of hot rock) and fast velocities under stable continental shields (underlain by cooler material), had been predicted in a general way from the present understanding of plate tectonics. Seismic tomography, however, also discovered some unexpected or previously undetectable characteristics of the Earth. For example, the hot material beneath mid-ocean ridges does not consist of simple vertically rising sheets of magma; instead, lateral flow of magma from a few source zones is involved. For example, the material responsible for both the East Pacific Rise and the Southeast Indian Ocean Ridge comes from a single region under the South Pacific at depths of several hundred kilometers.

Global and transportable arrays. In spite of the great success already achieved by seismic tomography in mapping the Earth's interior, the resolution attainable was limited by the geographic distribution of earthquakes and seismographic stations. Of course, nothing can be done about the distribution

Garcia—Gamma/Liaison

Mayon Volcano on Luzon Island in the Philippines erupts on Sept. 16, 1984. The volcano generated rivers of molten lava and mudflows that caused the evacuation of approximately 73,000 people.

of earthquakes, but major efforts were under way to increase both the quality and quantity of the seismic data. These data were expected to be useful not only for probing the Earth's interior but also for exposing the mysteries of the earthquake source. This latter result, in particular, would have important consequences for efforts in earthquake hazard reduction. As of 1985 few records of the ground shaking close to a large earthquake were available, yet it is such earthquakes that control the seismic design of such structures as nuclear power plants, dams, and hospitals.

The seismic network is the fundamental tool used by seismologists, just as telescopes and accelerators are fundamental tools for astronomers and physicists. Many of the seismic networks in use today are a direct result of the need to monitor underground nuclear explosions. The basic global network, termed the World-wide Standardized Seismograph Network (WWSSN), came into existence in the early 1960s and greatly stimulated seismological research by providing easily obtainable data from a network of well-calibrated similar instruments.

Although the WWSSN was still providing useful data in 1985, technological advances as well as demands for broader bandwidth with more dynamic range than it could provide led to several major programs to develop new seismological networks. Some of the programs were concentrating on dense arrays in seismically active regions, such as the Parkfield area of the San Andreas Fault in California. Others were global in scope. They included a network of long-period three-component instruments run by the Institut de Physique du Globe (IPG) in Paris and a major proposal for seismic recording by a consortium of more than 30 universities in the U.S. organized as the Incorporated Research Institutions for Seismology (IRIS).

The IPG network was planned to expand to 20–25 stations worldwide. These stations were designed to collect data in two quite different ranges of frequencies: very-long-period motions with periods greater than about 100 seconds and broadband data with periods between 1 and 100 seconds. Remarkably, both types of data will be derived from the output of a single seismometer. The long-period data were expected to yield information about large-scale features of the Earth (those with dimensions of hundreds to thousands of kilometers), and the data of shorter period would allow study of smaller features and of the mechanism involved in the rupturing of earthquake faults.

The IRIS proposal was ambitious. It included a global digital seismic array with real-time satellite transmission from 100 seismographic stations, a mobile array of 1,000 portable digital seismographs, a central data management and distribution facility, and a computational facility for analysis of the data. The use of satellite telemetry and real-time processing in the global array makes it possible to assess the potential of major earthquakes for generating tsunamis (giant sea waves), to plan disaster aid programs, and even to assess the size of volcanic eruptions in remote places (if explosive, these eruptions send out seismic waves just as do earthquakes). Even without those benefits the improved quality and distribution of the instrumentation was expected to improve the resolution of geologic structures and allow finer details of the earthquake source to be studied.

The other major part of the IRIS proposal called for many self-contained, portable, digital seismic recording units. These units were to be used in a wide range of studies, using both explosions and earthquakes as sources. The studies would include large-scale cooperative projects targeted on specific geologic structures, passive array deployments in which seismic waves from earthquakes would be used to study crustal and mantle structure in different regions, and specialized studies using smaller arrays formed with groups of 50–100 instruments. This latter application might include the detailed location of aftershocks from a large earthquake and the mapping of ground motion amplification in urban areas underlain by varying thicknesses of soil.

—David Boore

Hydrological sciences

Hydrologists during the year studied the interaction of decreased rainfall and increased human activities as the cause of desertification in some areas of the world. Among the noteworthy developments in oceanography was the discovery of the deepest known plant life in the world's oceans.

Hydrology. The year was one of extremes in hydrology. In Bangladesh heavy rains in August 1984 brought widespread flooding and loss of life. Northeastern India was similarly affected by floods in June and August. In contrast, widespread areas of Africa suffered continuing drought conditions that led to extreme famine, large-scale migrations, and large numbers of deaths from hunger and disease.

Droughts were also reported from large areas of western Europe; the east Ukraine and northern Caucasus in the Soviet Union; the southwest United States; the southern prairies of Canada; and Liaoning, Jilin, Fujian, and Hunan provinces of China. In England and Wales the rainfall total from April to August, amounting to 56% of the long-term average, was the second lowest of the 20th century, only 1976 being lower. The drought primarily affected the western part of Britain, which normally serves as a major water supply source. More than half of the population of England and Wales was affected by water restrictions, the imposition of water rationing being only narrowly averted by timely rains in September. Sufficient rainfalls and good weather in lowland Britain resulted in record cereal grain harvests.

However, problems of water supply that were experienced in western Europe paled into insignificance beside the problems of drought in Africa. The drought affected areas in both northern Africa south of the Sahara (the Sahel region) and in a band across southern Africa from Angola and South West Africa/Namibia to Mozambique. Nearly 30 countries were listed by the United Nations Food and Agriculture Organization as dangerously short of food.

The failure of rains in 1984 exacerbated the effects of earlier dry years in those regions. In fact some areas have suffered below-average rainfalls ever since the "Sahelian" drought of 1968–74, most notably parts of Mali and Burkina Faso (formerly Upper Volta). The levels of Lake Chad began to fall in the early 1960s with notable lows reached in 1972, 1974, and the years since 1980 when the level fell to such an extent that the Great Barrier ridge separated the lake into two pools. A major project planned in 1971 to distribute water from Lake Chad for irrigation purposes could not be used in recent years because the lake level was no longer above the intake. The period since 1981 is believed to be the worst drought in Chad in the 20th century.

In Kenya, until 1985 a grain-exporting country, the failure of the "long rains" in February 1984 led to widespread crop failures in the northern and central areas. Scientists feared that the water supply to the capital city of Nairobi, which received only 25% of its normal rainfall in 1984, would become critically short in 1985. In the Sudan food shortages caused by drought were exacerbated by influxes of large numbers of refugees from Ethiopia, Uganda, and Chad caused by both famine and civil war in those countries. In Ethiopia at least six million people were at risk of starvation because of crop failures caused by drought.

This brief summary can do no more than outline the appalling problems caused by drought in Africa. These problems appeared to be difficult to alleviate, despite the amount of aid that has been channeled into the countries of the Sahel since the drought of the early 1970s. These countries are situated in an area where rainfalls will always be uncertain and drought is to be expected as part of the normal course of events. The natural ecosystems of the marginal areas south of the Sahara are adapted to drought, including the traditional nomadic social systems of the human inhabitants.

Desertification. The seriousness of the current period of drought appears to have been magnified by the increasing pressures on the resilience of the vegetation and social systems caused by the growing populations of people and grazing animals. This conjunction of population growth and drought has resulted in desertification. The United Nations Conference on Desertification in 1977 was a direct result of the international concern generated by the Sahelian drought. Studies presented at the conference estimated that the area affected by desertification was increasing at a rate of 6 million ha (15 million ac) per year. A recent summary by Robert Lamb published in the World Meteorological Organization *Bulletin* of October 1984 suggested that this rate has not decreased since 1977 and that 35% of the world's land surface and 850 million people may be threatened. No change in the rate of desertification is expected before the year 2000; in fact, deterioration of agricultural areas may accelerate.

It has become an increasingly important scientific question as to whether the deterioration of the Sahel and similar areas is part of a long-term natural change or whether human influences may be the primary cause of desertification. A number of recent research papers addressed the question of whether long-term trends in rainfall can be determined. Peter Lamb of the Illinois State Water Survey examined annual rainfall data for sub-Saharan West Africa and showed that lower than normal annual totals persisted from 1968 to the present time. J. C. Olivry, a French researcher, similarly demonstrated that there has been no sign of recovery of rainfall in the western Sahel. However, a study in the Sudan suggested that the run of dry years since 1968 may not be a statistically significant departure from normal conditions in that region. The study indicated that a drought from about 1900 to 1919 may have been more severe than that of the present time in terms of rainfall amounts. It also pointed out that annual rainfall totals may be somewhat misleading in regard to the impact of drought on human and natural systems. The annual statistics may conceal information about changes in the magnitude and frequency of short-term rainfalls and also about localized changes in rainfall. The researchers found that changes in the larger daily rainfall amounts closely paralleled the changes in annual totals but that this was not the case for daily totals of less than 10 mm. They also discovered that there may be significant variations in rainfall totals over short distances. This localization of rainfall was apparent over time periods ranging from 48 hours to five years with no clear spatial pattern. Such local variations may be important contributors to the desertification process.

There were also suggestions that the effects of human activity may be leading to a long-term desiccation of the Sahel. It was postulated that any reduction of the vegetation cover (such as might be caused by overgrazing) would lead to an increase in the reflectivity of the Earth's surface to solar radiation, which, in turn, would intensify subsidence in the overlying atmosphere. This would have the effect of suppressing convection and the formation of rainbearing clouds, thereby prolonging periods of low rainfall. A second feedback hypothesis proposed that overintensive land use in areas at risk might reduce the water-holding capacity of the soil, thereby limiting the amount of water available for evaporation and again suppressing the formation of rain-bearing clouds. Both hypotheses were supported by modeling studies such as those of J. G. Charney of the Massachusetts Institute of Technology.

These hypotheses assumed that much of the rainfall of the Sahel and similar areas is derived from local moisture, in which case desertification could indeed be self-sustaining. However, in a review of desertification in the World Meteorological Organization *Bulletin*, F. Kenneth Hare of the University

John Chiasson—Gamma/Liaison

Parched earth in northern Mali reveals the severity of the drought that has afflicted the Sahel region in Africa, an east-west belt extending from Senegal to the Sudan and lying south of the Sahara.

of Toronto suggested that most climatologists would look to global causes for the widespread droughts experienced in Africa and elsewhere. Over the long term the global atmospheric circulation demonstrates relative stability. This view therefore implies that the present drought is due to a natural fluctuation in climate and that there will eventually be a return to wetter conditions.

Carbon dioxide. One effect of human activity with truly global consequences is the possible climatic effect of increasing concentrations of carbon dioxide (CO_2) in the Earth's atmosphere. Some of the hydrological implications of such changes were addressed at a joint meeting of the Royal Meteorological Society and the newly formed British Hydrological Society by Tom Wigley, director of the Climatic Research Unit at the University of East Anglia. The warming effect of increased CO_2 concentrations should lead to increased evaporation from the oceans. Increased evaporation should lead to increased rainfall. Wigley suggested that the resulting changes in the atmospheric circulation should lead to increased precipitation in higher latitudes and a general lengthening of the dry season in middle latitudes.

Changes in precipitation, however, might be more than offset by the direct effects of changes in CO_2 levels on plants. A doubling of atmospheric CO_2 levels can lead to a 30% increase in the efficiency with which plants use water. This could result in a reduction in evapotranspiration and consequently increase runoff of water even in areas where precipitation might be expected to decline.

Thus, globally, through increasing CO_2 levels, and regionally, through overintensive use of marginal lands, man may be having a direct effect on atmospheric circulation and hydrology. A proper understanding of the complex processes involved has not yet been achieved, but it is clear that without further hydrological research and planning in areas at risk the impacts of human activities on drought and desertification will be increasingly severe.

—Keith Beven

Oceanography. Application of high technology ranging from satellites to high-speed computers to fiber optics continued to reveal new features of the ocean. Oceanographers in all disciplines achieved a deeper understanding of the causes of environmental change as the interactions among physical, chemical, geological, and biological processes became clearer. Plans were laid for new global programs to study the ocean's role in climate change.

El Niño and global climate change. As the major El Niño event of 1982–83 disappeared into the natural "noise" of more normal climate behavior, oceanographers turned their attention to assembling an accurate description of the episode so that it could be compared to previous events and understood dynamically. Effects of the El Niño could be seen as far north as the eastern Bering Sea, where a significant warming lagged the major event by 6 to 15 months. The warm water and storms associated with the El Niño also constituted a catastrophic-scale disturbance to California kelp communities. During the summer and fall the layer of warm water extended to depths below that of the kelp forests, leading to depleted nitrogen in plant tissues and deterioration of the major species.

Biological productivity was observed to decrease by a factor of 20 in the coastal and island regions where the normal upwelling of cold water did not occur and by a factor of four in the open ocean. This was caused by increased mixing in the surface layers of the tropical oceans as well as by the movement into the regions of nutrient-poor waters.

Satellite data were used to advantage in the study of the recent El Niño, particularly for biological studies. The dramatic changes in the patterns of ocean color around the Galápagos Islands revealed a redistribution of food resources that could have contributed to the reproductive failure of sea birds and marine mammals.

Field studies revealed new evidence in both the Pacific and Atlantic tropical oceans for the existence of long internal waves that are propagated westward along the Equator during an El Niño event. Such evidence was found in data from instruments that measure heat content, profiles of current, and sea level. These waves are fundamental to the changing dynamics of the tropics during an El Niño, and so the information was expected to be an important component of the predictive models being constructed.

Following several years of planning and field studies, a major international conference was held in Paris in September to assess present knowledge about ocean-atmosphere interactions in the tropics, as particularly exemplified by the 1982–83 El Niño event, and to present a scientific plan for a program to begin in 1985. (*See* Atmospheric Sciences, *above*.)

Global circulation. Studies of ocean circulation continued throughout the world. For example, the distribution of chlorofluoromethanes used as chemical tracers, taken during the Long Lines expedition of January–February 1984, showed that the flushing time (time required to change its characteristics in an essential way) of the Weddell Gyre (a giant surface current in the Weddell Sea) is on the order of a decade or less. Similar measurements were used to reveal the extent of exchange between the Indian Ocean and the South Atlantic Ocean in the region of the Agulhas Current.

Along longitude 152° W in the Pacific Ocean between Hawaii and Kodiak, Alaska, scientists studied large-scale currents, density, nutrients, and tritium. The data they obtained were consistent with a condition of upwelling of water driven by gradients in the surface wind stress, as predicted by theory. Additional support to the theory was provided by work carried out with data on the southeastward shift and anomalous strengthening of the Aleutian low-pressure system in the 1982–83 winter when compared with data gathered from moored current meters at latitude 42° N and longitude 152° W. New results from numerical models run on high-speed computers showed that by including the ocean circulation and heat flux it is possible to obtain a realistic prediction of the seasonal movement of the edge of the ice in the Greenland Sea.

Sea level as an indicator of global climate change received increased attention during the year. Time-scale variations on the order of ten years were shown to extend deep into the ocean in contrast to seasonal and interannual variations that are concentrated in the upper layers. The interactions among changes in the amount of carbon dioxide, the chemistry of the ocean, and sea level were shown to be important as influences on climate.

Planning continued for the World Ocean Circulation Experiment (WOCE), which focuses on understanding the ocean's role in climate. Core programs included a global survey of the circulation and a study of the dynamics of the North Atlantic/Gulf Stream system. Central to WOCE were to be satellite measurements of winds and currents.

As part of the growing interest in large-scale circulation, the study of trans-Pacific sections was planned for March and September 1985 along latitudes 24.5° N and 47.8° N. Scientists expected these sections to provide data on the structure of currents, heat transport, and chemical tracers.

Real-time data distribution and forecasting. With the increasing recognition of the importance of the ocean in a variety of practical areas has come the demand for real-time (that is, within a short time after the data have been collected) forecasting. Although as of 1985 this could not be done with any degree of accuracy, studies during the past two years revealed that under certain circumstances and in limited areas such forecasts might be feasible. For help in forecasting El Niño-type events the Intergovernmental Oceanographic Commission and the World Meteorological Organization established a project to provide monthly maps of sea-surface topography based on information from tide gauges. Data came from Chile, Ecuador, Fiji, Kiribati, Nauru, New Caledonia, New Zealand, Papua New Guinea, Peru, the Solomon Islands, and the U.S.; the maps were to be issued within four weeks after the end of the respective month.

Acoustic techniques for providing near real-time descriptions of the internal structure of the ocean continued to be developed. Real-time data assimilation into regional circulation models was achieved for an area off the coast of California. During this study hydrographic data were collected in a region of about 200 km (120 mi) on a side, and a numerical model was used to forecast changes in the density of the seawater, which is then used to calculate ocean currents.

Open-ocean biology and chemistry. A multidisciplinary study of warm-core rings identified four major processes that determine the evolution of these eddy-like structures that are detached from such currents as the Gulf Stream. They include overturning caused by winter storms, seasonal restratification in the spring, catastrophic interactions with the Gulf Stream, and slow decay by turbulence. The rings are useful for biological study; they initially have the low levels of phytoplankton typical of their parent waters, but then the biomass rapidly grows. This occurs because the dynamics of the rings produces upwelling in the center, bringing more nutrients to the upper layers. Nine years of satellite observation proved important for establishing the frequency, distribution, and lifetimes of the rings as they grow, decay, and coalesce.

As demonstrated by the warm-core rings program the physics of the ocean must be measured together with the biology; for example, measurements made from a stable research platform in the Pacific provided some of the first direct observations of three-dimensional flow within the surface mixed layer of the ocean. There was some evidence that the observed narrow regions of downwelling jets transport heat and phytoplankton within the mixed layer. Direct measurements of velocity as a function of depth in a region in the Atlantic seaward of Cape Hatteras, N.C., showed current-induced upwelling that could be responsible for increased amounts of phytoplankton found there.

During the spring and fall of 1984 the Subarctic Pacific Ecosystem Research project conducted two month-long cruises to the Gulf of Alaska. The goal of the program was to examine the interactions among mixing, nutrient availability, light, phytoplankton growth and standing stock, and grazer (of phytoplankton) abundance and feeding. The aim was to understand why the phytoplankton stocks are nearly constant in the subarctic Pacific during all seasons.

Strong interest continued in the study of hydrothermal vents, as high temperatures typical of such vents were discovered near the Juan de Fuca Ridge off the coast of Washington. Further studies of the chemistry near the vent fields on the East Pacific Rise and in the Guaymas Basin of the Gulf of California were carried out. A distinct hydrothermal plume was discovered 200 m (656 ft) over the vent field area of the Endeavor Ridge in the northeast Pacific Ocean at latitude 48° N, longitude 129° W.

Evidence from deep-sea drilling in the southeast Pacific Ocean demonstrated that hydrothermal activity fluctuated widely in the geological past. Evidence was presented in support of the hypothesis that increased hydrothermal emissions in the past caused a global warming.

The study of fluxes of biological and geological aggregates from the surface to the bottom of the ocean continued to be an important activity. Natural radioactivity was used to calibrate the efficiency of the sediment traps used for these studies, and a strong seasonality was observed in the total flux.

The deepest known plant life was discovered on an uncharted seamount near The Bahamas by scientists using a submersible vehicle. The living marine algae, large enough to be seen with the naked eye, were found at a depth of 268 m (879 ft); it had been thought that plant life could not develop below about 200 m (650 ft) because of low light intensity. The great abundance of algae in this previously unknown deep-sea realm underscored their potential widespread importance to productivity, the marine food web, sedimentology, and reef biogenesis in clear tropical oceans.

Coastal regions. A joint U.S./China study of the water structure and circulation in the Yellow Sea used both field measurements and satellite imagery to show the effects of the Kuroshio current system. Detailed measurements were also made on the Alaska Coastal Current, in the South and East Coast Middle Atlantic Bight, and off the Australian coast. In the latter study measurements were made of both currents and winds in order to test coastal trapped wave theory.

During October 1984 the International Court of Justice resolved the dispute between the United States and Canada over mineral and living resources at the two countries' common Atlantic maritime border. The disputed area was roughly the top one-third of the Georges Bank area; the court awarded each country approximately half of the area. The border had been a source of conflict between the two countries for several decades, but the issue was not forced until each country passed legislation delineating 200-mi (322-km) fishery zones in 1977.

Processes at the seafloor. The High Energy Benthic Boundary Layer Experiment showed that currents at a depth of almost 5,000 m (16,400 ft) had significant effects on the deposition of silt on the seafloor. The processes responsible for deposition and erosion were being studied with a recirculating flume that was deployed on the seafloor at the site. The flume was able to impose a given shear stress for periods similar to those of deep sea storms, about five days, and thus allowed controlled experiments. The experiment also documented the effects of surface currents caused by winds on the global patterns of sediments on the ocean floor.

Fluxes of chemicals from the seawater near the ocean bottom into the seafloor continued to be a subject of great interest. The biogeochemistry of such processes is complex because of interactions both with turbulent currents and with the deep-ocean fauna. The first direct determinations of fluxes of the trace metals manganese, copper, nickel, vanadium, molybdenum, and silicon dioxide were made from sediments of the San Clemente Basin off the coast of southern California. The instrumentation that was used was a free vehicle that gently settled to the seafloor and sequentially collected samples at the site from an isolated parcel of bottom water over an undisturbed sediment surface. The data suggested that a large fraction of the particulate metal and silicate rain to the sediments is recycled from interfacial sediments to overlying waters.

Deep-sea drilling. Significant progress was made toward the conversion of the drilling ship "SEDCO/BP 471" into a scientific drilling vessel. The new ship, which was to replace the "Glomar Challenger" used from 1968 to 1983, was designed to provide capabilities not previously available to the marine geology and geophysics community. They included the ability to drill in 8,180 m (27,000 ft) of water and a drill string of 9,100 m (30,000 ft), enabling sampling in the world's deep ocean basins and trenches. The ship was strengthened to withstand ice so that it could operate in higher latitudes, heavier seas, and stronger winds than could the "Challenger." The ship was to be equipped with 930 sq m (10,000 sq ft) of scientific and laboratory space and an up-to-date array of geophysical instrumentation and computers, providing scientists with the best possible environment for their work while aboard for the two-month-long expeditions.

Ocean Drilling Program, Texas A & M University, College Station

Deep-sea drilling ship "SEDCO/BP 471" is maintained over a specific location by a computer system while drilling in water up to 8,180 meters (27,000 feet) deep.

The new ship was to be known as the "JOIDES Resolution," JOIDES being the acronym for Joint Oceanographic Institutions for Deep Earth Sampling, the international group that has guided the program since 1968. Members of that group in 1985 included West Germany, France, Canada, Japan, the United Kingdom, the United States, and the European Science Foundation representing The Netherlands, Norway, Sweden, Italy, and Switzerland.

The new ship, to be operated by Texas A&M University, began drilling near The Bahamas in mid-January 1985 in order to study carbonate bank development and to collect information on young sediments. During the remainder of the year the ship was scheduled to work in the North Atlantic, the Norwegian Sea, and the Labrador Sea. In 1986 and 1987 it was expected to move through the Panama

Canal to the eastern South Pacific Ocean and then south to the Weddell Sea.

New technology and design. Renewed interest was shown in designs for research vessels, and new instruments continued to appear. The U.S. Navy announced its intention to build a new research vessel for operation by the academic research community. The design for the new ship may be of the Small Waterplane Area Twin Hull (SWATH) type, which uses twin submerged, streamlined hulls to support a structure that rides above the surface on struts that are attached to the hulls. Thus the deck shape is entirely unconstrained by the hull shape. Moreover, the ship motion is greatly reduced and can be made nearly independent of wave direction relative to ship heading. For example, for a given displacement a SWATH ship can provide twice the amount of deck space that a single-hulled vessel can offer, and the ship motion is only 10% of that experienced by a single-hulled craft. Although the SWATH concept is not new, the renewed interest was due to the design's ability to meet many of the required multidisciplinary characteristics of a research vessel. A second design that was under consideration was a sail-assisted ship, either with one or two hulls. As a general-purpose ship, the twin-hulled design would provide less wave-induced motion, less vibration and noise, and more flexibility.

Microprocessors and fiber optics dominated instruments development during the year. The inverted echo sounder, a device that measures the time for a sound pulse to travel from a bottom-mounted instrument to the surface and back again and thus monitors thermal changes in the water structure, was designed to be controlled by microprocessors in order to achieve greater flexibility. The device was being used routinely for monitoring changes in the heat content of ocean layers to accuracies of a few percent. Modern low-power electronics enabled the instrument to be deployed for 18 months at a time and to collect 10 million bits of information on a self-contained tape recorder.

The Canadian Hydrographic Service developed a microprocessor-controlled tide gauge that both collects data and then transmits it by radio or telephone. The microprocessor controls all the electronics and external circuits, turning on the radio and sensors only when required for a measurement. The system can operate for 47 days with a single battery.

The use of fiber optics also became more prevalent. Because those fibers, which transmit optical signals, provide signal transmission rates up to 1,000 times greater than the insulated wires normally used for transmitting information, they were rapidly replacing the older conventional wires. New well-logging cables using fiber optics and electrical conductors can provide data transmission with a capacity in excess of 10 megabits per second. This type of cable is of importance for transmitting information from sonar and seismic arrays, and from moored buoys to shore or island stations. The development of new armored designs for such cables made them increasingly more useful for oceanographic purposes.

—D. James Baker

Electronics and information sciences

During the past decade computers and communications have become so closely entwined that it is increasingly difficult to distinguish one from the other. This trend continued, as evidenced by the entrances of the largest computer company (IBM Corp.) into the communications business and the largest communications firm, American Telephone and Telegraph Co. (AT&T), into the computer business. Both areas of endeavor employ electronic devices that have become smaller and faster with each passing year. Also, each year fewer people have been dealing with "products" and more people have been dealing with "information."

Communications systems

The speed and capacity of communications systems increased dramatically during the past year. Just as dramatic was the appearance of new applications

Words are displayed on a computer terminal screen in response to spoken commands. IBM developed this experimental speech recognition system.

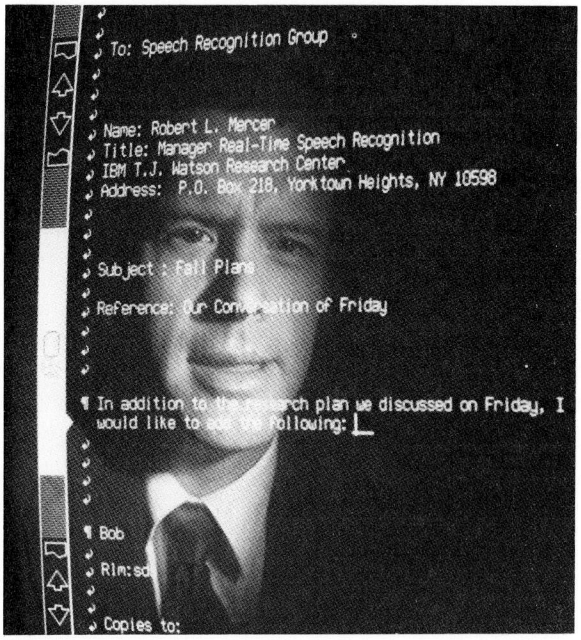

IBM

for those systems and the improvement in the technologies that they employed. In a number of cases during the year technologies such as fiber optics and cellular radio moved out of the laboratory and into the real world.

Central office switching. One of the most significant results of the breakup of AT&T and the Bell System was the sudden opening up of the market for telephone equipment. As of 1985 approximately 80% of the telephone subscribers in the United States were served by one of the 22 Bell operating companies. Previously those companies purchased almost all of their equipment from Western Electric Co. (the manufacturing arm of AT&T, now renamed AT&T Technologies), and hence 80% of the U.S. market for such equipment was essentially closed. As a result of the breakup, however, these telephone companies may purchase their equipment from any supplier. Consequently, there was a great deal of activity on the part of suppliers throughout the world.

This activity was especially marked in regard to central office switching equipment. (Central office switching equipment is used to connect one subscriber to another. Modern switching equipment is "digital" in nature; that is, a voice is sampled many times each second, and the result of each sample is converted to a coded character. This character is transmitted to the distant central office, where it is decoded and converted to an analog "voice" signal.) The market for central office switching equipment, for the Bell operating companies alone, was approximately $366 million in 1983. This was expected to increase to more than $2.5 billion in 1987. It was understandable, therefore, that manufacturers from throughout the world were competing for some of this market with U.S.-based companies such as AT&T Technologies, GTE, and ITT. They included Northern Telecom (Canada), Plessey (U.K.), CIT-Alcatel (France), Siemens (West Germany), NEC (Japan), and L. M. Ericsson (Sweden).

Equal access. Until recently the vast majority of all long-distance telephone traffic in the U.S. was handled by AT&T Long Lines. This long-distance network was accessed by dialing "1," then a three-digit area code, and then the seven-digit telephone number. The move to deregulation in the U.S. permitted other long-distance carriers to operate, but accessing the network of those carriers was extremely tedious. Often it required dialing 22 or more digits. And, also, the connecting facilities were inferior. As a result, access charges—those fees paid to the local telephone company by the long-distance carrier for the privilege of accessing the local network—were set lower for the other carriers than for AT&T.

However, one of the requirements of the consent decree that separated the Bell operating companies from AT&T was that these operating companies provide "equal access" to all carriers. Thus, by 1985 equipment was being designed and installed to make this possible. Industry observers expected that the entire program would be substantially completed in 1986 at a cost of about $1.5 billion.

Corning Glass Works

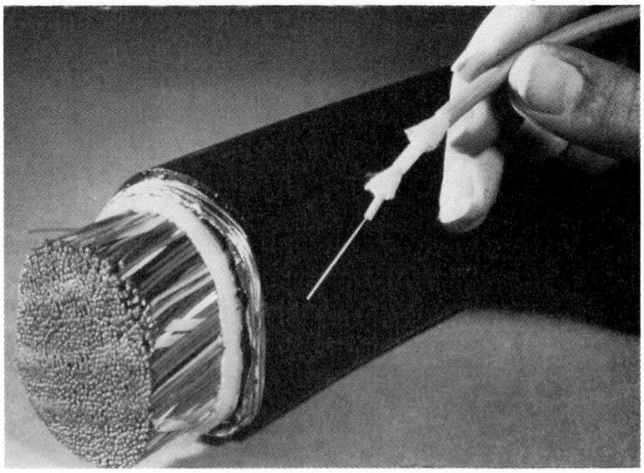

Single glass fiber (right) can carry more than 1,000 telephone messages simultaneously, the same number as can the standard wire cable at left.

When the program is completed, each subscriber will preselect his "preferred" carrier, and thereafter by dialing "1" he or she will gain access to it. Other carriers may be accessed by dialing a slightly more complex series of numbers.

ISDN. During the year many companies from around the world were hard at work on the Integrated Services Digital Network (ISDN). This was to be a fully digital telecommunications network providing a variety of services. As of 1985 no full-fledged ISDN system or subsystem existed, and it seemed likely to be another decade before substantial implementation could be achieved. However, segments of the network were being tested; one in Venice, Italy, by L. M. Ericsson, a Swedish firm that provided much of the equipment, and by SIP, the Italian telephone operating company.

The overall ISDN program was being guided by the International Telecommunication Union, based in Geneva. When implementation is finally achieved, subscribers will have the following capabilities: digital end-to-end connectivity; support of a wide range of services via a limited set of standard interfaces; enhanced telephone service; high-speed data service; packet-switched data transmission service; and global connectivity between users' equipment.

Specialized networks. Until a few years ago nearly all communications within the U.S. were transmitted through the standard telephone network. This network was arranged in a hierarchical fashion. Subscribers were connected to what were known as Class 5 offices (or End Offices); these were connected to other Class 5 offices (and thus to other subscribers) by way of a Class 4 office, and so on up to Class 1. This "pyramid" was still in existence in 1985 and remained an effective system. However, specialized networks were beginning to evolve, due both to the rapid advance of technology and to the advent of competitive long-distance networks. Although the majority of traffic continued to be carried over the standard telephone network, largely the property of AT&T Communications (formerly AT&T Long Lines), companies such as MCI and GTE Sprint were establishing their own long-distance networks. These networks were connected with local subscribers at the Class 5 office, which belonged to the local telephone operating company.

Another long-distance network structure was used for packet switching of data transmission. Data transmitted from a subscriber's place of business at relatively low rates of speed would be collected and briefly stored at a switching center of the packet network. Then this would become part of a "packet" of data that would be transmitted at very high speed to a switching center near the destination, at which point it would be stored and then transmitted at a slow rate of speed to the intended recipient.

Specialized networks were not used only for long-distance traffic; in fact, much attention was being given to local area networks. The main question in this regard involved how a particular company could transmit data at high speeds between various departments within one building or between buildings. Several possible solutions were suggested. In the Star system all points on the network would be connected, perhaps by fiber optics or by coaxial cable, to a central point. This central point, a voice/data private branch exchange (PBX), would perform all switching. This approach was advocated by manufacturers of telephone equipment. In the Ring arrangement all points would be connected in a high-speed ring. Sophisticated electronic techniques would be employed to prevent collision of messages sent simultaneously by several stations. Xerox, with its Ethernet system, espoused this method. In the Tree system branches would extend from a backbone communications channel to one or more stations. Again, techniques would have to be developed to prevent the collision of messages.

It appeared that no one configuration would be the "winner." All were technically feasible, and it seemed likely that all would survive. Indeed, hardware and software that would marry the different systems and allow message transmission from one to another began to appear in recent months.

Fiber optics. "Photonics" is the name applied to the transmission of information at frequencies so high that the carrier signal is in the infrared or even the visible light range. The medium for such transmission is a glass fiber no thicker than a human hair, thereby giving rise to the name "fiber optics." Perhaps the most interesting activity in fiber optics during the past year was the establishment of a number of fiber optic networks using the rights-of-way of railroads. It seemed certain that the U.S. soon would be laced with such networks. In light of this, one might wonder about the future of satellite and microwave transmissions. It seemed certain that both would continue to be widely used, but it should be noted that fiber presents an equally viable alternative.

The use of fiber optics for communications has a number of distinct advantages. First, and very important, is cost. Glass, made from silicon dioxide, is cheap as compared with the copper required for transmission by wire. But the greatest advantage probably is the ability of a fiber system to carry large amounts of information owing to the high frequency of the light being used as a carrier. In fact, a single strand of glass is able to carry thousands of voice conversations. (For additional information about fiber optics, see *Electronics,* below.)

Cellular radio. Cellular radio systems have operated in Norway, Sweden, and Denmark for several years, and the rush to implement cellular radio

Telephone calls from airplanes to the ground could be made from several airlines. The cordless system operated by means of radio wave transmissions.

Airfone, Inc.

continued in the United States. Many systems were under construction, and each month more were becoming operational.

The greatest advantage of cellular radio systems was their capacity to serve a large number of subscribers. A conventional radiotelephone system was prone to severe overload, a condition that worsened as more and more subscribers attempted to obtain service. This occurred because a metropolitan area was assigned a limited number of frequencies, and each could carry only one conversation.

With cellular systems the power and, therefore, the range of the transmitter at the base station was quite low, and so the particular set of frequencies used by the base station in its relatively small geographical area ("cell") could be reused by another base station several miles away. Therefore, the size of a metropolitan area was not a limiting factor; more cells could easily be added (although this was far from inexpensive), and additional groups of subscribers could be brought into the system.

A cellular system offered two distinctive features. The first was "hand-off." As an automobile in which a passenger received a telephone call moved from one cell to another, it was not acceptable to force the call to be terminated; somehow it had to be transferred (or handed off) to the transmitter and receiver of the new cell. A cellular system acomplished this by careful monitoring of the automobile's signal strength in both cells. When it became obvious that the subscriber had moved out of one cell and into another, the master computer mechanism (normally located in a main office called the Mobile Telephone Switching Office—MTSO) through the active cell instructed the automobile radio to change its transmit and receive frequencies to those assigned to the new cell. This was done automatically and rapidly, and the subscriber (even if he or she was talking on the telephone at that moment) was unaware of the transfer.

The second feature was the ability to locate a subscriber anywhere in the service area. This was done by programming the MTSO to send a query to the particular mobile unit from each cell in the serving area. When a response was received, the call was assigned to that particular cell.

In the U.S. it was decided that two companies would be allowed to establish a cellular system in each metropolitan area. One would be a wire-line company (normally a telephone company or a consortium of telephone companies), and the other would be a non-wire-line firm (usually a radio common carrier or consortium of radio common carriers). Some interesting applications were being considered for cellular systems, including some nonmobile applications that, for instance, in a rural area could make telephone poles and wire unnecessary.

A somewhat different application using radio (although not cellular per se) provided telephone communications from commercial airplanes operated by several major carriers. The passenger used a cordless telephone from his or her seat on the airplane. Charges, however, were relatively high.

Other services. A number of developments involving the world's communications systems resulted as much from political pressures as from technological advances. In both the U.K. and the U.S. deregulation of the telecommunications industry was making it possible for entrepreneurs to enter the marketplace and was stimulating the development of various kinds of hardware and software.

Interactive videotex, now several years old, was an example. Subscribers, by using a telephone and a television monitor, "called up" information from huge data banks and even purchased items by dialing additional digits. In many parts of France telephone books were being replaced by such equipment. A second example was shared-tenant services. Entire building complexes were wired so that a "communications landlord" could handle all of the telecommunications needs of the tenants—from installation of equipment to selection and provision of the least expensive long-distance telephone service.

Yet another service was concerned with fire and burglar alarms, using telephone wires. And still another involved automatic reading of water, gas, and electric meters. Each of these applications resulted from the realization that a telephone was used, on the average, only 20 minutes per day. Thus, it seemed certain that they would be only the first of many additional uses for this resource in the months and years ahead.

—Robert E. Stoffels

Computers and computer science

The past year in computer science was marked by a major scientific advance in algorithms for optimization problems. Also during the year the Unix operating system, once used only by computer scientists, gained wide acceptance throughout the industry. The pace of technological developments increased, prompting many new computer designs and business ventures by new and existing companies.

Algorithms. An algorithm is a method for solving a particular computational problem. Algorithms are analogous to blueprints in the construction industry—they tell how to solve problems without referring to the specific details. When programmers construct software, they work from algorithms, either those that they invent themselves or those that computer scientists have invented and written down, filling in the details needed to make a particular computer solve problems. Algorithms to solve trivial problems

Courtesy, AT&T Bell Laboratories

Narendra Karmarkar, a scientist at AT&T Bell Laboratories, demonstrates the algorithm he developed for solving a large class of computational problems called linear programming problems.

are obvious, and some programmers can work for days or months without ever using complex algorithms; they merely use the computational methods they learned in grammar school. However, programmers responsible for complex scientific software cannot rely on conventional computational methods. Programmers who work on computationally intensive problems such as weather predictions or airplane-wing design need special-purpose algorithms. They must choose algorithms to solve their problems carefully because poor algorithms can make their programs uselessly slow.

A young scientist at AT&T Bell Laboratories named Narendra Karmarkar surprised computer scientists during the past year when he announced an important new algorithm for a large class of computational problems called linear programming problems. These are mathematical optimization problems, and they arise in such diverse industries as telecommunications, automotive manufacturing, oil exploration, and airline traffic routing.

Until the recent discovery linear programming problems had been solved with a method developed nearly 40 years ago by George Dantzig of Stanford University. Dantzig's algorithm, called the simplex method, considers possible solutions one after another until a correct solution has been found. To make the search for a solution more efficient, the simplex method tries to avoid examining candidate solutions based on knowledge of mathematical properties of the linear programming problem on which it is working. With simplex, many linear programming problems required hours or days to solve on the world's fastest computers. Other problems were simply too large for current computers.

Karmarkar's algorithm will change much of that. The crux of his discovery is a new way to avoid considering candidate solutions unnecessarily. Because the new method eliminates candidate solutions much faster, its search for the optimum solution moves along rapidly, and it finishes more quickly than the simplex method. In an early experiment Karmarkar compared a program using his algorithm to one of the best-known implementations of the simplex method. For typical problems having about 5,000 variables, the simplex algorithm took 50 times longer than Karmarkar's to find the solution. More important, the speedup was related to problem size; the new algorithm was expected to work much faster than the simplex method on larger problems. Thus scientists and engineers will be able to solve problems that had been too large for even the fastest computers.

Karmarkar's algorithm has another surprising advantage over simplex; it can be divided into pieces and run on a computer with multiple processors in order to solve problems even faster. Although it might seem that all algorithms should run faster on a multiple-processor machine, they do not. Many, including simplex, must perform one computation at a time to work correctly. When simplex is programmed for a multiple-processor machine, it does not run faster than the single-processor version because each processor must duplicate the computations performed by other processors. Karmarkar's algorithm, however, can be divided in a way that does not duplicate effort.

Both Karmarkar and Dantzig used mathematics to derive their algorithms. For an extended discussion of this aspect of the subject, *see* Year in Review: MATHEMATICS.

Unix operating system. Most computers contain software called an operating system that controls the execution of programs, manages resources, and maintains permanent storage files on disks or tapes. Early operating systems were crude and large. They depended on particular machines and only worked with particular printers, disks, and tapes. In the mid-1960s a group of computer scientists met at the Massachusetts Institute of Technology to study and build a new type of operating system. They called the system Multics. Two participants in the Multics project, Kenneth Thompson and Dennis Ritchie, returned to Bell Laboratories, where about 1970 they developed their own operating system based on some of the principles from the Multics project. They observed that Multics was too slow and that it required some special-purpose hardware. They concentrated on keeping their system efficient and independent of any particular computer hardware.

By the mid-1970s more researchers at Bell Laboratories had joined with Thompson and Ritchie, and the group began to use the system for its day-to-day computing. A colleague, Brian Kernighan, coined the name Unix (to stand for the uniprogrammed version of Multics). Bell Laboratories released Unix to universities, where it became immensely popular among computer scientists. Universities began exchanging software that used Unix as well as improvements to the system itself. Several universities started major research projects based on Unix.

The decision to keep Unix independent of particular computer hardware paid off handsomely. Replacing old computers with new models had always been painful because old programs had to be rewritten for the new machines. With Unix, however, researchers were no longer dependent on computer manufacturers. Once they moved Unix onto the new machine, their old programs worked. Unix, thus, was the first portable operating system.

By 1980 the Bell System was developing Unix, using it internally and selling it commercially. Bell Laboratories had begun using Unix for many of its development projects, while Western Electric Co. sold it as a commercial product. Researchers continued developing new versions and incorporating improvements discovered at universities. Among colleges and universities Unix was universally accepted. More than 90% of the computer science departments in the U.S. ran Unix on at least one computer; this was far more than any other operating system. Many computer science departments adopted Unix as their primary computer system.

Unix PC, developed by AT&T, can stand alone as a single-user workstation but also has the processing power to handle eight additional users at the same time.

Courtesy, AT&T Bell Laboratories

After years of apathy toward Unix the major commercial manufacturers in 1984 began introducing Unix-based hardware and software products. As expected, in the wake of the telephone system divestiture AT&T entered the computer market with a variety of Unix-based systems. It was somewhat surprising, however, that the two leading computer manufacturers, IBM Corp. and Digital Equipment Corp., both moved into the Unix marketplace. Under license with AT&T, Digital Equipment Corp. introduced its own version of Unix and announced that it would be available for most Digital Equipment machines. Other leading computer firms such as Hewlett-Packard Co., Honeywell, Sperry Corp., and Apple Computer, Inc., also announced the development of Unix-oriented products. Finally, because Unix can be adapted to new machines easily, many computer makers offered it as the basic software on their new products. The result was that Unix became one of the most widely used operating systems.

New computer designs. Rapid improvements in computer technology continued to fuel the development of new computers. The price of silicon chips continued to decline, making it possible to build computers that are more powerful but cost less.

One of the most surprising developments in 1984 was an announcement from AT&T that it had been able to build a one-million-bit random access memory (RAM) chip that could be in mass production within about one year. IBM and the Japanese firms NEC Corp., Fujitsu Ltd., and Hitachi Ltd. also had built one-million-bit RAM chips, but they were still in experimental stages and much further from mass production. The capacity of RAM memory chips had been doubling every few years, but the increases announced in 1984 were expected to have an especially profound effect on the capacity of computers. The industry standard 64K-bit chip can hold roughly 64,000 bits (zeros or ones). When chips that hold 256,000 bits appeared in 1984, some scientists believed that silicon technology would reach a limit near 512,000 bits per chip. The announcement of more than one million bits on a chip surprised many of them and stimulated research into machines that have memories far larger than present systems.

The succession of technological improvements was so rapid that it kept the computer market relatively unstable. Customers found it difficult to choose between increased computational power and dramatically lower costs. New companies often found that their products had become obsolete even before they were delivered.

During 1984 sales of the least expensive home computers fell far below predicted levels, but sales of personal computers to small businesses increased. As a result several manufacturers of small home systems, such as Coleco Industries, Inc., announced that they would withdraw from the market.

Selling computers in an unstable market is difficult even for large companies. Those firms with a large customer base, most notably IBM, followed a conservative approach, producing computers that were either smaller, faster, or cheaper but which were otherwise copies of older products. Other manufacturers tried to find a market niche by producing computers with radically new architectures. While new architectures did not produce a large share of the total revenue, they showed some interesting possibilities for the future.

Workstation architecture dominated the new products. The term workstation refers to a personal computer, usually with generous amounts of random access memory (1–4 megabytes) and private disk storage (30–300 megabytes; one byte equals eight bits). Workstations are usually distinguished from large personal computers because they have more powerful processors and high-speed network connections to other workstations or mainframe computers. The workstation uses its network connection to access files of data or to have the mainframe computer perform special-purpose functions. The idea behind workstations is economic; it costs less to give each employee a small computer than it does to purchase a large central computer with equivalent power.

Workstation prices continued the decline that began after they were first introduced a few years earlier. In 1984 SUN Microsystems Inc., which manufactured a workstation based on the Motorola 68010 microprocessor, introduced a repackaged system that cost only about 60% of the 1983 model but delivered the same computational power. Other manufacturers such as AT&T and Hewlett-Packard announced workstations for under $10,000 that would have cost twice as much in 1980.

Another development of the past year was the reemergence of an interesting computer configuration, the multiprocessor. Most computers are uniprocessors; they contain a single processing element. A multiprocessor is a computer that uses multiple processors to perform computation faster than a uniprocessor machine.

Computer scientists had been interested in multiprocessors for many years, but no single multiprocessor design had emerged as the best. Also, even when multiple processors were available, existing computational methods were not usually able to take advantage of them. Thus, the problem had two complementary aspects: computer scientists designing algorithms dreamed of advanced multiprocessor machines capable of performing their computations, and engineers built multiprocessor machines from existing technology and hoped that computer scientists would discover ways to use them.

During the past year engineers made the greatest advances in multiprocessor design by using new technologies to produce faster systems. These new systems have a variety of architectures. The new designs differ from one another in the ways that they connect processors to memory, to other processors, and to input/output devices. The challenge now passes to computer scientists, who must develop software that controls and exploits the speed of the new hardware.

—Douglas E. Comer

Electronics

The electronics industry enjoyed a period of rapid growth during 1984. In the U.S. revenues increased by more than 14% to over $200 billion, representing almost 40% of the worldwide figure. There were signs, however, that the trend might not continue at the same brisk pace through 1985 but might slow to perhaps a 7–8% rate because of overproduction and a cooling of the general economic growth rate.

Integrated circuits continued to be at the forefront of research and development efforts as evidenced by a 55% increase in sales to more than $21 billion in 1984. It is interesting to note that even with the increased interest in IC's the discrete (individual electronic components) market also increased by almost 30% to some $6 billion. Experts forecast, however, that the interest in IC's would continue to grow so that they would eventually absorb the major share of the discrete market.

There was pressure on the IC industry during the past year to increase the density and speed of commercial units. The result was a 300% increase in wafer yields (number of good dies from a silicon wafer) and a more than 60% reduction in the space required for the standard transistor configuration. By 1985 there were some semiconductor memory units available with access times of 10 nanoseconds (1 nanosecond = 0.000000001 sec); such a capability makes possible 100 million operations per second.

Motorola's 16-bit MC68000 microprocessor, one of the most successful on the market, contains approximately 70,000 transistors on a single chip that measures 240 by 281 mils (1 mil = 0.001 in). This is an area that is less than one-quarter the size of a postage stamp.

Courtesy, AT&T Bell Laboratories

In an area smaller than a dime, a new computer memory chip developed by AT&T can store more than one million digital bits of information.

Silicon compilers received increased attention by IC designers as a way of achieving custom designs in a period of days or weeks rather than months or years. A silicon compiler is a software approach to translating a request for specific circuit operations to a design from which masks (design patterns) can be made for the production cycle. It is a structured approach to design that utilizes packages of previously designed systems of all sizes and functions in concert with its awareness of design limitations and the host facility.

CMOS technology. The field of electronics continued to experience growth and advances in manufacturing technology and in product sophistication and complexity. Considerable work was taking place in the area of complementary symmetry metal-oxide semiconductor (CMOS) circuitry. This form of circuit seemed certain to become increasingly popular because it does not dissipate power easily.

A CMOS circuit contains both an n-channel and a p-channel metal-oxide semiconductor field-effect transistor (MOSFET) connected to a common input as the active element. Thus, turning the n-channel device on also turns the p-channel device off (or vice versa). Therefore, one of the two is always off, resulting in zero current through the CMOS circuit. Because there is no current, no power is drawn from the power supply and no power is dissipated by the circuit.

Of the projected $40 billion IC market in 1990, at least $15 billion was expected to be in CMOS products. These products cover a wide area of electronic and computer circuits, although as of 1984 and 1985 the computer area was undergoing the greatest growth. CMOS was used in operational amplifiers, analog-to-digital converters, telecommunications, data communications, and, of course, digital computers. Most knowledgeable experts expected that CMOS would be the dominant technology during the last years of the 20th century.

Fiber optics. As the need to transmit greater amounts of information at faster rates continued to grow, the use of fiber optics was becoming increasingly important. Scientists at various laboratories in both the U.S. and Japan attempted to extend the limits of this technology. In Japan transmission of information at 445.8 megabits (million bits) per second over 134.23 km (83.41 mi) of fiber-optic cable was achieved—without the need for an amplifying repeater. At Bell Laboratories in the U.S information was sent at a rate of 420 megabits per second over 161.5 km (100 mi) through a fiber-optic cable. It is interesting to consider that at those information rates the entire text of the *Encyclopædia Britannica* could be transmitted from Los Angeles to San Diego in only one second.

Considerable effort was also in progress between Great Britain, France, and the United States to develop an underwater cable across the Atlantic. The second test of an undersea fiber-optics system in deep ocean waters was carried out by a Japanese company (the first was performed by AT&T in 1982). AT&T was preparing to install a system between two islands in the Canary Islands as a test for future longer distance ventures. In July and August 1984 fiber-optic links were used at the Olympic Games in Los Angeles. The links transmitted digitized video information at a speed of 90 megabits per second. Several long-distance carriers set up transmission systems that operated at speeds of more than 400 megabits per second.

Wafer-scale integration. Considerable effort took place during the past year to produce superchips, wafer-scale integrated circuits. A single wafer ranging in diameter from 5 to 20 cm (2 to 8 in) would contain an electronic system having a capacity of about 25 to 100 of the present microprocessor units. Already available is a 4-megabit ROM (read-only memory) that holds the entire Kanji alphabet. While some success has been achieved, other attempts to package an entire mainframe computer on a single wafer have thus far been unsuccessful. The future of wafer-scale integration probably lies in the semiconductor memory area.

Medicine. Efforts in medical electronics were directed primarily toward the imaging process (obtaining two-dimensional cross sections of vital organs), improving the available biosensors (transducers that convert a biological condition or process into an electrical signal), developing implants and assist systems, and improving the quality and capability of pacemakers.

Within the past year nuclear magnetic resonance (NMR) scanning was approved by the U.S. Food and Drug Adminstration to obtain the important two-dimensional cross-sectional images. In early 1984 there were only 14 such units in the U.S. One year later there were more than 90 in the U.S. and about 140 throughout the world. Patients are placed within a large magnetic coil that causes the magnetic dipoles of the hydrogen protons in the individual's body to align with one another. A radio pulse with an alternating magnetic field is then applied at right angles to the original magnetic field, twisting the dipoles of the hydrogen protons out of their original alignment. When the applied pulse ends, the protons return to their original state and emit a radio signal that is picked up by the antennas of the system. The results are then processed by a computer and converted to a two-dimensional display.

Ultrasonics is a method for obtaining two-dimensional scans that does not employ nuclear radiation. Physicians use it for such purposes as measuring the velocity of blood flow in high-risk pregnancies, taking motion pictures of a patient's heart, and detecting cancerous tumors in blood vessels. Ultrasound signals are generated by a piezoelectric transducer, which then converts electrical pulse signals into mechanical vibrations. These vibrations are coupled to the skin of the patient and echoed back to receiver transducers in order to provide a view of the area being scanned. The level and frequency of the ultrasound signal are adjusted for the particular body structure being studied. For example, abdominal and cardiac studies require penetrations of 10 to 20 cm (4 to 8 in), and so frequencies of 2–3.5 MHz are used. Proportionally higher frequencies are used for smaller structures.

During the year pacemakers that adjust the heart rate during exercise periods were under development. Through biosensors the blood temperature or respiratory rate is monitored and the pulse adjusted accordingly. Improved batteries permit long periods without maintenance and small units due to the reduced weight and size.

Microprocessors were also being used in portable programmable insulin injection systems. They permitted a flow of insulin at a relatively slow rate over a comparatively long time in order to remove the discomfort associated with other techniques.

Music. Personal computers were rapidly becoming standard items in recording studios. Such computers were used for a variety of functions, including creating sounds not otherwise possible. Quality recording on digital discs was becoming prevalent throughout the industry. A plastic disc of about 12 cm (4.7 in) in diameter, called a compact disc, can store more than one hour of music in micrometer (millionth of a meter)-sized pits that are read by a laser beam in the record player—all without touching the surface of the disc. This method of recording can record a dynamic range of 90 decibels (compared with 70 decibels for the best available audio discs).

By 1985 compact disc players were available for home use, and they were expected to be available soon in automobiles. Compact disc technology incorporated such electronic devices as optical lenses and high-precision servomechanism systems.

Displays. The length required for the electron beam of the cathode-ray tube (CRT) has been a major deterrent to the miniaturization process in a wide range of applications. Japan appears to be leading in the development of commercial units to solve this problem, and U.S. revenues from such products are expected to reach $1.25 billion in 1987, as compared to about $40 million in 1983.

Industry spokesmen agreed that the flat panel was the direction of the future. It seemed only a matter of time before most displays would be 2.5 centimeters (one inch) or less in thickness. The three prominent technologies in flat-panel design included liquid crystal displays (LCD's), gas plasma displays (GPD's), and electroluminescent displays (ELD's). The liquid crystal of the LCD is a material that will flow like a liquid but has some molecular structure characteristics usually associated with solids; confined between two parallel transparent plates, it will display different characters when particular regions are energized. In GPD's a gas such as neon or argon is ionized between a pair of electrically charged electrodes. In ELD's light is emitted by the stimulation of a luminescent material between a pair of charged electrodes.

As of 1985 LCD's accounted for about 70% of the market, GPD's about 20%, and ELD's 8%. Recent trends suggested that by 1987 LCD's would account for the same percentage but that the percentages for the other two would reverse. Major shortcomings of the flat displays include limited viewing angle, brightness, readability, speed of response, and lack of color. The color problem was being challenged by introducing dye guest molecules with the host molecules of the LCD or by using colored filters in front of the panels.

Robotics. The U.S. robotics industry experienced a 50% growth in 1984 to revenue levels in excess of $300 million. New developments in microprocessor and mechanical design resulted in robots capable of delicate maneuvers, permitting their use in areas such as electronic assembly and semiconductor manufacturing. The largest number of robots continued to be used in such traditional areas of application as welding, spray painting, and materials handling. However, as robots become more sophisticated and

© Brian R. Wolff 1985

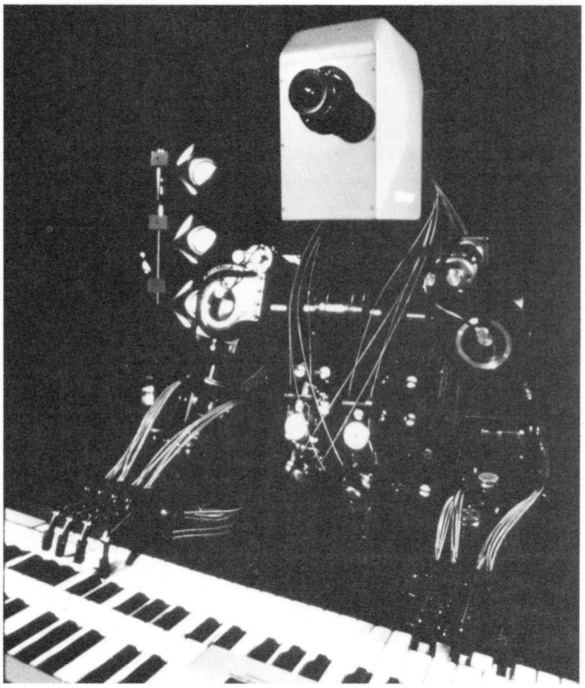

Robot developed by Sumitomo Electric of Japan can play an organ. It has 50 joints and is controlled by 17 16-bit microprocessors.

enter the electronic assembly area in larger numbers, revenues in the robotics industry are expected to grow substantially.

Robots can be separated into three major categories: turtles, arms, and mobiles. As the name suggests, turtles are small armless boxes usually controlled by a local computer. If equipped with a drawing instrument, they can be used to produce a hard copy of a display pattern. For arm robots the base of the arm is usually fixed in place, but the arm itself can have four, six, or more degrees of movement. Mobile robots include all those that are free to move and that have sensors to provide one or more of the senses.

Sumitomo Electric of Japan developed a robot having a general human shape that has the capability of playing an organ (and using the foot pedals) through memory control or by reading sheet music. The android has 50 joints and is controlled by 17 16-bit microprocessors.

—Robert Boylestad; Louis Nashelsky

Information systems and services

The information industry continued to to grow in size and importance during the past year. According to John Diebold, president of the Diebold Institute for Public Policy Studies Inc., more than half of the U.S. work force in 1985 was made up of information workers. The industry was responding to this growth by creating data processing equipment with greater capacity, higher quality, and lower cost, and with software that was not only easier to use but also was designed to enhance human creativity and improve decision making.

Advanced information technologies were being used to modify existing business procedures and to create new services. In addition to the credit card, banks devised a debit card, which supports the immediate transfer of funds between a retailer and a financial institution. An even more advanced variation was the "smart card," a plastic card with an embedded microprocessor and memory chip. In addition to functioning as a credit card, the smart card could be used to store personal information such as one's medical record and for a variety of other personal services that require identification.

Fraud and computer crime continued to be matters of concern, with private industry reporting annual losses ranging between $145 million and $730 million; losses in government departments may be even higher. To deal with this problem a number of laws were introduced in the U.S. Congress. However, laws were not enough; computer crime was difficult to detect, and the computer criminal was difficult to apprehend. Donn B. Parker, a recognized authority on the topic, claimed that in order to reduce computer crime, society would need a new code of computing ethics as well as new laws.

Universities were meeting the challenge of the information age by devising programs for teaching technical skills connected with computers and telecommunications. College students were learning how to manage information and how to access and use information to solve problems in school, industry, and society. Administrators added new subjects to the curriculum, reorganized traditional departments, and created new schools so as to bring together the curricula of computer science, communication studies, information resource management, and information systems development. Computer manufacturers provided computer hardware and software valued in the millions of dollars to establish a network of instructional workstations and to enable both faculty and students to determine how best to cope with the information society.

U.S. information systems. The U.S. National Commission on Libraries and Information Science issued a statement pointing out the need for federal, state, and local officials to support effective library and information services as they developed plans to improve the quality of education. Officials involved with the problems of rural America were concerned that many such communities lacked adequate information resources. The National Rural Information Development Program, working with the National

Bud Lee

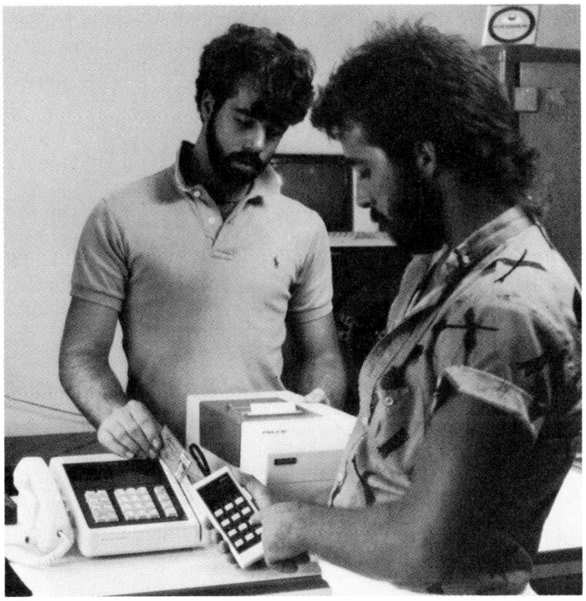

Shopper using a debit card presses his personal identification number. Debit cards facilitate the immediate transfer of funds between a store and a bank.

Agricultural Library, sought to strengthen and improve rural libraries and provide more effective information services to local community residents.

The familiar library card catalog was in many places being converted into an on-line electronic data base (one in direct communication with a computer). Users of libraries that had converted their catalogs in this way no longer had to thumb through printed cards while searching for specific books, authors, or subjects. Instead, they sat in front of a terminal or perhaps an inexpensive microcomputer and quickly determined whether the library had the desired document.

To expedite requests for interlibrary loans and photocopies the American Library Association created Alanet, an electronic mail and information service. Subscribers wishing to borrow books or obtain photocopies from member libraries could complete the request forms on-line by interacting with the system. Alternatively, participants with access to microcomputers or word processors could prepare the requests off-line. In either case the completed forms were transmitted instantly to the electronic mailbox of the lending library while automatically sending the borrower an acknowledgement.

Many additional data bases, covering a wide variety of subjects, were available in 1985 for computerized on-line searching. The U.S. Department of Labor developed a national computerized job bank to help people throughout the nation find jobs. Computerized job lists prepared by state employment services were sent to a center in Albany, N.Y., for conversion into a national master list. The service was supported by the Department of Labor, and no fees were charged to a state for using it. When a person found a job that he or she was interested in, that individual submitted a résumé to the state employment office, which then forwarded it to the employer. If the employer was interested, the job seeker was contacted and an interview was arranged. When a job was filled, the listing was removed from the system, and the master list was updated. Joyce Kaiser, the associate assistant secretary for employment and training, explained that the main advantage of the system was that it provided individuals who wanted to relocate with the opportunity to research job markets in specific areas of the country before they made commitments.

The Aircraft Owners and Pilots Association organized a Forum that served as a message center, an on-line conferencing facility, and a reference library for members. The Forum could be used to leave and receive messages and to communicate with fellow members and with the association's staff. The on-line reference library contained news items, an aviation calendar of events, general aviation documents, information on state and federal legislative matters, and a question-and-answer section on aviation-related topics.

The American Chemical Society Committee on Professional Training created a Directory of Graduate Research File (DGRF) and made it available for computerized searching. The DGRF offered career guidance to students by providing information about colleges and universities offering degrees in all specialties of chemistry as well as individual biographical records of faculty members and their specific areas of interest.

The American Medical Association (AMA) during the year cooperated with the federal Centers for Disease Control (CDC) and the John F. Kennedy Institute for Handicapped Children to create an on-line bulletin board of disease alerts to be beamed to physicians via Med/Mail, the electronic messaging service of Minet (GTE Telenet's Medical Information Network). The CDC planned to report disease occurrences throughout the U.S. as well as recent developments about such diseases as AIDS (acquired immune deficiency syndrome) and the toxic shock syndrome. The John F. Kennedy Institute was to supply a directory of resources, accessible by geographic region, for serving disabled persons.

Using programs prepared by Massachusetts General Hospital and Harvard Medical School, the AMA added to Minet a data base that enabled physicians to test their clinical skills through simulated patient encounters. The simulated patient maladies ranged from hypertension to meningitis, and the cases were designed to convey the essentials of patient management in regard to priority, safety, cost, and sequencing of decisions.

Eyenet was introduced during the year as a data base designed to help ophthalmologists keep up with the latest information about eye disease treatment, new and rare eye diseases, eye care products and instruments, and drug information. The data were obtained from the ophthalmology section of *Excerpta Medica* and were updated weekly. The service was made available by the American Society for Contemporary Ophthalmology through General Electric Information Services Co. The American Society of Hospital Pharmacists computerized its drug information service and offered it on-line to hospitals, medical libraries, retail pharmacists, and physicians.

News of a general nature was also available on-line. More than 200 national and international news, analysis, and feature stories were filed daily by Associated Press reporters and were available on-line from ITT Dialcom Inc. The full text of the *Wall Street Journal,* including news stories, editorials, and other features, was made available at 6 AM on the day of publication to subscribers of the Dow Jones News/Retrieval Service. The data base, updated every business day, served as a valuable reference tool for corporations, investors, students, and the general public.

International information systems. The Commission of the European Communities, headquartered in Luxembourg, sought during the year to stimulate the European information industry. According to Jansen van Rosendaal, the director for information management, third-generation data base services would be designed to expand the market greatly from the present 400,000 users and to respond to the information needs of European professionals by helping select, organize, translate, and deliver documents to managers, farmers, lawyers, and students for use in decision making. To accomplish this goal it was necessary to overcome existing impediments, including an inadequate communications infrastructure and a lack of a clearly defined and unified communications strategy across national boundaries. As an example the Commission pointed out that with the introduction of national networks ten different procedures were needed for accessing data bases in 14 European countries; this required a multiplicity of passwords. Recognizing these problems, the Council of Ministers of the European Communities adopted a plan for transnational development.

Intelpost, a service of the government-owned *Australia Post,* used facsimile machines to transmit exact copies of documents or graphics across the telephone network to similar machines in *Post* centers throughout Australia. *Australia Post* claimed that the complete transaction never took more than two hours. Intelpost's speed and the variety of material it could handle were its biggest assets and contributed to the success of the service.

In the United Kingdom the Scottish College of Textiles was developing a computer graphics system for industrial designers that would simulate patterns in tweeds and other woven textiles. Patterns would be designed by specifying yarn colors and other parameters and then adding threads sequentially in a simulation of the way cloth would be woven from individual colored yarns on the loom. Although the computer program would not eliminate the need to weave samples of fabric from which buyers made their final choices, it would enable artisans to adjust designs rapidly and economically.

Oxford University Press began a multimillion-dollar project to digitalize the 500,000 entries and 60 million words in the authoritative *Oxford English*

The 16-volume, 60-million-word Oxford English Dictionary *is being digitalized by the Oxford University Press so that it will also appear in the form of a computer-readable data base.*

Ric Gemmel—Discover Magazine © 1984 Time Inc.

Dictionary so that it would be transformed into a computer-readable data base. When the task is completed, it will include every word in the English language since 1150 AD. Linguists and other scholars thus will be able to conduct a computer search for the first known appearance of a word in print and its subsequent changes in meaning. The project is supported by IBM Corp. in the form of a 4300 computer system.

Information science research. An important branch of information science research is termed expert systems or knowledge engineering. Its goal is to apply the results of artificial intelligence research in order to improve ways that computers can store information and assist in solving complex problems. As of 1985 one of the great bottlenecks in the construction of expert systems was the process of transferring information between expert and system. In a project at Stanford University researchers analyzed a model for constructing knowledge-based programs by transferring information, one item at a time, from the human expert to the data base stored in the computer. Researchers at New Mexico State University investigated the structure of information in knowledge-based systems while paying special attention to the representation of judgmental and commonsense knowledge.

The U.S. National Library of Medicine supported research at Tufts University to develop an expert system for medical disorders of sodium and water balance in kidneys. This project was designed to test the hypothesis that cause-and-effect reasoning requires the use of a richly structured data base that can be derived from relatively few individual experts. The researchers expressed the belief that the methodology would be readily transferable to other knowledge domains that also depend upon causal analysis as part of the problem-solving process.

The relationship between the use of scientific and technical information and an organization's, or nation's, growth and prosperity was under investigation during the year. King Research Inc. was studying the economic conditions, communication patterns, and national policies affecting information production and transfer in energy research and development activities. Researchers at the University of Pennsylvania constructed mathematical models of organizations and contracting rules in order to compare benefits of alternative information and organizational structures. At Yale University researchers investigated the formal properties of information in economic organizations. Problems being studied included information relationships among economic agents, performance evaluation, and centralized versus decentralized organizational structures. By representing and comprehending the resulting patterns of information behavior in organizations, managers, it was hoped, would be helped to understand how economic mechanisms operate, or can be made to operate, productively and efficiently.

—Harold Borko

Satellite systems

Earth-orbiting satellites that utilize their vantage points in space for economic benefit and military purposes are termed applications satellites. Three general classes of satellite systems are: communications, Earth observation, and navigation. Satellite systems are developed and operated by individual nations, groups of nations, and by private industrial concerns. Although the United States and the Soviet Union continued to dominate such activities throughout 1984, other countries, among them Japan and China, were using satellites increasingly for such purposes as communications and remote sensing.

During 1984 the Ariane expendable launch vehicle came under the control of Arianespace, Inc., a group of European companies and banks. Once under the auspices of the 11 Western European countries that compose the European Space Agency (ESA), Ariane was now to be built, launched, and marketed by Arianespace.

Communications satellites. By 1985 applications of communications satellites had expanded to include mail delivery, banking, printing, and teleconferencing. In the United States alone, communications and information markets were projected to be $100 billion in 1985; world markets were expected to reach $325 billion by 1988.

Comsat General Corp., a subsidiary of Communications Satellite Corp. (Comsat), and Holiday Inns Inc. formed a partnership to distribute video programming and data services via satellite to 1,500 Holiday Inns in the United States. Satellite Business Systems (SBS), owned jointly by IBM Corp. and Aetna Life and Casualty, increased its SBS Skyline system of satellites. Twenty-three SBS Earth stations located in major U.S. metropolitan areas direct the electronic traffic point-to-point in the U.S., Puerto Rico, the Virgin Islands, and to 80% of the telephones in Canada.

SBS announced plans to sell data and video communications services in Belgium and Mexico. The organization also reached agreements with the national telecommunications administrations of France and Switzerland to provide digital videoconferencing links between the U.S. and their respective countries, to begin in mid-1985. Public videoconferencing rooms were to be located in Paris and eight other cities in France. In Switzerland a similarly designed room was to be located in Zürich.

A decision made by a U.S. Court of Appeals panel, confirming an earlier Federal Communications Commission ruling, disallowed local and state government regulation of television programming distributed by satellite to apartment complexes and individual homes. This decision could lead cable operators to expand into satellite services in order to compete economically.

In the U.S., Direct Broadcast Satellite (DBS) service to the home was delayed. Most of the U.S. firms that had once indicated active plans for DBS systems stopped or slowed down such plans in 1984, although four firms continued to proceed with system construction. The newly formed Direct Broadcast Satellite Association announced that it would select and adopt in 1985 a "single industry standard" for home DBS systems. The group hoped to standardize home DBS equipment.

Internationally, satellite television services brought the 1984 Summer Olympic Games to more than two billion people around the world. Live broadcasts from the Games in Los Angeles were transmitted over 14 channels on six satellites. The satellites were owned and operated by the 108-member International Telecommunications Satellite Organization (Intelsat). During the Games a revolutionary Torus antenna was set up about 35 km (22 mi) outside of greater Los Angeles to communicate with a number of satellites in varying orbits. A conventional antenna can operate only with the satellite at which it is pointed.

Late in 1984 U.S. Pres. Ronald Reagan approved the use of private satellites for international communications. The president's endorsement was seen as weakening the near-monopoly held by Intelsat for providing international satellite communications. Cable television operator Ted Turner fought both Comsat and Intelsat and won the right to use the Soviet Union's Intersputnik system for receiving in the U.S. direct video transmissions of the U.S.S.R.'s "Friendship Games."

GTE Sprint Communications Corp. stated its intent to provide international long-distance satellite service to the United Kingdom, other major European countries, Australia, Japan, and Hong Kong. GTE anticipated that by early 1985 approximately 1.4 million of its customers would be able to call anywhere in the U.K. at discount rates that would be nearly half those charged by American Telephone & Telegraph Co. (AT&T).

In 1984 Japan launched its first operational direct broadcast satellite, BS-2A. Designed to provide color programming for homes, the satellite experienced electronic problems that forced a curtailment of direct broadcast to homes.

The U.S. National Aeronautics and Space Administration (NASA) launched several communications satellites from the space shuttles. In February space shuttle "Challenger" deployed a Palapa B-2 telecommunications satellite for the Indonesian government and an advanced Western Union communications satellite, Westar 6. Due to faulty kick motors both satellites failed to reach their proper orbits. In early November, however, the space shuttle "Discovery" rescued both the Palapa B-2 and Westar 6. Using special equipment, shuttle astronauts rendezvoused with the errant spacecraft, retrieved them from orbit, and returned the satellites in the shuttle cargo bay to the Earth for repairs.

Competitor to the U.S. space shuttle for payload customers, Europe's Ariane launcher achieved a string of successful flights in 1984. In March an Intelsat 5-F8 was launched by Ariane from a jungle site in Kourou, French Guiana. This flight marked the last "developmental" launch by Ariane under sponsorship of ESA. The first commercial endeavor of Arianespace Inc. took place in May, when Ariane placed in orbit the U.S.-built GTE Spacenet 1. This geostationary-orbit satellite was the first U.S.-owned spacecraft to be launched by a non-U.S. booster rocket. (In a "geostationary," or "geosynchronous," orbit a satellite remains over the same point above the Earth at all times.) Spacenet series satellites are in use by GTE Sprint for its long-distance telephone, video, and data customers.

In April China announced the launch of its 15th satellite into geosynchronous orbit by a new CZ-3 booster. As China's first experimental communications satellite, the one-ton spacecraft was designed to relay transmission of radio, television, telegraph, telemetric, and radiophotographic signals during its lifetime.

NASA used the expendable Delta launch vehicle to orbit successfully the Galaxy C communications satellite. From its geosynchronous orbit the satellite relayed video, voice, data, and facsimile communications in the continental U.S. for business customers. Another Delta launched the fourth in a series of NATO-111 satellites in October, providing secure communications links for nations in the North Atlantic Treaty Organization (NATO).

Both a European Communications Satellite (ECS 2) and the French satellite Telecom 1 were launched by an Ariane in August. Telecom 1 was Europe's first domestic telecommunications satellite. It provided wideband data transmission for business users, communications with French territories abroad, and cable television distribution, and it also relayed governmental data.

Arianespace launched the GTE Spacenet 2 satellite and an ESA Maritime Communications Satellite, Marecs B2, in November. Leased by the International Maritime Satellite Organization (Inmarsat), the Marecs B2 provided the equivalent of more than 50 simultaneous ship/shore telephone circuits from geostationary orbit. Inmarsat is a cooperative organi-

GTE Spacenet Corporation

The GTE corporation's Spacenet 1 communications satellite is launched by the European-built Ariane. Spacenet provided telephone, video, and data services.

zation involving 41 member countries. The Inmarsat system is used by commercial shipping, passenger liners, offshore drilling rigs, icebreakers, fishing fleets, and seismic survey ships. By the end of 1984 there were nearly 3,000 users equipped with ship/shore access to the Inmarsat system.

Earth observation satellites. This category of applications satellites provides data on Earth resources, meteorological conditions (weather), and military reconnaissance.

Earth resources satellites. On March 1 NASA launched the U.S. National Oceanic and Atmospheric Administration's (NOAA) Landsat 5 from the Western Space and Missile Center at Vandenberg Air Force Base in California. In a 700-km (435-mi) orbit the satellite replaced Landsat 4, which had become inoperative. Landsat 5 orbited the Earth from pole to pole every 100 minutes, crossing the same spot on the globe every 16 days. Its main instruments, a thematic mapper and multispectral scanner, viewed the Earth in varying spectral bands. Data were made available to all nations for applications to requirements of forestry, water utilization, and land use.

Weather satellites. In July a premature failure of a U.S. Geostationary Operational Environmental Satellite (GOES 5) left weather-watchers "blind in one eye," with no accurate weather data for parts of the eastern U.S. and the Atlantic Ocean. The malfunction led to the move of a GOES 6 from its location in Equatorial orbit over the Pacific Ocean to cover both coasts of the U.S. A nine-year-old GOES 1, turned off in 1978, was reactivated and pressed into service for coverage of Hawaii and the Gulf of Alaska. The failure of the GOES 5 was a reminder of how valuable weather satellites had become.

In August Japan's National Space Development Agency launched a Himawari 3 into geosynchronous orbit. The country's third such meteorological satellite, it relayed data for tracking typhoons over Asia and the Western Pacific.

Military reconnaissance satellites. Predominantly owned by the United States and Soviet Union, military reconnaissance satellites remotely observe military activities on a global scale. For the U.S. the long staytime-in-orbit KH-11 series of satellites was in polar orbits, covering areas on the Earth twice a day. Weighing an estimated 13,500 kg (30,000 lb), a KH-11 spacecraft relays digital signals in real time. A new generation of shuttle-tended KH-series photoreconnaissance satellites was anticipated. They were to be equipped with enhanced sensors in order to image military targets during the night and in poor weather conditions.

A majority of the reconnaissance satellites orbited were Soviet. Launched once every two to six weeks, they provided electronic intelligence, early warning, and land and ocean surveillance. In 1984 China's 16th satellite was believed to be for military reconnaissance. Placed into orbit in September, it ejected a recoverable film pod to the Earth.

Navigation satellites. Evaluation of the U.S. Global Positioning System (GPS), known as NavStar, continued during the past year. Plans called for 18 large, powerful satellites to be circling the Earth in six orbital planes by the late 1980s, transmitting three-dimensional navigational data. Uses of NavStar include air traffic control, search and rescue operations, and worldwide navigation.

Since September 1982, nearly 350 lives have been saved by an international search and rescue program. Called SARSAT (Search and Rescue Satellite-Aided Tracking), the program included as participants the U.S., Soviet Union, Canada, France, United Kingdom, Norway, Sweden, Finland, and Bulgaria. U.S. and Soviet satellites carry instruments for receiving distress signals from aircraft and ships outfitted with special radio beacons. When a SARSAT satellite receives an emergency signal and relays it to land, the site can be pinpointed within an accuracy of 21 km (13 mi). An agreement signed in Leningrad, U.S.S.R., in October committed both the Soviet Union and the U.S. to maintain two satellites equipped with SARSAT transponders. A U.S. meteorological satellite carrying SARSAT equipment was orbited in December.

—F. C. Durant III

Energy

After a decade of high oil prices it appeared by 1985 that their impacts were at least as surprising as the original price increases. Oil price behavior has proved difficult to predict. A few reasonably clear conclusions can be reached about the behavior of the members of the Organization of Petroleum Exporting Countries (OPEC). Such knowledge, however, cannot fully explain past behavior or forecast future actions.

Supply and demand. By the mid-1980s it had become apparent that oil was not as scarce as believed by either OPEC or those outside observers who predicted continued price rises. OPEC prices weakened considerably from their 1980 peaks. A resumption of price increases was being forecast for the late 1980s at the earliest, but many questioned whether such price rises would occur even then.

The price weakness also indicated that the OPEC countries did not operate a perfectly functioning organization for monopolizing the oil market. It was not until the end of 1984 that OPEC agreed to establish even the most rudimentary elements of a formal cartel. The members then agreed to permit representatives of the organization to observe actions in each country. Doubts immediately arose about whether the surveillance would succeed in detecting or discouraging price cutting.

The entry of new producers into the oil market proved less extensive than some had expected, and the newcomers did not topple OPEC. The rise of new producers, nevertheless, was a major restraint on the cartel. For example, production decisions by Britain and Norway had to be considered by OPEC. Conversely, however, the failure of oil prices to return to their pre-1974 levels refuted the view that oil price rises were the result of transitory forces that would be overcome.

The relative importance of oil scarcity and monopolization as influences on oil pricing remained unsettled. Such members of OPEC as Saudi Arabia and Kuwait played major roles in helping to limit price declines, while at the same time other members, most notably Nigeria, acted in ways that weakened prices. Tension arose among the OPEC members about sharing the output cuts needed to hold up prices. The predictable result was a price somewhere below what a tight monopoly would charge but well above the level that would be reached if competition were free.

Saudi Arabia continued in 1984 to assume a major role in output restraint. The Saudi output decline pattern that emerged in the 1980s persisted. Kuwaiti output rose somewhat, although it remained well below prior highs. While production in Saudi Arabia was about half as high as in the late 1970s, Kuwait was only producing about one-third of its previous peaks (*see* TABLE I).

Table I. World Crude Oil Production
(000 bbl per day)

Country	1973	1979	1983	August 1984
United States	9,208	8,552	8,688	8,781
U.S.S.R.	8,465	11,460	12,034	11,870
Total OPEC	30,989	30,928	17,562	16,350
Saudi Arabia	7,596	9,532	5,086	4,490
Iran	5,861	3,168	2,426	1,800
Iraq	2,018	3,477	1,005	1,200
Kuwait	3,020	2,500	1,064	1,090
Libya	2,175	2,092	1,076	980
Nigeria	2,054	2,302	1,241	1,100
Venezuela	3,366	2,356	1,768	1,820
OPEC less Saudi Arabia	23,393	21,396	12,476	11,860
Mexico	465	1,461	2,686	2,680
Non-OPEC total	24,685	31,607	35,419	36,070
World	55,674	62,535	52,981	52,420

Source: U.S. Department of Energy, *Monthly Energy Review*.

The reaction to the policies of OPEC involved considerably more reduction in energy consumption and less development of coal and oil from non-OPEC sources than had been previously predicted. To be sure, a sizable part of the slowdown in consumption (perhaps as much as half) was the result of a reduced growth of economic activity around the world. Some of this slowdown may have been caused either directly or indirectly by the oil price increases. For example, resources may have been diverted from promoting growth to mitigating the consequences of the oil price rises. Moreover, the rises took place while the world was affected by inflationary pressures, and the policy responses to the further effects of the oil price increases may have ultimately retarded growth.

Efforts to determine the effects of OPEC foundered, however, on the near impossibility of knowing what would have happened otherwise. The role either of growth-retarding policies adopted for other reasons or of market developments cannot be determined at least without inordinately expensive efforts to gather and analyze data.

Energy consumption also declined because of the reduced use of energy per unit of total output. This was particularly notable in North America, Western Europe, and Japan. Energy consumption continued to rise in those areas until 1979; subsequently, use tumbled below 1973 levels. Again the exact causes of the decline were difficult to determine. The possibilities included making do with less, employing facilities that utilized energy more efficiently, and restructuring the economy to reduce the role of industries that use energy heavily.

Clearly some of all of this occurred. A move toward less winter heating and summer cooling became evident. So were efforts to improve home insulation and the design of boilers. There was a shift to cars that could travel farther per unit of gasoline. At least part of the slowdown in the production of metals must be attributable to higher energy costs since energy is a major element in the production costs of many metals, notably steel and aluminum.

However, reduced oil consumption unquestionably was a major element in the changed situation. During the last decade North America, Western Europe, and Japan all lowered their use of oil by amounts that substantially exceeded the decline in total energy use. (The drops from 1973 to 1983 for North America were 87 million oil equivalent metric tons for energy and 135 million for oil; for Western Europe, 19 million for energy and 163 million for oil; in Japan, 7 million for energy and 63 million for oil.) The reduction in oil use largely took place outside the transportation sector. Electric utilities shifted to coal and nuclear power. The remainder of the economy reduced direct use of mineral fuels but tended to use more electricity.

In regard to other fuels, coal use rose significantly in North America and Japan but did not do so in Western Europe. A long-running Western European decline in coal use did not begin to reverse itself until the late 1970s. Consumption in Western Europe rose from its 1975 low of 309 million metric tons to nearly 340 million in the early 1980s. By contrast, 1970 consumption was 393 million. The Japanese increase in coal use was more broadly based than that in North America. Both the U.S. and Canadian increases resulted from the rising use of coal for electricity generation.

Table II. Consumption of Major Fuels in the United States
(000,000,000 BTU's)

Year	Coal	Natural gas	Petroleum	Water	Nuclear	Total
1947	15,824	4,518	11,367	1,326	0	33,035
1960	10,110	12,390	19,920	1,650	10	44,080
1970	12,600	21,790	29,520	2,650	240	66,830
1973	12,895	22,512	34,840	3,010	910	74,212
1974	12,655	21,732	33,455	3,309	1,272	72,479
1975	12,615	19,948	32,731	3,219	1,900	70,485
1976	13,519	20,345	35,175	3,066	2,111	74,297
1977	13,863	19,931	37,122	2,515	2,702	76,215
1978	13,841	20,000	37,965	3,141	3,024	78,039
1979	15,049	20,666	37,123	3,141	2,776	78,845
1980	15,336	20,391	34,202	3,118	2,739	75,900
1981	15,843	19,926	31,931	3,105	3,008	73,940
1982	15,268	18,507	30,232	3,577	3,115	70,807
1983	15,861	17,535	30,076	3,880	3,235	70,721
1984*	17,192	18,031	30,004	3,783	3,546	73,730

*Preliminary figures.
Source: U.S. Department of Energy, *Monthly Energy Review*.

In Japan, however, coal use rose in every sector of the economy except residential and commercial. In particular, Japan was still using significantly more coal for coking in the early 1980s than in the early 1970s. Such usage declined elsewhere with a particularly severe drop in the U.S. (from 75 million to 80 million metric tons to about 35 million). European increases in coal use for electricity generation were not sufficient to restore coal use to the 1970 level.

The situation concerning gas was varied. U.S. consumption continued to decline from its 1973 peak. Canadian and Western European consumption grew until the late 1970s and then began to drop, while Japan's continued to increase through 1983.

Various forces produced these trends. Coal has many disadvantages compared with oil and gas. It generally is more difficult to extract, transport, and burn. The pollution caused by coal burning is more difficult to mitigate than that caused by oil and gas. Additionally, the coal industries of Western Europe and Japan have extremely high costs. It is cheaper for these nations to import coal from the United States, Australia, and South Africa or to use other fuels. Western European coal output, therefore, contracted sharply, and the decline would have been far greater were it not for massive subsidies.

The disadvantages of coal use decrease for large-scale users such as electric power plants. Their scale lowers the unit cost of every element of coal utilization, from shipping from the mine to waste disposal. The coal can be shipped directly in train or shipload lots, saving the expenses of rehandling necessary with smaller shipments. Receiving facilities, coal handling at the plant, and air-pollution control equipment all have lower unit costs in large facilities.

A more critical and more technical consideration is that unit capital costs also tend to fall as plants become larger. In part, this is due to the fact that the need for labor does not rise in proportion to the increase in size. Another reason is that a given increase in volume requires a smaller increase in area. A doubling of output thus can be achieved with less than a doubling of area and, therefore, of cost. Doubling the output without having to double the total cost reduces the cost per unit.

Prospects for coal. Many efforts have been made starting at least with Germany during World War II to find ways to overcome the problems in coal use. Work has been directed at both the fuel processing and combustion stages of the process. A broad range of transformations of coal has been considered. The simplest proposals were to increase the extent to which the washing and crushing processes could remove the impurities that cause difficulties in burning coal. More complex approaches involve synthesizing methane (the main component of natural gas), gasoline, or fuel oil from coal. Much stress was given in

E. I. du Pont de Nemours and Company

Tension leg platform in the North Sea, an innovation of Conoco Inc., permits deepwater reserves of petroleum to be exploited economically.

the 1970s to less elaborate transformations such as finding an alternative or an upgrading of the known technologies for gasifying coal. Such gases have less heat content per cubic meter than does natural gas, but producing them was considered to be much simpler than synthesizing methane.

In practice, neither improved gasification nor any other change in coal processing proved commercially successful. The costs of all the transformations far exceeded the increase in fuel value that was produced.

Similar problems were encountered in efforts to improve combustion technology. Long-standing efforts to develop an alternative type of boiler called the fluidized bed still have not produced a clearly competitive process. Interest in the late 1970s shifted to mixes of coal with either oil or water. Boilers built to burn oil and gas have tubes too small to allow the passage of coal but could possibly accommodate mixtures. Again, however, difficulties were encountered in making the processes profitable to adopt.

Major changes in the steel industry have reduced the need for coke. First, technology reduced the coke requirement per ton of final steel produced. This involved numerous developments. The blast furnace can produce pig iron using less coke per ton of pig iron. More scrap and less pig iron may be used to produce a given amount of raw steel. New methods of casting hot metal require less raw steel production per ton of final output. Thus, coking coal is not an area of major growth opportunities for the coal industry.

As of 1985 electric power was the major prospective market for coal. About 85% of U.S. coal consumption was for electricity generation. The share was 58% in Western Europe but only 15% in Japan. (The range within Western Europe was considerable. Electric power accounted for 75% of coal use in Britain but only 35% in Italy.) The U.S. generated slightly more than half its electricity from coal; Western Europe, about one-fourth; and Japan, about 10%.

Nuclear power. In the area of electrical generation a significant challenge to coal has come from nuclear power. In the late 1960s and early 1970s the U.S. made a heavy commitment to nuclear power.

By the late 1970s and early 1980s disenchantment had arisen. The ordering of new plants ceased, and the expansion of nuclear power was due entirely to those earlier orders that were carried to completion.

All sides of the debate agreed that practical economics made nuclear power uneconomical. The disagreements arose about the causes. One view attributed them to industry problems; another, to public policy. The first view held that nuclear power would be uneconomical in the U.S. under the best of circumstances. It maintained that the performances of reactor manufacturers, plant engineer contractors, and utilities all had been unsatisfactory. The second view was that the regulatory climate was highly unfavorable. Its proponents maintained that the Nuclear Regulatory Commission imposed so many barriers to the construction and operation of nuclear plants that their costs exceeded those of coal plants. A further problem was the unwillingness of state public utility regulators to allow rates to rise high enough to cover the cost of new capacity.

Evidence exists that the regulatory restraints would suffice by themselves to kill nuclear power in the U.S. Even critics of nuclear power concede that the regulatory process is too cumbersome. Enormous effort is expended without satisfying fears about nuclear power. The open questions are, first, whether these regulations were the best possible to meet concerns about nuclear safety, and, second, whether in the absence of regulatory difficulties, other problems would still have killed nuclear power.

Japan and Western Europe generally have had a more favorable view of nuclear power than do the United States and Australia. Clearly, the economic situation in both areas is more attractive for nuclear power than it is in the U.S. As noted above, the cheapest sources of coal are imports from the United States, Australia, and South Africa. The transportation costs involved make coal generation more expensive in Japan and Western Europe than in Australia or the United States. Nuclear costs, in contrast, are, if anything, lower in Western Europe and Japan than in the U.S. With the notable exception of Great Britain, where efforts to develop a uniquely British reactor technology floundered, Western European countries and Japan proceeded smoothly in building and operating nuclear plants.

Future outlook. The halt in the development of nuclear power leaves coal as the fuel to meet the expansion of electric power generation capacity in the United States. In other places coal is challenged by nuclear power. Even in the United States the slowdown in power growth and the completion of nuclear plants ordered prior to 1975 limit increases in coal use. Moreover, Canadian utilities have begun to argue that they can supply power more cheaply than can new coal-fired plants in the U.S.

Thus, the major changes in energy involved a reduced use of energy rather than either the rise of numerous new suppliers of petroleum or a shift to nonpetroleum energy sources. The retreat from a major commitment to developing synthetic fuels from coal or oil shales continued. New technology, particularly that to utilize coal more satisfactorily, had not yet become well established.

—Richard L. Gordon

Environment

Two events affecting the environment created great havoc over the past year. In December 1984 there was a massive escape of the poison gas methyl isocyanate from a pesticide manufacturing plant at Bhopal, India. Within one week more than 2,500 people living near the plant died, an additional 1,000 were expected to die, and more than 100,000 were injured. The other occurrence was not a single event but a long-term development. Famine and starvation affected large parts of Africa, with Ethiopia receiving the most attention. Perhaps the single statistic that best captured the degree of misery in that area was a life expectancy of 39.4 to 42.6 years in Ethiopia.

Research to identify the root causes of both these events uncovered a tangled web of economic, governmental, social, and cultural factors at work, as in most environmental problems. It appeared certain that the immediate causes of the Bhopal disaster were the inadequacy of a scrubber for neutralizing the gas when it escaped and the failure of an incinerator for burning the gas. However, it soon became apparent that there was at least one important feature in this incident that had implications for toxic and hazardous chemical plants and storage depots throughout the world. Investigation revealed that even in the United States the primary goal of regulations for such chemicals is to try to ensure the safety of the workers, not the public living close to the plant. Thus, the federal standards set by the U.S. Occupational Safety and Health Administration did not address the issue of community or public access to information about the nature of chemicals manufactured or stored nearby. As of 1985 20 states and more than 40 cities in the U.S. had hazardous-chemical disclosure laws that required companies to label hazardous chemicals and to inform workers about potential dangers. However, most of those laws did not give the public the same access to information that could be important for emergency planning. The scale of the disaster at Bhopal might change this situation, however.

Confidence in the technology for minimizing the risk of such incidents was shaken not only by Bhopal but also by a liquified gas explosion in Tlalnepantla,

© Sandro Tucci—Gamma/Liaison

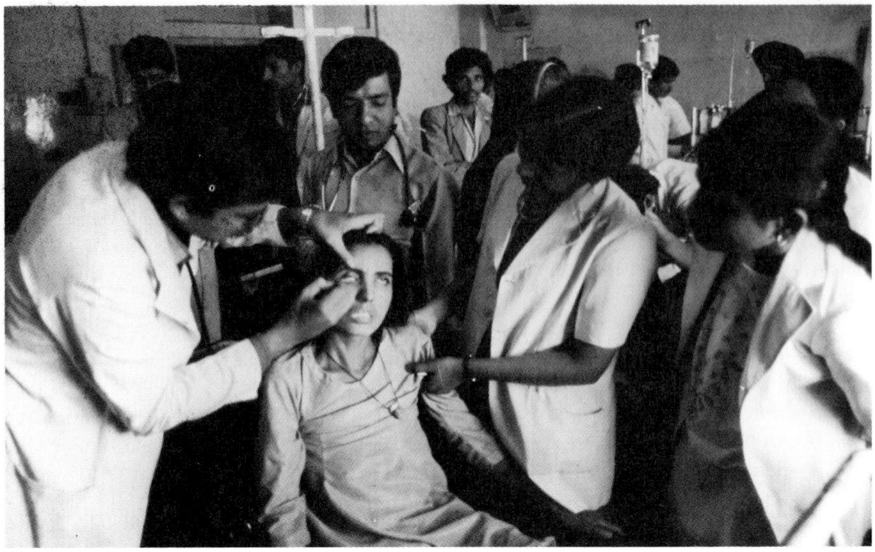

In Bhopal, India, the poison gas methyl isocyanate escaped from a pesticide manufacturing plant in December 1984. More than 2,500 people living near the plant were killed, and over 100,000 others suffered injuries, often to their eyes.

Mexico, a suburb of Mexico City, that killed 452 people and injured 4,248 on Nov. 19, 1984, and a gasoline explosion in Cubatão, Brazil, on Feb. 25, 1984, that killed more than 500.

While most of the attention in regard to starvation in Africa was focused on Ethiopia, the problem was widespread throughout the continent. It affected a wide belt of sub-Saharan (Sahel) countries, a row of tropical countries below those, and parts of the eastern and southeastern coasts. The root causes were many and complex; they included drought, overgrazing, mismanagement of water and soil resources, and overpopulation. However, there appeared to be a consensus that mismanagement of the economies of most African nations by their political leaders was the problem that was basic to most of the others.

Acid rain. Acid rain continued to be an important issue, attracting widespread concern in Europe and North America and much scientific investigation that was gradually exploring more and more ramifications of the subject. One question that arose was concerned with whether acid rain really originates primarily from the activities of civilization or whether natural emissions, as from volcanoes, play a primary role. A problem in answering this question is that little data can be found for the years before 1950. Accordingly, Gene Likens of the New York Botanical Gardens and James Galloway and Mark Hawley of the University of Virginia used another approach; they compared the chemical composition of precipitation at five remote sites with that at five sites in eastern North America. They discovered that many areas of North America receive precipitation with substantially greater concentrations of acid chemicals than do the remote areas. In particular, SO_4 is 2 to 16 times as concentrated in eastern North American precipitation as in precipitation from the remote sites in Australia, the Indian Ocean, Bermuda, Venezuela, and Alaska. Thus, it seems clear that acid rain is caused primarily by the activities of man.

A great deal of information about the effect of air pollution on soil and plant life was contained in a new book edited by Michael Treshow. For example, casual visitors to a nation may wonder why there is so much alarm about the impact of air pollution on trees there, when they see so little evidence of damage. One explanation is that most traveling is done in warm, low altitudes, whereas, as pointed out in an article by Jan Materna, a given concentration of air pollutants has a much more toxic effect on trees at an altitude of 1,200 m (3,900 ft) than on those at 400 m (1,300 ft).

In the same article it is mentioned that increased soil acidity implies that an increased proportion of soil aluminum will be present in a soluble form. The relationship between acid rain and the leaching of aluminum out of the soil was perceived as important because aluminum is toxic to living organisms. The most intriguing implication of the relationship was expressed in hearings before the Select Committee on Aging of the U.S. House of Representatives. Researchers discovered that there are more aluminum atoms in the brains of people with Alzheimer's disease, a degenerative disease of the central nervous system, than in normal brains. Daniel Perl reported to the committee that the brains of individuals suffering from Alzheimer's disease have two characteristic microscopic alterations: the neurofibrillary tangle and the senile plaque. The former consists of an accumulation of abnormal fibers in certain nerve cells. Perl and his associates used a sensitive electron microscope to identify accumulations of the element aluminum in nerve cells containing neurofibrillary tangles. Normal cells from those patients or from nondemented elderly control cases did not reveal comparable accumulations of aluminum.

Every step of this case has not been proved; it must be demonstrated precisely how the soluble aluminum released from soils by acid rain can work its way into the brains of patients with Alzheimer's disease. However, it appears likely that there is a connection between the two. Indeed, local incidence of Alzheimer's disease may be a useful signal for scientists to investigate the immediate environment for soluble aluminum and acid rain concentrations.

Carbon dioxide and climate. A related topic of continuing cause for concern in 1984 was the possible effect upon world climate of carbon dioxide concentration in the air. Some scholars believe that higher concentrations of the gas will increase temperatures throughout the world; others think that it will increase temperatures at some latitudes but cause glaciation at the poles.

Concern about this problem was leading to a variety of research efforts. For example, many researchers were concluding that the effect of deforestation in increasing the carbon dioxide concentration of the air ranges from half as important to equally as important as combustion of fossil fuels. Accordingly, there was a major new concern with monitoring the amount of terrestrial vegetation worldwide, using remote sensing by satellites.

"Grain" of a landscape. One new environmental issue became prominent during the year. For some time theoretical ecologists have been interested in what they refer to as the patchiness, or "grain," of an environment, or landscape. By this they mean the extent to which the landscape is either one homogeneous surface, as when vast tracts are planted in a single crop such as wheat, corn, or rice, or is a checkerboard of small squares or polygons of differing character, as when apple orchards, forests, plots of vegetables, and small ponds are interspersed. During the past year a number of books and papers appeared pointing out how this environmental characteristic can be related to a variety of other matters, including the management of agricultural systems and the survival of endangered species.

A number of other characterisics of the landscape are related to grain. Coarse-grained landscapes have low species diversity and require a high fossil fuel energy input; fine-grained landscapes have high species diversity and require a much lower energy input. There is no clear relationship between grain and either the level of economic activity or standard of living. Switzerland has very fine-grained landscapes but also maintains a high standard of living and a high level of gross national product per capita.

In a new book on agricultural systems John Krummel and Melvin Dyer pointed out that in highly managed systems in which the landscape pattern is relatively homogeneous, carnivorous animals are managed by such strategies as chemical control or physiological changes in animal breeding strategies. Use of chemical pesticides in large contiguous tracts of cotton, wheat, or corn, or management of physiological factors and nutrition in confined populations of hogs, cattle, or poultry are examples. In less managed systems the management of carnivores focuses on factors related to the landscape pattern. For example, in semiarid rangeland there is less management effort than in a cattle feedlot; there is also a greater diversity of herbivorous animals, such as cattle, deer, mice, rabbits, and grasshoppers. Consequently, carnivores can utilize the herbivore species in a fine-grained manner; that is, the diet in each locality is a specific response to the food available there. Or, in a fine-grained forest landscape carnivorous animals can be manipulated by the selective cutting of trees. In general, the fine-grained landscape is associated with a lower energy input.

Larry Harris pointed out another aspect of landscape grain in a new book on the relationship between the theory of island biogeography and strategies for preserving biotic diversity and endangered species. Two important rules emerged from the study of islands (the term "island" meaning both an island in an ocean of water and an "island" of one type of habitat surrounded by an ocean of landscape of a different type, as where a small forest is surrounded by agricultural land). First, in the absence of compensating colonization, species become extinct in small islands faster than in larger ones. Second, islands distant from the nearest similar habitat support fewer species than islands closer to the nearest similar habitat. Both of these rules are expressions of laws of probability. Large islands support larger populations than do small islands; therefore, the likelihood that a species will become extinct on an island due to chance alone becomes greater as the island becomes smaller. This principle indicates that one would expect to find a consistent relationship between the size of islands and the number of species they support. Indeed, the observed relationship has been so consistent that it can be described by a mathematical relationship: the logarithm of the number of species on an island is a straight-line function of the logarithm of the size of the island area. Similarly, researchers have determined that the farther an island is from the nearest similar habitat the more likely it is that a series of accidents in flight, swimming, or drifting in air or water will prevent it from being colonized.

One more finding from ecology is required to understand the basis of a strategy for maintaining endangered species. In stands of Douglas fir it has been discovered that the number of species of animals supported by a stand is related to the particular evolutionary stage of the forest. Such forests can be classified along a continuum comprising regeneration, seedlings and saplings, pole timber, sawtimber, large sawtimber, and old growth. These six categories represent the following series of tree ages: 0–10 years, 10–50, 50–100, 100–150, 150–250, and 250–450. It has been discovered that the total number of species of amphibians, reptiles, birds, and mammals using these six forest stages for feeding, reproducing, escape, and overwintering is greatest during the second and sixth stages and least for the third stage. This provides a justification for maintaining very old Douglas fir forests.

From the facts just given, a strategy for maintaining endangered species of animals can be developed. Harris points out that, reasoning from the nearest habitat discovery in regard to island biogeography, remnant patches of old-growth forest salvaged from much larger continuous stands will support greater numbers of species than will comparably sized replacement stands developed in isolation from contiguous old-growth forests. This notion provides an incentive to conserve old-growth forest stands, in the interests of preserving endangered animal species, rather than developing replacement stands.

Furthermore, an integrated system of islands for preserving wildlife must consider not only the nature of the islands themselves but also the positions of the islands with respect to other islands being used for the same purpose. Thus, one is led to the notion of preserving endangered wildlife species by planning island archipelagoes in which deliberately saved habitat islands would be sufficiently close together to allow migration of individuals from one to the other with minimum loss in transit.

Landscape management and history. During the last year new research by William Cronon appeared relating the notion of landscape management to history. This work created great interest among scholars because of the new insights it reveals, its profoundly interdisciplinary nature, and, most important, because it provides a new type of model for research on the consequences of interrelationships between societies and their environments over long periods.

The unusual nature of the work by Cronon can be emphasized by pointing out that a great deal of classic historical research would never mention concepts such as trees, forests, landscape, soil, water, or wildlife management. Cronon, by contrast, related the long-term fate of the Indian and European settlers in New England to the interactions between both of them and to the ways in which they managed the landscape. The story is particularly complex and fascinating, because he reveals that the way in which the Indians managed the landscape had an effect on the subsequent success of the colonists in establishing themselves in surroundings that differed markedly from those they were used to exploiting.

Three key features characterized the relationship between the cultural belief system of the Indians and the effect they had on the landscape. First, they understood that controlled burning would prevent the development of large unbroken forests with a

Lead-iron phosphate glass (disk at center of various lead phosphate glass wedges) is considered a promising material for the storage of high-level nuclear waste. By melting lead-iron phosphate glass with simulated nuclear waste, scientists were able to form at a relatively low temperature a stable glass in which the radioactive nuclei of the waste material are chemically incorporated.

Oak Ridge National Laboratory, Oak Ridge, Tennessee

dense tangle of huge trees and nearly impenetrable underbrush. The edges between forest and intervening open areas increased the total supply of food for herbivores, which both attracted game and allowed it to increase in numbers. Thus, the first European settlers in the 1600s discovered remarkably open, almost parklike woods extending from the Saco River in Maine south to the Hudson River in New York. In the language of Krummel and Dyer the Indians had produced a fine-grained, dynamic, heterogeneous landscape rather than a coarse-grained, static, homogeneous one. Second, the Indians achieved a balance between the productivity of land and their own population by restricting their birthrate. Third, the Indians lived in a symbiotic relationship with the beaver, another critical element in determining the character of the landscape. The extraordinary fertility of the soil under beaver ponds that had been drained by the Indians was ideal for primitive farming. Gradually, the fine-grained, dynamic, heterogeneous landscape of the Indians shifted to one that was coarse-grained, static, and homogeneous. The Indians had lived in a world in which they fostered high species diversity of plants and animals. By contrast, the white settlers gradually moved toward monoculture: vast tracts of wheat or corn were the characteristic use of the land.

Soil and water resources. During 1984 increasing attention was focused on topsoil and water resources throughout the world. Lester Brown and Edward Wolf of the Worldwatch Institute concluded that each year the world is losing about 25 billion more tons of topsoil than it is forming. This represents about 7% of the world's topsoil per decade. Brown and Wolf concluded that governments around the world must devise appropriate soil conservation programs.

The Ogallala aquifer underlies eight western states of the U.S., from South Dakota to Texas. The aquifer brings life and prosperity to 6 million ha (15 million ac) of farmland that would otherwise be arid. However, water supply from the aquifer is falling. During 1984 the Texas Water Resources Development Board unanimously recommended adopting a state water plan amendment calling for discussions with neighboring states on exporting water to Texas.

Concern about the ability of the Ogallala aquifer to meet water demand in the future exemplifies growing concerns about whether there will be enough water in the future for cities, farms, and industries. This concern is justified in the U.S. because of the population shift from water-rich northern states to the dry southwest.

The Ogallala aquifer has been drained heavily since about 1945. Irrigated acreage rose from 1.4 million ha (3.5 million ac) in 1950 to more than 6 million ha (15 million ac) in 1980. During the same period the amount of water pumped from the Ogallala aquifer increased from 4.2 to 12.8 trillion gallons. Increasingly, wells drawing on the aquifer are either running dry or producing a fraction of the water they once did. One possible solution is huge projects to move water around North America. A suggestion for accomplishing this is a $20 billion pipeline from Texas to the mouth of the Mississippi River, 860 km (535 mi) away.

—Kenneth E. F. Watt

Food and agriculture

The principal developments concerning food and agriculture during the year included near-record levels of world food production coupled with severe famine in some areas; breakthroughs in genetic engineering and biotechnology making possible disease-resistant plants and animals and early sex identification of embryos; and nutrition and health research that could lead to better diets and improved health for humans.

Agriculture

World output of agricultural products in 1984–85 was expected to be at or near record levels. Favorable growing conditions brought a sharp increase in the production of wheat, coarse grains, rice, soybeans, and cotton, particularly in Europe and China, offsetting expected declines in the Soviet Union, Canada, and Australia. World meat output, although less favorable, was expected to continue to increase in 1985, primarily because of greater poultry production. Milk output might be up slightly, principally as a result of rising Soviet production and continued growth in many of the smaller milk-producing countries.

U.S. crop production in 1983–84 was more than 25% below that of the preceding year. Yields were reduced by the worst drought in 50 years and by the record amount of land taken out of production under a government acreage-reduction program. Crop prices rose sharply, resulting in higher feed prices and reduced returns to livestock producers.

U.S. agricultural exports were expected to reach $38 billion in 1984–85, compared with $34.8 billion a year earlier, despite record crops abroad and the strong U.S. dollar. U.S. grain exports should expand as Soviet purchases increased the levels of international trade. Overall, however, U.S. crop prices in 1985 were expected to remain under pressure. Meat supplies were expected to remain plentiful, though slightly below the 1984 level, and average livestock prices should be higher.

Sub-Saharan Africa. Poor 1984 harvests in several African countries were the natural result of the worst

Tass/Sovfoto

In the Soviet Union harvested flax stems are mechanically wound into tight rolls that are easy to handle and transport. This newly developed procedure saves significant amounts of labor in harvesting and preprocessing the crop.

drought since the mid-1970s. More than two dozen countries in sub-Saharan Africa were affected, with the largest declines in the East African countries of Kenya, Sudan, and Ethiopia. The coastal countries of West Africa experienced good rains and thus a significantly improved 1984 harvest. The situation was mixed in the Sahel, where drought-related food shortages were most severe in Niger and Chad. In southern Africa food supplies in Mozambique were precarious, but the region's import needs decreased slightly because of a better than expected corn crop in Zimbabwe.

During the winter of 1984–85, worldwide attention focused on the food crisis in Ethiopia. Thousands of hunger-related deaths were reported, and the Ethiopian government estimated that at least seven million people were at risk of starvation. Famine developed as a result of a continued drought in several parts of the country, compounded by civil war and overwhelming logistical problems that made distribution difficult. Initial reports suggested the 1984 grain harvest was down by 80% in some districts, and people flooded into feeding camps as farm food supplies dwindled. Food aid, vehicles and equipment, medical supplies, and other support poured into the country. However, the situation was exacerbated by the fact that some of the worst affected areas were under rebel control. There were accusations that the Ethiopian authorities had no interest in sending food to those stricken provinces.

In Sudan severe drought affected much of the northern part of the country. The sorghum crop was estimated to be two-thirds its average size and the wheat crop perhaps half of normal. As with many other countries, Sudan had little foreign exchange and therefore relied on food aid to meet its requirements. Sudan was also faced with an influx of refugees fleeing famine and/or war in the surrounding countries. It was estimated that one-third of the more than three million refugees in Africa were in Sudan. Kenya's crops were also ravaged by drought, but food supplies were likely to remain adequate because the country had a larger commercial import capacity and a better distribution system. Grain import requirements in the landlocked Sahelian nations of Chad, Niger, and Burkina Faso (formerly Upper Volta) were estimated at 700,000 to 800,000 tons. Chad's food aid needs in 1984–85 would be significantly higher than in the preceding year because of below-normal rainfall in the major producing areas. In Mozambique continued action by guerrillas hampered food distribution, and overall food supplies were extremely tight. Even with adequate rainfall, no improvement could be expected until after the spring 1985 harvest.

Technology and food production. Before 1900 most of the increase in U.S. farm production came from expanding the agricultural land base, but this changed with the closing of the land frontier. By the 1950s and 1960s most of the gain came from increased productivity resulting from the application of science and technology, primarily labor-saving mechanization, agricultural chemicals (fertilizers, pesticides, feed additives), and modifications in biological systems (use of improved crop varieties and animal breeds). Much of this new technology had its origin in public research and was developed and delivered through the private sector.

The supply-demand situation in the 1980s suggested that farm profitability, rather than yield improvement, should receive more attention from the agricultural research system. This would require the identification of technologies that improve productivity by means of cost reduction, increasing the resistance of crops to disease and harsh environments, and enhancing product quality. Other factors requiring attention included linking farm production to human nutrition requirements; maintaining the integrity of soil and water resources; and providing the scientific foundation for a safe and efficient regulatory system.

There was concern about the ability of the agricultural sector to sustain the rate of productivity growth that prevailed during the first three decades after World War II. As agriculture becomes more dependent on sophisticated technologies, more effort is needed to sustain the effectiveness of those technologies. This area, called "maintenance research," applies to many aspects of agricultural research, from plant breeding to pest control. Like buildings and equipment, technology depreciates and must be constantly updated.

Scientific research. Scientists at South Dakota State University were developing technology for restructuring meat products. The eating quality of lower-priced meat cuts, such as chuck, can be improved dramatically through such processes. Chunked and formed "steaks," the second generation of restructured meat products, contain much larger lean particles than the earlier flaked products. They simulate more closely the texture of intact-muscle steaks such as the T-bone or sirloin. The object of restructuring is to offer the consumer a product of value and quality intermediate between ground beef and steak. These products are also desirable to the hotel, restaurant, and institutional trade because composition and portion size can be controlled precisely. Restructured steaks that are 88% lean can be retailed profitably at twice the price of boneless chuck.

The American chestnut, a premier hardwood forest tree of the eastern United States that was virtually eliminated five decades earlier by fungal disease, was experiencing a comeback. Using new biotechnological techniques, scientists had manipulated a chemical known as double-stranded ribonucleic acid (dsRNA) to change fungus strains lethal to the trees to benign ones.

Despite such major accomplishments biotechnology continued to be attacked by those who feared possible unforeseen and disastrous consequences. Work at the University of California involving the release of genetically altered organisms into the environment was temporarily halted pending the outcome of a lawsuit charging that the effects on the environment are unknown and that the National Institutes of Health violated the law by not fully assessing the impact before approving the testing. The research involved spraying genetically altered organisms onto potato plants in outdoor fields to determine if the organisms could delay the formation of frost. Such

Stan Grossfeld—The Boston Globe

Driven from their homeland by a severe drought, Ethiopian refugees cross the border into Sudan.

work could clear the way for producing frost-resistant crops and increase food production in areas where such weather factors are a problem.

The specific experiment in question called for spraying potatoes with a bacterium, *Pseudomonas syringae,* which exists on almost all plants. In its natural state it helps trigger the formation of ice on vegetation, causing millions of dollars of frost damage annually. The bacterium was altered in the laboratory by removing the section of its DNA responsible for helping ice crystals form on plants. The altered bacteria would be sprayed on the plants as soon as they sprout so that the "innocent" bacteria could take over the niche on the plant where the damaging bacteria used to preside. In greenhouse versions of the experiment, the altered bacteria gave potatoes resistance to frost damage down to −5° C (23° F). They usually suffer frost damage just below freezing.

A second lawsuit sought to halt genetic-transfer experiments being conducted through the use of public funds. Filed against the U.S. Department of Agriculture (USDA) and two of its research agencies, the suit focused on the transfer of human growth hormone genes into pigs but aimed to stop all such experiments. If successful, it would halt research on similar projects in many U.S. universities that received USDA funding.

The research project cited in the suit was being conducted at the USDA's Beltsville, Md., facility. The human version of the growth hormone gene was used because it had been studied most frequently by scientists conducting gene-transfer studies. In the experiments a piece of nucleic acid is added to genes that are then inserted by microinjection into fertilized eggs removed from sheep and pigs. The eggs are then implanted into the uteri of surrogate mothers to continue through a normal gestation. Besides testing whether such genes can be transferred, the researchers hoped to learn how to control the growth rate of the animals involved. This could lead to genetically engineered farm animals that grow more quickly and efficiently to a standard size. The experiments also were prototypes for introducing other desirable genes, such as those conferring disease resistance, into valuable species.

Scientists at land-grant universities in California, Iowa, New York, Texas, Washington, and Wisconsin were mapping the location of livestock and poultry genes that control resistance to specific diseases. California scientists developed a procedure for early identification of the sex of embryos, using an antibody technique. The procedure is nearly 85% accurate. Scientists also were investigating procedures for splitting cattle and sheep embryos to produce genetically identical twins. One laboratory produced identical quintuplets from a single sheep embryo. This technology would be useful for increasing the number of offspring from valuable females and would provide a potent research tool for distinguishing between genetic and environmental effects.

—John Patrick Jordan

Nutrition

The nutritional impact of food aid was assessed by Shlomo Reutlinger, senior economist in the Agricultural and Rural Development Department of the World Bank, and Judit Katona Apte, nutrition adviser with the UN Food and Agriculture Organization. Essentially, food aid can be provided in two ways. One is to give food as an addition to the meager and deficient supply available to people suffering from a disaster such as famine, war, or flood. In such cases the nutritional effect of the donated food is predictable. The nutritional cost effectiveness of different foods that are supplied is directly related to the nutritive value and cost per unit of each food.

The other option is to provide foods that can be substituted for foods that are available for purchase by the aid recipient but are not sufficiently varied to ensure an adequate diet. The aim is to enhance the quality of the food eaten; an example would be to substitute a quality protein source for some of the rice or other carbohydrate food being eaten by the recipient. Foods not commonly available or chosen at the market are likely to be consumed if they are given as food aid, thereby increasing the chance for a better balanced diet. This approach also provides the recipient with increased purchasing power. Donors are sometimes disappointed if this additional discretionary income is used for nonfood items, but clothing, housing, and education, for example, also improve the recipient's health and well-being.

Numerous studies have suggested that the nutritional effect of food aid is not limited by the nutritional contents of a food-aid package. A study of 302 families in Peru showed that the food-aid product was substituted for normally consumed foods of lesser nutritional and health value. Researchers in Colombia, who made similar findings, stressed that the effects of food aid could not be assessed solely in nutritional terms because the supplement freed family resources to satisfy other needs. Education programs did not improve diets unless they were accompanied by increased resources.

Unsatisfactory nutritional status is a very complex condition among individuals and families in any given socioeconomic group. Particular circumstances can affect utilization of food and nutrients in any one individual. Conditions can change the amount and kind of food eaten so that results fail to reflect the food aid. The best possible nutritional-cost-effective food-aid program requires a new approach that will separate requirements for disaster relief programs,

Photos, USDA

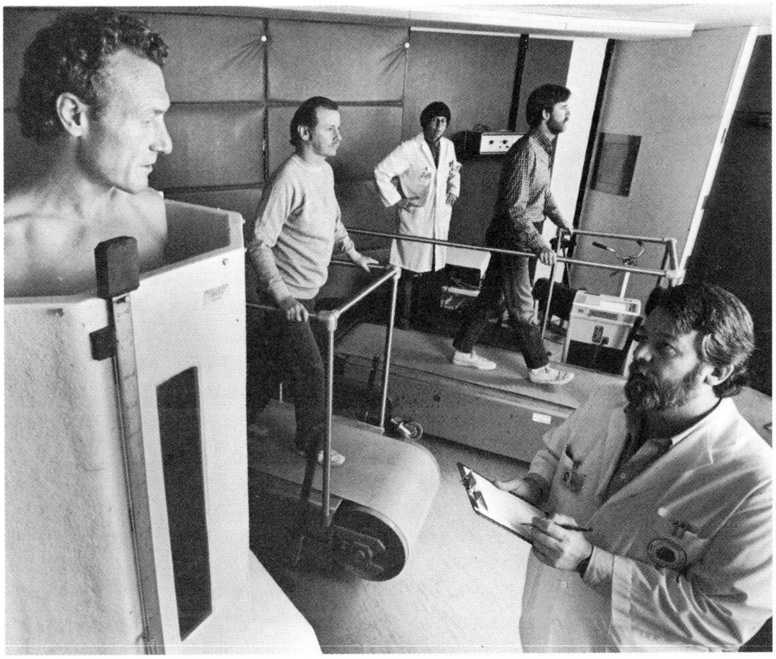

Head dietitian Alice Fong plans meals for nutrition studies at the Western Human Nutrition Research Center in San Francisco (left). Two men on treadmills and one in a water displacement tank (above) are monitored in order to determine the effect on blood pressure of increasing the amount of polyunsaturated fats in the diet.

in areas where food supplies are deficient, from programs in areas where food is available in the local market but households lack the purchasing power to utilize it in effective eating patterns.

Dietary fiber. Throughout history enthusiasts have promoted whole grain breads and other products. In the United States Sylvester Graham and W. K. Kellogg are familiar names. Denis Burkit of Great Britain's Royal Society has been called "the modern father of fiber." In numerous medical reports he identified absence of malignant tumors in people living on high fiber diets, as well as absence of cancer of the colon, constipation, hemorrhoids, varicose veins, appendicitis, hiatus hernia, obesity, diabetes, and coronary heart disease. In 1971 Neil Painter, an Oxford surgeon, demonstrated that diverticular disease of the intestinal tract was associated with prolonged bowel transit time and increased pressure in the sigmoid colon, caused by low fiber intake.

Dietary fiber and the role of the component types of fiber were being investigated. Dietary fiber intake depends on the nature and amount of vegetables, fruits, cereals, and bread products consumed regularly. Strict vegetarians may eat as much as 40 g of fiber daily; in general, people eat about 25 g, and some eat as little as 10 g. Dietary fiber affects the whole length of the gastrointestinal tract and influences the absorption of essential nutrients.

Trace elements in nutrition. Dietary imbalances involving numerous trace elements have been identified, probably starting in the 1800s with iodine deficiency. This was noted in certain Alpine regions of Europe and the northern midcontinent area of the United States, where geologic changes had reduced the iodine content of the soil so that foods grown there supplied no iodine in the diet. Since then changes in food processing and dietary practices may have reduced or eliminated other trace elements.

Controlled experiments have shown that marginal deficiencies of specific elements result in impaired or suboptimal functioning of specific enzyme systems, derangement of carbohydrate or lipid metabolism, or imbalances between metabolic pathways during the synthesis of cell components. Walter Mertz, director of the Human Nutrition Research Center of the USDA, reports that significant impairment of various immune systems can remain even when a severe clinical deficiency has been treated and changes in the development of cognitive and emotional functions have been demonstrated.

The enrichment of processed food with trace elements and the use of nutrient supplements for the individual have become controversial issues. These questions need to be answered for each nutrient: Is there a need to add it to the diet? What are the risks in doing so? Can it be done?

Diet fads and fallacies. Despite repeated explanations by nutritionists and dietitians, food fads and fraudulent claims and promotions continued to abound. The terms foodism and foodists were proposed as being useful to designate a wide range of questionable food doctrines, systems, and practices and the persons who promote them. The "foodist" may be a misled scientist, well informed in another area but ignorant concerning nutrition. "Foodism" may be founded on valid scientific evidence that has been misinterpreted, and it may create anxiety and distort people's perceptions of food and/or food use. Proponents of the new terminology hoped that it would help promote the acceptance of facts about nutrition, as opposed to fantasies.

A recent report of a survey of nutritional knowledge and food practices of 943 high-school athletes in Connecticut indicated that young people have a reasonably accurate knowledge of what constitutes sound food choices but are not well informed about nutrition as it affects the functioning of the body. Young women in the study did not appear to utilize their nutritional knowledge as well as young men in making food choices, a finding that was supported by the results of other studies. Generally, researchers have concluded that knowledge about nutrition alone may not be sufficient to convert people to sound eating habits that contribute to health and well-being. Rather than change their eating habits, people in general expect that the medical and pharmaceutical professions will restore body functions that have deteriorated after years of dietetic abuse and neglect.

Nevertheless, there were signs that many people were becoming more aware that food has an important influence on health, that some ingredients in food have specific functions, and that "enough of a good thing is enough"; one example was sodium. In the fall of 1982 random telephone interviews were made with the goal of asking 4,000 adults about their awareness of salt and sodium information on food labels and their need for better labels with more helpful information.

Of the respondents 12% had been advised by a health professional to reduce salt or sodium intake and another 49% had some degree of concern, ranging from wanting to use less salt to strict, self-imposed, low-sodium diets. Most respondents recalled the types of salt and sodium information available, but most were unclear about the difference between "salt" and "sodium," how much sodium was needed as compared with the "milligrams" listed on the label, and the meaning of "without added salt," "reduced sodium," and "moderately low sodium." Not surprisingly, 65% felt confused and thought too many different terms were used.

The number of consumers who looked for and tried to use label information about salt and sodium had increased from 14% in 1978 to 40% in 1982, a greater rise in awareness than for any other type of label information on foods or medications. However, most respondents who had been advised to reduce salt or sodium intake did not know how to determine what foods they should avoid other than those that tasted obviously salty, such as ham, bacon, pickles, and chips. They could not tell from a label whether a product had a high or low salt content, partly because they were unaware that ingredients were listed in decreasing order of concentration. Apparently many consumers would use nutrition information if they realized that it could have an immediate application in their lives and if they could interpret the statements on labels.

Control of nutrition behavior. In November 1984 a group of researchers at the University of Illinois Department of Nutrition and Medical Dietetics, Health Science Center, published a report in the *Journal of the American Dietetic Association* entitled "Perceived Control of Nutrition Behavior: A Study of the Control Theory Among Healthy Subjects." The report described research that was designed to gain insight into the perceptions of individuals concerning their willingness and ability to change their nutrition behavior. A questionnaire, designed to examine beliefs about internal and external factors that determine dietary habits and health, was given to 105 healthy young Jewish men and women from different areas of Israel. The questionnaire was given personally to each one of the subjects with written and oral explanations, but there was no discussion of nutrition or health.

A distinction was found between the attitudes of religious and nonreligious subjects indicating a difference in learning and training that has relevance for nutrition education. Nonreligious/nontraditional Jewish subjects were more willing than religious/traditional subjects to change their nutritional behavior. They perceived themselves as more able to make such changes and to take responsibility for choosing "what is right" or "better."

—Mina W. Lamb

Life sciences

Among the major developments of the year in the life sciences were the discovery of a gene segment related to body formation in a wide range of animals, the cloning of DNA (deoxyribonucleic acid) from an extinct animal, the emergence of a new bacterium that caused canker in citrus trees, and an improved understanding of such cellular phenomena as the transmission of impulses by nerves and the recognition of one cell by another through their surface molecules.

Botany

Steady progress in many areas of botany took place during the past year. Among them were forestry, genetics, paleontology, physiology, and parasitology. In addition, a special plant was honored by the American Museum of Natural History.

Forest problems. Evidence continued to build that acid rain was causing serious harm to forests throughout the Northern Hemisphere. Some observers contended that many forests in 1985 were receiving as much as 30 times more acidity than they would if rain fell through clean air. The pollutants involved in acid rain included sulfates and nitrates, sulfur dioxide, ozone, and heavy metals. Their effects on the trees included yellowing and premature loss of needles and leaves, thinning of their crowns, and, eventually, death.

Nowhere were these effects more visible than in central Europe. The forests of West Germany were particularly affected. It was reported that only 52% of the forest in the area including the famous Black Forest remained healthy. Other countries reporting damage included East Germany, Austria, Poland, Yugoslavia, and Czechoslovakia. According to a study in the state of Baden-Württemberg in the southwestern part of West Germany, forests had declined very rapidly there since 1980. Damage to white fir trees was just beginning in 1980, but by 1982 there was a marked increase in mortality. Norway spruce, apparently healthy in 1980, showed evidence of widespread damage by 1982.

In the United States a decline of certain eastern forest species was noted as long ago as 1965. Several investigators studied tree rings of affected species and found that growth had been suppressed since at least that time, a condition most evident in red spruce, pitch pine, and shortleaf pine. An alarming increase in red-spruce mortality in the Appalachian Mountains was noted. These findings suggested that damage recently evident might be the result of long-range stress to forest trees, the nature of which was subject to much speculation and research. It was known, for example, that the 1960s were years of prolonged drought and unusual cold throughout the East. There were also pronounced increases in air pollution with resultant acid rain.

One example of research was reported during the year by C. F. Baes III and S. B. McLaughlin of Oak Ridge (Tenn.) National Laboratory. Their detailed study of tree rings of the shortleaf pine (*Pinus echinata*) in eastern Tennessee revealed increasing concentrations of trace metals in annual growth rings since the 1950s. Beginning in the 1970s metal content increased, while ring growth decreased. Because it had been proposed that acid precipitation increased the concentrations of aluminum and iron in the soil, greater accumulation of these elements, and perhaps others, in plant tissues might serve to indicate the effects of acid rain. A similar suppression of growth rings and increased metal concentrations were associated with a period of uncontrolled smelter emissions at Copperhill, Tenn., from about 1863 to 1912. In addition to the xylem of annual rings, these metals were also found in the cambium and phloem of current growth, where damage was potential. (Xylem, cambium, and phloem are types of tissues in the vascular systems of higher plants.) If they were found to produce harmful effects on trees, not only would there be a method to trace the history of certain kinds of air pollution, but there also would be evidence of the primary cause of growth-ring suppression.

Other general theories were being explored to explain the cause of tree damage. According to one concept acid rain damages needles or leaves, thus cutting down carbohydrate production and impairing the plant's root system. The latter could also be affected by a change in soil chemistry.

Tree rings. The world's longest continuous tree-ring chronology has been derived from the bristlecone pine (*Pinus aristata*) in California. The discovery of trees more than 1,000 years old as well as partially fossilized remains provided for a tree-ring sequence of 8,681 years. During the past year researchers reported an unbroken European tree-ring sequence spanning 7,272 years based on studies of two oaks, *Quercus petraea* and *Q. robur,* in Northern Ireland and West Germany. J. R. Pilcher and M. G. L. Baillie of Queens University (Belfast), B. Schmidt of the University of Cologne in West Germany, and B. Becker of the University of Hohenheim in West Germany synchronized sequences of ring widths by visual and statistical means. They established site chronologies first by cross-dating the oak remains in each place to provide a sequence typically 500–1,000 years long. Because these oaks do not live longer than 250 years, the task was much more complicated than with bristlecone pines, which record the most recent 1,000 years in living trees.

Once local chronologies were available, they were cross-dated over the distances between them by the same process of overlapping identical ring sequences used for local sequencing. Thus the development of enough local sequences made it possible to work out the whole chronology, match the most recent part of it to living trees, and provide a calendar to give absolute dates to each of the sites.

Plant genetics. The advent of cell-culturing techniques eventually showed that genetic variability for a number of morphological, biochemical, and chromosomal characteristics would occur in the cloned cells. Plants grown from cell culture retain some of this genetic variability, indicating that cell culture may be used to introduce variation into seed-crop species. D. A. Evans, W. R. Sharp, and H. P. Medina-Filko of DNA Plant Technology Corp. reviewed progress in this area during the last year and reported on their research with the tomato, *Lycopersicon esculentum*. They were able to reproduce tomato plants in tissue culture by removing young leaves from healthy plants and placing them on the appropriate culture medium. Shoots appeared within four weeks from calluses that formed on the leaves. Such shoots were rooted in rooting hormone and potted for growth in the greenhouse. Seeds from the resulting fruit were sown to detect recessive mutations in the second generation. Among the mutations that appeared was "jointless pedicel," a trait desired for mechanical picking of tomatoes because no stem remains on the fruit. It was thus an example of a desirable mutant trait that could be introduced into a crop-plant strain by tissue culture. The advantage of such a technique over conventional detection of mutants, which involves cross-breeding plants for generations to develop the useful strains, is the time saved.

The foregoing is an example of somaclonal variation, so called because nonreproductive tissues (soma) are cultured. In such tissues recessive mutations are not evident and cannot be detected until the next generation. Evans and his colleagues also experimented with gametoclones; these are plants regenerated from microspores, the cells that develop into pollen. Such cells and the resulting plants are monoploid—mutations appear in the first generation. The chromosomal number must be doubled, however, to produce genetically healthy plants. Treatment with colchicine, a crystalline alkaloid, is most frequently used to accomplish this.

It was becoming increasingly apparent that such changes occur because DNA is remarkably unstable. The recent 11th Annual Minnesota Plant Biology Symposium at the University of Minnesota explored the matter. Scientists there concluded that one cause of the instability might be due to changes in the large number of repeated sequences in plant DNA. Christopher Cullis of the John Innes Institute in Norwich, England, reported that different stressful conditions might induce changes in repetitive DNA. Some sequences increase in number, but more decrease so that there may be as much as a 15% difference in the DNA in a plant following the stressful condition. Cullis discovered these changes when trying to determine why different concentrations of nutrients such as potassium, nitrogen, and phosphorus could produce variations in plants of an inbred line. What is more, the variations were inheritable, as shown by breeding sequences.

Fossil plants. Some researchers during the year concluded that they had found evidence of a fossil parasitic fungus on a fossil gymnosperm (a woody seed plant that produces naked seeds not enclosed in an ovary, such as a conifer). Over the years a number of fossil fungi have been identified, but evidence for the parasitic relationship has been difficult to obtain. For this reason most investigators have assumed a saprotrophic relationship whenever a fungus is found with plant material in a medium such as coal. Because a saprotroph is an organism that lives on the dead remains or wastes of another organism, these researchers are indicating by means of this theory that it is difficult to prove that the host was living at the time that the fungus was there and that the fungus actually penetrated living cells.

Sara P. Stubblefield and Thomas N. Taylor of Ohio State University, Charles E. Miller of Ohio University, and Gary T. Cole of the University of Texas reported their find from a coal bed of the Upper Pennsylvanian Period near Berryville, Ill. In it they found cones of the gymnosperm *Lasiostrobus polysacci* that showed definite tissue destruction and other evidence of fungal activity. Of particular significance was the indication that the cells of the host were reacting to the fungus, thus proving that the host was alive and that the relationship was not saprotrophic. Some host cells apparently had produced small rounded bodies contiguous with their cell walls, a reaction noted in some present-day cases. Other cells had similar structures filling their entire volume, and others had what appeared to be fungal reproductive structures in them. The authors speculated that the rounded bodies were made of resinous material that might have served to impede the progress of the fungus and thus provide some host resistance.

Fossils of a more recent origin led to increased insight into domesticated plants in eastern North America as long as 2,000 years ago. It has been well

Carbonized seed of a species of goosefoot (Chenopodium berlandieri)*, about 2,000 years old, was discovered in Russell Cave in Alabama. Its characteristics reveal that it was a domesticated variety of goosefoot.*

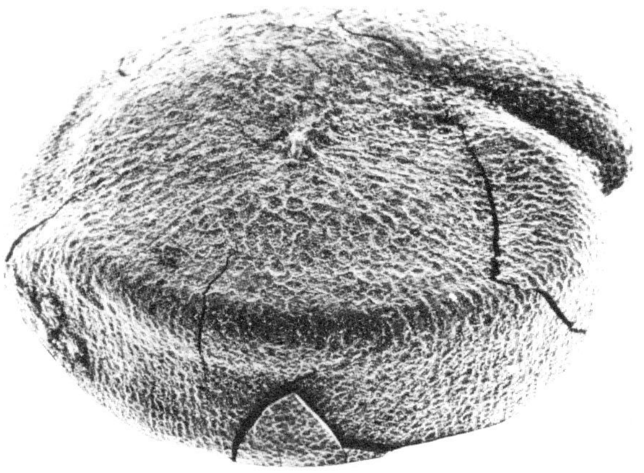

Smithsonian Institution

Stylites andicola, a relative of ferns that grow in the high Andes Mountains, is the first vascular land plant to be found that absorbs carbon dioxide through its roots for purposes of photosynthesis.

established that four domesticated food plants were being utilized at that early time: bottle gourd (*Lagenaria siceraria*), squash (*Cucurbita pepo*), marsh elder (*Iva annua*), and sunflower (*Helianthus annuus*). Others might have appeared in garden plots, but their seeds did not bear the distinguishing marks of domesticated plants. It is generally accepted that the lack of the distinctive hard outer coat and the presence of a truncated margin are characteristic of seeds from plants that have been domesticated for some time. The latter trait is associated with a cross section of the seed taking on a rectangular shape, which later provides the seedling with greater food reserves. Bruce Smith of the National Museum of Natural History in Washington, D.C., recovered such seeds of the plant *Chenopodium berlandieri* (a species of goosefoot) from Russell Cave in Alabama. The dating of those seeds at approximately 2,000 years before the present time and their apparent domestication indicated that yet another food plant was present at that time in garden plots of what is now the eastern United States.

Fossil evidence was causing some change in ideas about the history of tropical forests. Barbara Leyden of the University of Florida studied the pollen from two lake profiles in Guatemala to determine the vegetation changes that had taken place over the past 10,000 years (the Pleistocene Epoch). She discovered that, instead of a lush tropical forest as found in Central America now, there was a cool scrub and savanna landscape. Thus the common idea that tropical forests were preserved in such places as Central America and northern South America during the successive periods of glaciation that made more northern or southern areas inhospitable must be modified. It had been thought that such refuge forests expanded again when glaciers retreated. Leyden's work, added to scattered data from other places, indicated instead that tropical forests might have been "resynthesized" from various scattered components rather than preserved as a whole. The speed with which the highly diverse and ancient-looking modern forest was reestablished is impressive.

Mosses and light. Phototropism is the tendency of plants to grow toward a light source. While it seems of obvious value for plants to do this, the mechanism for detecting and responding to the direction of light is not well known. Thus far it is known that all plants do not use the same method. An example of phototropism in the horn-tooth moss *Ceratodon purpureus* was reported during the year by three researchers from West Germany. They found that phytochromes (certain pigments) are not only used to detect light but may be arranged in cells to detect direction of that light. One phytochrome, designated P_r, absorbs red light and is converted to P_{fr}, another phytochrome. The latter absorbs far-red light and is converted back to P_r. This combination makes it possible for plant cells to detect the presence of daylight because it contains red light, while there is far-red light even at night. The researchers grew the early algalike stage of *Ceratodon,* called the protonema, and noted that it was phototropic. They also found that the tubular cells acted as a lens system to interpret the light direction. Only when the cells are oriented certain ways does the light fall on the phytochromes and trigger the phototropic response. The orientation of pigments must be fixed in the cells, probably on certain membranes, a fact verified by polarized-light experiments.

CO_2 through the roots. It is commonly found that nonvascular aquatic plants absorb carbon dioxide (CO_2) through their roots in order to make it available for photosynthesis. Vascular land plants have stomata on their leaves to allow for the direct exchange of such gases as CO_2 between the air and the internal leaf. Aquatic plants usually do not have stomata, nor do terrestrial vascular plants absorb CO_2 through their roots. These two clearly different structural-functional relationships seem to mark the difference between aquatic and terrestrial vascular plants. It would be interesting to find a terrestrial vascular plant that absorbs CO_2 through its roots and thus illustrate a possible link.

Jon E. Keeley of Occidental College, C. Barry Osmond of the University of Nevada, and John A. Raven of the University of Dundee in the U.K. studied just such a plant. Its name is *Stylites andicola,* and it is a relative of ferns that grow on hummocky ground in bogs in the high Andes Mountains of Peru. Very little of the plant appears above ground, and only about 6.6% of the plant weight is green leaf tissue. Root-absorbed CO_2 is supplemented by crassulacean acid metabolism (CAM), a modification of photosynthesis very common to stomata-bearing terrestrial plants that need to keep stomata closed in order to preserve moisture. CO_2 gained by nighttime metabolism is stored chemically and then released internally during the day to be recycled by photosynthesis. The researchers concluded that it was too early to determine whether *Stylites* is a link between aquatic and terrestrial plants. If it is, CAM may have been present in vascular plants much longer than previously supposed.

Plant parasites. Some red algae give evidence of interesting host-parasite relationships. These are widespread eukaryotic (having true nuclei) plants in tidal communities of the oceans. More than 15% of all known genera in one major family are parasitic, and these always parasitize other red algae. It is in the parasitic relationship that interesting things happen, as illustrated by the parasite *Choreocolax* on its host, *Polysiphonia.* The parasitic filaments grow between the cells of the host and proceed to grow structures called secondary pit connections between parasite and host cells. These structures begin as conjunctor cells budded off the *Choreocolax* filament cells and are completed by connection to the host cells. A small nucleus, present in each of the conjunctor cells, is passed into the host cell, which may eventually contain many parasite nuclei as well as parasitic ribosomes, mitochondria, proplastids, and endoplasmic reticulum.

Lynda J. Goff of the University of California at Santa Cruz and Annette W. Coleman of Brown University, who reported confirmation of the process during the year, were able to use fluorescence microscopy to distinguish host from parasite nuclei. They concluded that it was likely that the parasitic nuclei might take over host cells and might represent a regulatory mechanism for this form of parasitism.

Peonies. A familiar flowering garden plant was honored during the year by an exhibit at the American Museum of Natural History in New York City. The exhibit celebrated man's use of the peony as reflected in art and illustration. Although prized today for its beauty, the peony was once used as a medicinal plant. As early as the 1st century AD recipes containing peonies were purported to dissolve kidney and bladder stones. Pedanius Dioscorides, well known for his encyclopedia of medicinal plants, believed that peony root promoted menstruation and that peony seeds in wine would help combat nightmares, hysteria, and cramps of the uterus. Into medieval times herb gatherers sold peonies to physicians for their prescriptions.

Today the peony is respected for the beauty of its large blossoms. The familiar herbaceous form, *Paeonia lactiflora,* originated in Asia. Tree forms also exist, having been known in the Orient for at least 1,500 years.

—Albert J. Smith

Microbiology

Among the highlights in microbiology during the past year were the identification of the virus believed to cause acquired immune deficiency syndrome (AIDS); the discovery of marine life at a depth of 3,000 m (10,000 ft) in the Gulf of Mexico similar to that found in hydrothermal vents in the Pacific Ocean; the emergence of a new bacterium that caused an outbreak of canker in citrus trees; and evidence that animal feeds containing antibiotics may lead to antibiotic-resistant bacteria that, in turn, may infect humans.

Health-related microbiology. Scientists at the National Cancer Institute in Bethesda, Md., and at the Institut Pasteur in Paris independently isolated viruses that were shown to cause AIDS. Later work established that the two viruses were identical.

The AIDS virus is closely related to two known viruses, human T-cell lymphotropic viruses (HTLV) I and II. Consequently, the AIDS virus was designated as HTLV III. Owing to the experience acquired from work with HTLV I and II, scientists were rapidly able to gain an understanding of the AIDS virus. The full nucleotide sequence of the viral genome was expected to be known in 1985, and experimental vaccines would soon follow.

Scientists recently demonstrated that chimpanzees are susceptible to the AIDS virus. Chimpanzees, therefore, can be used as an animal model for the study of AIDS. With the causative virus identified and characterized and an experimental animal model found, major obstacles in the study and prevention of AIDS were removed. These developments are important because in 1985 about 8,500 new cases were expected in the United States alone.

In Western Europe and the U.S. AIDS victims generally have been male homosexuals, intravenous drug abusers, and hemophiliacs who receive numerous blood transfusions. The AIDS virus has been found in blood, saliva, and semen of apparently healthy male homosexuals. The situation appeared to be different in Zaire, a nation in central Africa. A team composed of French and Belgian workers found that about 5% of the population of Zaire carried antibodies for the AIDS virus, indicating some type of exposure to it. Moreover, the same team found evidence for heterosexual transmission of AIDS. This led to the fear that AIDS could spread to the general population wherever the virus is present.

For more than 30 years antibiotics have been added to feed in order to promote growth in cattle, poultry, and hogs. This has fueled a debate as to whether this practice might lead to an increase in antibiotic-resistant bacteria in the animals so fed and, by animal-to-man transmission, to an increase in human infections by antibiotic-resistant bacteria. For this reason the use of antibiotics to promote growth in animals was banned by Western European countries. A similar proposal was made by the U.S. Food and Drug Administration in 1977, but the U.S. Congress failed to approve it.

This debate was rekindled during the past year owing to 18 cases of salmonellosis, a severe intestinal disease that in this situation was contracted by eating hamburger contaminated with the causative organism. The origin of the organism (and the hamburger) was traced back to animals which had eaten feed that contained antibiotics. Thus, there seemed to be a direct connection between the use of antibiotic feed additives and the occurrence of the 18 cases of salmonellosis.

As is often the case, however, all was not as clear as it seemed. For example, 70,000 lb of hamburger were involved, but there were no other cases reported. Of interest, too, was a report by scientists from the Bronx-Lebanon Hospital Center in New York City. These workers studied more than 40 million bacterial samples from 242 hospitals over a 12-year period. They found no increase in bacterial resistance to 15 different antibiotics and therefore concluded that the commonly held concept of an increase of bacterial antibiotic resistance at an alarming pace is a myth.

To everyone's surprise yeasts were found to produce steroid hormones that are identical to those of humans. Estrogen and corticosteroid are two such hormones. Yeasts also possess receptors that can bind both human and yeast steroids. As of 1985 the role of steroids in yeasts was not clearly understood, but there was evidence that the steroids may regulate the transformation of yeast from the threadlike mycelial form to the globular form. Yeasts may also be a source of environmental steroids.

Some yeasts are infectious. In some types of infections men are more likely to be infected than women, and the reverse is true in other types of yeast diseases. Microbiologists speculated that the infectious yeasts may select their hosts on a sexual basis by binding to the predominant steroid hormone (such as estrogen) produced by the host.

A perennial health problem is hospital-acquired, or nosocomial, infection. Between 5 and 10% of patients acquire infections after admission to a hospital. A scientist from the London (England) Hospital Medical College recently found that bacteria which occur normally on human skin produce an antibacte-

Grant Heilman

Hereford cattle eat commercial feed. Antibiotics have been added to feed for many years in order to promote growth in livestock. Some scientists, however, believe that this practice may cause an increase in antibiotic-resistant bacteria in the animals and, when the meat of such animals is eaten, may lead to an increase in human infections that are generated by such bacteria.

Doug Cavanah—Tampa Tribune

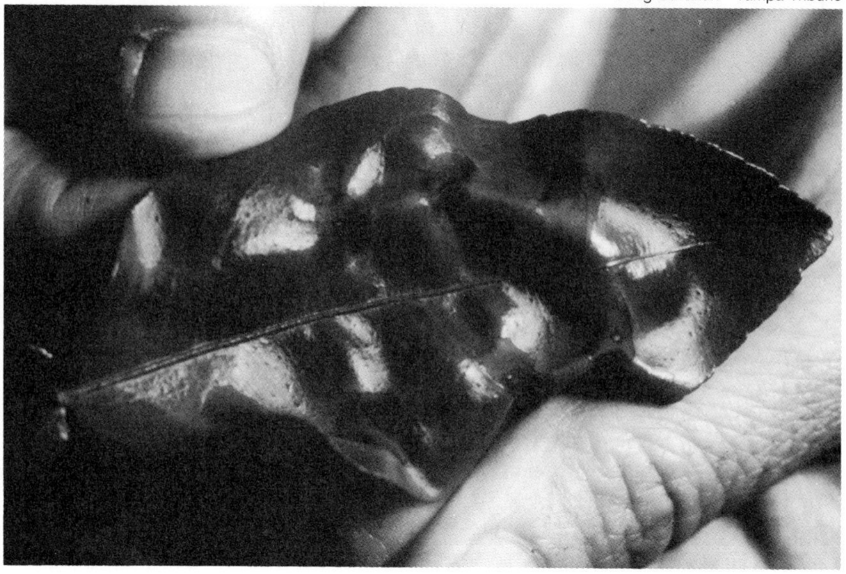

Dark spots on a citrus leaf indicate the presence of citrus canker, a disease that quickly spreads to destroy the entire tree.

rial substance(s) that inhibits some but not all infectious bacteria. These findings suggested the need for a series of "narrow-spectrum" skin antiseptics that leave intact the hospital staff's normal skin bacteria and their natural antibacterial factors but that kill those bacteria resistant to these factors.

Environmental microbiology. During recent months scientists from the Scripps Institution of Oceanography in California discovered biological communities composed of tube worms, clams, mussels, and other forms of marine life living at a depth of 3,000 m (10,000 ft) in the Gulf of Mexico. This was an unexpected and accidental discovery, because the purpose of the expedition was to study the effects of erosion on the continental shelf of Florida. The forms of life were similar to those found in the hydrothermal vents in the East Pacific. Because there is no sunlight at 3,000 m primary production cannot take place by photosynthesis. (A primary producer is the first form of life in the food chain.) Instead, analogous to the Pacific hydrothermal vent communities, it is thought that primary production is carried out by chemosynthetic, autotrophic bacteria that gain energy for growth by the oxidation of hydrogen sulfide or, perhaps, by oxidation of other reduced inorganic compounds. The only requirements would be a source of hydrogen sulfide or reduced inorganic compounds, oxygen, and carbon dioxide.

The microscopic marine alga *Prochloron* lives in symbiotic association with species of marine ascidians (sea squirts) among tropical reef habitats. The alga actually resides in the cloacal cavity of the sea squirt where its profuse growth may impart a green color to its semitransparent host. *Prochloron* was shown by a scientist from the University of North Carolina to fix nitrogen. This ability to convert molecular atmospheric nitrogen into organic nitrogen is important, because the amount of nitrogen-containing compounds in tropical waters is vanishingly low. Interestingly, this was the first report of nitrogen fixation attributable to symbiotic algae residing in either an aquatic or a terrestrial animal.

Associative nitrogen fixation is carried out by bacteria living on the roots of nonleguminous plants. In contrast, symbiotic nitrogen fixation is carried out by bacteria living in root cells of leguminous plants. A great deal is known about how symbiotic nitrogen-fixing bacteria adhere to roots of leguminous plants prior to invasion of the root cells; namely, adhesion relies on plant substances called lectins that recognize and bind to carbohydrates on the bacterial surface. Workers at Helsinki University in Finland recently reported that *Klebsiella,* an associative nitrogen fixer, attaches to plant roots by fimbriae (or pili), filamentous proteins that protrude from the bacterial surface. It appears that the fimbriae may have lectinlike activity that enables them and, thus, the bacteria to attach to some specific substance on the root surface.

Scientists in New Hampshire demonstrated that phosphorus is tightly conserved within hardwood forest ecosystems. A large part of this phosphorus retention is due to microorganisms associated with the root system of the trees.

Apparently the hot springs in Yellowstone National Park in Wyoming are a never-ending source of new and interesting bacterial forms. Recently a photosynthetic purple bacterium was isolated that differs from all other known purple bacteria.

Not only have the Florida citrus growers suffered considerable losses during the past five years because of cold weather, they also have been hit by a double "whammy" in the form of a new type of bacterium that causes a disease called citrus canker. The organism belongs to the same genus as four other known canker-causing bacteria, but it is a new variety. Moreover, while the other four varieties, or strains, are specific with respect to the type of citrus tree they can infect, the new strain is able to infect a wide variety of citrus trees.

As of 1985 the new canker-causing organism had infected only nursery plants. The only "cure" was to burn all plants in the infected areas. Thus, about three million seedlings, representing 21% of the total nursery stock in Florida, were destroyed. For that reason there will not be enough nursery plants available to replace citrus trees killed by cold weather.

The origin of the new canker-causing bacterial strain remained unknown. Scientists thought that it might have been brought into Florida from another country, perhaps by a traveler carrying contaminated fruit.

Miscellaneous. Studies on the production of methane by anaerobic (in the absence of air) bacteria continued during the past year. Anaerobic microbial digestion of sewage, in which methane is simultaneously produced, has long been an established technology. As of 1985, however, commercial production of methane itself had not been successful because the production rates and yields were too low to make the process competitive with conventional energy sources. Continued efforts to commercialize microbial methane production through innovative digester designs and other improvements were in progress in several countries, particularly the U.S., Belgium, West Germany, and China.

There was also a renewed interest in immobilized microbial enzymes. Enzymes are biological catalysts composed of proteins, and they are produced only by living cells. Isolated enzymes may be immobilized by adsorption onto a solid substance such as a resin, by chemically binding to a solid substance, by entrapment in a gel, or by encapsulation in membranes. Until recently immobilized enzymes were unable to compete economically with chemical processes mainly because the desired enzymes were too expensive to produce or too scarce. By 1985, however, this picture was changing owing to genetic engineering techniques that can change microbial cells into efficient producers of almost any desired enzyme at reasonable prices. Thus, it was predicted that the use of enzymes as catalysts was on the verge of remarkable growth. Dozens of enzymes already were in use: in the food industry; as components of detergents; in the pharmaceutical industry; in the treatment of certain diseases; in the manufacture of various products other than pharmaceuticals; and to detect or to detoxify a variety of toxic pollutants.

—Robert G. Eagon

Molecular biology

The flood of information flowing from the study of cloned genes continued during the past year. The incredibly powerful tools for analysis of genes and their products made it possible for scientists to propose reasonable models for a variety of cellular phenomena, including the transmission of impulses by nerves, the recognition of one cell by another through their surface molecules, and the orderly distribution of proteins to different cellular compartments.

Nerve transmission. The transmission of electrical impulses in the cells of the nervous system, called neurons, is accomplished by selective, abrupt changes in the permeability to ions of the membrane that surrounds the neuron. Cell membranes consist of a molecular bilayer of lipid molecules analogous to a soap bubble; the membrane prevents water, other molecules and ions from escaping the cell. Transport across the membrane is mediated by protein molecules embedded in it. In neurons there are two classes of membrane proteins that transport ions into the cell in response to specific external signals. One class responds to small molecules called neurotransmitters. Proteins of this class are found on the receiving side of a synapse. When a suitable (electrical) signal is transmitted to the sending end of a neuron, that end releases the chemical transmitter, which diffuses across the synapse and signals the receiving neuron to open its membrane protein to the flow of cations (positively charged ions). The best characterized system of this class utilizes the neurotransmitter called acetylcholine; the postsynaptic protein that binds acetylcholine and responds to that binding by opening a cation channel into the cell is called the acetylcholine receptor. Investigators have learned much about the structure and function of the acetylcholine receptor protein during the past year.

A second class of proteins in the membranes of neurons does not respond to neurotransmitters. Instead, this class can measure a voltage difference between the inside and outside of the cell. When that voltage exceeds a certain value, the protein changes its conformation in a way that, again, opens a channel for the flow of ions. The best studied example is called simply the sodium channel. The protein involved in this reaction has been the object of detailed study, and some clues to its structure and function have come to light.

In the cases of both the acetylcholine receptor protein and the sodium channel protein conventional biological materials contain too few copies of the specific protein to be useful sources of material for analysis. Fortunately, neurophysiologists have for many years studied specialized organs of animals that have turned out to be rich sources of those proteins. For the acetylcholine receptor protein the electric organ of the stinging ray *Torpedo californica* is the starting material of choice. The sodium channel protein is most easily obtained from the electric organ of the eel *Electrophorus electricus*, although it has also been purified from rat brain, rat muscle, and other tissues.

Purification and characterization of these proteins depends on two different kinds of functional assay. Each of the proteins, when in its native state in the nerve cell membrane, can be poisoned by the binding of specific neurotoxins, which are themselves proteins. These neurotoxins are also obtained from animal sources, which are believed to manufacture them for defense against predators. Specific toxin binding is a necessary but not sufficient indication of a functional ion channel protein. The final test is reconstitution of the purified protein in a synthetic membrane and the demonstration of acetylcholine-dependent ion flow across the membrane, in the case of the acetylcholine receptor, or voltage-dependent ion flow, in the case of the sodium channel protein. For both proteins this was accomplished, and some information on the structure of the proteins thus became available.

The acetylcholine receptor protein is a complex containing five polypeptide chains, of which two are identical; the structure is written $\alpha_2\beta\gamma\delta$. Workers in the laboratory of Shosaku Numa at the Kyoto University School of Medicine in Japan prepared recombinant DNA (deoxyribonucleic acid) clones containing genes that encoded each of these proteins. The complete sequence of the repeating units of DNA (adenine, guanine, cytosine, and thymine) was determined for each of these genes, and from that sequence it was possible, using the known genetic code, to predict the complete sequence of amino acids in each of the receptor protein subunits.

At first glance it would seem that knowledge of amino acid sequence alone gives little insight into the three-dimensional folding of the protein and thus into its functional domains. However, the three-dimensional structure of a significant number of proteins has been determined by X-ray crystallography, and it is now possible, by comparing known structures with known amino acid sequences, to develop empirical rules that correlate types of local protein domains with particular sequences. For example, a given stretch of amino acids may be found, four out of five times, to form an α-helix. In addition to these empirical "rules" for structure prediction, the 20 different amino acids can be classified as hydrophobic (lacking affinity for water), hydrophilic (having an affinity for water), or indifferent, depending on the chemical nature of each. When a protein known to be located in a membrane is found to have a stretch of α-helical hydrophobic amino acids, it is a reasonably good bet that that stretch of amino acids is located within the membrane bilayer.

Applying these considerations to the amino acid sequence of the α-subunit of the acetylcholine receptor protein led to the development of a model for this protein. Roughly half of the protein chain (the amino-terminal half, first to be synthesized) is on the outside of the membrane; the remaining half includes five α-helical segments, each of which spans the membrane; the final small carboxy-terminal piece is inside the cell. Four of the α-helical segments are hydrophobic. The fifth is called amphipathic because, looked at from the top or side, one side of the helix displays hydrophobic residues while the other displays electrically charged residues. It is this column of charged residues spanning the membrane that is thought to provide the channel for ion movement across the membrane.

This model was tested by the Numa laboratory in Japan in a fantastic way. Starting with the cloned α-subunit gene, the researchers prepared a series of engineered derivatives in which each of the five α-helical segments was individually deleted. Other "mutants" were created in which single amino acids were replaced with others. Then each of these man-made altered receptor proteins was tested for its ability either to bind specific neurotoxin or to pass a current when activated by acetylcholine. In order to prepare cells with membranes that contained the modified receptor proteins, Numa's colleagues first used their synthetic genes to direct the in vitro synthesis of messenger RNA (ribonucleic acid). (Messenger RNA carries genetic information outside the cell nucleus.) This RNA was then injected into oocytes of the toad *Xenopus laevis*. These large shell-less eggs are favorite objects for such work, because they faithfully make proteins according to the information encoded in the injected RNA. The complete acetylcholine receptor protein, it should be recalled, has five subunits, and so it is necessary to inject, as well as RNA encoding the modified α-subunit, messenger RNA encoding the unmodified β, γ, and δ subunits. When this was done, the oocyte manufactured all of the subunit proteins and assembled fully functional receptor protein in the cell membrane. Deletion of parts of the α-subunit protein predicted to be inside the cell had no effect on function. Deletion of all or part of four of the five helical segments resulted in loss of ability to allow ions to pass, but neurotoxin could still be bound. Deletion of the helical segment closest to the N-terminal resulted in loss of toxin binding as well as of ion permeation. The same effect was obtained by replacing the amino acid cysteine at positions 128 or 142 with serine. While much remained to be done, this work represented a great stride forward in understanding the relationship of the structure of a protein to its function.

As if this were not enough, Numa and another team of co-workers also elucidated the amino acid sequence of the sodium channel protein. Unlike the multi-subunit acetylcholine receptor, the sodium channel consists of a single very long protein chain containing 1,820 amino acid residues. Again, the structure predicted from the amino acid sequence contains unusual features, as shown in the figure. The protein contains four sections that are related to one another by sequence homology (chemically similar but differing one from the next by a simple structural unit) and by similarity of predicted structure. Within each section there are six elements; of these four are likely to be α-helical, spanning the membrane, and two are thought to be located on the inner side of the membrane. One of the latter two is rich in amino acids with positively charged side chains that occur every third residue. This sequence can form a structure in which the positive charges line up along one side of a helix, possibly providing the ion channel. Although this structure has not yet been tested by the site-directed mutagenesis applied to the acetylcholine receptor, such experiments have now become possible.

Targeting proteins. During the year researchers addressed the question of how a cell assures the proper location of the proteins it manufactures. Most proteins, in cells with nuclei, are made in the cytoplasm (the protoplasm of a cell that is exterior to the nuclear membrane). Yet some of these proteins belong in the nucleus, some belong in such non-nuclear cell structures as mitochondria or chloroplasts, some belong in the cell membrane, and some are to be exported from the cell. It is a general rule of biology that the way in which a protein folds up (its three-dimensional structure) is determined by its sequence of amino acids. It is this rule, in fact, that permits the structure predictions described for the nerve cell proteins above. Amino acid sequences also determine where a protein goes in a cell.

The critical step in protein traffic is crossing or entering a membrane. The mitochondria, for example, have an outer membrane that defines the organelle (a collective term for all membrane-bounded structures within the cell cytoplasm) and a highly folded inner membrane containing the protein complexes that couple the flow of electrons with the synthesis of ATP, the chemical that supplies energy to drive many biochemical reactions. Some of the mitochondrial proteins are made in the mitochondrion itself, but most are made outside. These must pass through the outer membrane; some will insert into the inner membrane, while others must also pass through the inner membrane to the interior matrix.

For some years it has been known that many proteins destined to be secreted by cells are synthesized with extra amino acids at the amino-terminal, or beginning, of the protein chain. These extra amino acids were termed a "leader" sequence; they were shown to be essential for transport of the rest of the protein across the membrane and were removed in that process. Thus, the mature protein is shorter than when it was originally made and, lacking the leader sequence, is incapable of returning through the membrane by which it exited the cell. Subsequently it was discovered that proteins destined for locations in mitochondria or chloroplasts also have leader sequences that are processed or removed during membrane crossing or membrane insertion. The detailed mechanisms by which membrane crossing or insertion is accomplished differ: some work during synthesis of the protein, while others work on the finished proteins. The names given to the part of the protein responsible for targeting to a particular location also differ: a "leader" sequence is removed during synthesis of the protein; a "transit" sequence is removed after synthesis is completed; and a "signal" sequence is responsible in any way for protein targeting, whether it is removed or not.

Many signal sequences for a wide variety of proteins were determined recently. This work was facilitated by gene cloning and determination of the nucleotide sequence of the DNA corresponding to those proteins, as described above for the nerve cell

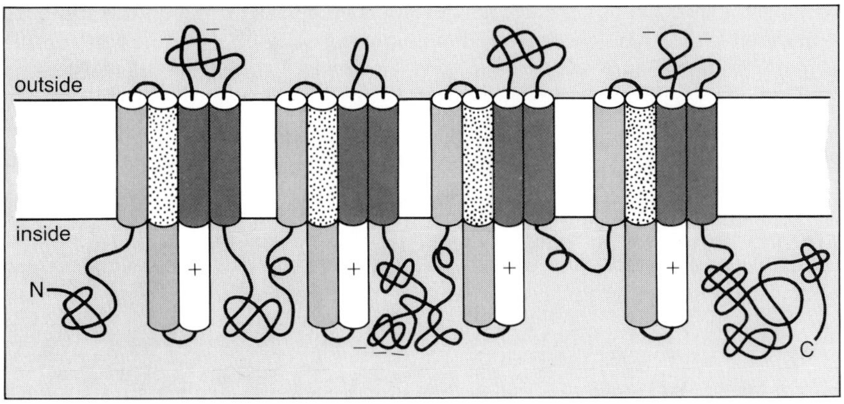

The sodium channel protein contains four sections that are chemically similar but differ one from the next by a simple structural unit. Of the six elements that comprise each section, one is rich in amino acids with positively charged side chains that occur every third residue.

Reprinted by permission from *Nature*, vol. 312, no. 5990, pp. 121–127. Copyright © 1984 Macmillan Journals Limited proteins. Comparison of the amino acid sequence predicted from the gene sequence with the actual sequence of the mature protein led in many cases to the discovery of possible signal peptides. These were later confirmed by biochemical experiments designed to detect short-lived intermediates in the synthesis of the finished proteins.

Once the genes encoding various signal peptides were known, it became possible to ask detailed questions about targeting. The most striking results of the past year showed that a given signal peptide can direct other proteins to the customary location of that peptide's usual client protein. The implications of these results for genetic engineering are profound. The following examples illustrate the power of these procedures.

One of the major enzymes present in chloroplasts catalyzes the conversion of carbon dioxide to carbohydrate. This enzyme contains two kinds of protein subunit, termed "large" and "small." The gene for the large subunit is located in chloroplast DNA, but that for the small subunit is in the cell nucleus. Small subunit protein is made in the cytoplasm (external to the nuclear membrane) in the form of a precursor containing a transit peptide. The precursor is transported across the chloroplast membrane, during which process the transit peptide is removed. Inside the chloroplast the small and large subunits associate to form the active enzyme.

The question arises as to whether the transit peptide can be used to transport other proteins into chloroplasts. The answer is yes, at least for some proteins. In one dramatic example workers in the laboratories of Marc van Montagu at the University of Ghent in Belgium and Anthony Cashmore at Rockefeller University in New York City fused a part of the pea gene encoding the small subunit transit peptide to a bacterial gene encoding the enzyme neomycin phosphotransferase. This enzyme adds a phosphate group to the antibiotics neomycin or kanamycin, rendering cells carrying the enzyme resistant to those antibiotics. It was shown that the small subunit transit peptide targeted the neomycin phosphotransferase to the interior of chloroplasts in whole tobacco cells or, in vitro, to the interior of chloroplasts isolated from pea leaves. Transport required the transit peptide. These results suggested that many other proteins can be targeted to the chloroplast by the small subunit transit peptide. Since herbicides, in general, work by poisoning specific proteins in chloroplasts, this procedure is likely to find wide application in tailoring plants that are resistant to specific herbicides.

Other experiments of the kind just described have shown that foreign proteins can be targeted to the nucleus, to the outer membrane of mitochondria, to the matrix of the mitochondria (crossing both outer and inner mitochondrial membranes), and to the endoplasmic reticulum, the intracellular membrane system that initiates the process of protein secretion from cells.

Lest it be concluded that leader or transit peptides are responsible for all cellular targeting of proteins, it should be added that many proteins, of which the sodium channel protein is a prime example, do not have processed signal sequences. In that case the finished protein is synthesized as such. The structure of the finished protein is sufficient to dictate how and where the protein inserts into a membrane. In all these cases there is much to be learned about the components of the membranes that process the transit sequences or accept the finished protein.

—Robert Haselkorn

Zoology

During the past year a 20-year research project describing the developmental position and fate of every cell in a small nematode was completed, and a gene segment related to body formation was found in fruit flies, annelids, frogs, and humans. Neurobiologists continued to find that the brain is a dynamic organ, and olfactory imprinting of salmon was used commercially in ranching operations. DNA from an extinct zebralike animal was cloned, and other researchers compared DNA from many different birds to determine evolutionary relationships. In the U.K. badgers were implicated as repositories of bovine tuberculosis.

Developmental biology. After 20 years of research biologists at the Laboratory of Molecular Biology, Cambridge, England, described the complete development of a 1-mm-long nematode worm, *Caenorhabditis elegans*, from the fertilized egg to the 995-cell adult. The exact cell lineages are now known, including which cells give rise to all other cells at any stage of development. This work laid the foundation for answering a fundamental problem in biology, namely, what events determine how cells become different from each other during development.

During development in this type of nematode worm about ⅛ of the cells die. Hilary Ellis of the Massachusetts Institute of Technology identified mutants in which cell death does not occur. The study of these mutants should shed light on the role of certain genes in programming cellular death and how death results in the formation of structures in the embryo.

Researchers also identified about 400 other genes and thousands of mutations. Of great interest was the complete description of the nervous system. It has 302 neurons that make 8,000 synapses. (A synapse is the point at which a nervous impulse passes from one neuron to another.) Though seemingly large, it is infinitely more simple than the human brain with its 100 billion or more neurons. Mutations identified with changes in behaviors were to be studied to see if the behaviors result from changes in the number, structure, or association of cells in the nervous system. Genetic analysis of the role of the nervous system in determining inherent behavior now appears open to direct observation.

During development, nerves grow along tortuous paths to reach targets. How cells follow these paths has puzzled biologists. Recent research appears to indicate that the pathways are marked by a protein called nerve cell adhesion molecule (NCAM) that occurs on the surfaces of tissues along the way. Joshua R. Sanes of Washington University in St. Louis demonstrated that embryonic muscle has large amounts of NCAM on its surface but innervated (supplied with nerves) muscle has none. If an axon (a nerve-cell process) leading to a muscle is cut, the muscle produces NCAM to guide future innervation. Urs Rutishauser and Jerry Silver of Case Western Reserve University showed that nerves in a developing chick eye follow routes marked by NCAM. If NCAM is blocked by administering an antibody, the nerve cells do not follow normal paths and randomly elongate. No such NCAM is found in areas where axons do not grow. Essentially NCAM is a molecular equivalent to a highway marker, indicating nerve routes to target areas.

A great stir occurred among developmental biologists during the past year when similar genes regulating the formation of the segmented body during early development were found in fruit flies, annelids, frogs, mice, and humans. The original interest in these genes came from analyzing bizarre mutant fruit flies that have legs in place of antennae. William McGinnis, Walter J. Gehring, and colleagues at the University of Basel in Switzerland cloned these developmental regulating genes, which are called homeotic genes. Using a DNA hybridization technique, they showed that DNA from fruit flies would bind to DNA from several other species providing that the latter had segmented bodies during development; the DNA from fruit flies would not bind to DNA from species lacking body segmentation. There was great hope that this discovery would help unlock some of the secrets of gene regulation during development.

During the past year successful embryo transfers were carried out between two species. At the Louisville (Ky.) Zoo a quarter horse gave birth to a zebra, and at the Cincinnati (Ohio) Zoo a common African antelope gave birth to a bongo, a rare species of antelope. Both of these were the world's first offspring from an embryo transfer between different species. The researchers then attempted to freeze embryos before the transfers in an effort to determine whether they could maintain the embryos for long periods of time. These techniques, considered impossible only a few years ago, are becoming routine. They may be used to save species that are becoming extinct, and they offer a simple way to provide genetic input into inbred zoo populations from wild species.

Neurobiology. For many years the brain was considered a fixed organ incapable of cell divisions. By 1985, however, evidence was accumulating that this is not the case. John A. Paton and Fernando N. Nottebohm of Rockefeller University demonstrated that cells in a vocal control area of the canary brain divide in adults when the birds are given injections of testosterone to stimulate singing. Adult birds were

AP/Wide World

Quarter horse mare nuzzles the baby zebra to which she has just given birth. The zebra was one of the world's first offspring that resulted from an embryo transfer between two species.

given a radioactive chemical, thymidine, that would only be taken up by dividing cells, thereby marking those cells. The canarys' brains were then penetrated with an electrode to record electrical activity in cells of the vocal area. Cells were found that were electrically active to sound stimuli. These cells were injected with a marker chemical, the animals were killed, and the brains were examined by microscopy. Cells that were active also had taken up the thymidine, indicating that the brain produced new cells in response to hormone treatment and that these cells were incorporated into functional circuits.

Carl Cotman and Manuel Nieto-Sampedro of the University of California at Irvine showed that neurons respond to growth factors produced when the brain is injured. When brain cells are damaged, the synapses, which can be regarded as communications junctions between adjacent cells, degenerate. Over a few weeks undamaged cells nearby may form new synapses with the ones that dead cells had contacted, restoring the number of synapses formerly found. However, these new junctions are usually not farther away than 30 micrometers from any cell. Neurons are stimulated to make these contacts by the formation of neurotrophic chemicals when there is brain injury. The factors that promote the synapse formation can be demonstrated by taking fluids from injured brains and putting them on nerve tissue cultures, causing the neurons to grow. Such extracts could be used to help brain grafts become functional by stimulating contacts with the host's tissue.

Behavior. The magnetic sense organ of yellowfin tuna may have been located by Joseph L. Kirschvink, a geobiologist at the California Institute of Technology, and Michael M. Walker of the University of Hawaii. Their studies indicated that crystals of uniform magnetization are concentrated in a sinus in the dermethmoid skull bone, where approximately 85 million crystals 50 nanometers long occur. It is thought that groups of these crystals deform mechanoreceptors when in a magnetic field. Histological studies revealed nerve axons in the dermethmoid region with connections to the supraophthalmic trunk nerve. Others found similar crystal arrangements in the dermethmoid bones of chinook salmon, further implicating this region as a magnetic sense organ.

The Kemp's Ridley sea turtles are an endangered species whose nesting sites are confined to a beach in Mexico that is about 15 km (9 mi) long. Joint efforts between the U.S. and Mexico were under way to establish breeding colonies for the turtle on Padre Island, Texas. Because sea turtles were thought to imprint (a rapid learning process that takes place early in life) on olfactory clues of home beaches, turtle eggs collected in Mexico were put in Padre Island sand and brought to the United States, hatched, and placed into the surf at Padre Island. The juveniles were then collected and raised for 9–12 months at the National Marine Fisheries Laboratory in Galveston, Texas, before being released. If those animals were imprinted, females should favor returning to the beach on Padre Island in several years to lay eggs. M. A. Grassman and D. W. Owens of Texas A&M University, R. Marquez of the Instituto Nacional De Pesca, Mexico, and J. P. McVey of the U.S. National Marine Fisheries Laboratory obtained experimental evidence of imprinting. Young turtles raised at the National Laboratory were put into a chamber where water collected from four geographic locations was circulated. The turtles spent significantly more time in an area of the chamber that had water coming from Padre Island. The researchers interpreted this to mean that the animals had imprinted on olfactory clues from the island and that the stocking program should be successful. This was the only evidence that sea turtles might chemically imprint and implied that imprinting is a necessary step in restocking programs.

National Marine Fisheries Service

Kemp's Ridley sea turtle was one of many that took part in an experiment at the U.S. National Marine Fisheries Laboratory. When placed in a chamber containing water from different areas, the turtles spent the most time in the water from the area where they were raised.

Olfactory imprinting in salmon was being put to commercial use. Young hatchery-raised salmon, imprinted with the characteristics of local waters, have been released into the ocean to grow and eventually return for harvest. In Chile 800 chinook salmon returned from a total of 300,000 that had been released into the ocean when young. The Weyerhaeuser Corp. began experimenting with salmon release in the Pacific Northwest but remained uncertain as to whether the operation would be profitable. During their five-year maturation in the sea many salmon are killed by predators and others are caught by commercial and recreational fishermen. Nonetheless, the idea that fish can be raised for a short time, released, and then later return to a processing plant located at the rearing site has appeal. Oregon has permitted 20 private salmon ranchers to release 180 million young salmon annually.

Organismal zoology. Peter Moran of the Australian Institute of Marine Science in Queensland observed a second outbreak of an infestation of crown-of-thorns starfish, *Acanthaster planci,* on a segment of the Australian Great Barrier Reef. These animals, which can reach sizes of 60 cm (24 in) across, consume living corals. The reef thus loses its regeneration capacity and erodes under wave action. Moran studied the recovery of the reef from an invasion of *Acanthaster* 12 years ago. In 1982 he noted that 35% of the reef had recovered, but visits in 1984 indicated that living corals had been reduced to 2% of the reef.

The *Acanthaster* outbreak was not confined to Australia but seemed to be occurring throughout the Pacific. In Micronesia outbreaks appeared to be correlated with heavy runoff of water from land, which increases nutrients in the sea. Plankton blooms result from the nutrients, increasing the food supply to larval stages of the starfish. The movements of adult starfish were not well understood, and Moran and his colleagues were attempting to track them. Moran microinjected miniature radio transmitters into the body cavities of the starfish, allowing him to track their movements and study individuals.

In the Caribbean Sea populations of the long-spined black sea urchin, *Diadema antillarium,* were decimated by a waterborne pathogen. H. A. Lessios and colleagues of the Smithsonian Tropical Research Institute in Balboa, Panama, witnessed massive mortalities of the sea urchin, starting in 1983 on the Venezuelan coast and spreading north and east to the coast of Florida and the Caribbean. This was of concern because the sea urchins, which can reach sizes of up to about 30 cm (12 in) in diameter, are effective grazers of algae on coral reefs. In their absence scientists feared that the algae would out-compete the coral and begin to slow reef growth and replacement.

Other researchers from the Smithsonian Institution's National Museum of Natural History observed echinoderms in their natural habitats using a submersible, the Johnson-Sea-Link II, as deep as 750 m (2,460 ft) in The Bahamas. David L. Pawson and colleagues made a total of 41 dives, videotaping echinoderm behavior and collecting specimens. They found 5–10 new species, but, more important, they had the unique opportunity of seeing how these animals live. A startling discovery was that sea lilies, echinoderms that were thought to be attached to the seabed by a stalk, were found to be mobile. When sea lilies were removed from the seabed and placed on the ocean floor, they used their featherlike arms to crawl. Brittle stars were seen to use their spine-studded arms to swim in whiplike motions or to use them to form a basket, perhaps for catching small prey.

Marine fish produce a large excess of offspring, but relatively little is known about predation of the larvae. Cannibalism of young by adults of the same species was thought to account for most losses. Heino Moller of the University of Kiel in West Germany reported that, in the Kiel Fjord, jellyfish, *Aurelia aurita,* are significant predators of larval herring. He noted correlations over four years between the size of the jellyfish populations and the number of larvae found. Close examination of the jellyfish showed that on average they contained from 0.4 to 4.4 larvae per stomach in samplings at different times. Thus, predation of vertebrates by invertebrates may be a major consideration in how many of the former survive.

Entomologists during the year perfected a method to study a habitat that had only been sampled sporadically in the past. That habitat was not beneath the sea but instead was 60 m (200 ft) above the Earth in the canopy of tropical forests. Terry Lee Erwin of the Smithsonian Institution's National Museum of Natural History, working in the Peruvian Tambopata Wildlife Reserve in the Amazon basin, sampled insect populations in the treetops by using a gun to shoot a rope into the canopy. An insecticide fogging device was then hauled into the canopy, and from it was released a biodegradable insecticide. The insecticide only kills arthropods; it does not harm vertebrates and has a two-hour life. Dead insects are caught as they rain down from the treetops into nets that look like inverted umbrellas. The yield of insects was incredible. By far the most common insects in this aerial habitat are ants, but the diversity of other varieties is great. Based on his work in this forest, Erwin estimated that there may be 30 million to 50 million species of insects on the Earth, most not yet described.

The DNA from the mitochondria (structures in the cell outside the nucleus) of a quagga, a zebralike animal that last lived about 100 years ago on the steppes of South Africa, was cloned by Russell Higuchi and Allan Wilson and colleagues at the University of California at Berkeley. The tissue samples from which the DNA was extracted came from a skin preserved with salt in a German museum. The cloned DNA fragments were compared to fragments from zebras and horses in order to determine evolutionary relationships and changes. Work such as this suggested that

DNA fragments from mummies could eventually be cloned. Wilson's group also tried to isolate material from mammoths found frozen in Siberia but had limited success. Bacterial contamination accounted for most of the DNA found in such corpses, but some DNA of mammoth origin was recovered. This pioneering work may mean that someday scientists will be able to retrieve DNA from bones of other animals and to create banks of extinct genes. These will be useful for determining evolutionary rates but will not result in recreation of extinct species because whole, intact DNA in nuclei is not found.

Ecology. A massive hydroelectric project was completed in Canada in the James Bay area of Quebec. At first there were fears that the project would be an environmental catastrophe. Northern habitats, especially the taiga (the area between forests and the tundra), are considered fragile environments, and the magnitude of this project was forecast to have profound effects, possibly changing the climate and triggering earthquakes. Begun in 1971, two reservoirs were created to divert water into the La Grande River, and three more reservoirs were created on the La Grande itself. These inundated an area half as large as Wales. Earthworks were built that, if placed end to end, would stretch 125 km (77.5 mi) with the tallest equaling a 40-story building. Environmental inventories indicated that animal populations seemed to have adapted to the project and had not suffered great losses.

Sperm whales are an endangered species, and in 1981 the International Whaling Commission (IWC), a volunteer regulatory agency of 38 nations, decided to suspend whaling in the northwest Pacific by not setting catch limits. Japan, where whale products are part of the culture, objected and was granted special concessions to whale in the area in 1982–84. At the commission's meeting in 1984 in Buenos Aires, no quotas were set and no concessions granted. Japan's fleet left port despite the moratorium and caught at least two sperm whales. The IWC cannot take action to protest Japan's action, but member nations can. For example, the U.S. was considering limiting Japan's rights to fish within the U.S. 360-mi (580-km) territorial limit. This would be a substantial sanction because Japan's fleet takes 74% of the bottom fish caught off the coast of Alaska, and this amounts to 10% of Japan's total catch.

In the mid-Atlantic states an outbreak of rabies occurred in raccoon populations. First reported in raccoons in Florida in 1953, there were 1,608 cases reported in mid-Atlantic states in 1983. The growing prevalence of raccoons in urban areas raised concern at the U.S. Centers for Disease Control in Atlanta that rabies may spill over into the human population either by direct contact or through transmission of the disease by nonimmunized cats and dogs.

In Great Britain badgers were suspected of being repositories of bovine tuberculosis. In the 1930s more than 35,000 people died there annually from tuberculosis, about 10% of those from the bovine form. Pasteurization, immunization, and removal of infected animals from herds all but wiped out the disease among humans and in cows, but isolated outbreaks still occurred. In southwest England the incidence of bovine tuberculosis in cattle in 1984 remained 5–10 times greater than the national level. Testing of badger populations in the areas showed that about 10% of them also had the disease, and it was presumed that they transmitted it to cattle. Badger eradication programs were undertaken, including gassing of the animals in their burrows and trapping them. But there appeared to be a lack of evidence showing that control of badger populations lessens the disease in cattle, leading conservationists to oppose the control plan.

Evolution. A nearly complete skeleton of a human ancestor was found on the shore of Lake Turkana in northern Kenya by Kamoya Kimeu of the National Museums of Kenya under the direction of Richard Leakey. The 1.6 million-year-old bones were from a young male about 12 years old who stood about 5 ft 6 in tall and would have been 6 ft at maturity. The skeleton, missing the entire left arm, the lower right arm, and most of the feet, was classified as *Homo erectus,* but a detailed analysis remained to be done. (*See* Year in Review: ANTHROPOLOGY.)

Charles Sibley and Jon Ahlquist of Yale University completed 20,000 DNA comparisons among birds of the world to determine their relatedness. They took nuclear DNA, broke it into fragments 500 base pairs long, and mixed it with DNA from another species. When the DNA was heated, the strands of the double helices that comprise the DNA separated; then, when the DNA was cooled randomly, they reassociated. Reassociation depends on complementary base-pairing between the strands, and the more similar the two DNA's the more tightly they bind. New double helices were formed that contained strands from both species. When the DNA was reheated in the range of 100° C (212° F), the melting temperature of the hybrid double helices was related to the strength of bonds, that is, the number of base pairs that match each other. Strands with mismatched bases melt before those with complete matches, and so the DNA that melts soonest is composed of strands of both species. By looking at the differences in temperature where 50% of the DNA is dissociated, scientists were able to obtain a measure of genetic difference for any two mixed DNA samples.

If it is assumed that changes in DNA are linear with time and if it is known when two species evolved from each other, the technique can be calibrated so that a melting point difference can be related to a divergence date in years. Sibley and Ahlquist used this technique to demonstrate that the theories on origins of birds in Australia were incorrect. The diverse bird life there was previously thought to have originated by waves of migration from Eurasia, but Sibley and Ahlquist showed that all birds evolved from one lineage about 60 million years ago. The two researchers then turned their attention to human evolution. They reported that gibbons diverged from the hominoid family 18 to 22 million years ago followed by the orangutans at 13–16 million years. The gorillas diverged some 8 to 10 million years ago, leaving humans and chimpanzees sharing a common ancestor until they split 6–8 million years ago. This finding is contrary to the evolutionary pathway suggested by fossils.

—Warren D. Dolphin

Materials sciences

Research highlights during the past year included reports on advanced glasses for a variety of applications, novel uses for piezoelectric ceramics, and methods to lower the power requirements in the electrolytic production of aluminum. Ingenious treatments being developed for the iron and aluminum smelting industries promised significant reductions in impurities in the extracted metals.

Ceramics

Their attractive appearance, transparency, and the economy with which they can be formed into complex shapes have made silica-based glasses one of the most useful ceramic products. Although early civilizations learned by trial and error to make these glasses by melting silica sands with various fluxing and stabilizing oxides, their underlying nature is still a subject of intense study. For example, research in the rapid solidification of molten materials has shown experimentally that many materials will freeze into a glassy (noncrystalline) state if they are cooled quickly enough. A kinetic theory of glass formation recently put forth by Donald R. Uhlmann of the Massachusetts Institute of Technology suggests that any molten material can be made glassy by exceeding some critical cooling rate. Today, in more and more cases, that rate is being estimated from the basic properties of the materials.

Another aspect of glass under study is its physical structure. Most theories have proposed and experimental evidence has supported the idea that glassy structures contain a significant amount of randomness and disorder, as opposed to the perfect regularity of ideal crystal structures. This concept was further developed by a number of researchers. For example, James Phillips at AT&T Bell Laboratories in the U.S. suggested the existence of partially polymerized clusters in silica glass, and S. Dembovsky and D. Chechetkina of the Kurnakov Institute in the Soviet Union proposed that "quasimolecular" defects in the overall structure affect viscosity and the critical cooling rate required for production and retention of the glassy state.

Among interesting specific developments were advances in heavy-cation fluoride glasses that transmit medium-wavelength infrared light, fast-ion conducting glasses, and glasses formed by gel processing. Glasses that transmit well at longer wavelengths (to about ten micrometers) in the infrared part of the spectrum are important both for efficient fiber-optic communication and for various electro-optic applications in defense technology. Heavy-cation fluoride glasses, discovered serendipitously by Marcel Poulain in 1974, appear well-suited to these requirements, and more than a hundred research papers dealing with their development have since appeared. In these glasses the majority of the structure consists of atomic polyhedra of zirconium, hafnium, and thorium fluoride linked together by barium fluoride or other constituents added as glass formers. The high atomic mass of the heavy zirconium, hafnium, and thorium ions are conducive to the low structural vibration frequencies needed for good long-wavelength transmission. Practical applications for such products as optical fibers, however, will probably require solutions to several difficult problems related to production and serviceability. These glasses tend to be very fluid and to recrystallize easily on reheating, making fiber drawing difficult. They also are weak and very susceptible to surface corrosion.

Calcium aluminate glasses also appear to have promise for transmission in the infrared, at least at wavelengths of three to five micrometers. They are therefore particularly attractive for use as window materials in infrared-guided tactical missiles, but they tend to be weak and susceptible to erosion when flown through rainstorms. GEC Power Engineering Ltd. in England showed that crystallizing the surface of these glasses improves both their strength and their erosion resistance.

Much work has been done on fast-ion conducting materials for use in advanced batteries, the best known material being beta-alumina solid electrolyte. Although most of these materials are crystalline, recent research has shown that glasses also can be made into fast-ion conductors. Such glasses typically contain lithium, silver, chlorine, and iodine. Uhlmann and co-workers demonstrated that the chlorine and iodine in these glasses enhance the movement of the lithium ions, and that the lithium content of the glass helps prevent physical attack of the glass by the molten lithium in the battery.

Synthesis of glasses by the sol-gel route has become an extremely popular topic of investigation. Though known for a long time, sol-gel processing has rarely been pursued outside the laboratory, a situation likely to change as a result of recent advances in sol-gel technology and some important new potential applications. Sol-gel processing begins with the raw materials in a liquid solution or a colloidal suspension (a sol). If diffusion between reactants is required, it can be made to occur at this stage without the high heat needed in conventional high-temperature ceramic processing. The solution or sol is then reacted to form a gel, and finally excess water or other solvents are removed to obtain the desired composition, structure, and in some cases even the final shape of the glass product.

Two major problems with sol-gel processing to date, however, have been the long drying times required in forming thick, solid components and the cracking that is generally encountered during drying. Larry Hench and co-workers at the University of Florida showed that both problems can be reduced by controlling the pore structure, shrinkage rate, and evaporation rate during the gelation and drying processes. Pore size can be controlled by adjusting the processing variables, and the evaporation rate by adding an organic solvent, termed a drying-control chemical additive, to the sol. This approach allowed the investigators to dry large transparent glass bodies, without cracking, in much shorter times than had been possible previously.

Sol-gel processing can be used to form a wide variety of products. It was shown, for example, to be an inexpensive way to deposit adherent coatings of indium-tin oxide on glass. Because of their high transparency to ultraviolet and visible wavelengths but excellent reflectance in the long-wave infrared, these coatings are superb heat mirrors and can be used on insulating windows to convert them to passive solar collectors. It was also demonstrated that sol-gel processing can be used to synthesize such natural materials as olivine that are difficult to reproduce by conventional processing techniques as well as calcium silicate glasses that do not expand or contract with temperature and that cannot as yet be made by conventional melting procedures. Sol-gel glasses also showed the potential for unique bonding arrangements that could lead to properties not observed in melt-derived glasses.

Piezoelectric materials, both natural ones like quartz and tourmaline and synthetic ones like barium titanate and lead zirconate titanate, have long been known. Their atomic structures are polarized, and they therefore can generate large electrical voltages when subjected to mechanical stresses. Although an applied voltage produces the opposite effect, an expansion or contraction of the material, the dimensional changes are small and difficult to exploit in practical devices. One emerging use for the latter property is in highly energy-efficient actuators. Working very much like an insect's wings, for example, a resonant piezoelectric driver can be made to flap a flexible blade and move sizable amounts of air. Unlike motor-driven fans, which become very inefficient in small sizes, piezoelectric blowers can be scaled down to centimeter and even millimeter sizes. They also generate no magnetic or electrical interference or mechanical noise and appear to offer long, maintenance-free service lives. Since little current flow is required for their operation, they are quite efficient; a piezoelectric blower consuming only a few milliwatts can remove several watts of generated heat. They thus appear to offer advantages for use as cooling devices in the microminiature world of solid-state electronics. Another emerging application for piezoelectric actuators is in large, deformable space mirrors, where the small deformations would be well suited to the critical shape adjustments needed to maintain precise focusing accuracy.

Piezo Electric Products, Inc.

Small efficient fan incorporates two winglike blades driven by the vibrations of a piezoelectric material. The new device generates no magnetic or electrical interference and promises a long, maintenance-free life.

Development of ceramics for use as biomaterials made significant advances. Though the field has been dominated by metals and plastics, clinical studies have shown that ceramics have important advantages in many biomedical applications, particularly as bone implants. The Battelle Memorial Institute's laboratory in Columbus, Ohio, has had excellent results with dense, hydrostatically pressed polycrystalline alumina (aluminum oxide) as dental implants in animals, and current tests in humans at Ohio State University look promising. For some years

Kyocera Corp. in Japan has successfully used single-crystal alumina (sapphire) dental implants given a special chemical surface treatment to increase their strength. Their material, called Bioceram-S, is undergoing testing in the U.S., where it has had a success rate of more than 90%. Kyocera also has had excellent success, both in Japan and in initial U.S. testing, with a polycrystalline alumina material, Bioceram-P, for use as inert vertebral spacers to replace malignant or severely fractured spinal material and as inert bone extender in total knee and hip replacement operations.

Another family of useful bioceramics is based on various calcium phosphates, materials that more closely resemble the composition of bone itself. For example, Calcitek, Inc. is making a dense hydroxyapatite that can be injected under the skin of the jaw to form a very stable replacement for damaged or deteriorated jawbone.

Bioglasses, glasses designed especially for biomedical use, offer several promising advantages as implants. They are generally soda-lime silicate glasses with additions of phosphates and fluoride ions. Leaching of calcium and phosphorus from the glass surface promotes the formation of a silica-rich gel coating on the glass. This process in turn generally leads to in-situ crystallization of hydroxyapatite on the surface and chemical bonding of the implant to surrounding bone and tissue. In fact, the surface reactivity of these glasses probably can be tailored by adjustment of their composition to match the specific tissue response needed for a given application. Such flexibility could allow their use, for example, as anchors for implants to soft tissues such as tendons and ligaments. One major limitation of the bioglasses, however, has been their relatively low strength and brittleness. Applications to date therefore have been primarily as coatings on ceramic or metal implants. Recent studies showed that reinforcing bioglasses with fibers, which results in stronger and tougher composite materials, may overcome these difficulties and permit their use as the implant material itself.

Meanwhile, progress continued in the fabrication of whole ceramic prostheses. For example, Taylor Ceramic Engineering, working with the University of New South Wales in Australia, developed and demonstrated an artificial knee composed entirely of ceramics.

—Norman M. Tallan

Metallurgy

Recognition that the properties of steels can be improved by decreasing the concentrations of the impurity elements sulfur and phosphorus to very low levels has stimulated the search for economical ways to remove these elements by means of injection metallurgy. Metallurgical coke, used as the reducing agent in the iron blast furnace, is the major source of sulfur occurring in hot metal, the impure liquid iron produced in the blast furnace; although 80–90% of the sulfur charged to the blast furnace is absorbed by the slag, the hot metal typically contains about 0.03% by weight of sulfur. Iron ore is the major source of phosphorus, and although phosphorus does not present a major problem to North American steelmakers, hot metal made in Europe and Japan can have phosphorus levels as high as 1% by weight.

Before the introduction of a special treatment of the hot metal prior to its conversion to steel, sulfur and phosphorus were taken out during the steelmaking process. The phosphorus was removed by oxidation to phosphate, and the sulfur by reduction to sulfide. Steelmaking, however, which is a process of selective oxidation, provides less than ideal conditions for efficient removal of sulfur. The need for very low concentrations of sulfur led to the introduction of so-called external desulfurization, in which the hot metal is treated in the vessel in which it is transported from the blast furnace to the steelmaking furnace. External desulfurization involves injecting desulfurizing agents into the hot metal in a stream of carrier gas blown through a tube immersed in the melt. In North America two types of reagent are in use: calcium carbide and magnesium granules coated with magnesium salts. External desulfurization can decrease the concentration of sulfur in the hot metal from as high as 0.07% by weight to levels of 0.01–0.005%. Both reagents have their disadvantages,

Bone tissue engulfs microscopic spheres of a biocompatible metal alloy applied to the surface of a metal hip implant. The porous coating, created by a process called Porocoat, encourages formation of a durable bone-to-metal bond.

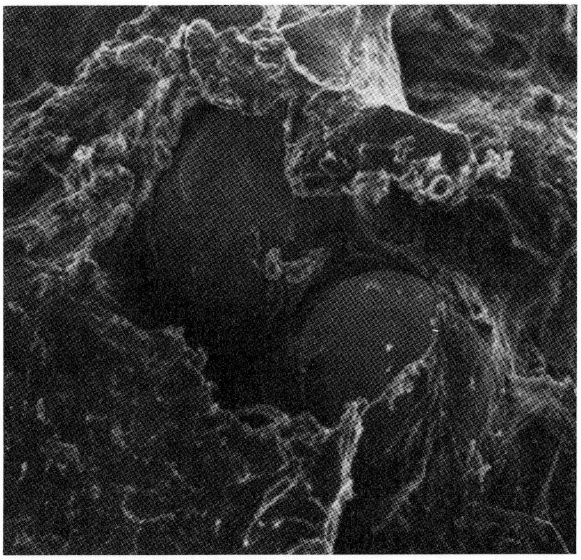

DePuy

Sandia National Laboratories

Kay Hays of Sandia National Laboratories in New Mexico demonstrates her new plasma-deposition method of plating parts with corrosion- and wear-resistant noncrystalline (glassy) metal alloys. In the technique, gases carrying the desired metal elements in the required proportions are converted into a plasma, from which the alloy deposits onto the workpiece as a noncrystalline film.

however. The cost of producing calcium carbide is increasing, and the use of salt-coated magnesium is losing favor because of the violence of the reaction that occurs when the magnesium comes into contact with the hot metal.

Because of these disadvantages Japanese researchers have been investigating methods of external dephosphorization and desulfurization making use of a mixture based upon calcium oxide (CaO) and containing calcium carbonate ($CaCO_3$), calcium fluoride (CaF_2), and carbon. Upon injection of these powders, thermal decomposition of the $CaCO_3$ into CaO and carbon dioxide (CO_2) causes the formation of very fine particles of chemically active CaO, while the reaction of the CO_2 with carbon dissolved in the hot metal to form carbon monoxide (CO) produces the desired degree of turbulence in the vessel. The use of $CaCO_3$ in the reagent mixture eliminates the expense of calcining (roasting) $CaCO_3$ to form CaO, and the heat absorbed by the decomposition after injection is almost entirely replaced by that released from oxidation of silicon dissolved in the hot metal by the CO_2 produced by the decomposition. The substitution of sodium carbonate (Na_2CO_3) for $CaCO_3$ in the mixture produces even better desulfurization and dephosphorization, but industrial use of this process must await development of a means of recovering and recycling the Na_2CO_3. The use of Na_2CO_3 can decrease the concentrations of sulfur and phosphorus to less than 0.002% and 0.01% by weight, respectively.

Recognizing the differing chemical requirements for the removal of sulfur and phosphorus, Japanese researchers are developing a method in which the two elements are removed at different locations in the hot metal. In this procedure powdered mixtures of CaO, CaF_2, and iron oxide (Fe_2O_3) are injected with an oxygen-nitrogen carrier gas deep into the hot metal, and CaO-CaF_2 powders are added to its surface. The injected powders melt and begin to rise through the hot metal as liquid globules. The presence of Fe_2O_3 and oxygen gas provide the oxidizing conditions necessary for the removal of phosphorus, and during their ascent the globules oxidize and absorb silicon and phosphorus and lose their iron content. Reaching the surface of the hot metal, the globules coalesce to a CaO-CaF_2-SiO_2-P_2O_5 slag, which dissolves the CaO and CaF_2 that have been added to the surface. Because the Fe_2O_3 has been removed, the reducing conditions necessary for the removal of sulfur are restored. The flexibility of the process arises from the ability to vary the proportions of the three reagents in the injected powders as well as the ratio of oxygen to nitrogen in the carrier gas.

Aluminum is produced by the century-old Hall-Héroult process, in which alumina (aluminum oxide) dissolved in molten cryolite (Na_3AlF_6) is electrolyzed to produce liquid aluminum at a carbon cathode and a mixture of carbon monoxide and carbon dioxide at a carbon anode. Because liquid aluminum does not wet carbon, a thick layer of liquid aluminum must be maintained at the bottom of the electrolysis cell in order to prevent contact between the cryolite electrolyte and the carbon cathode, which lines the cell bottom. Interaction between the electrically induced magnetic field in the cell and the layer of liquid aluminum causes undulations in the layer such that the depth of the aluminum at the center of the pool can be as much as five centimeters (two inches) greater than at the periphery. In order that contact between the electrode products be avoided, the distance between the anode and the cathode is maintained at four to six centimeters. Consequently as much as 40% of the electric power required for the process is consumed by resistance heating of the large depth of electrolyte.

Because a decrease of 1–1.5 volts in the cell voltage could be achieved by decreasing the anode-to-cathode distance to one to three centimeters, investigators have searched extensively for a material that is wetted by liquid aluminum and that could be used as the cathode. This research has identified titanium diboride (TiB_2) as a possibility. TiB_2 is wetted by liquid aluminum, has a low solubility in both liquid aluminum and liquid cryolite, and has a high electrical conductivity compared with liquid cryolite. In one proposal for its use a TiB_2-coated cathode is placed in the cell at an angle so that the thin film of deposited aluminum drains continuously into a holding well; the negligible interaction between the thin film and the magnetic field allows the anode–cathode distance to be reduced.

Fabricated TiB_2 parts, however, tend to be brittle and sensitive to thermal shock. Moreover, because of the difference between the thermal expansivities of TiB_2 and graphite, it is very difficult to produce a bond between a TiB_2 coating and a carbon substrate that can be maintained between room temperature and the cell temperature of about 1,000° C (1,830° F). One approach to improve bonding applies the technique of graded metal-to-glass bonding in the form of a three-layer composite comprising an upper layer of TiB_2, an intermediate layer of bonding material that accommodates the difference in thermal expansivity, and a lower layer of graphite provided with recesses to increase holding power.

In another approach a wet paste of TiB_2 powder, carbon cement, and carbon fibers is applied to a carbon substrate and then is cured and carbonized by a complex heat-treatment procedure. The presence of the carbon fibers prevents the spread of cracks that form during the curing process, and the mixture contains only enough TiB_2 (35–60%) to achieve complete wetting of the cured surface by liquid aluminum. At present, cathodes formed by this process are being tested in commercial electrolysis cells.

Introduction of the use of electrolytes containing lithium fluoride (LiF) in Hall-Héroult cells has increased the concentration of lithium occurring as an impurity in the aluminum, leading to the development of a process called TAC (treatment of aluminum in crucibles). In this process the molten aluminum is stirred and reacted with aluminum fluoride (AlF_3) in a special crucible after its removal from the cell. Adding a half kilogram (1.1 pounds) of AlF_3 per metric ton of aluminum decreases the lithium concentration from 20–25 ppm (parts per million) to 3 ppm. The reaction product, which has the general formula $(Li_xNa_yCa_z)AlF_6$, floats to the surface of the metal where it is easily removed. The reaction product is then recycled to the electrolyte in the Hall-Héroult cell.

—David R. Gaskell

See also Feature Article: NONDESTRUCTIVE EVALUATION—COMING OF AGE.

Mathematics

The pace of mathematical discovery during the past year seemed to be increasing at a surprising rate, both in pure and applied mathematics. Two important breakthroughs were confirmed in 1984, and there were signs of near-success on other important problems.

Linear programming. The most widely used mathematical technique of all time is the so-called simplex algorithm for solving linear programming problems. (In this context "programming" means "planning.") In 1984 Narendra Karmarkar, a young Indian mathematician working at Bell Laboratories in Murray Hill, N.J., devised an ingenious new algorithm for solving these problems. (An algorithm is a procedure for solving a mathematical problem in a finite number of steps; it frequently involves repetition of an operation.) On early test cases Karmarkar's method seemed to be about 100 times as fast as the simplex method.

Linear programming problems arise in a wide variety of industrial situations, from blending fuel to scheduling airlines, from preparing livestock feed to routing telephone calls. The common feature is a large number of variables (often as many as 10,000) limited by various constraints that can be expressed by linear equations such as $y = mx + b$ or the related inequalities $y > mx + b$ or $y < mx + b$.

The objective of linear programming is to determine a feasible solution (one satisfying the constraints) that minimizes operating costs. The difference between an optimal solution and a manager's best guess can be of enormous importance to a company's economic fortunes. Since timely solutions to linear programming problems can make the difference between success and failure in the marketplace, it is not surprising that enormous efforts are devoted to solving those problems.

The traditional simplex algorithm, developed during World War II by George Dantzig of Stanford University, uses algebraic manipulations to move around the edges of a high-dimensional polyhedral figure called a simplex. This simplex method searches for the optimal vertex by a slow, inexorable uphill climb. On average, it works reasonably well for typical problems of modest size. But it can fall into the trap of computational intractability if the number of variables is too large or the conditions particularly difficult. The simplex algorithm is what mathematicians call an "exponential" method, because in the worst case the time of solution can increase exponentially with the number of variables in the problem.

In 1979 a Soviet mathematician, L. G. Khachian, introduced a new "ellipsoid" method, the first known algorithm for linear programming that was not exponential; instead, the time of solution increased in "polynomial time." The following example reveals the difference between exponential and polynomial times. If the time required for a computer to solve a problem shows polynomial growth according to, for example, the cube (degree 3) of the number of variables, then increasing that number from 20 to 50 increases computing time from about 1 second to 12 seconds. However, if the growth is exponential, the exponent increases from 20 to 50 and the time required increases by about 1,000 centuries. In the ellipsoid method the degree of the polynomial is sufficiently high (degree 6) and the computations sufficiently difficult, so that on typical problems the simplex method is still faster.

Like the ellipsoid method, Karmarkar's algorithm operates in polynomial time. But the degree of the polynomial is lower in Karmarkar's method (degree 3.5), and realistic comparison trials on major problems suggest that his method significantly reduces the computation time required for solving linear programming problems. What formerly took hours could now be done in minutes, making possible "real-time" calculations to influence decisions.

Karmarkar's method, like Khachian's, plunges into the interior of the simplex and identifies a feasible (but not optimal) solution. To do this it uses techniques of projective geometry to transform the simplex into a nearly regular polyhedron; once that is accomplished, a relatively easy calculation (also based in ellipsoids —three-dimensional ellipses) produces a direction to move in which one can find a better approximation. Successive applications of this technique rapidly converge to the optimal result.

The task of transforming Karmarkar's algorithm into a commercially useful tool was complicated by copyright and trade-secret laws. Many of the major linear programming problems arise in classified military applications or in highly competitive industrial situations. To put the new method to a thorough test would require running it on many large realistic problems. At least a year probably will be needed to fine-tune early computer programs into productive industrial computer code. But early efforts continued to produce favorable results, suggesting the possibility of a significant breakthrough in computation speed.

Bieberbach conjecture. The most important theoretical result of 1984 was the surprising solution by Louis de Branges of Purdue University, West Lafayette, Ind., of the Bieberbach conjecture, a famous unsolved problem in the field of complex analysis. Many mathematicians believed that a solution would require significant new mathematical tools; some actually thought that the conjecture was false. Yet by employing old results in clever new ways de Branges accomplished what many believed to be impossible.

The problem that de Branges solved was proposed in 1916 by the German mathematician Ludwig Bieberbach in his study of coefficients of power series—infinite polynomials—that represent what are called analytic functions of a complex variable. These functions form the central core of much of modern analysis and mathematical physics and have innumerable applications in engineering.

Analytic functions are very smooth: they have neither sharp corners nor abrupt changes. Bieberbach's conjecture, in essence, sets limits on the growth of the coefficients of analytic functions, thereby describing inherent limits on their curvature.

Analytic functions act on complex numbers of the form $a + bi$, where i signifies the imaginary square root of -1. Complex numbers arise commonly as solutions to quadratic equations, given by the famous "quadratic formula" of high school algebra. Like ordinary "real" numbers, complex numbers can be added and subtracted, multiplied and divided. Unlike real numbers, however, they cannot be represented on a one-dimensional "number line." Because each complex number includes two ordinary numbers—the "a" and the "b"—a two-dimensional plane is required to represent the geometry of complex numbers.

Complex numbers are traditionally denoted by the letter "z." The magnitude (or absolute value) of a complex number, denoted by $|z|$, represents its distance from the origin (the complex number 0). Polynomials in complex numbers work just like ordinary polynomials; the same is true with power series, which look like infinite polynomials. For example, the expression $3z^3 + 2z + 7$ is a polynomial in the complex variable z, whereas the expression $z^2 + 2z^3 + 3z^4 + \ldots$ is a power series that represents an analytic function.

An analytic function establishes a relationship between the complex number z and its value, say w, at z; this relationship is usually expressed by the shorthand notation $w = f(z)$. If the numbers z are selected from a certain region in the complex plane, the "image" points w form a new region, one that usually has a different shape than the original. In this way the analytic function is said to "map" the original region to the new one, as if the old region were a city and the new region a street map for the city. The most common region is the "unit disk,"

Purdue University

Louis de Branges of Purdue University in 1984 solved the Bieberbach conjecture, a problem in the field of complex analysis that had challenged mathematicians since it was proposed by Ludwig Bieberbach in 1916.

the circular disk with a radius of one centered at the origin. The behavior of an analytic function on the unit disk provides a total picture of its behavior on the entire plane, the only difference being a matter of scale.

The nicest analytic functions are those in which two different points are never mapped to the same point. (That's the way all good road maps work too.) These are called "univalent" functions, since each value is derived from just one original point. Bieberbach's conjecture is just this: every univalent analytic function defined on the unit disk has the property that the magnitude of the coefficient a_n of the nth term in its power series can be no larger than n. In terms of symbols this can be expressed as $|a_n| \leq n$.

Bieberbach proved this only for the first and second coefficients. It took more than 60 years for others to verify the conjecture for the third, fourth, fifth, and sixth coefficients; proving it for all coefficients simultaneously seemed quite beyond reach. Recently, using a different approach, the Soviet mathematician I. M. Milin of the University of Leningrad proved that the coefficient of the nth term of such functions is never more than 1.24 times n. Others reduced this factor to 1.07. De Branges proved an even stronger conjecture of Milin, from which the Bieberbach conjecture follows as a direct consequence.

De Branges had worked on this particular problem for more than seven years without much encouragement from knowledgeable colleagues who believed the problem to be too difficult to solve. Most recently he reduced the issue to a certain inequality that he began testing by computer to determine if it could possibly be true. Then, fortuitously, he discovered that the inequality he needed had been proved in 1976 by Richard Askey of the University of Wisconsin at Madison and George Gasper of Northwestern University, Evanston, Ill.

Pulling together all the pieces, de Branges produced a massive first draft of the proof, more than 300 pages in manuscript form. Shortly thereafter he spent three months in Leningrad, where Milin and his students tested the proof in their seminar. Their work verified the result and reduced its essence to about 15 pages.

Riemann hypothesis. Without question one of the most important unsolved problems in all of mathematics concerns the solutions to an equation called the "zeta function" introduced by the German mathematician Bernhard Riemann in the middle of the 19th century. This equation relates prime numbers to complex numbers and thus forms a fundamental bridge between the discrete structures of number theory and algebra on the one hand, and the continuous structures of real and complex analysis on the other. (A prime number is any integer other than 0 or 1 that is not divisible without remainder by any other integers except 1 and the integer itself.)

In 1859 Riemann published an analysis of this function in which he conjectured that all solutions (or "zeros") of the function lie along the vertical line of the complex plane for which the real part of the solution is equal to one-half. Mathematicians ever since have tried to prove this but with little success. They have attacked the problem by every conceivable means including number theory, group theory, differential equations, probability theory, and complex analysis. But no one has been able to marshal sufficient evidence to settle the issue.

At the end of 1984, however, two tentative solutions were under serious discussion. One was an attack via group theory announced in Paris by the Japanese mathematician Hideya Matsumoto. The other, less firm, was a strategy announced by Louis de Branges at a meeting of the American Mathematical Society. Both approaches were only in outline form, and there were many gaps that remained to be filled. But both were taken seriously, and there was much hope that within a year the Riemann hypothesis would become a theorem.

Awards. The major mathematics award of 1984 was the National Science Foundation's annual Waterman Prize, honoring the most outstanding U.S. scientist under 35 years of age. It was given to Harvey Friedman, a logician at Ohio State University, for his work in the foundations of mathematics that continued the great tradition of Gottlob Frege, Bertrand Russell, and Kurt Gödel. In recent years Friedman has proved a great variety of important theorems that establish the independence of certain mathematical results from the traditional axioms of logic and set theory, thereby providing new philosophical perspective on the fundamental nature of mathematics.

Gödel's work in the 1930s showed that axiomatic mathematical systems must be incomplete, that there will always be some statements in the system that cannot be proved or disproved by axiomatic means. Until recently all known examples of such "undecidable" propositions have been disguised forms of logical consistency statements, more part of logic itself than of mathematics.

Friedman's great discoveries were of concrete facts in traditional mathematics that lie beyond the reach of axiomatization. His work challenged the centuries-old paradigm of founding theoretical science on the axiomatic method. It suggests, said MIT's Gian-Carlo Rota in his Waterman Prize address, "that the axiomatic method itself will soon be rendered obsolete."

—Lynn Arthur Steen

Medical sciences

During the past year medical scientists achieved considerable advances in their understanding of acquired immune deficiency syndrome (AIDS). Among other developments was the discovery that many people who suffer from chronic, recurrent flulike illness show evidence of infection with the Epstein-Barr virus and the finding that a new class of antibacterial agents, the quinolones, is effective against many organisms that are resistant to other drugs. Dental researchers believed that they were close to developing a vaccine to prevent tooth decay, and veterinary scientists in late 1984 produced a vaccine effective against feline leukemia.

General medicine

Intensive worldwide research efforts at last produced some clues to the origin of AIDS. Two research groups, one headed by Robert C. Gallo of the U.S. National Cancer Institute in Bethesda, Md., and one from the Institut Pasteur in Paris headed by Luc Montagnier, reported the isolation of a virus believed to be responsible for the condition. The virus, a member of a large group known as retroviruses, is variously termed human T-cell lymphotropic virus III (HTLV III); lymphadenopathy-associated virus (LAV); and AIDS-associated retrovirus (ARV).

After some initial controversy over whether HTLV III and LAV were in fact the same, researchers concluded that they are indeed variants of the same virus. Four research groups subsequently determined complete nucleotide sequences for various isolates of the virus. Gallo's group reported the complete nucleotide sequences of two viral isolates on Jan. 24, 1985, while Paul Luciw of Chiron Research Laboratories in Emeryville, Calif., with Jay Levy of the University of California (San Francisco) School of Medicine and associates reported the sequencing for another isolate on Feb. 1, 1985. Another report was expected early in 1985 from a group of researchers at Genentech, Inc. in San Francisco, while Montagnier's Institut Pasteur research group published the sequence for LAV in the January 1985 *Cell.* Although the sequences differed, indicating that the viral genomes of the isolates vary in structure, investigators hoped to identify the gene segments conserved from isolate to isolate that might be most important in the activity of the virus.

The work raised hopes that it might soon be possible to produce an AIDS vaccine. Investigators warned, however, that more must be learned before an effective, safe vaccine can be manufactured.

Efforts to produce a blood test for AIDS were intensified after the AIDS-linked virus was discovered. Montagnier's group reported the development of a "sensitive and specific radioimmunoprecipitation test . . . for the detection of antibodies to the major core protein of LAV," and the Institut Pasteur licensed a Seattle-based company to market it in the U.S. and Canada. Gallo's group developed a different enzyme-linked immunosorbent assay for antibodies to the virus's surface proteins and made virus lots available to five U.S. companies competing to produce a test kit. The U.S. Department of Health and Human Services was expected to approve at least one test early in 1985 and said that it would take steps to make the test widely available even for those who could not afford to pay for it.

The possibility that a drug used for sleeping sickness might help AIDS victims also emerged in 1984. Gallo's group found that in a test-tube environment the drug surinam suppresses activity of the virus linked with AIDS and began tests to determine whether patients might benefit from it.

In the meantime, however, AIDS continued to spread unchecked throughout the world. In central

Africa rapid transmission was reported among sexually promiscuous heterosexuals as well as homosexual men. In the U.S. cases had climbed to more than 8,000, with 48% dead, by the end of January 1985. James Curran, director of the AIDS task force at the Centers for Disease Control in Atlanta, Ga., predicted another 8,000 U.S. cases in 1985, with between 250,000 and 300,000 people showing antibodies to the virus.

Cholesterol and heart disease. At a U.S. National Institutes of Health conference an expert panel recommended that everyone over age 2 have cholesterol levels taken and follow a diet espoused by the American Heart Association (AHA) that cuts back on dietary fats. Some termed the panel report "the last word on cholesterol," but others objected to the commission report on the grounds that existing data do not justify recommending that people with normal cholesterol levels and even children should go on low-fat diets. The American Academy of Pediatrics continued to oppose feeding the AHA diet to growing children, arguing that it is low in calcium and iron and may interfere with normal growth.

The panel, chaired by Daniel Steinberg of the University of California at San Diego, urged "a shift from the current typical American diet to one that is lower in total fat, saturated fat, and cholesterol. . . . The evidence justifies for men, women, and children ages 2 years and older the reduction of calories from fat from its present average level of 40% to 30%, calories from saturated fat to 10% or less, and dietary cholesterol to no more than 250 to 300 mg daily. We recommend that calories from polyunsaturated fat not exceed 10% of total calories."

The panel also urged individuals, health professionals, and health agencies to "recognize the need to control obesity both to aid in controlling blood cholesterol levels and to reduce the other health risks of obesity. Other elements important to the prevention of cardiovascular disease, including avoidance of cigarettes, control of high blood pressure, and maintenance of reasonable levels of physical activity are recommended."

The panel recommended educational services that would teach adults and children how to make informed choices concerning their eating habits. Programs for physicians, dietitians, and other health professionals, and changes in fast-food and restaurant menus and in government and school food programs were advised. Panel members also called for more research efforts in such areas as lipid metabolism; the mechanisms of atherosclerosis; and the effects of "precisely defined diets and pharmacologic interventions" in individuals under controlled conditions.

Exercise-induced amenorrhea. The recent widespread efforts to achieve physical fitness may have harmful effects on female athletes, according to a number of researchers. On the average women have a larger percentage of body fat than do men, about 26% as compared with 12%. The difference, equivalent to about 300,000 kilojoules, is approximately the energy needed to make a full-term human fetus, according to exercise physiologist Craig Sharp of the University of Birmingham, England.

When women engaging in sports training drop below a body fat threshold of about 16–18%, they often develop amenorrhea (cessation of menstruation), perhaps because fat converts androgen produced by

Robert C. Gallo (left) and his co-workers at the U.S. National Cancer Institute in Bethesda, Maryland, and a research team from the Institut Pasteur in Paris (right) each reported that they had isolated the virus believed to be responsible for acquired immune deficiency syndrome (AIDS).

(Left) National Institutes of Health, Bethesda, Md.; (right) Christian Vioujard—Gamma/Liaison

Tardy-Perrin—Gamma/Liaison

Women athletes who drop below a level of 16–18% body fat often develop amenorrhea (cessation of menstruation) and suffer from a significant loss of mineral density in their lumbar vertebrae.

the adrenal glands into estrogen that supplements the estrogen produced by reproductive organs. Although the amenorrhea such women experience can be quickly reversed if training is discontinued, Barbara Drinkwater and her associates from the University of Washington found that a group of 14 amenorrheic young women athletes suffered significant loss of mineral density in the lumbar vertebrae, to a level commonly seen in 51-year-old women. Bone loss was not detected in the radius of the forearm when measurements for the amenorrheic group were compared with a group of 14 menstruating female athletes who were following the same training regimen.

Robert Marcus and his associates at several research facilities in California reported similar findings in a study of 17 women distance runners, 11 of whom had experienced amenorrhea for one to seven years. However, the amenorrheic women who trained hardest had higher spine mineral density than did those whose training was less severe.

Marcus and his colleagues also found that running-related fractures were more frequent in the amenorrheic women and that thyroid hormone levels were significantly lower than levels in either control subjects or menstruating runners. Dietary intake was low in both groups, and, although the women studied consumed large quantities of vitamin and mineral supplements including potentially toxic amounts of vitamins D and A and large quantities of iron, they did not take calcium supplements.

Marcus and his colleagues also found that their data supported earlier suggestions that girls who begin serious athletic training before the onset of menstruation (menarche) may experience a delay in that onset and also increase their risk of secondary amenorrhea (later cessation of menstruation). Those who were amenorrheic had started training within one year of menarche, with four beginning before its onset, while those who continued to menstruate had not begun training until about five years after the onset. Jill Lindberg and her associates at the Letterman Army Institute of Research in San Francisco also observed decreases of bone density in 11 amenorrheic runners and noted that nearly half had reported stress fractures within the year before their study was conducted.

Investigators suspected that estrogen deficiency causes the bone density loss and warned about the potential long-term effects of intense athletic training. Hunter Heath III of the Mayo Clinic suggested that calcium supplements be provided to female athletes and that the administration of estrogen might be a logical step.

Alzheimer's disease. Although the cause of Alzheimer's disease, a devastating degenerative disorder of the nervous system, remained unknown, Stanley Prusiner and his colleagues at the University of California at San Francisco raised the possibility in 1984 that an infectious protein particle much smaller than a virus might be implicated in the condition. Prusiner and his colleagues suggested that aggregates of the particles, termed prions, may constitute the amyloid plaques seen in the brains of people with Alzheimer's disease and other degenerative neurological conditions, such as kuru, Creutzfeldt-Jakob disease, and Gerstmann-Straussler syndrome. Prions have already been identified as the cause of a neurological disorder of sheep and goats called scrapie.

The researchers found that scrapie prions contain one major glycoprotein termed PrP and aggregate into rods that are structurally similar to amyloid plaques. It is possible that prions, which are thought to be about 100 times smaller than a virus, may insert a small piece of nucleic acid into a host-cell chromosome, thereby promoting or enhancing gene expression by the host cell.

The investigators demonstrated prions in amyloid plaques from brains of scrapie-infected hamsters and in the brains of both animals and humans that were infected with the agent which causes Creutzfeldt-Jakob disease. Since amyloid plaques are considered a nearly definitive sign of Alzheimer's disease and are linked to the degree of mental dysfunction shown by patients, "the possibility that Alzheimer's disease might be caused by prion infection is intriguing," Prusiner observed.

Epstein-Barr virus. Two groups of investigators linked the Epstein-Barr virus, which has been generally associated with infectious mononucleosis, with recurrent, chronic illness. A team headed by Stephen E. Straus of the U.S. National Institute of Allergy and Infectious Diseases reported that 23 of 31 adult patients complaining of chronic illness with fatigue after infectious mononucleosis showed evidence of chronic Epstein-Barr virus infection; in some cases the infection appeared to have persisted for years after the patients suffered bouts of mononucleosis from which they were thought to have recovered. Most complained of a chronic or recurring "flulike" syndrome characterized by muscle aches, mild sore throat, tender lymph nodes, and low-grade fevers as well as severe fatigue. They had difficulty working; most reported visiting numerous physicians because their symptoms were considered unimpressive and nonspecific; and many had become angry or distrustful of traditional medicine and had turned instead to such remedies as special diets, megavitamins, meditation, and acupuncture in an effort to gain relief.

Another group of 26 adults and 18 children with similar complaints was seen at the University of Arizona College of Medicine at Tucson. James F. Jones and his colleagues found that 39 were positive for Epstein-Barr virus antibody with levels indicating active infection for at least one year, although immune function studies in 11 patients did not show the changes often associated with acute viral infections or immunodeficiency diseases. Many in this group had missed as much as six or more months of school or work due to their complaints, and also reported rejection by physicians, self-doubt, and family and social conflicts.

"The disorder we are considering to be chronic Epstein-Barr virus infection is not rare and warrants further investigation. For now, it is of immeasurable benefit to patients with this disorder to document an organic basis for their complaints," Straus and his colleagues reported in the *Annals of Internal Medicine* in January 1985.

Fragile chromosome sites. University of Minnesota geneticist Jorge J. Yunis proposed that the discovery of 51 "fragile sites" prone to breakage in human chromosomes may make possible the development of a blood test assessing susceptibility to some forms of cancer. Yunis initiated a large-scale study of cancer patients, cancer-prone families, and healthy people in the hope of linking cancer and high chromosome breakage rates.

Fragile sites are regions of chromosomes thought to be segments that do not undergo normal compaction during mitosis, according to Michelle M. LeBeau and Janet D. Rowley of the University of Chicago department of medicine. (Mitosis is a pro-

José Moré—Chicago Tribune

Equipment at the Hillfarm Dairy in Melrose Park, Illinois, is inspected for evidence of Salmonella *bacteria. Some 15,000 people in the Middle West suffered from* Salmonella *poisoning caused by tainted milk processed at the dairy.*

cess that takes place in the nucleus of a dividing cell; it results in the formation of two new nuclei, each of which has the same number of chromosomes as the parent nucleus.) "The remarkable concordance between the chromosomal location of human cellular oncogenes and the breakpoints involved in certain consistent chromosomal rearrangements associated with various cancers has enhanced the excitement" surrounding this area of study, they reported. (An oncogene is a gene capable of producing a tumor when activated.)

Yunis found that 20 of the 51 fragile sites correspond to chromosome defects already associated with some leukemias, lymphomas, and malignant solid tumors. Of the 17 oncogenes located on human chromosomes 7 were found at or near fragile sites.

The sites are believed to be sensitive to folic acid. The expression (extent to which they have an effect) of many of them is enhanced by the absence of folic acid in a culture medium, and according to Yunis the addition of folic acid to a culture medium prevents chromosomes from rearranging in patterns characteristic of cancer chromosome defects. Yunis also reported that chromosomes from one cancer patient and two healthy people who were given folic acid supplements resisted breakage. Although no direct evidence indicates that dietary supplementation with folic acid can prevent cancer, such a possibility was being tested in clinical trials sponsored by the National Cancer Institute. Yunis stated his belief that people with a family history of cancer and those exposed to carcinogens (cancer-causing agents) might consider eating diets rich in folic acid.

Drugs and vaccines. A new class of antibacterials, the quinolones, appear to be effective against a wide range of organisms resistant to other drugs and cause only minimal side effects. The quinolone derivatives work by blocking an enzyme, DNA gyrase, that is essential for replication of bacteria and were expected to be useful for severe infections. As many as 18 of them were under development in Japan, Europe, and the United States. One, norfloxacin, was approved for use abroad, and by 1985 clinical trials of the quinolones had involved thousands of patients in Europe. Studies of ciprofloxacin in the U.S. were expected to involve about 100 research centers.

Organisms sensitive to the quinolones include *Escherichia coli, Proteus, Pseudomonas aeruginosa, Klebsiella,* and *Hemophilus influenzae.* Results also were good in the treatment of gonorrhea, urinary tract infections, shigellosis, pneumococcal pneumonia, staphylococcal cellulitis, hemolytic strep infections, streptococcal erysipelas, and anaerobic infections.

Caffeine reexamined. After years of bad publicity caffeine received some positive attention in 1984. One group of investigators headed by Allan B. Becker of the University of Manitoba conducted a study comparing its effects to those of theophylline, a drug commonly given to asthmatics, in 23 asthmatic children between the ages of 8 and 18. They found that caffeine was metabolized to paraxanthine, theobromine, and theophylline in the children, as had previously been shown in adults, and was as effective a bronchodilator as the theophylline. Although the researchers noted that their findings may not be applicable to commercial caffeine formulations and did not recommend it for regular use, they suggested that caffeine may be valuable for temporary use when prescribed antiasthma medications are not immediately available.

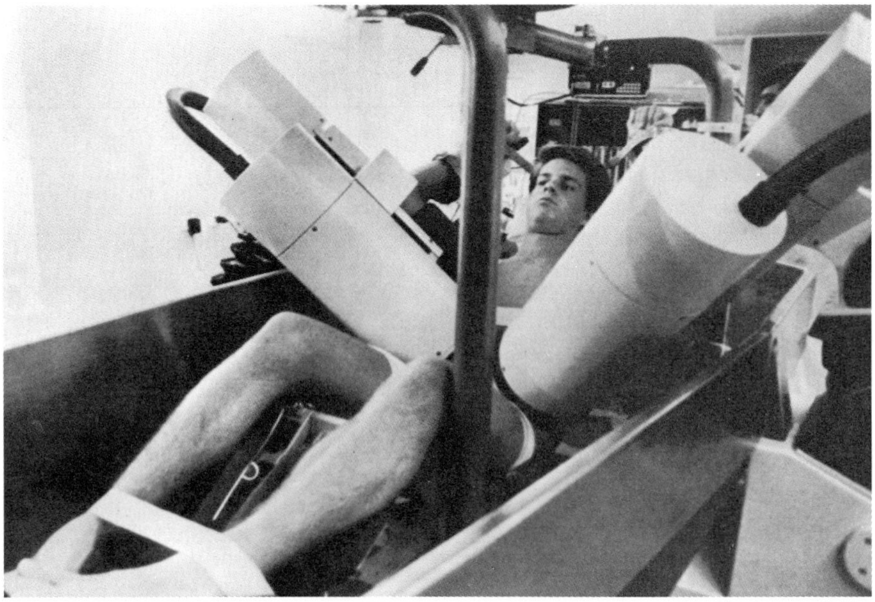

Volunteer is strapped into a newly developed machine that generates shock waves to pulverize kidney stones. The stones are shattered into particles small enough to be passed spontaneously through the urinary tract.

UPI/Bettmann Archive

Eugene M. Laska of the Rockland Research Institute, Orangeburg, N.Y., and his colleagues analyzed raw data from 30 clinical studies involving more than 10,000 patients over the past 20 years to assess the value of caffeine in analgesic preparations. Although such preparations had been available for many years, in 1977 a U.S. Food and Drug Administration advisory panel questioned the practice of adding caffeine to aspirin, acetaminophen, phenacetin, and salicylamide, stating that data showing that caffeine improved the efficacy of such painkillers were insufficient. Laska and his colleagues, however, concluded from their data review that "on the average, when an analgesic contains no caffeine, a dose that is 40% larger is required to obtain the same degree of pain relief as is obtained when the analgesic is given in combination with caffeine." The researchers suggested that in most of the studies that showed no analgesic-enhancing effect for caffeine the doses studied were too low. Although the mechanism by which caffeine enhances pain relief remained unknown, Laska and his colleagues said that adding 65 mg of caffeine to an analgesic tablet "seems reasonable" and "provides an important therapeutic option for treating the patient with pain." In a third set of caffeine studies researchers at the Long Beach Veterans Administration Medical Center found that 17 men with coronary heart disease and exercise-induced angina were able to walk for longer periods without chest pain after drinking coffee than when they drank a decaffeinated substitute.

Researchers did not know whether the methylxanthines in coffee, primarily caffeine, have coronary vasodilator effects (dilate the passageways in the coronary arteries) or whether coffee offers a mild analgesic effect. They cautioned, too, that coffee can cause heart arrhythmias, although no increase in premature ventricular contractions was seen in their study. However, one cup of coffee allowed their subjects to increase the length of pain-free exercise by 8% and two cups, by 12%, the investigators found.

Malaria vaccine. A malaria vaccine moved a bit closer to reality in 1984, when two groups of researchers used molecular biology techniques to describe the structure of the gene for the principal antigen of the sporozoite form of two strains of the protozoan *Plasmodium falciparum* that cause human malaria. (A sporozoite form of such a protozoan is capable of motion and is infective.) John B. Dame and Thomas F. McCutchan of the National Institute of Allergy and Infectious Diseases and their colleagues at the National Institutes of Health and the Walter Reed Army Institute of Research, cloned the gene for the circumsporozoite protein that covers the surface of the malaria sporozoite and described the nucleotide sequence of the gene and the amino-acid sequence of the protein. On comparing it to the circumsporozoite protein of *Plasmodium knowlesi*, a malaria parasite that affects monkeys, they found two closely similar sections of the nucleotide sequence, raising hopes that immunization with one section might give protection against several species of human malaria.

Similar findings for another strain of *P. falciparum* were reported by Vincenzo Enea, Ruth Nussenzweig, and colleagues at New York University. In 1983 this group described the structure of the antigen for *P. knowlesi*. According to Victor Nussenzweig of New York University the repeated amino-acid sequences identified in *P. falciparum* may be a good candidate for a synthetic vaccine.

Such a malaria vaccine might eventually be delivered in a cowpox virus, according to Anthony S. Fauci, director of the National Institute of Allergy and Infectious Diseases. His agency was inserting a number of different antigens into the virus, which would make it possible to offer protection against several diseases at once. Among the possibilities, according to Fauci, were hepatitis B, influenza, herpes, rabies, and malaria. Cowpox virus vaccines, which made the eradication of smallpox possible, are inexpensive and need no refrigeration. They also can be administered by scratching the skin rather than by a sterile hypodermic needle, making their use in less developed countries especially attractive.

Gonorrhea vaccine. Another long-term public health problem, gonorrhea, may soon be minimized by a vaccine as well. In 1984 gonorrhea was the most common reportable communicable disease in the United States, and it was epidemic in Southeast Asia and western Africa; in parts of Africa it caused infertility in up to one-quarter of the women by age 25.

Gary Schoolnik and his colleagues at Stanford University School of Medicine developed a vaccine based on a bioengineered protein fragment that appears to prevent the pili (tiny hairlike structures) on gonococcal bacteria from adhering to cells in the urogenital tract. The vaccine effectively blocked gonococcal pili from adhering to human cells in tissue culture and was to be tested in human volunteers for safety and effectiveness in 1985. The Stanford School of Medicine was attempting to patent the fragment discovered by Schoolnik. Should the patent be approved, Stanford planned to give a license for production to the Institut Mérieux, in Lyon, France, a company that funded its development.

Another potential component of a gonorrhea vaccine was also developed at Stanford by Stanley Falkow, who cloned a gene that codes for a protein that gonococcal bacteria use to protect themselves from the immune system during infection. The cloning of this gene may make it possible for researchers to learn how to inactivate the protein.

—Susan V. Lawrence

Dentistry

The question of whether there would be enough dentists to serve future needs of the U.S. population continued to be a major issue of the dental profession during 1984. A new manpower report by the federal government found that declining dental school enrollment appeared to be "a corrective marketing factor" that should be allowed to exert its influence on the supply of dentists. "Even so, the downward trend should be monitored and objectively related to future requirement forecasts so as to determine a trigger point below which enrollment would indicate a future supply problem," the report said. This finding underscored concern expressed for years by the American Dental Association (ADA). A similar warning flag was raised by Harvard School of Dental Medicine researchers. Their study indicated that current improvements in preventive dentistry and treatments, coupled with the large increase in the number of older adults who were retaining their teeth, may lead to greater, not lesser, dental care needs during the next 20 years.

The Harvard study, conducted by Chester W. Douglass, noted that there was a decline of 30–50% in the rate of tooth decay in children and young adults during the ten years from 1974 to 1984. In the older population, however, between the early 1960s and the early 1970s, the number of decayed, missing, and filled teeth remained fairly stable. Douglass projected that in the year 2000 approximately 42 million adults aged 18–74 would be at risk and that there would be an increased number of persons over 75. Because of the growing elderly population the study projected a marked increase in dental restorations by the year 2000, which could change dental manpower needs.

Meanwhile, an ADA survey revealed a slight decline of dental visits during the second half of 1984, which was seen as "coincident with the recent slowing in the rate of growth of the over-all economy to a more sustainable long-term rate." On a more positive note Donald E. Bentley, 1984 ADA president, reported that the indigent, elderly, and handicapped were finding dental care more accessible than ever before. Latest ADA figures showed that 125 state and local dental societies were conducting comprehensive dental programs that were designed to serve those special population groups.

Decay vaccine. A dental researcher from Washington University in St. Louis, Mo., predicted that a vaccine to prevent tooth decay could be available within the next three to ten years. Roy Curtiss III analyzed the surface proteins of *Streptococcus mutans*, the strain of bacteria most prominently identified with the development of cavities, to discover the components that are essential to allow this organism to adhere to teeth. His goal was to find a way to prevent *S. mutans* from colonizing tooth surfaces. Antibodies against the organism are present in saliva but not in sufficient strength to prevent colonization on tooth surfaces.

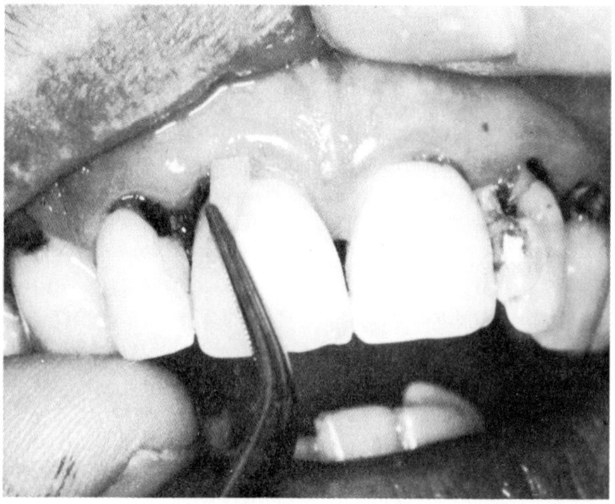

To combat gum disease, a small, porous plastic strip is inserted into a gum pocket that was formed when the gums retracted from the teeth. Containing antibiotics, such strips reduced the depth of the pockets by several millimeters.

Curtiss identified the surface protein antigen A (SpaA protein) as the principal cell surface protein on *S. mutans*. In rat tests he found bacteria lacking S antigen A protein to be totally nonvirulent. Through gene cloning techniques he made pure preparations of antigen A protein and of another surface protein, glucosyltransferase (GtfA protein). Recombinant clones expressing GtfA protein and SpaA protein may prove the basis for an effective vaccine against *S. mutans*-induced dental caries (decay).

Curtiss recommended developing a capsule as the most effective way of administering a caries vaccine and giving it to three- or four-year-old children for the best results. He emphasized that the vaccine should be used with other preventive measures such as water fluoridation, good dental hygiene, and routine dental care.

Plastic strips. A dental researcher at the Dental School in Cardiff, Wales, developed an innovative technique to combat periodontal (gum) disease. Hanah Hassan inserted small, porous plastic strips containing an antiseptic or antibiotic drug in the periodontal pockets that develop in patients when the gums retract from the teeth. The most commonly used drugs in those plastic strips were tetracycline and metronidazole. Even very deep pockets responded well to the treatment, shrinking the pockets by several millimeters. Monitoring of the bacteria in the pockets revealed considerable reductions, and clinical signs of inflammation were also reduced.

The effects of the treatment lasted for several months. Although the bacteria did recolonize the areas, the reduction in the pocket depth was maintained—particularly after metronidazole therapy.

Gum disease. The oral bacteria associated with periodontal disease may be transmitted from person to person, suggested an Emory University dental researcher. Bacterial samples from 14 married couples indicated a "statistically significant match" between the bacteria of spouses, according to Steven Offenbacher. "While the results of the study are by no means conclusive, they do suggest the possibility that bacteria linked to gum disease can be transferred by mouth-to-mouth contact." These bacteria behave somewhat like dominoes because the presence of one type paves the way for the appearance of the next. "The longer dental plaque is left in the mouth, the more types of bacteria emerge and the more easily they move from site to site within the mouth. The bacteria may be transferred from person to person in much the same way," he explained.

The bacterial "matches" between couples could be due to the fact that couples often share similar oral hygiene habits. "If a person has plaque build-up, chances are his or her spouse will too," he said. Bacteria may be transferred indirectly via shared drinking glasses and other objects. "There is still much to learn about bacteria's role in gum disease, but with further research, bacterial analysis may become a useful diagnostic tool," Offenbacher concluded.

Dentists as information sources. For help with everything from choosing floss to deciding whether or not to undergo orthodontic treatment, people turn first to their dentists, according to a survey in the U.S. by Opinion Research Corp. of Princeton, N.J., conducted on behalf of M&M/Mars, Inc. Almost two-thirds of the public said that they counted primarily on their dentists for dental health information.

Though the findings may seem obvious, dentists themselves did not seem to be fully aware of how heavily their patients relied on them for dental information. Only 39% of 103 dentists surveyed thought that the public relied primarily on the dental profession for dental information, with 46% citing television as the public's leading source. By contrast, only 11% of the public named television as their main dental information source.

The study also found marked differences between the professional dental community and the public regarding the link between "sticky" foods (such as raisins and toffees) and dental decay. Though almost all dentists surveyed agreed that food adhesiveness is at least fairly important in determining a food's ability to cause decay (64% labeled stickiness "very" important), only 1% of the public made any connection between sticky foods and decay. The study also found a large gap between dental professionals and the public in awareness of sodium fluoride's role in decay prevention. While dental researchers named sodium fluoride as the single most important factor in controlling tooth decay, the public named "not brushing teeth" as the leading cause of decay, with very few mentioning lack of sodium fluoride.

Courtesy, Norman L. Corah, State University of New York at Buffalo

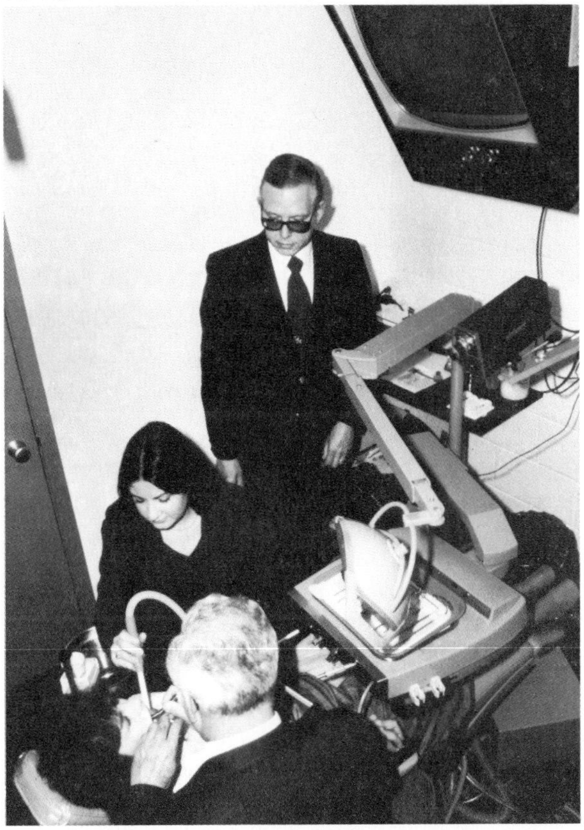

Dental patients who played video games or watched a comedy on television experienced reduced anxiety levels and reported significantly less discomfort than those for whom no distractions were provided.

Invisible tooth fillings. Tooth-colored dental fillings of the future may be undetectable to anyone but a dentist, according to Karl Leinfelder of the University of Alabama. Once scientists develop materials hard and durable enough to withstand the pressure of biting, grinding, and chewing on back teeth, where most cavities develop, even patients with mouthfuls of fillings will look as though they never had a cavity. The materials, combinations of plastic and ceramic substances known as composite resins, are expected to replace the standard, dark-colored metal restorative materials.

In 1985 composite resins were being used in the procedure called cosmetic bonding to repair chipped, cracked, or badly stained front teeth. Though such resins have worked well on front teeth, they have not been durable enough for the heavy chewing and grinding that take place on back teeth. As research yields more wear-resistant materials, however, color-matched fillings should come into widespread use for these back teeth as well.

Videos and dental visits. According to a study from the State University of New York at Buffalo, patients who play video games or watch comedy programs while in the dental chair can successfully reduce their anxiety levels over impending treatment. Reducing anxiety also has been shown to reduce perception of pain, according to Norman L. Corah. Both the video games and the television comedy programs provided the distraction and relaxation necessary to reduce a dental patient's anxiety during treatment. Less effective, the researchers found, were listening only to an audio portion of a comedy special and listening to recorded music.

The study, conducted at the university's dental clinic, involved 80 college-aged adults—40 males and 40 females—who required at least two dental restorations. Prior to treatment each patient was rated for dental anxiety and fearfulness of the dental experience. on the Corah Dental Anxiety Scale developed by Corah several years earlier. At the first dental visit electrodermal responses or changes in skin resistance using painless electrodes attached to the patient's hand were recorded to obtain a baseline against which anxiety exhibited during future visits would be compared. At the start of the second dental visit the 80 patients were divided into four groups: one played a video ping-pong game during treatment; the second watched and listened to a television comedy program; the third listened to the audio portion of a television program featuring well-known comedians; and the fourth received no distraction or relaxation aids.

Overall results of the study showed that the video-game participation and watching and listening to the comedy program were nearly equal in effectively reducing dental anxiety levels among the patients. But while both groups reported significantly lower discomfort ratings, only the video-game group did not present a significantly increased pain rating at that time.

Following completion of their dental treatment, the patients in three groups were asked to rate the effectiveness of the experimental techniques. Those who played the video game or watched and listened to the comedy program reported that they believed the technique they used to be "more effective" than only listening to the audio comedy presentation. All three experimental techniques were rated higher than scores given by those in the control group, who received no distraction during treatment.

Dry mouth remedy. Many people suffer from "dry mouth." Those especially afflicted by this condition are patients irradiated for oral cancer and those receiving drug therapy in which dry mouth is a side effect. Researchers in The Netherlands developed a mucin-containing artificial saliva with properties that simulate those of natural saliva.

Reporting to the general session of the International Association for Dental Research, Hans Gravenmade said that the use of artificial saliva applied by dripping, sucking, or spraying the liquid would be contrary to normal function and could be disturbing to the patient during routine social contacts. The new technique, however, offers a more natural version of artificial saliva application by using an intra-oral reservoir as a container. The reservoir is easily filled and cleaned and can be worn by patients with natural teeth or by denture wearers.

New treatment for bone loss. Can broken bones be made to heal faster? Can the "irreversible" bone loss seen in some dental diseases be reversed? These questions indicate some of the possible future applications of a substance recently discovered at the U.S. Army Institute of Dental Research in Washington, D.C. A still unnamed substance stimulates the incorporation of calcium into living bones in culture. Furthermore, it blocks the loss of calcium from the bones, a loss that is caused by some known stimulators of bone degeneration. Produced by human white blood cells in response to stimulation by an extract from mung beans, this substance belongs to a class of biological compounds known as lymphokines. Although it is produced in very small quantities, it was reported to be potent and effective at extremely low concentrations.

Current research was aimed at producing sufficient quantities of the lymphokine to allow studies of its chemical nature as well as its biological effects. Extensive research will be required before this substance can be used in patients.

Oral pain. A pain in the mouth could mean much more than a toothache, and the dentist often is able to determine its cause. Brain tumors, multiple sclerosis, circulatory diseases, and a host of other ailments are often spotted in a dentist's office, according to a University of Florida dental researcher.

"Just as head, neck and shoulder pain sometimes stem from dental disorders, mouth and jaw pain sometimes stem from medical disorders. What at first seems like a dental problem could really be something quite different," said Parker E. Mahan of Gainesville. "The pain of migraines, for example, often manifests itself in the jaw area, where it is mistaken for a dental ailment known as TMJ (temporomandibular joint or jaw joint) syndrome." By precisely identifying the sources and causes of patient's pain, dentists are often the first health professionals to spot nondental problems and refer patients to physicians.

—Lou Joseph

Veterinary medicine

A major breakthrough in veterinary medicine was the development of a feline leukemia vaccine announced in late 1984 by a team of immunologists at the Ohio State University College of Veterinary Medicine, headed by Richard G. Olsen. In 1945 the disease was discovered to be caused by a virus, which was first isolated from cats in 1964. Transmission was once thought to be only from parent to offspring through the genes, but in 1973 a veterinarian at Memorial Sloan-Kettering Cancer Center, William D. Hardy, demonstrated that spread was by virus in saliva, urine, feces, or milk, probably most often by mother cats nursing or grooming their kittens. For a number of years the development of a feline leukemia vaccine was considered unlikely, because killed-virus vaccines were ineffective and attenuated live-virus vaccines caused leukemia. The problem was solved by using proteins extracted from virus-infected feline cells, and marketing of the resulting vaccine, by Norden Laboratories, was announced in January 1985.

As many as one million of the 52 million pet cats in the United States can be expected to die of feline leukemia, and the virus has worldwide distribution. More than half of the free-ranging cats become infected, at least transiently, with the virus, which can wipe out a population of susceptible cats in a household. Most cats exposed to the virus develop some immunity and eliminate the infection. Only about 20% of chronically infected cats develop leukemia, but most others die of related diseases.

The feline leukemia virus resembles that of human acquired immune deficiency syndrome (AIDS) in that it suppresses the immune system and makes cats susceptible to diseases they would ordinarily resist. Myron Essex, a veterinarian in the Harvard School of Public Health, recently reported that feline leukemia virus causes a feline version of AIDS, which he calls FAIDS, and that the feline virus is closely related to a human T-cell lymphotropic virus considered the most likely cause of AIDS. However, there is no evidence that the feline virus can infect humans.

In 1979 a new disease of horses was recognized in Maryland and named Potomac horse fever. Since that time it has been reported throughout the U.S. Also called acute equine diarrhea syndrome, even with appropriate treatment it killed about 30% of the affected animals and consequently stimulated intensive research on its cause and mode of transmission. Because the disease was most prevalent during warm weather and had characteristics suggesting an insect-transmitted viral infection, most investigators looked for such an agent but were thwarted in their efforts. After a two-year study involving a number of false leads researchers at the Virginia-Maryland Regional College of Veterinary Medicine demonstrated a rickettsia, a small bacterium having some viruslike features, in lesions of the colon of affected horses.

Meanwhile, an entirely independent investigation by two researchers at the University of Illinois College of Veterinary Medicine led to positive identification of the rickettsia after a study of only about four months. This remarkable success story began with Jean Sessions, a veterinarian in the area of Maryland where the disease was first recognized; she had begun her career with the U.S. Army Veterinary Corps doing research on rickettsial diseases of dogs. In the spring of 1984, after noting certain

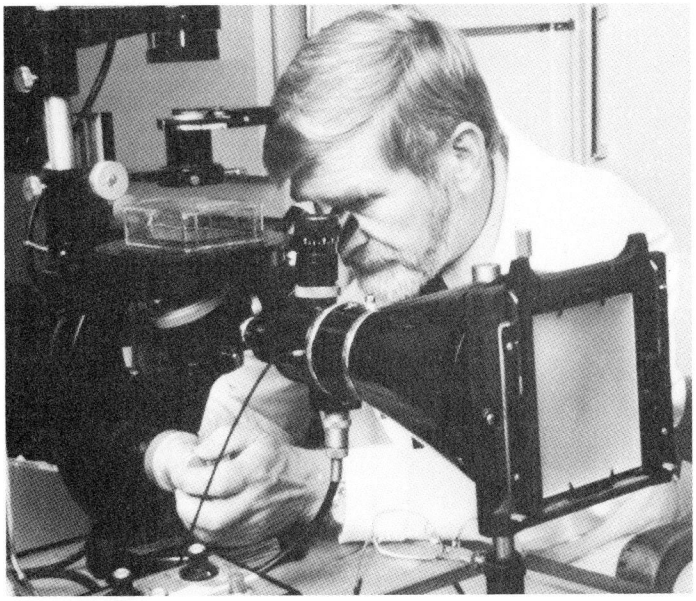

Norden Laboratories

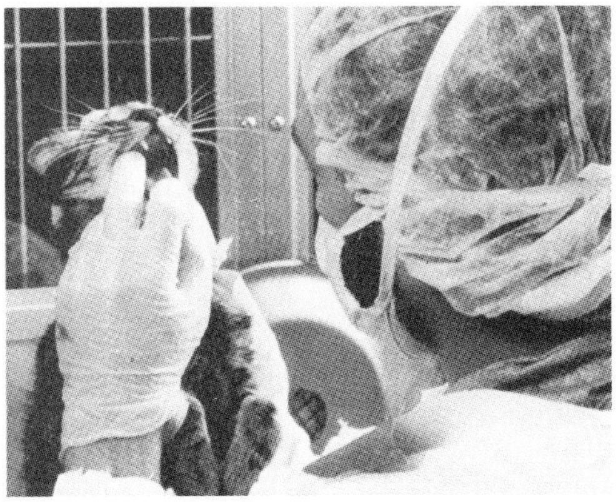

Richard Olsen (left) and his co-workers developed an effective vaccine against feline leukemia by using proteins extracted from feline cells infected with the leukemia virus. Young kittens (above) and cats over the age of eight are most susceptible to the disease, from which chances of recovery are poor.

features that the equine disease had in common with ehrlichiosis of dogs, she took her suspicions to her former adviser, Miodrag Ristic at the University of Illinois, who in 1968 had identified *Ehrlichia canis* as the rickettsial agent causing a new disease of military dogs in Vietnam. Using coded blood samples, Ristic and graduate student Cynthia Holland isolated an organism similar but not identical to *Ehrlichia sennetsu,* a rickettsia causing a mononucleosis-like disease in persons, mainly in Japan. A pure culture of the organism, since named *Ehrlichia risticii,* produced diarrhea and fever identical to that of the naturally occurring equine disease, now termed equine monocytic ehrlichiosis.

The centennial of the Bureau of Animal Industry, the veterinary arm of the U.S. Department of Agriculture until 1953, was celebrated on May 29, 1984, with day-long ceremonies in Washington, D.C. Among the notable early achievements of the Bureau was the first isolation of *Salmonella* organisms (now commonly identified in food poisoning) in 1885 by its first chief, Daniel E. Salmon. In 1886 Salmon was the first to demonstrate that vaccines (bacterins) produced from killed cultures of certain organisms would protect against disease. In 1892 contagious pleuropneumonia of cattle was eradicated by the slaughter of affected and exposed animals, allowing the U.S. to become the first nation to eliminate this disease completely. Discovery of the difference between human and bovine tuberculosis organisms in 1895 led to programs resulting in near-eradication of the disease in cattle and people. The discovery by Bureau veterinarians in 1893 that the organism causing Texas fever of cattle was carried by ticks led to control of that disease and others transmitted by insects, thus making possible the building of the Panama Canal through control of the yellow fever mosquito. In 1953 the functions of the Bureau were absorbed by the U.S. Department of Agriculture's Agricultural Research Service and the Animal and Plant Health Inspection Service.

Four state veterinary associations, those of Connecticut, Indiana, New Jersey, and Ohio, also celebrated their centennials in 1984, as did the School of Veterinary Medicine at the University of Pennsylvania. The latter was the second of the existing U.S. veterinary schools to be established, following that at Iowa State University in 1879.

Robert Phillips, a physiologist at Colorado State University and the first veterinarian to be recruited by the U.S. National Aeronautics and Space Administration (NASA), was selected for a January 1986 space shuttle mission devoted to studies on weightlessness. Phillips, whose special field was the control of fluid and electrolyte balance, anticipated more opportunities for veterinarians in space. "Space stations will provide an opportunity for longer-range biological experiments and presumably the utilization of a wider range of experimental animals. Following that, there will be permanent colonies on the moon. As such colonies become self-sufficient, there will be a requirement for veterinarians," he said.

The U.S. Food and Drug Administraion (FDA) Center for Veterinary Medicine (CVM) banned the use of the antibiotic chloramphenicol in animals used for meat, egg, or milk production and warned that the restriction would be rigorously enforced. According to CVM director Lester Crawford, "The unique toxicity of this product to humans has been so well established that a comprehensive enforcement program by federal and state authorities will continue until diversion of chloramphenicol to food animal use ceases." Since 1950 more than 700 cases of chloramphenicol-induced aplastic anemia, with about 70% mortality, have been reported in the U.S., in some instances after exposure to minute amounts of the drug. In one recent case a rancher developed aplastic anemia four months after he began treating his cattle and died a month later.

In September 1984 the *New England Journal of Medicine* published an article purporting to show that meat from antibiotic-fed cattle had been responsible for human illness diagnosed as antibiotic-resistant *Salmonella* infection. This generated numerous stories in the public press, to which the Council on Public Health of the American Veterinary Medical Association responded, "The AVMA seriously questions the credibility of studies that suggest that serious human health problems have been the direct result of the use of antibiotics in livestock feeds." In 1978 the FDA had proposed banning the use of antibiotics in feeds, but reaction from the livestock industry prompted Congress to study the matter further. A report by the U.S. National Academy of Sciences in 1980 concluded that the potential hazards to human health which might result from such use had been neither proved nor disproved.

Robert Phillips, the first veterinarian to be recruited by NASA, was chosen to be a member of a January 1986 space shuttle mission devoted to studies on weightlessness.

NASA

After the FDA proposed in 1983 to allow additional uses of antibiotics in feeds, a committee of scientists including former FDA commissioner Jere Goyan petitioned against further delay in banning their use. Goyan said, "Unless we take action now to curb the use of these drugs in the livestock industry, we will not be able to use them to treat human diseases in the future." This argument was disputed by Fred H. Holt, president of the Animal Health Institute, who claimed, "No substantial evidence has been developed to demonstrate that the risk exists." He also pointed out that the lowered efficiency of weight gain without medicated feed would increase the cost of meat to consumers, one estimate being an additional $3.5 billion per year.

—J. F. Smithcors

Optical engineering

Continued growth in the engineering of electro-optical devices took place during the last year. The total laser and electro-optical devices market continued to grow in parallel with the international growth in the economy. Japan remained the primary source of new optical devices, but other nations were beginning to emerge as producers.

Optics in the factory. As usual, defense projects dominated the electro-optics industry in the United States. Industrial requirements and the consumer industry were not ignored, however. The application of laser machining and optical gauging in production areas reached maturity. A listing of machine shops using laser cutting or welding revealed several dozen in the U.S. Because little capital investment was required, even small shops were beginning to install laser workstations; there were some estimates that the investment could be recovered in less than a year for many applications. Some recent developments of importance in the laser machine shop included optical fibers that can convey several kilowatts of laser power over a distance of about 1 m (3.3 ft). This greatly increased the range of possible types of machining operations that could be carried out on relatively inexpensive machines.

The role of optics in automated factories continued to increase. Some assembly lines in the automobile industry were using laser sensors to determine if parts such as dashboards were too large when they came from a molding press. If they were, the excess material was then automatically cut away by a high-power laser. A major revolution in the semiconductor business was the introduction of laser trimming for increasing the yield of acceptable microcircuit chips. In this process a large memory chip is produced by photolithography on a silicon disk having several extra circuit components. After being sliced from the disk and mounted, the assembled chip is tested to locate any defective components. A computer-directed laser is then used to cut through some of the connections on the chip in order to disconnect the bad regions and replace them with new connections to acceptable circuit components. The use of these techniques made possible huge increases in the yield in chip manufacturing.

Lasers and medicine. Medical applications of optics also increased. Surgery can be carried out by using a variety of lasers in different wavelengths. This permits tailoring the procedure to the problem to be solved. The conveying of several watts of high-power laser radiation along a fiber permitted surgeons to carry out surgical procedures within the body without the necessity of cutting the skin for access. Such previously major operations as the removal of tumors in the stomach or bladder can be accomplished by using a fiber endoscope to view the interior of the body cavity through a normal body opening. When the tumor is identified, the surgeon can point the high-power laser fiber at the offending tissue and burn away the tumor. This technique is especially interesting for treating surgical problems within the eye, as the laser can be imaged directly through the lens of a patient's eye. Lasers thus have allowed many complex surgical operations that once required long hospitalization to be reduced to outpatient procedures.

Consumer products. Electro-optics was used increasingly in consumer products. Photographic cameras became more electronic, with "talking cameras" and automatic focusing features becoming standard on many models. The most popular approach to automatic focusing was the use of projection of light

Minolta Maxxum 7000 is the first 35-mm single-lens reflex camera with a built-in automatic focusing system. It employs two microcomputers to achieve automatic focusing in less than one second.

Minolta Corporation

Hewlett-Packard

Computer output printer (right) uses lasers to produce text and images. When connected to the HP 150 touchscreen computer of Hewlett-Packard, it prints eight standard pages of text per minute. It is also much quieter than non-laser printers.

from an infrared diode, with the return light measured by a detector array. The use of this technique made it easy to set the optimum focus into a series of acceptable zones that could be read and set by the camera as the operator pressed the shutter release.

Most of the operations required for use of the camera can be communicated to the user by employing a voice-synthesizing computer in the camera. Although this does not appear to be within the field of optical engineering, the camera designer can use this approach to provide an overall integrated system, including the camera user, in making the decisions under guidance of the camera. The exposure is determined by a photodetector that often uses the same lens used for image formation; this permits the introduction of some rather elegant methods for allowing the camera to find the best average exposure for a particular scene.

The all-electronic still camera using electronic detectors for recording the image probably will not arrive for a number of years. Prototypes were demonstrated, the most impressive being the use of electronic still cameras to provide instant recording and transmission of events at the Olympic Games in Los Angeles. However, even the best electronic array detector can record only about 10,000 picture elements, whereas a 35-mm color photograph can contain easily 50,000 or more picture elements. The image quality provided by photographic film still well exceeds that available from video detectors and displays and seemed likely to continue to do so for several years.

A reasonably successful attempt at miniaturization of cameras is the Eastman Kodak Disc camera, introduced in 1982. Although the camera gained favorable comment, sales were somewhat disappointing, probably because the image quality available on such a small-frame format is not as good as normal 35-mm photography.

Consumer video continued to grow in importance, as the price for a videocassette recorder came within the reach of anyone who could afford a television set. The image quality of video systems was adequate for scenes in which motion was being recorded, as the human eye does not have the ability to examine moving scenes for the detail required in still photography.

A major change in the recorded video market was the announcement by RCA Corp. that it was dropping its capacitance videodisc system because of huge financial losses in attempting to market the system. The capacitance videodisc player could not be used to record in the home, and it offered images of lower quality than those of the optical or laser videodisc. This left only videotape and laser videodiscs as the media for storing images electronically. The laser-read videodisc achieved a prime market in industrial and educational applications and showed promise as a mass storage device for computers. The present difficulties in erasing and reusing space on an optical disc did not seem too important when it was realized that several hundred million words of program or data could be written on a single such disc 10 cm (3.9 in) in diameter.

Optics also gained importance as the primary medium in printing. There was continued growth of the use of laser scanners for reading documents and producing color separation plates. Relatively expensive but high-quality computer output printers have used lasers for writing text and images. During the past year the laser printer took a large step toward common use through the introduction of a low-priced laser printer module of Japanese origin which was accepted by several computer manufacturers for incorporation into office-type systems. A word and graphics processor driven by a microcomputer seemed likely to become commonplace in offices during the next few years.

Astronomical instruments. Some important milestones were reached in regard to astronomical optical instruments. The Space Telescope structure and optics were completed by Perkin-Elmer Corp. when that company finally integrated the precision fine-guidance trackers into the optical system. The telescope was then shipped from Connecticut by air to Lockheed Corp. in Sunnyvale, Calif., where it was to be integrated with the entire spacecraft and detector packages. A launch in 1986 on the space shuttle was contemplated.

Other new space programs of importance to optical engineers included the High-Energy Astronomical Observatory and the Large Deployable Reflector. The first of these consisted of a far-ultraviolet and X-ray telescope with an aperture of about one meter. Grazing incidence (incidence of radiation at a very small glancing angle) optics that are difficult to fabricate were to be used in this telescope. The latter project involved a plan to build a telescope made up of several segments and having an overall aperture of 30 m (98 ft); it would operate in the long-wavelength infrared region. The first of these projects may begin in 1985, the latter probably not until 1990.

Ground-based astronomy progressed in several areas. An announcement was made by the University of California and California Institute of Technology that they had obtained funding to initiate the construction of the California optical telescope. This instrument, with a primary mirror 10 m (about 400 in) in diameter, would use 36 mirror segments that would be maintained in precise alignment by the use of a servomechanism system. If all goes well, by 1991 the telescope should be located at an altitude of 4,180 m (13,800 ft) on Mauna Kea, an inactive volcano in Hawaii. Many innovations will be required for the fabrication of this telescope, including a lightweight structure and dome and, of course, the servomechanism system to keep the segments of the primary mirror in alignment.

Another astronomical project of interest during the year was the National New Technology Telescope. The decision on the form of this telescope was made in mid-1984. It was to be 15 m (590 in) in diameter, with four 7.5-m (295-in) telescopes joined together in a manner similar to the Multiple Mirror Telescope already in operation. The project was to be under the direction of the National Optical Astronomy Observatories in Tucson, Ariz.

The basic telescope design and also the concept for the primary mirrors resulted from a suggestion by the University of Arizona: Pyrex-type low-expansion glass would be cast close to the required shape in a rotating furnace. The effects of gravity and centrifugal force would then combine to form a surface within a few millimeters of the desired shape. This would reduce the amount of excess glass that would have to be removed in the process of fabricating the optical surface of the mirror.

Roger Angel, University of Arizona, Tucson

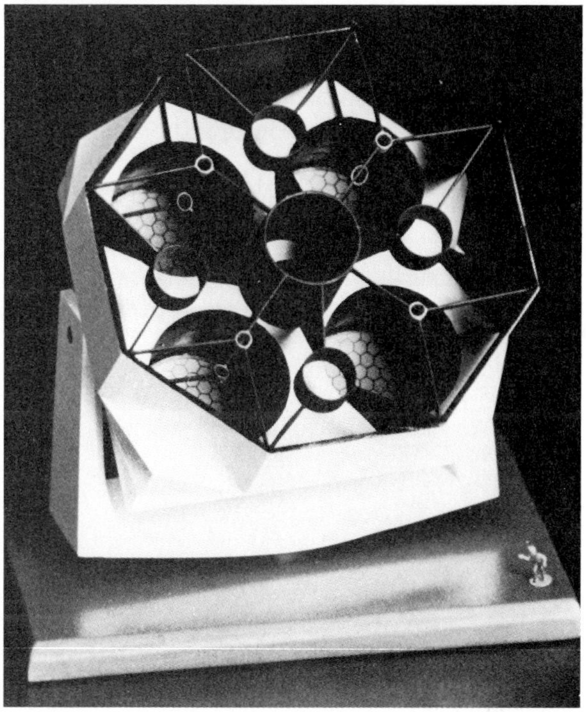

The U.S. National New Technology Telescope will consist of four telescopes joined together. Each of the four will be 7.5 meters (295 inches) in diameter and will use Pyrex-type low-expansion glass for its mirror.

By 1985 the University of Arizona had already installed a new optical surface generator capable of cutting the surface of a glass blank that is eight meters in diameter to within a few wavelengths of the desired shape. Astronomers hoped that some large mirrors for a new generation of eight-meter telescopes would be available within the next two years. However, as of 1985 no firm funding for the construction of those projects had been approved.

In the U.S. considerable attention was directed toward the Strategic Defense Initiative proposed by Pres. Ronald Reagan, popularly called the "Star Wars" proposal. High-power lasers in space or on the ground would be an essential part of this missile defense system. Other necessary electro-optical equipment would include the pointing, tracking, and high-power beam-handling devices necessary for such a weapon system. The requirements were challenging. Pointing a high-power laser beam to an accuracy of better than 0.01 of an arc second would be required. This is at least a factor of ten better than could be achieved in 1985. *See* Year in Review: DEFENSE RESEARCH.

—Robert R. Shannon

Physics

Research in physics during the past year provided investigators with a broad range of unprecedented abilities and tools as well as a wealth of new knowledge. Techniques to slow down, trap, and image individual atoms were refined; continued improvements to the transmission electron microscope brought scientists to the brink of seeing directly the atomic structure of materials; and new methods of investigating the atomic nucleus over an increasingly wide energy range were allowing the study of previously unsuspected phenomena taking place deep in the nuclear interior. Physicists found evidence for the long predicted *t*, or top, quark and discovered a new magnetic material that promised to outperform the best presently available permanent-magnet alloys.

General developments

In recent months scientists learned to handle individual atoms and bring them to a halt without "touching" them, found a new use for "pure" mathematics in real-world experiments, and developed new magnetic materials that may soon find their way into the home.

Hold that atom. Scientists at several institutions including the U.S. National Bureau of Standards (NBS) recently perfected techniques to slow down, cool, trap, and "photograph" individual atoms. This ability should lead to better knowledge of the interaction of atoms with each other, to the development of improved time standards, and ultimately to the study and use of atoms made of antimatter.

Atoms in a gas are moving in all directions with velocities near that of a jet aircraft. Furthermore, they constantly disturb each other by collisions. Scientists can make a beam of nearly undisturbed atoms by allowing a jet of gas to flow through a series of small holes into a high vacuum. Since the atoms in the beam are all moving in nearly the same direction, collisions are infrequent. Atomic beams allow scientists to study atoms uninfluenced by their neighbors. The best clocks available today use beams of atoms as their basic reference. The atoms in such a beam, however, are still moving at nearly the speed of sound. Even better experiments could be done if the atoms were motionless.

To slow down atoms in a beam, scientists train a laser beam on them from a direction opposite to that of their flight. They also tune the frequency of the laser so that its photons interact only with those atoms that are moving rapidly toward the laser. When an atom in the beam absorbs a laser photon, the momentum of the absorbed photon causes the atom to slow down. The atom soon reemits the photon, but in a random direction. Thus, on the average, the "kicks" the atom receives from a large number of reemitted photons cancel each other out.

After absorbing hundreds of laser photons, all coming from one direction, while reemitting the photons in all directions, the atom in the beam finally is brought to a halt. Once the atom is no longer moving toward the laser, the photons from the laser stop interacting with that atom but continue to work on other atoms that are still moving forward. After all the atoms have been brought to a halt, they can be put into a trap.

A trap suitable for charged atoms, or ions (atoms that have lost or gained at least one electron), is a Penning trap. As configured for positively charged atoms the trap has two conical, positively charged electrodes at the top and bottom of the trap and a negatively charged ring electrode around the middle. A strong magnetic field running from top to bottom completes the trap. The positively charged atom is repelled both upward and downward by the positive conical electrodes and starts to move sideways toward the negative ring electrode, but the strong magnetic field makes the charged particle travel in circles, never allowing it to reach the ring. If the energy of the atom is small enough, the trap is very stable. A Penning trap set up for negatively charged particles has kept a single electron trapped for more than nine months.

Scientists took a "picture" of a single barium ion in such a trap. By shining a laser beam at the barium ion and causing it to reemit many photons, they made a "time exposure" photograph of a single atom. The size of the image was used to determine how much the atom moved. From its motion its effective "temperature" was determined to be less than 0.02 K, or a fiftieth of a degree above absolute zero.

Such trapped atoms can be used as clocks. The internationally accepted definition of the second is "the duration of 9,192,631,770 periods of the radiation corresponding to the transition between the two hyperfine levels of the ground state of the cesium-133 atom." The atomic clocks presently used by the NBS as time standards, which employ cesium atoms moving at high speed, are accurate to one part in 10^{13}. It is expected that trapped atoms will ultimately produce clocks having accuracies of one part in 10^{15} or better.

A more exotic application for Penning traps is as containers for atoms of antimatter. Plans are under way by a collaboration of scientists from Los Alamos (N.M.) National Laboratory (LANL) and a number of universities to capture a quantity of antiprotons and carry out experiments with them.

Antiprotons are presently being made and stored in quantity at the European Laboratory for Particle Physics (CERN) near Geneva. The antiprotons are generated by smashing ordinary high-speed protons into a dense metal target. Some of the proton energy is converted into matter and antimatter, including antiprotons. The antiprotons are captured and stored in a ring until there are enough for experiments.

CERN also has a machine that will slow down the antiprotons—a particle accelerator that is run backward as a decelerator—and put them into its Low Energy Antiproton Ring. Because the antiprotons are still moving fairly rapidly, the LANL collaboration plans to build another decelerator to bring the particles' velocity close to zero so that they can be caught in a Penning trap. The scientists hope to capture about a million antiprotons and take them off in their "bottle" to another laboratory for investigations. One of the first experiments will be to see if antimatter falls up or down in the Earth's gravity.

Mathematical chaos in the laboratory. Mathematicians like to play games with numbers. The further removed the mathematical game is from reality, the better they like it. How many colors, for example, does it take to paint a map of n dimensions? (The pursuit of that question is still pure mathematics.) What kind of space allows two parallel lines to meet twice? (The answer lies in the realm of spherical geometry, which is used every day by airline pilots.) How does one find a prime number without dividing it by every number smaller than itself? (That mathematical technique is now used to find large prime numbers so that business people can encrypt important messages.) It is becoming increasingly more difficult for mathematicians to find something "impractical" to tinker with.

One day a mathematician at LANL was exploring a strange equation using his pocket calculator and found some interesting behavior that had been missed by people using big computers. Sometimes the equation gave a single answer; sometimes it gave two answers, or three, or four; and sometimes it acted chaotically and gave what seemed to be random answers that varied widely. The strange equations had to be "nonlinear," in that the unknown variable (x in most equations) is either raised to some power like x^2 or is embedded in some mathematical function like sin x or e^x. One way of solving these equations is to make a guess at the answer, substitute it for the x in the nonlinear parts of the equation, and solve the rest of the equation for x. The answer for x is then taken as the next guess and the process repeated until x settles on a value. For some types of equations x did not always settle down.

Mathematicians who began studying these types of equations identified "attractors" that seemed to "attract" the value of x toward them. They found equations with "strange attractors" for which the values of x seemed to vary randomly while still remaining clustered in a "chaotic" region around the value of the strange attractor. They studied the chaotic regions and found that they were not truly random. There were patterns in the sequences of numbers, but patterns very difficult to unravel. To pursue their studies the mathematicians had to switch from pocket calculators to the most sophisticated computers since a small round-off error in the calculation would send the sequences of numbers wildly off in a different direction.

An elementary example of a "strange" equation that the reader can try out on a pocket calculator is the iterative equation $x_{n+1} = 4Ax_n(1 - x_n)$, in which x_n is the nth estimate for the ultimate value of x, and x_{n+1} is the next estimate. The value of A determines

Ten tubes of laser amplifiers each 137 meters (449 feet) long (left) join together in an aluminum target chamber almost 5 meters (16 feet) in diameter (right) to form Nova, the world's most powerful laser system. Dedicated in April 1985, the $176-million facility will be used in weapons-simulation and laser-fusion studies.

Photos, Lawrence Livermore National Laboratory

Delco Remy Division/General Motors Corporation

Redesigned automobile cranking motor (right), scheduled for introduction in the 1986 model year, is smaller and lighter by half than its predecessor (far right) thanks to a new permanent-magnet material discovered by U.S. and Japanese scientists. The high-performance alloy—a mixture of neodymium, iron, and boron—replaces the motor's bulky copper-coil electromagnet.

the strangeness of the behavior of the equation. First, choose some value from 0 to 1 for A. Then choose any value from 0 to 1 for x_n and use that to calculate x_{n+1}. Then use x_{n+1} for the next guess, and repeat the process several times while watching what happens to x.

If one chooses $A = 0.4$ and $x_1 = 0.4$, then $x_2 = 0.3840$, $x_3 = 0.3785$, $x_4 = 0.3764$, $x_5 = 0.3755$, $x_6 = 0.3752 \ldots$, and $x = 0.37500000000$. All other starting values for x_1 will converge ultimately to the same value. If one chooses $A = 0.8$, however, eventually the answers for x will be found to alternate between 0.7995 and 0.5130. The equation now has two answers. Other values of A give results that alternate among four or more answers. But for any value of A greater than 0.891, the results vary in an apparently random, chaotic sequence.

Although strange and seemingly only of interest to mathematicians, chaotic behavior has been found in many physical systems such as chemical reactions (*see* Feature Article: PENDULUMS IN THE TEST TUBE), electronic circuits, and fluid dynamics. Recently a team of scientists at the University of Texas found that the turbulence in a fluid between two counterrotating concentric cylinders showed period-doubling behavior that can be described by the theory of strange attractors. In another recent example scientists at Hughes Research Laboratories, Malibu, Calif., working with high-power lasers and nonlinear crystals to make "magic mirrors" (see *1985 Yearbook of Science and the Future* Year in Review: PHYSICS: *General developments*) found that under certain conditions the nonlinear crystals they used showed chaotic behavior. The output of the optical cavity containing the nonlinear crystals varied with time even though the laser power into the crystals did not vary. If they changed the length of the optical cavity slightly, the period of the output variations would be twice as long or four times as long, and for some conditions the output signal became chaotic and produced seemingly random noise. Since their crystals reacted to the laser light in a nonlinear fashion and there was period doubling and quadrupling, they suspected that they were observing chaotic behavior in the system.

The scientists then created a mathematical model for the system and found that the nonlinear equation in the model was similar to some of the equations that the mathematicians had been investigating. The changes that they had made in the length of the optical cavity corresponded to different values of A in the iterative equation above. Once they finally understood the random behavior of their apparatus, they were able to choose a value for the optical parameters that resulted in stable operation—yet another accomplishment owed to mathematics that had been developed "just for fun."

A new magnetic material. When people think of magnets, they usually think of iron. The best presently available magnetic material, however, is a samarium-cobalt alloy, which contains no iron. Samarium-cobalt is used in a wide range of devices from compact electric motors to lightweight earphones for portable tape players. The high price of cobalt, however, has kept samarium-cobalt from being used more widely. Recently scientists in Japan and the U.S. found a new group of magnetic materials comprising a rare earth element (R), iron (Fe), and boron (B) in the general compound $R_2Fe_{14}B$. The rare earths neodymium and praseodymium seem to produce the best results.

By 1985 magnetic materials based on this formulation had been made that have 1.5 times the energy density of the best cobalt-based magnets and more than 10 times that of the conventional ferrites used to hold notes on refrigerators. The strength of a magnet also depends on the way in which the material is prepared. Scientists have made the materials as rapidly cooled metallic ribbons, powders, and aligned microcrystals while subjecting them to magnetic fields and various heat treatments to improve the final properties. These new magnetic materials are allowing the design of motors and speakers that are smaller and more powerful than those formerly available. And someday soon they may even be mundanely decorating refrigerator doors.

—Robert L. Forward

High-energy physics

As in the previous year the most important advances in our knowledge of elementary particles in 1984 came from the proton-antiproton colliding-beam facility at the European Laboratory for Particle Physics (CERN). This machine, located near Geneva, was built and used originally as a 400-GeV (billion-electron-volt) proton synchrotron. Recent modifications have allowed it to circulate protons and antiprotons each of 270 GeV in opposite directions, making possible two-beam collisions with a total energy of 540 GeV. This energy is equivalent to that released by a proton of about 200 TeV (trillion electron volts) striking a stationary proton. Although the collision rate is low, the very great energy available has permitted the discovery and study of phenomena and particles never before seen.

Thus it was that in 1983 the intermediate vector bosons, the W and Z particles, were observed at CERN. These particles, having rest masses of 81 GeV and 93 GeV, respectively (almost one hundred times greater than the rest mass of the proton), had long been predicted. They are understood as the field quanta that carry the weak interaction, that force which governs the radioactive decay of nuclei, and they play the same role with respect to the weak interaction as the photon does with respect to the force of electromagnetism. In 1984 Carlo Rubbia of CERN and Harvard University and Simon van der Meer of CERN were awarded the Nobel Prize for Physics for developing the CERN collider and for the discovery of the W and Z particles (*see* SCIENTISTS OF THE YEAR).

Table I. The Elementary Particles of Physics

	Electric charge[1]	Generation I	Generation II	Generation III
quarks	$+2/3$	*u* (0.3)[2]	*c* (1.5)	*t* (~50)
	$-1/3$	*d* (0.3)	*s* (0.5)	*b* (5)
leptons	-1	e (5.1×10^{-4})	μ (0.106)	τ (1.784)
	0	ν_e (0?)	ν_μ (0?)	[ν_τ (0?)][3]

[1] Electric charge is in units of the magnitude of the charge of an electron.
[2] The numbers in parentheses are masses in units of GeV/c^2 (energy in billions of electron volts divided by the square of the speed of light). Quark masses are approximate.
[3] The ν_τ (tau neutrino) has not been experimentally observed.

Table II. The Basic Forces or Interactions of Physics

Force	Field particle and rest mass	Strength relative to strong force at 10^{-13} cm distance	Particles that experience force
electromagnetic	γ (photon)[1]; 0	10^{-2}	all electrically charged particles; all quarks and all charged leptons
weak	$W^{\pm}$, Z° (intermediate bosons); ~90 GeV	10^{-13}	all particles
strong	g (gluon)[1]; 0	1	quarks and hadrons (particles composed of quarks)
gravitational	G (graviton)[2]; 0?	10^{-38}	all particles with mass

[1] The gluon has been identified indirectly but has not been, and perhaps cannot be, isolated.
[2] The graviton has not yet been experimentally observed.

Evidence for the *t* quark. As summarized in Table I all matter is presently understood in terms of six quarks and six leptons. These fundamental particles interact through four forces—the strong, electromagnetic, weak, and gravitational interactions—summarized in Table II. Each of the three "generations" of quarks and leptons seems characterized by a different mass scale, and the discoveries of the *b* quark and the τ (tau) lepton in the 1970s suggested that the third generation includes particles considerably more massive than the others. Whereas some theories could be forced to accommodate only five quarks, the *t* quark would round out the symmetry of the particles and result in a theoretical model much cleaner and simpler. Nevertheless, there were no theoretical predictions of the *t* quark mass, and early, intensive searches for evidence of the *t* quark, particularly at the PETRA electron-positron collider in Hamburg, West Germany, had proved fruitless.

Fortunately, studies of mesons composed of other heavy quarks had reached a state of sophistication such that the signal expected from a *t* quark was quite clearly defined. These other studies of meson states, in which the *b* quark and the *c* quark are components, had been carried out at electron-positron colliders in Hamburg as well as at Cornell and Stanford universities in the U.S. Other studies at the CERN 400-GeV proton synchrotron and a similar machine at the Fermi National Accelerator Laboratory (Fermilab) near Chicago had further clarified the lifetimes and decay modes as well as production characteristics of the *c* quark states.

In 1984 at the CERN proton-antiproton collider six events were observed that correspond to the production of a W particle followed by its decay to a *t* quark and a *b* antiquark. The *t* quark in turn gave rise to a neutrino, a charged lepton (electron or muon), and a *b* quark, which produced a cascade, or jet, of more common mesons. The detector used to observe and interpret these events is an elegant and remarkable device in its own right, requiring time, skills, and money on a scale quite comparable to the accelerator itself. Nevertheless, the presence of undetected neutrinos, which carry away a certain amount of energy, in the final state of the *t* decay leaves the mass of the *t* quark quite uncertain; Rubbia's group was only willing to bracket the figure between 30 and 50 GeV.

Anomalous collision events. Other results from the CERN collider are worth noting. Earlier it had been feared that the debris from a proton-antiproton collision at such high energies would be so complex as to make the clear observation of rare phenomena impractical. Fortunately this proved not to be the case. The proton (or antiproton) is now understood to be composed of three quarks bound by gluons, the field quanta of the strong force. When a proton and an antiproton collide at very high energies, one quark or gluon from each particle may experience a very "hard," or head-on, collision, while the rest of the constituents of both particles continue in their original directions, only somewhat disturbed. The products of the hard collision often will fly off at large angles and carry very high transverse momentum, whereas the remaining constituents lead to a stream of many particles (mostly mesons) all moving in nearly the original direction of the proton or antiproton. If the hard collision is "elastic," the final state will reveal two jets of particles moving opposite to each other and both at large angles to the original particle directions. From observation of these jets a great deal is being learned concerning quark-quark, quark-gluon, and gluon-gluon scattering processes. Of course, on rare occasions a hard collision results in the production of a W or Z particle instead of simpler debris.

Among the millions of proton-antiproton collisions observed and studied, some cannot be understood in terms of any known processes. Some events contain only a single, large-angle jet among the debris; others contain only a single electron having a very large transverse momentum; and yet others are equally difficult to interpret, although more complex. At the very least it appears that new phenomena beyond scientists' present understanding are made manifest at these energies, and it is a challenge for the future to study them.

By 1985 a second proton-antiproton collider, at Fermilab, was approaching its full operating potential. This facility will provide higher collision energies than CERN (as high as 2 TeV). Furthermore, for future studies both at CERN and Fermilab the collision rate will be much higher than was available at CERN for the important discoveries cited above.

Testing the standard model. A great deal of other research at high-energy accelerators around the world has increasingly confirmed what is now termed the "standard model." This model, which treats all strong interactions in terms of the exchange of gluons, differs from the quantum theory of electromagnetism chiefly in that there are three aspects (called colors) of the particles that experience the strong interaction whereas there are only two aspects, or signs, of electric charge (positive and negative). The theory of the strong interaction is referred to as quantum chromodynamics, or QCD. The standard model also includes the unified theory of electromagnetism and the weak interaction, wherein the two forces are understood as different facets of a single electroweak interaction.

Some experimental results are still beyond the present ability of the standard model to explain. At Brookhaven National Laboratory on Long Island, N.Y., a 30-GeV proton synchrotron has been modified to accelerate polarized protons; that is, protons aligned so that their spins (the direction of their intrinsic angular momentum) are in the same direction. When a beam of polarized protons is elastically scattered on a target of polarized protons, the angular distribution of the scattered protons may be a function of this polarization. Whereas at low energies these polarization effects have been observed, simple application of QCD theory suggests that they would decrease with increasing energy. During 1984 a group led by Alan Krisch of the University of Michigan carried out the first scattering experiments using the new polarized beam at Brookhaven. To their surprise certain polarization effects appeared to increase rather than decrease with energy. Clearly more experimental and theoretical work is needed to resolve this discrepancy.

An interesting class of experiments that relate intimately to high-energy physics, inasmuch as they deal with elementary particles in a fundamental way, does not make use of accelerators. Foremost is a group of ambitious experiments seeking evidence for the spontaneous decay of the free proton. In 1984 experiments were in progress in India (the first serious study), Japan, Italy, and the U.S. The latest report, by the collaboration using the largest detector, concludes that no convincing evidence exists for proton decay. Their detector is a tank of 8,000 tons of pure water viewed by 2,048 phototubes and located in a salt mine near Cleveland, Ohio. This collaboration, which included physicists from the University of California at Irvine, the University of Michigan,

Courtesy, Fermi National Accelerator Laboratory

Line of particle-beam magnets awaits full installation in the antiproton source tunnel at Fermilab. Once generated, antiprotons will be directed into the main ring, the Tevatron, for acceleration to energies as high as 1 TeV (one trillion electron volts) and subsequent collision with 1-TeV protons.

and Brookhaven, observed several events that might be interpreted as proton decays but which are more readily explained as the expected interactions of cosmic-ray neutrinos. The Japanese and Indian groups felt that some of their own observed events are more difficult to explain as neutrino interactions. In any event, as established from these experiments, the experimental lower limit to the lifetime of the proton against spontaneous decay is greater than 10^{32} years, which is 10–100 times greater than that predicted by the most straightforward application of current theory.

As with the proton decay experiments, negative results may often be as significant as positive results. In 1982 there appeared a published report of evidence for a magnetic monopole. This particle, allowed by theory and suggested to play a significant role in the early evolution of the universe, is understood as a single north or south pole of magnetism. It is predicted to have a very great rest mass, perhaps 10^{16} times the mass of a proton, and a unit of magnetic "charge." During the past year several experiments, including a new experiment implemented by Blas Cabrera of Stanford University, who made the 1982 observation, extended the sensitivity of the original experiment by more than a hundredfold with negative results. Thus, it appears that the earlier observation was an experimental fluke and that the magnetic monopole must once again be considered a hypothetical particle. If found and confirmed, however, it would certainly be a momentous discovery.

Although the standard model has systematized and clarified a broad spectrum of experimental data on the interactions of elementary particles at high energies, it has led to challenges. One still unverified prediction of the standard model is the existence of a particle more massive than the W or Z, the so-called Higgs particle, which is responsible for the breaking of the simple underlying symmetries of the model. Scientists do not understand the origin of mass, much less the basis for the particular masses of the known particles. It is also conceivable that quarks have a substructure. To all indications these and other tantalizing problems should be resolvable with particle experiments conducted in an energy range much higher than that available at the CERN or Fermilab colliders but still technically realistic.

Fortunately the technology of superconducting magnets, which make possible the generation of magnetic fields more than twice as intense as those of the strongest conventional magnets while using negligible electric power, has been perfected. Indeed, the most recent accelerator facility at Fermilab, the Tevatron, is a ring of such magnets. The Tevatron was brought into successful operation for physics research during 1984 for the first time and will be the ring in which Fermilab will operate its proton-antiproton colliding-beam experiments.

During the year U.S. physicists under the leadership of Maury Tigner at the Lawrence Berkeley Laboratory were developing a detailed design and proposal for a superconducting proton-proton colliding-beam accelerator capable of bringing together 20-TeV protons and thus making available 40 TeV for experiments. The collision rate, or luminosity, will be thousands of times greater than in the CERN collider. From a preliminary study in early 1984 the basic design features, costs, and operating characteristics were defined. Throughout the year physicists convened to explore various specific aspects of this Superconducting Super Collider, or SSC. By early 1985 the excitement and enthusiasm were high for moving forward with the detailed design and construction of the machine.

—Lawrence W. Jones

For an account of the interrelationship of high-energy physics and the early evolution of the cosmos *see* Feature Article: THE INFLATIONARY UNIVERSE.

Nuclear physics

For nearly the past half century nuclear physics has been limited primarily to the study of phenomena in the nuclear surface and phenomena governed almost entirely by the behavior of neutrons and protons in the nucleus. In the past few years major new frontiers have opened up in the study of the nuclear many-body system. Workers in the field are confident of major advances based both on a new understanding of the nucleus and on new experimental techniques and facilities.

Inside the nucleus. Essentially all studies of shell and collective phenomena in nuclei have focused on the behavior of the nuclear surface or on that of nuclear neutrons or protons (nucleons) very close to the surface. With the availability of highly precise beams of electrons, mesons, and heavier, strongly interacting particles over wide ranges of energy, it has now become possible to probe deeply into the nuclear interior and to study previously unsuspected phenomena that cannot be understood solely in terms of neutron and proton constituents. Specifically, recent high-precision measurements by Bernard Frois and his collaborators of the Center for Nuclear Research, Saclay, France, extended physicists' detailed understanding of neutron and proton orbital motion throughout the complete nuclear interior and showed that shell-model concepts retain their validity; *i.e.,* to the first order the nucleons do not change their character when in the presence of many others. At the same time measurements by Raymond Arnold and his collaborators at the Stanford Linear Accelerator Center (SLAC) in California showed conclusively that the scattering of high-energy electrons from light nuclei, such as helium-3 and helium-4, cannot be understood even qualitatively without taking into account the interaction of the electrons with currents of mesons inside the nucleus in addition to their interaction with neutrons and protons. At higher energies measurements by the European Muon Collaboration, working at the European Laboratory for Particle Physics (CERN) in Geneva, demonstrated that some of the quark constituents of the nucleons spend at least part of their time adventuring outside of their parent nucleons and that entities made of six quarks, in addition to the normal three-quark nucleons, are present in the nuclear interior.

This rich spectrum of phenomena, only glimpsed as yet, now lies open to experiment. Furthermore, it shows promise of being understandable on the basis of quantum chromodynamics, the fundamental theory of the strong nuclear interaction.

High-energy electron beams are making it possible to examine systematically the change of behavior that occurs in the nuclear many-body system as its temperature is increased. At low temperatures, near the nuclear ground state, nuclei are characterized by phenomena in which neutrons and protons play the dominant roles while small admixtures of mesons provide the "glue" holding the nucleons together. As the temperature is raised, however, the neutrons and protons absorb energy and are promoted into excited quantum states—the so-called nucleon isobars—which begin to play major roles in determining observable nuclear characteristics. At still higher temperatures the nucleons begin to melt, liberating their constituent quarks and gluons, which mediate the strong force holding the quarks together. Finally, at very high temperatures not yet accessible, it is anticipated that an entirely new form of matter, quark-gluon plasma, will result.

Although electron beams permit systematic exploration of these changes, study of the quark-gluon plasma awaits the availability of ultrahigh-energy beams of heavy nuclei. In collisions with heavy targets such beams can create quark-gluon plasmas in volumes large enough to contain many participants and, in so doing, recreate conditions as they existed in the earliest instants of creation.

In early 1985 major new electron accelerators were either under construction or approved for construction in Moscow, in Paris, at the Massachusetts Institute of Technology, and in Newport News, Va., where the Southeastern Universities Research Association has planned the CEBAF 4-GeV (billion-electron-volt) electron facility. In parallel with this work additions were being made at SLAC specifically to enable nuclear experiments to be carried out at that well-known particle physics facility. With respect to ultrahigh-energy heavy ions, the only two facilities presently in operation are the Synchrophasotron at Dubna in the Soviet Union and the Bevalac at the University of California at Berkeley, which provide ion beams as high as 4 and 2 GeV per nucleon, respectively. An interim program at much higher energy was under way at CERN, and ultrahigh-energy facilities have been designed for installation at GSI in Darmstadt, West Germany, and at Brookhaven National Laboratory in the U.S. As of early 1985 both of these latter two facilities awaited approval at their respective federal levels. Scientists worldwide have been planning for these new facilities and for their use once they become operational.

Nuclear decay studies. While attention is frequently focused on physics done with accelerators, it bears emphasis that nuclear physics began with the study of radioactivity and still uses radioactive decay processes as a powerful source of information on nuclear behavior and fundamental questions involving both nuclear and particle physics.

The neutrino mass. In 1980 V. A. Lubimov and his associates at the Institute for Theoretical and Experimental Physics in Moscow reported on the results

Adapted from information obtained from D. Allan Bromley

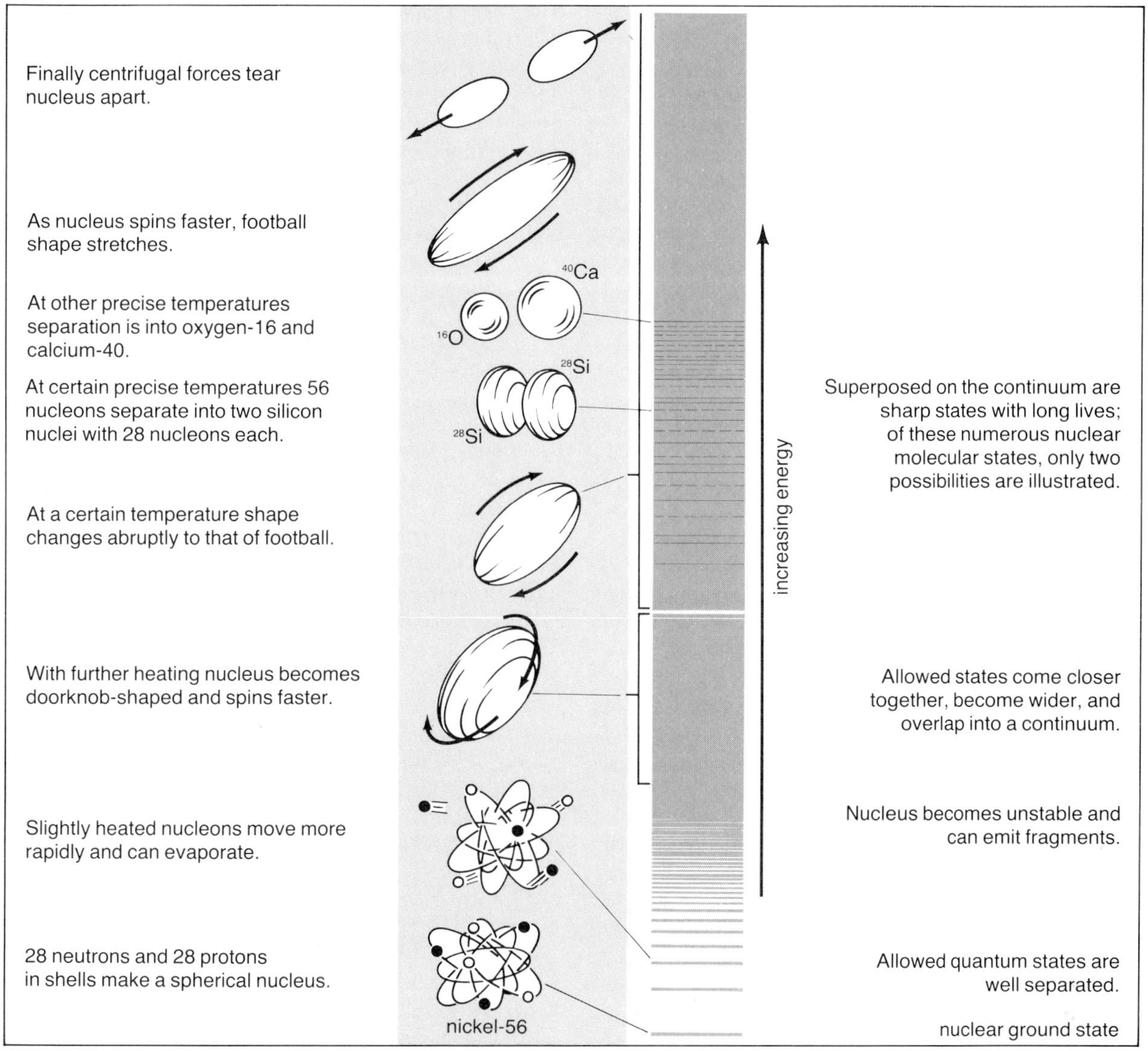

Diagram summarizes the large-scale systematic changes seen in the nickel-56 nucleus as more and more energy is added. An unexpected feature is the reassembly of its nucleons into nuclear molecular configurations.

of their measurements of the beta-decay spectrum of tritium (hydrogen-3), which indicated that the electron neutrino, long thought to have a zero rest mass, has a rest mass between 14 and 46 eV (electron volts). These measurements have continued, and in 1984 the Moscow group reported that the mass is greater than 20 eV at a 95% confidence level.

Reflecting the enormously important consequences of a nonzero neutrino mass in fundamental physics, in astrophysics, and in an overall understanding of the universe, many groups around the world have undertaken to repeat precision measurements on the endpoint of the beta-decay spectrum of tritium; in other words, on the mass difference between the mass-3 isotopes of hydrogen and helium. Whereas all of the results are not yet in, recently a group at the Institute of Chemical and Biological Physics of the Estonian S.S.R. Academy of Sciences in Tallinn, led by E. Lippmaa, reported an atomic mass difference of $18{,}599 \pm 2$ eV, to be compared with a value of $18{,}594 \pm 25$ eV obtained by W. R. Dixon and his collaborators at the National Research Council Canada in Ottawa and of $18{,}603.6 \pm 7$ eV recently reported by Lubimov. All of these results appear to require a nonzero neutrino mass for their interpretation, and great importance will attach to the results from other experiments expected during the coming year. If a nonzero mass is confirmed experimentally, it will be one of the most far-reaching discoveries in fundamental physics in many decades.

A new kind of radioactivity. In 1914, after searching for radioactive emission products heavier than the alpha particle in the decay of heavy nuclei, Ernest Rutherford was able to show that if any such particles were present they were certainly there at a level of less than one per 10,000 alpha particles. In 1980 Walter Greiner of the Institute for Theoretical Physics at the University of Frankfurt in West Germany, with A. Sandulescu and D. N. Poenaru, returned to this question as part of their study of the nuclear fission process within the framework of modern nuclear shell models. They predicted that heavy radioactive nuclei that fission spontaneously should, on rare occasions, divide very asymmetrically, even to the extreme case in which one of the fragments is an alpha particle, making this case synonymous with familiar alpha-particle radioactivity. In 1984 H. J. Rose and G. A. Jones of Oxford University reported the discovery of a decay mode of radium-223 (^{223}Ra) involving emission of a 29.9-MeV (million-electron-volt) carbon-14 (^{14}C) nucleus and having a frequency of occurrence, relative to alpha particle decay, of one in 85 billion. S. Gales and his collaborators of the University of Paris at Orsay confirmed this discovery and, in addition, obtained a single event that they interpreted as radioactivity involving emission of an oxygen-16 nucleus. Very recently Buford Price at Berkeley and his collaborators observed ^{14}C radioactivity in the decay of ^{222}Ra and ^{224}Ra in addition to ^{223}Ra.

This new information can be used to calibrate the earlier shell-model calculations for extreme configurations and thus provide every valuable new insight into the minute structure of heavy nuclei as well as more confident extrapolation of these shell models to predict the characteristics of transuranium and supertransuranium species (actual and hypothetical elements with atomic numbers greater than that of uranium). From a qualitative point of view the observation of ^{14}C radioactivity is consistent with the existence of stable pear-shaped deformations of the radium nuclei with the ^{14}C being emitted from the small end of the pear and the lead partner from the large end.

Heating the nuclear many-body system. As noted above, almost all of nuclear physics to date has focused on relatively cold nuclei and on peripheral phenomena. One of the major open frontiers only now being explored is that of higher temperatures obtained as more and more energy is added to individual nuclear systems. In addition to the systematic internal transformations discussed above, there are more macroscopic systematic changes, which are shown schematically in the diagram on p. 369. After only about 8 MeV are added to an ordinary nucleus, it becomes possible to evaporate single nucleons from the nucleus, and valence nucleons can be promoted into previously unoccupied shell-model orbits. It is characteristic of the scattering processes whereby energy is added to nuclear systems that increasing energy carries with it increasing angular momentum, so that the nucleus spins faster as it is heated. Reacting to this increased angular momentum, most nuclei go through an oblate spheroid (doorknob-shaped) phase before switching to a prolate (football) shape.

What was completely unexpected, and what Russell Betts and his collaborators at Argonne National Laboratory in Illinois and at Yale University showed to be a systematic feature of essentially all nuclei, is that at very specific resonant energies—as much as 70 MeV above the ground state—the nucleons present reassemble themselves into two or more separate nuclear fragments or clusters that exhibit very simple motions with respect to one another in a nuclear molecular configuration. There are many such configurations, and the transition from one to another occurs over a very narrow energy or temperature range; between them the nuclear system reverts to its normal football shape, in which all the nucleons present participate.

As the angular momentum and temperature increase, the nucleus is stretched by centrifugal forces. Although the phenomenon has not yet been observed experimentally, it is confidently expected that at angular momenta in the range of 50–70 units, the nucleus will be torn apart by these forces. In the nickel-56 system shown in the diagram, which was initially studied by Stefanie diCenzo and her collaborators at Yale, existing accelerator facilities permitted the system to be studied to angular momenta of 44 units; current theories suggest that disruption will occur at about 50 units. New electrostatic accelerator facilities will allow study of higher angular momenta in the near future.

—D. Allan Bromley

Solid-state physics

The past year was characterized by continued vigorous research in solid-state physics and its many technical applications. This field of physics, which is now frequently called condensed-matter physics, has probably the shortest connection path between its basic science and its technical applications, each providing stimulus to the other. One such connection is between new materials and novel devices, the most famous case being pure single-crystal silicon and the transistor. Underpinning much of the progress in solid-state physics is the continued development of sophisticated instrumentation to study new physical phenomena and to characterize material systems. Recent developments in high-voltage transmission electron microscopy have given scien-

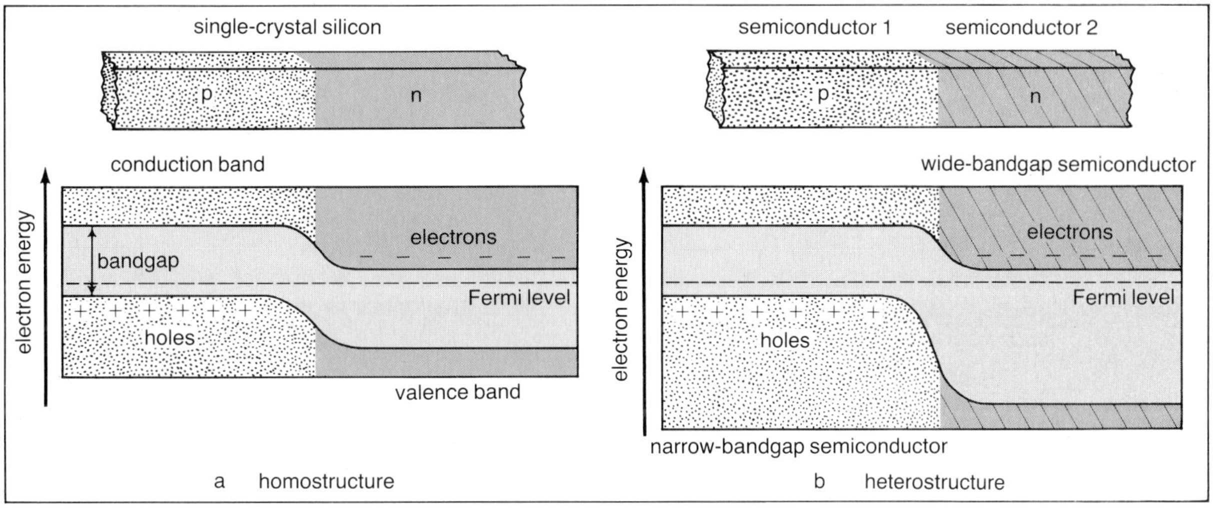

Figure 1. In a homostructure (a) differing electronic properties between adjacent regions are achieved by doping the same single-crystal semiconductor. In a heterostructure (b) the effect is accomplished by joining different semiconductors, but with an atomic alignment that permits the flow of charge carriers across the interface.

tists an unprecedented ability to study materials on the atomic building-block level.

New superconducting state. An example of the coupling between new materials and the discovery of unpredicted new physical phenomena is the ongoing work on so-called heavy-fermion materials. These are metal alloys containing elements with partially filled *f* electron shells (rare-earth metals or actinides). This class of materials exhibits unusually large low-temperature electronic specific-heat coefficients, indicative of large effective mass of the electrons at the Fermi level, thus the name "heavy fermion" systems. The discovery in 1984 by scientists at Los Alamos (N.M.) National Laboratory of superconductivity in the uranium-platinum compound UPt_3 seems to suggest strongly the existence of a new type of superconducting state and a new mechanism leading to superconductivity.

Artificially structured materials and quantum size effects. Most applications of semiconductor crystals in electronics are built on homostructures, such as the p-n junctions in transistors (*see* figure 1a). In such devices the same semiconducting material, most commonly single-crystal silicon, is used throughout the entire device, and the necessary change in the nature of the electrical conductivity (p-n junction) is achieved by doping the host crystal with small amounts of foreign atoms.

On the other hand, if a junction is formed by joining two different types of semiconductors having different electronic properties, the result is a heterostructure (figure 1b). The ability to fabricate such structures easily and in great variety would offer tremendous opportunities for conceiving novel devices. In fact, it would provide the materials foundation for the "microengineering" of tailor-made devices with predictable properties for specific applications. The requirements on such junctions, however, are very stringent. For the junction to pass current, an essentially perfect alignment of the atoms is needed at the interface of the two materials. A mismatch or numerous defects at the interface will cause the negative or positive charge carriers (electrons or holes) to be trapped or scattered.

Most heterostructures in current use are made between compound semiconductors such as gallium arsenide (GaAs) and semiconducting alloys; for example, gallium aluminum arsenide ($Ga_x Al_{1-x} As$), in which some gallium atoms have been replaced by aluminum atoms. The best known heterostructure device is the solid-state diode laser. The generation of laser light at a p-n junction was demonstrated in 1962, and the technical breakthrough for this device was achieved with the invention of the so-called double-heterostructure laser and its operation at room temperature. A schematic diagram is shown in figure 2 on p. 372.

The two semiconductors forming the double heterostructure are GaAs and $Ga_x Al_{1-x} As$. Since these two materials have different indexes of refraction for light, with GaAs having the higher one, the laser light created at the p-n junction by passing an electric current through the device is confined in the "active" region of GaAs, just as light is confined in a fiber-optic cable because its index of refraction is higher than that of air.

Progress in solid-state lasers has been determined primarily by the development of crystal-growth techniques. Spectacular and promising progress has been achieved by Robert D. Burnham and his co-workers

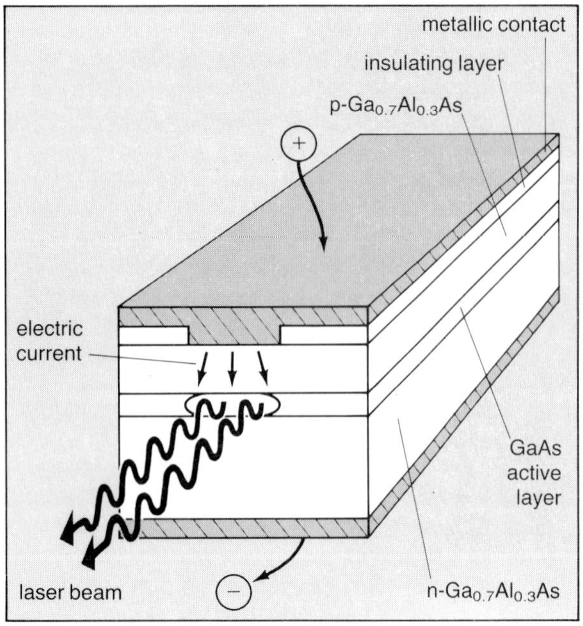

Figure 2. In the solid-state diode laser, light generated by the recombination of electrons and holes at the p-n junction is trapped in the GaAs active layer because its index of refraction is higher than that of the $Ga_x Al_{1-x} As$ layers above and below. The laser beam emerges only at the unconfined edges of the GaAs layer.

at the Xerox Palo Alto (Calif.) Research Center using a technique called metal-organic chemical vapor deposition (MOCVD), which allows fabrication of essentially defect-free heterojunctions. Crystal growth occurs in a heated reactor into which are fed the elements to be deposited (Ga, As, Al, and dopants). The elements are in gaseous form as organometallic compounds diluted in hydrogen as a carrier gas. The high temperature in the reactor causes the compounds to decompose; the metal atoms are deposited on a substrate to form the desired chemical compounds. The optical quality of the heterostructures produced by this technique now supersedes that of any other method of growth.

Altering the composition of the gases in a periodic way makes it possible to build up a crystal structure with alternating layers of different composition. Such a structure is called a superlattice. By accurately controlling the crystal-growth parameters Burnham and his research group have grown superlattice structures having layers only a few atoms thick and exhibiting atomically smooth and abrupt interfaces. An example of such a superlattice consisting of alternating layers of GaAs and $Ga_x Al_{1-x} As$ is shown in figure 3.

The thickness of the active GaAs layer in a solid-state laser like that illustrated in figure 2 is typically larger than 500 angstroms (Å), and the charge carriers behave as they do in a bulk crystal (one angstrom is 10^{-8} centimeters). If the active layer thickness is smaller than 500 Å, however, the de Broglie wavelengths of the electrons and holes are comparable to the layer thickness. (In quantum mechanics all particles of matter can be represented as waves with their own associated wavelengths.) This represents one of the very few cases of a man-made structure that exhibits quantum size effects (QSE) corresponding to a textbook example of quantum mechanical behavior. Whereas the explanation is technically demanding, the net effect is that the light emitted from the active layer is shifted toward shorter wavelengths. By exploitation of QSE, the emission of a GaAs laser, which normally is in the invisible, infrared portion of the spectrum, can be pushed into the visible region. In practice several layers in the form of a multiple quantum well heterostructure (QWH) have been employed as the active region. This seems to be the most promising approach to a visible-light solid-state laser, which

Figure 3. Electron micrographs made at two levels of magnification reveal the structure of a superlattice built up of alternating layers of GaAs and $Ga_x Al_{1-x} As$. Each layer is only about ten atomic layers thick. As is apparent in the more highly magnified image, the interface between the layers (marked with arrows) is atomically smooth and completely abrupt.

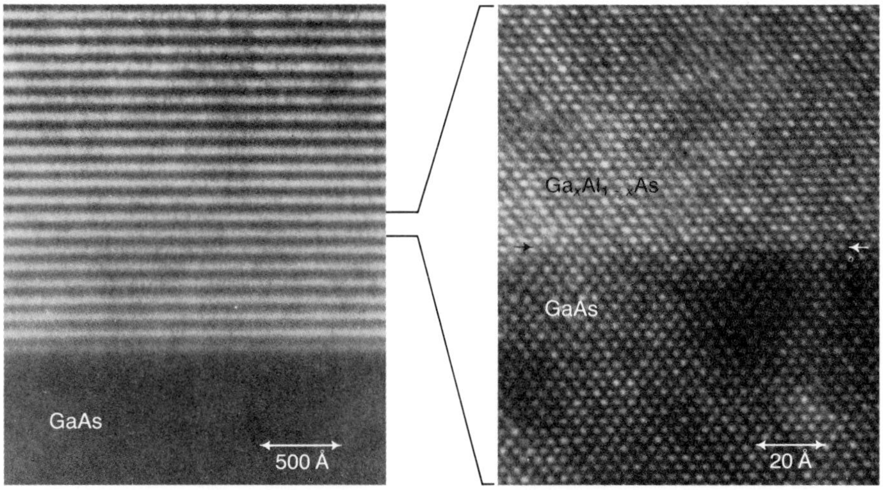

Photos, courtesy, Fernando Ponce and Joe Tramontana, Xerox Palo Alto Research Center

Courtesy, Fernando Ponce, Xerox Palo Alto Research Center

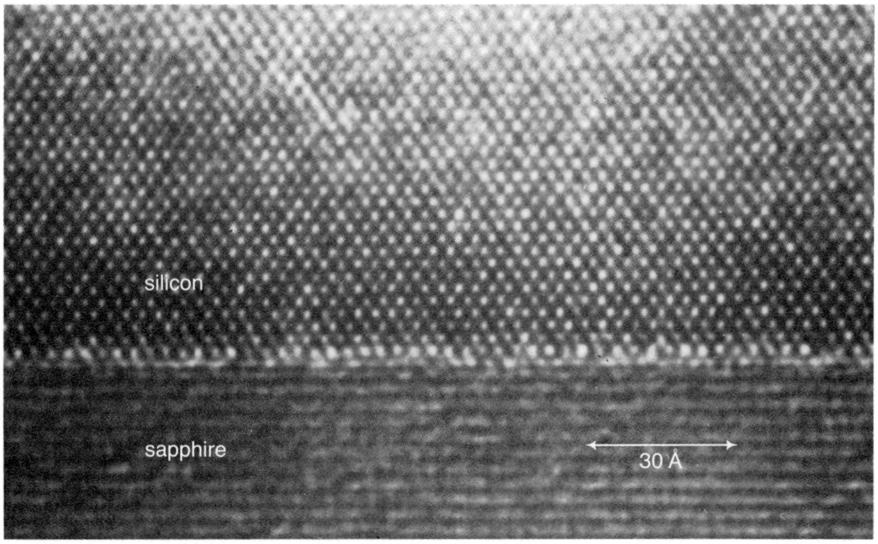

Figure 4. The electronically important interface between silicon and sapphire is revealed at nearly atomic resolution in the electron micrograph at left. Advances in high-voltage transmission electron microscopy are providing scientists detailed information about the atomic structure of materials.

could replace costly and bulky gas lasers and provide new technical opportunities.

Impurity-induced disorder. Even though many superlattices are quite stable, they can be disordered at high temperatures. This treatment, however, would not allow a laterally selective disordering of a superlattice or of a multiple QWH, which would be desirable when building device structures by the commonly used photomasking technique. During the past year Nick Holonyak, Jr., and his students at the University of Illinois and Burnham and co-workers at Xerox developed a new technique to achieve disordering of a QWH by diffusion of impurity atoms at low concentrations, for example, silicon in a $GaAs/Ga_x Al_{1-x} As$ QWH, at lower temperatures than thermal disordering. By depositing a thin layer of silicon on a QWH wafer in a pattern determined by photomasking and then heating the wafer, the researchers created a lateral pattern of regions with different bandgaps. This achievement has immediate implications for new and improved laser structures and should assist in future efforts to create integrated circuits from superlattice structures. Various mechanisms for this type of disorder-inducing diffusion have been proposed, but the exact mechanisms of it are still unknown.

Atomic-resolution electron microscopy. The ability to produce artifically structured materials such as superlattices on an atomic-layer scale requires not only very precise synthesis tools like MOCVD but also highly sophisticated tools for analysis and characterization. In many cases scientists must resort to indirect methods; *e.g.,* measuring such properties as the electrical and optical characteristics of a device and then deducing from its performance the quality of the component materials. More direct methods are desirable, and in the ideal case scientists would like to be able to "see" the structure on an atomic scale. The recent development of atomic-resolution high-voltage transmission electron microscopy, has come close to achieving this goal.

Theoretically the electron microscope should be able to resolve structure even on a subatomic level. To date, however, practical limits have put its resolution at a few angstrom units, not quite sufficient to separate individual atoms. Recently a sophisticated research high-voltage microscope with a near-atomic resolution was developed by a Japanese instrument company for the National Center for Electron Microscopy at Berkeley, Calif. This development led the way to more generally available instruments with similar performance. During the past year the first of this new generation of electron microscopes was delivered to the Center for Solid-State Science at the Arizona State University, Tempe.

Figure 4 shows an electron micrograph, made with such an instrument, of an interface between silicon, a semiconductor, and sapphire, an insulator. Projected atomic columns are clearly visible as bright spots, and the figure shows that the interface is abrupt within one atomic layer and shows no defects. The full use of these new electron microscopes, however, requires a sophisticated technique for simulating the images in a computer. By comparing a simulated image with a real one, investigators can deduce the atomic arrangement. This development offers yet another example of the way in which the branch of physics that spawned the largest technical revolution in modern times, the electronics revolution, is now benefiting from the technical applications of its scientific discoveries.

—Stig B. Hagstrom

See also Feature Articles: CURVES, KNUCKLE BALLS, BREAKING PITCHES: THE KINEMATICS OF BASEBALLS; NONDESTRUCTIVE EVALUATION—COMING OF AGE.

Psychology

The discipline of comparative psychology has for the past two decades seemed to be in a moribund condition. Recently, however, there have been some clear signs of a resurgence of activity and interest in this field. During the past year in the G. Stanley Hall lecture given at the annual convention of the American Psychological Association, psychologist Duane Rumbaugh presented a strong case for a renewed and vigorous comparative psychology.

Drawing heavily on the recently published *Animal Cognition,* edited by H. L. Roitblat, T. G. Bever, and H. S. Terrace, Rumbaugh argued convincingly that a truly comparative psychology provides a useful perspective for the teaching of introductory psychology. Some of his strongest points were drawn from his own research with chimpanzees at the Georgia State University Language Research Center. For example, an ability to categorize by using geometric symbols ("lexigrams") was documented in two of the most highly trained chimpanzees, Austin and Sherman. They were able to classify lexigrams, which they had previously used in communication, as either tools or ingestibles (foods/drinks). They did so by selecting the appropriate superordinate lexigram (one indicating either a tool or ingestible) as each of the more specific lexigrams was presented. Only a single error was made in 33 trials, and even this error—classification of the lexigram for sponge as an ingestible—could be explained on the basis of the erring chimpanzee's earlier observed tendency to nibble at sponges when they were used to extract liquids from containers.

Evidence for generality of language functions among higher animal forms recently also came from research with marine mammals. Porpoises (dolphins) were found to be capable of accurately carrying out instructions—even new instructions—when they are conveyed either by hand signals presented by trainers or by computer-generated sounds. Sea lions revealed similar abilities. Thus, the study of language in a variety of animal species appears to offer great promise for providing the kind of conceptual framework upon which psychology can proceed to develop some truly comparative and illuminating principles.

Learning in utero. Experimental psychologists have recently turned their attention to the problem of prenatal learning. In so doing they have accumulated considerable evidence indicating that in both people and anthropoid apes (gorillas, chimpanzees, orangutans, and gibbons) unborn babies learn while still in the uterus. For example, psychologist Anthony DeCasper of the University of North Carolina at Greensboro demonstrated auditory learning during the prenatal period. In one experiment he and his associates hooked up earphones to human babies 1½ to 4 days old. Different consequences were arranged for the sucking response. Sucking during recorded beeps started a tape recording of their mothers' voices, while sucking during silence started a recording of another woman's voice. Overall, 21 of 26 babies sucked predominantly during beeps; this indicated a clear preference for the mother's voice, presumably heard and learned during the prenatal period. The babies did not prefer their father's recorded voice, however, over that of another male.

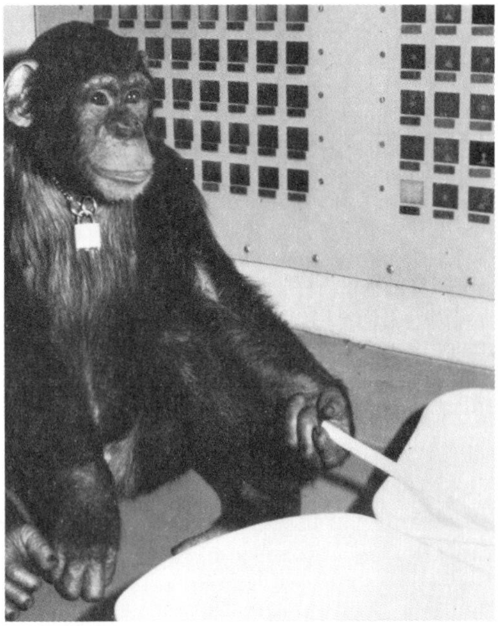

Photos, courtesy, Language Research Center, Georgia State University and Yerkes Primate Center, Emory University

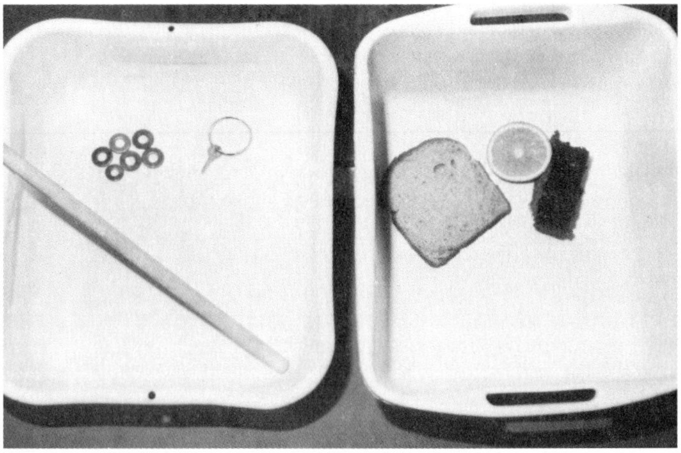

Chimpanzee (left) was one of three trained to sort three foods into one bin and three tools into another (above) and then to label each with a symbol indicating whether it was a food or tool. When additional foods and tools were introduced, two of the three chimpanzees were able to use the symbols to classify all but one of the objects correctly into the two categories.

In a similar experiment pregnant women read aloud from a nursery book twice daily during the last six weeks of pregnancy. Tested within two days after birth, 10 of the 12 babies of these women preferred the story that had been read by their mother over a new one.

Odor preferences developed by rats in utero were shown to exert subsequent powerful behavioral effects. Psychologists Elliot Blass of Johns Hopkins University, Baltimore, Md., and Patricia Peterson of Yale University exposed rats in utero to citral, a chemical with a pungent odor. After birth those rats were found to prefer to suckle rat mothers whose bedding had been saturated with citral. No such preference was observed in a control group not given in utero exposure to citral. A greatly diminished effect was found in rats given postnatal exposure to citral.

In other research the amniotic sacs of fetal rats were infused with both apple juice and lithium chloride, which causes intestinal discomfort. After birth these rats avoided nipples coated with apple juice, even those of their own mothers. These examples suggest the potential for research into prenatal effects on learning and memory.

Learning under anesthesia. A new approach to an old problem was introduced by a team of psychobiological researchers at the University of California at Irvine. They attacked the problem of whether learning occurs under anesthesia, a time when the brain appears to be unresponsive to external stimulation. The new variable used in this research was the administration of epinephrine (adrenaline), the hormone responsible for the fight/flight reaction in animals. Several groups of rats were anesthetized and then stimulated by white noise and shock. No changes were recorded in heart rate or body reflexes from either the administration of the epinephrine or the noise/shock stimulation.

All of the rats that had been given epinephrine and had had the white noise paired with shock during anesthesia subsequently showed inhibition of drinking behavior when stimulated by white noise. No drinking inhibition occurred in control rats, either those given a saline solution or those for whom the white noise and shock had been presented separately rather than together during anesthesia.

This was an interesting result, not only because it demonstrated for the first time that the brain can be responsive and mediate learning during anesthesia but also because the learning that was demonstrated did not seem to be dependent upon the experiencing of pain by the subjects. Perhaps there are more subtle signals within the brain to indicate potentially noxious stimulation.

Social psychology. Called the "Pygmalion effect" by Harvard University psychologist Robert Rosenthal, the self-fulfilling prophecy has been well established by a variety of empirical demonstrations. Now this research design has been extended to physical education by Rosenthal, working with Elisha Babad at Hebrew University in Jerusalem.

These researchers first determined the extent to which 26 high-school physical education teachers were influenced in their grading by extraneous information. Sixteen teachers were categorized as "high bias" and the remainder as "low bias." The researchers next provided all of the teachers with alleged information about which of their students had the greatest potential (actually names were selected randomly) and then observed their treatment and grading of both selected (named) and nonselected students. Relatively equal treatment of all students was reported to have occurred among the low-bias teachers, but the students described as having the most potential were not only graded more favorably by the high-bias teachers but also seemed to be treated with greater warmth by them.

Societal problems. Preliminary findings from a long-term, five-city survey of mental illness were announced from data collected in Baltimore, Md.; St. Louis, Mo.; and New Haven, Conn. Approximately one in every five persons questioned by trained interviewers reported at least one mental disturbance in the previous six months. Anxiety disorders (such as phobia, panic, obsession-compulsion) were the most common disturbances. Equal numbers (about 6% of the respondents for each) admitted to being substance abusers (mainly alcohol) or sufferers from some affective disorder (such as major depression and manic-depressive syndrome). Schizophrenia and antisocial personality were each reported in about 1% of the sample. The study is remarkable for its scale, as it involved a sample of 20,000 adults.

Internationalization of U.S. psychology. During the past year there were indications of increased awareness within U.S. psychology of international developments in the field. For some time U.S. psychologists have been notoriously oblivious to the contributions of those from other nations, as evidenced particularly in their failure to cite such sources in books and research reports.

Perhaps the most noteworthy single event suggesting the end of this insularity was the publication of a set of papers on international education in psychology in the November 1984 issue of *American Psychologist.* The flavor of these reports is suggested by their titles: "The Price of Success: Our Monocultural Science," "The World Beyond Our Borders: What Might Our Students Need to Know About It?", "Cross-cultural Psychology," "Psychology in Its World Context," and "American Psychologists and Psychology Abroad."

In another significant publication (*American Psychologist,* August 1984) University of California psychologist Mark R. Rosenzweig compared U.S. and world psychology and drew some interesting conclusions. He explained the pronounced insularity of U.S. psychology on the basis of the explosive growth of the discipline in the U.S. during the four decades after the end of World War II and its consequent self-sufficiency. That growth can be indexed by a number of statistics. For one example, in 1984 the American Psychological Association had almost as many members (approximately 50,000) as all of the other 43 member organizations of the International Union of Psychological Science put together.

Cross-cultural research. As an example of the kind of empirical cross-cultural research that is rapidly finding increased favor among U.S. psychologists, a University of Michigan comparison of U.S. schoolchildren with those in Japan and Taiwan may be cited. A total of 5,000 children in 120 first- and fifth-grade classrooms were studied over a two-year period. A major conclusion of this study was that improvement in reading and mathematics by U.S. schoolchildren will require some fundamental changes in homes as well as schools.

"In whatever way the data are summarized," reported developmental psychologist Harold Stevenson, "the poor performance of American children is evident." For example, U.S. first graders received 57 of the 100 lowest scores in reading comprehension (compared to a chance expectancy of 33) and 58 of the 100 lowest scores in mathematics. Even more dismal results were noted for the fifth graders: U.S. children received 67 of the 100 lowest scores. Moreover, only a single U.S. child was among the top 100 fifth graders in mathematics.

Some of this inferiority in performance can probably be attributed to the fact that substantially less time is spent in the classroom by U.S. children. They average 178 days per year in school compared with 240 days for Japanese and Taiwanese children, who attend one-half day on Saturday in addition to weekdays. However, observers noted that the Oriental children were much more attentive in the classroom, and so the difference in sheer numbers of days in attendance probably does not explain all of the U.S. inferiority.

The most important new finding of this study concerns the crucial contribution of parental expectations to school achievement. In general, in spite of the demonstrably poorer school performance of the U.S. children, their mothers were more satisfied with their children's performance than were Japanese and Taiwanese mothers. More than 40% of the U.S. mothers reported themselves as "very satisfied" compared with fewer than 10% of the Oriental mothers. Also, the U.S. mothers tended to rate ability as more important than effort in school achievement and, accordingly, placed a much lower value on homework than did the Oriental mothers, who rated effort as more important than ability.

—Melvin H. Marx

Developments in Europe. The European Federation of Professional Psychologists Associations (EFPPA), a new international organization, held its third general assembly in Vienna. The EFPPA was founded in 1981 to promote communication between member associations in Europe, to further the establishment of ethical codes of practice among psychologists' associations, and to promote the use of psychology to assist in the relief of distress of those to whom psychology offers a service. F. M. McPherson of the British Psychological Society (BPS) was the first president of the EFPPA. The Netherlands Institute of Psychology provides the secretariat.

In the United Kingdom the BPS adopted a revised code of conduct for psychologists. Unlike in the U.S. and many other Western countries, psychology is an unregistered profession in the U.K. Thus no license is required for anyone wishing to practice as a psychologist. The BPS was seeking government approval to maintain a voluntary register of bona fide practitioners under the terms of its Royal Charter of incorporation. This was seen as a means of protecting members of the public by ensuring that when they consult a psychologist the person concerned is qualified. Similar efforts to establish registers of adequately trained psychologists were taking place in other countries within Europe.

During the past year a new Group for Teachers of Psychology was set up within the BPS. This action reflects the concern of British psychologists to encourage high standards among those teaching psychology both to trainee psychologists and to other professional groups. The Council of the BPS also issued a major position statement entitled "Psychological Aspects of Nuclear War," which was published on a worldwide basis by John Wiley & Sons, Ltd. In 1984 the BPS added a new periodical entitled *Guidance and Assessment Review* to the list of learned journals it publishes.

—Colin V. Newman

See also Feature Article: PHOBIAS: FEAR OUT OF CONTROL.

Space exploration

The space shuttle program of the United States continued to perform successful manned missions during the past year, and three Soviet cosmonauts set a new record for time in space by a single crew. Several nations were developing probes to observe Halley's Comet in 1986, and the U.S. completed the first probe designed to enter the atmosphere of an outer planet.

NASA

Robert Stewart appears to hover over the ocean as he tests the nitrogen-propelled Manned Maneuvering Unit during the "Challenger" space shuttle mission that took place in February 1984.

Manned flight

The U.S. space shuttle orbiters "Challenger" and "Discovery" during the past year placed seven commercial satellites in orbit, and U.S. astronauts repaired in orbit the Solar Maximum Mission (Solar Max) solar physics satellite. Two communications satellites were captured in orbit and returned to the Earth for reflight. The Manned Maneuvering Unit (MMU) was flight-tested and proven during the satellite repair and retrieval missions. Three Soviet cosmonauts lived aboard space station Salyut 7 for eight months to set a new record for time in space for a single crew.

Space shuttle. Space shuttle orbiters "Challenger" and "Discovery" hauled commercial and scientific payloads into space during five flights in 1984. The Manned Maneuvering Unit was test-flown and then put to work as astronauts left the orbiter and went into space to repair an ailing solar physics satellite and retrieve two communications satellites that were not in the proper orbits. Techniques for refueling satellites in orbit were tested during one flight, which also included the first space walk by a U.S. woman. The first commercial payload specialist and the first Canadian payload specialist were among 1984 shuttle flight crews.

On Feb. 3, 1984, the "Challenger" was launched from the Kennedy Space Center in Florida. Aboard were Vance Brand, Robert L. Gibson, Bruce McCandless II, Ronald E. McNair, and Robert L. Stewart. The Western Union Westar 6 and Indonesia's Palapa B-2 communications satellites were deployed, but both failed to reach geosynchronous orbit (an orbit in which a satellite remains over the same place on the Earth) because of identical failures of their upper stage rocket nozzles. The West German Shuttle Pallet Satellite was not deployed because of problems with "Challenger's" remote manipulator arm. McCandless and Stewart test-flew MMUs untethered up to 90 m (300 ft) away from "Challenger." For the first time a shuttle orbiter landed at the Kennedy Space Center Shuttle Landing Facility on February 11.

The "Challenger" again went into orbit on April 6, 1984. The crew consisted of Robert L. Crippen, Francis R. Scobee, Terry J. Hart, James D. van Hoften, and George D. Nelson. Among their major achievements, was the deployment of the 9,550-kg (21,000-lb) Long Duration Exposure Facility, to be retrieved during a 1985 flight. Nelson and van Hoften captured the Solar Max satellite and brought it into the "Challenger's" cargo bay for replacement of the attitude control system and the coronagraph/polarimeter main electronics box. Solar Max was then released into orbit to continue solar physics measurements. The "Challenger" landed on April 13 at Edwards Air Force Base in California after a weather "wave-off" from Kennedy Space Center.

The shuttle orbiter "Discovery" began its first flight on August 30 when it lifted off from the Kennedy Space Center. Henry W. Hartsfield, Jr.,

George Nelson (lower left) makes a first excursion to the damaged Solar Max satellite in April 1984. Later Nelson and James van Hoften captured the satellite and brought it to the space shuttle for repairs.

NASA

NASA

Dale Gardner (left), holding a "For Sale" sign, and Joseph Allen stand near the Westar 6 communications satellite. They had helped retrieve it from a useless orbit and bring it to the space shuttle "Discovery" for return to the Earth, where it was to be repaired and resold.

Michael L. Coats, Judith A. Resnik, Steven A. Hawley, Richard M. Mullane, and Charles Walker—the first commercial payload specialist—comprised the crew. Satellites deployed included the Satellite Business Systems SBS-4, U.S. Navy Leasat, and AT&T's Telstar 3C. The astronauts also tested the 31-m (102-ft)-long solar array panel. "Discovery" landed at Edwards Air Force Base on September 5.

On October 5, the "Challenger" again went into orbit. The crew consisted of Robert Crippen, Jon McBride, Sally Ride, Kathryn Sullivan, David Leestma, U.S. Navy oceanographer Paul Scully-Power, and Canadian payload specialist Marc Garneau. The Earth Radiation Budget Satellite was deployed, and the Large-Format Camera and Shuttle Imaging Radar-B were operated for Earth survey mapping. Sullivan and Leestma spacewalked to connect valves and tubes in the cargo bay for tests of inflight transfer of hydrazine propellant, in anticipation of in-orbit refueling of satellites in the future. "Challenger" landed at the Kennedy Space Center on October 13.

"Discovery" was launched on November 8, this time with Frederick Hauck, David Walker, Anna Fisher, Dale Gardner, and Joseph Allen on board. The astronauts deployed Canada's Telesat H and the Leasat communications satellites. Allen and Gardner flew manned maneuvering units to retrieve the Westar 6 and Palapa B-2 satellites that had failed to reach geosynchronous orbit after deployment from the "Challenger" flight in February. The mission was the first to include the 3M Corporation's Material Science Laboratory, designed to investigate the growth of semiconductor crystals in space. "Discovery" landed at the Kennedy Space Center on November 16.

"Discovery" again lifted off from the Kennedy Space Center on Jan. 24, 1985. All five crew members, Capt. Thomas Mattingly, Lieut. Col. Loren Shriver, Lieut. Col. James Buchli, Maj. Ellison Onizuka, and Maj. Gary Payton, were U.S. military officers. The mission was undertaken for the U.S. Department of Defense, and no information on the payload was released to the general public. On January 27 "Discovery" returned to the Kennedy Space Center.

After a delay of more than a month because of problems with the satellites it was to deploy, "Discovery" went into orbit again on April 12. The crew included Air Force Col. Karol Bobko, Navy Capt. S. David Griggs, Navy Comdr. Donald Williams, M. Rhea Seddon, Jeffrey Hoffman, and Jake Garn, a U.S. senator from Utah. During the mission the astronauts successfully launched one communications satellite, but a second one failed to operate. Hoffman and Griggs made a space walk to the satellite in an effort to activate it, and Seddon used the shuttle's mechanical arm for the same purpose, but the satellite did not respond. The shuttle returned to the Kennedy Space Center on April 19.

Soviet manned flight. Soviet cosmonauts established a new spaceflight duration record during 1984 with a 237-day occupation of space station Salyut 7. Two cosmonaut crews and five unmanned resupply spacecraft joined the space station during the eight-month stay.

Soyuz T-10 with a crew of Leonid Kizim, Vladimir Solovyev, and Oleg Atkov was launched on Feb. 8, 1984, from Tyuratam Cosmodrome and docked with Salyut 7 the following day. A two-man crew had inhabited the space station for 211 days during 1983. On February 23 the Progress 19 resupply spacecraft carried fuel, scientific equipment, and mail to Salyut; undocked from the space station on March 31, it entered the atmosphere and vaporized on April 2. Four other Progress spacecraft serviced the Salyut 7 during the cosmonauts' stay.

On April 3 Soviet cosmonauts Yury Malyshev and Gennady Strekalov, and Rakesh Sharma of India docked with Salyut 7 in the Soyuz T-11. They returned to the Earth April 11 near Arkalyk, Kazakhstan, aboard Soyuz T-10.

Soyuz T-12 was launched July 17 and docked with Salyut 7 on July 18. The crew consisted of Vladimir Dzhanibekov, Svetlana Savitskaya, and Igor Volk. Savitskaya made the first space walk by a woman, during which she tested space welding and construction techniques. The three cosmonauts then returned to the Earth in Soyuz T-12 on July 29.

Kizim, Solovyev, and Atkov ended their long mission by returning to the Earth in Soyuz T-11 on October 2. They landed at Dzhezkazgan, Kazakhstan, in the Soviet Union after spending 237 days, 22 hours, and 10 minutes in space. The crew made six space walks during the eight months aboard the space station in order to install additional solar panels for generating electrical power and to repair a leaky fuel manifold.

—Terry White

Space probes

As 1984 dawned, major space probe activity around the world centered on preparations for observing Halley's Comet. Probes from the U.S.S.R., European Space Agency (ESA), and Japan were readied and tested before being launched into trajectories toward the comet when it makes its closest approach to the Sun in March 1986. The U.S. National Aeronautics and Space Administration (NASA) continued developing its core program for small- and moderate-scale planetary missions to be flown throughout the remainder of the century. The program also included planning for the preliminary assay of near-Earth resources on the Moon.

Targeting Jupiter. During the year construction of the Galileo space probe was completed. It was to be the first probe to enter the atmosphere of an outer planet, Jupiter. Galileo was scheduled to penetrate 675 km (400 mi) into the atmosphere of the planet. The probe is 1.2 m (4 ft) in diameter and weighs 2,555 kg (5,622 lb). It was to be carried into low Earth orbit in May 1986 by the space shuttle. From there it would be launched onto its Jupiter trajectory by the first "widebody" Centaur injection stage rocket which has 50% more propellant capacity than earlier models. The probe was due to arrive at Jupiter on Aug. 25, 1988.

Galileo's descent module, weighing 188 kg (414 lb), was designed to enter the atmosphere at 184,000 km/h (115,000 mph) and experience maximum g forces equal to 350 times the gravity of Earth. As the module descends by parachute into Jupiter's atmosphere, it will study the planet's three main cloud layers. In doing so it could well encounter hurricane winds up to 322 km/h (200 mph) as well as heavy rains at the base of the water clouds believed to ex-

Soviet cosmonauts Leonid Kizim, Vladimir Solovyev, and Oleg Atkov parachute to the Earth in the Soyuz T-11 after spending a record 237 days in space aboard the Salyut 7 space station.

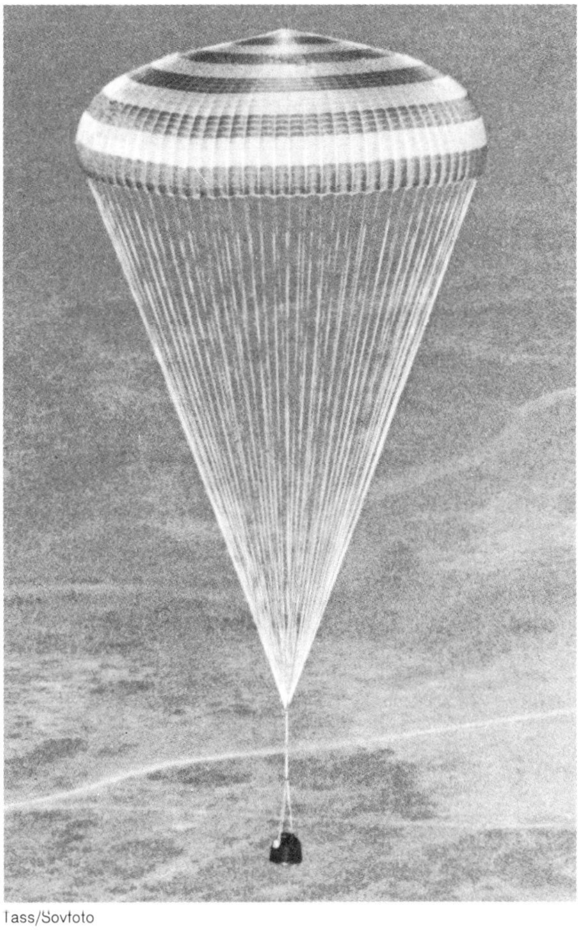

Tass/Sovfoto

UPI/Bettmann Archive

Rakesh Sharma of India (left) joined Soviet cosmonauts Yury Malyshev (center) and Gennady Strekalov in a mission to dock their Soyuz T-11 spacecraft with the Salyut 7 space station in April 1984. They returned to Earth in the Soyuz T-10.

ist on the planet. Seven instruments in the module will measure the chemical composition and physical state of the atmosphere and clouds.

The descent module will separate from the orbiter approximately five months before arrival at the planet and then follow its own trajectory into Jupiter's atmosphere. After entry the module's mission will be completed in 60 to 75 minutes. The seven instruments, weighing 28 kg (62 lb), will study the atmosphere of the planet and try to determine its chemical composition, to obtain a density profile from initial detection to the end of the mission, location and structure of clouds, and the existence and characteristics of lightning.

The orbiter, which will circle Jupiter at an altitude of 160,930 km (100,000 mi), will receive telemetry transmissions from the descent module and relay them to the Earth. Immediately after the descent module stops sending data to it, the orbiter will be oriented for a close flyby of Jupiter's satellite Io in an effort to gain information about its atmosphere. In addition to sending data to the Earth in real time, the orbiter will also store information for later retransmission.

The orbiter was designed to circle Jupiter for 20 months. During that time it will transmit to the Earth at least 50,000 high-resolution pictures of the planet and its satellites Io, Europa, Ganymede, and Callisto.

As the year ended, NASA was still trying to decide whether to divert Galileo to a flyby of the asteroid Amphitrite, an object the size of the state of West Virginia. While the change in trajectory to Jupiter would mean a slightly degraded mission at Jupiter, some scientists believe that the asteroid flyby would be worth the cost. A decision was not expected to be made until the scheduled launch date or within two months after launch.

In reviewing the Galileo mission Nick S. Vojvodich, program manager at the NASA Ames Research Center, Mountain View, Calif., said: "I think we will have learned more in the five-hour journey from the upper regions of the atmosphere to deep below the cloud layers than in all recorded history from all spacecraft that have preceded us."

Also during 1984 the Soviet Union evinced "great interest" in sending cosmonauts on a mission to Jupiter, where they would land on and explore one or more of the planet's satellites. The proposal was revealed by Konstantin Feoktistov, a spacecraft designer and former cosmonaut. He stated that his country could launch such a mission in 10 to 15 years if necessary.

Probing Venus. Early in the year the Soviets revealed more of the modifications made on their Venus probes Venera 15 and 16, launched in June 1983. Their primary mission was the radar mapping of the planet. Changes were made in the capacity of the propellant tanks, the area of the solar cell panels, and the diameter of the reflector of each probe's pencil-beam antenna. Additionally, special, airtight compartments were installed for the side-looking radars and other new instruments. It was also revealed that the radio altimeter of the Veneras had an accuracy of 50 m (164 ft).

The U.S.S.R. Academy of Sciences Institute of Radio Engineering and Electronics described the process for analyzing data from the radar mapping instruments of the probes. Each day the probes transmit displays of the Venusian surface in sectors that are 7,000 km (4,300 mi) long and 150 km (90 mi) wide. Scientists at the institute said that the amount of data that the Venera can acquire in 15 to 16 minutes requires 12 hours of processing at the institute plus another 4 hours for checking data from the radio altimeters aboard the probes. The radar mapping will make use of a map prepared by U.S. space scientists based on data from altimeter readings from U.S. space probes.

The Soviet probes Vega 1 and 2 were launched on Dec. 15 and 21, 1984, respectively, for a flyby mission to Venus and then a trajectory that will take them to a rendezvous with Halley's Comet in June 1986. As each probe passes Venus, it will drop off a descent module designed to land on the Venusian surface.

Each module, in turn, will release a balloon into the atmosphere to drift at an altitude of 54 km (33.5 mi) above the surface of the planet. Aboard each balloon will be a mephelometer developed by scientists at NASA Ames Research Center in cooperation with French space scientists. The instrument was designed to measure aerosol distribution. John A. Simpson at the University of Chicago developed a simple cometary dust detector that will also be aboard the Vega probe. In an agreement with NASA the Soviet balloons will be tracked by that U.S. agency's radar installations in Spain, Australia, and the U.S. These tracking instruments will receive telemetry signals from the balloons.

The U.S. also planned to assist the Soviets in analyzing data that will help them in targeting Vega landing sites. The two landers were aimed at points on the Venusian surface about 500–1,000 km (300–600 mi) apart.

During the year development of the U.S. Venus Radar Mapper (VRM) probe continued. The substitute for the previously planned Venus Orbiting Imaging Radar (VOIR) probe realized cost savings of some $335 million by a variety of means, including the use of hardware left over from other probe programs. However, the VRM retained the primary mission of the VOIR: a global map of Venus with a resolution sufficient to permit a geological analysis of the planet's surface.

The VRM was to be placed in an elliptical orbit around Venus ranging from 300 km (185 mi) to more than 8,000 km (4,900 mi) from the planet's surface. The planned angle of orbital inclination was designed to be between 86° and 88°. The closest approach to the planet was to be at about 10° N latitude in order to cover as much of the northern polar region as possible. Project scientists believe that more than 90% of the Venusian surface can be mapped by the VRM.

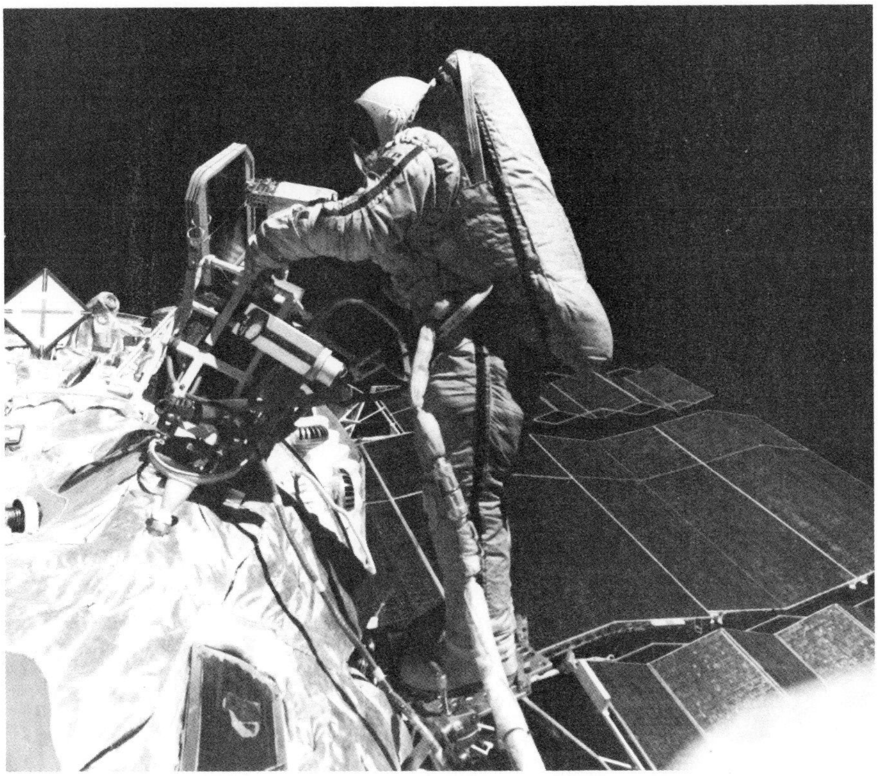

Tass/Sovfoto

In July 1984, during the first space walk ever made by a woman, Svetlana Savitskaya of the Soviet Union tests a newly developed hand tool designed to perform such operations as cutting, welding, and soldering in space.

Photos, NASA

Galileo space probe, the first designed to enter the atmosphere of an outer planet, Jupiter, was completed in 1984 (left). The drawing at the right is an artist's conception of Galileo with its Centaur injection rocket after being launched in 1986 from the space shuttle on its trajectory to Jupiter. Its 4.8-meter (15.7-foot)-diameter antenna (center) will then be unfurled.

Each 25 by 16,000 km (15.5 by 9,900 mi) swath from the probe was to be processed to produce images covering areas of approximately 20,200 sq km (7,800 sq mi). The VRM will have a radar resolution of about 200 m (656 ft) at the closest approach to Venus with obviously less definition as the probe moves farther away. Its altimeter was designed to provide data with a vertical resolution of approximately 30 m (98.4 ft).

Probing Mars. NASA began preliminary planning for a Mars probe, instrumented mainly to try to detect the presence of water on the planet. Cooperation with the Soviet Union was a possibility for the mission, scheduled to be launched in 1990. Meanwhile, the Soviets continued planning for a Mars probe to be launched in 1986 or 1988. It was to have an imaging system and be placed on a trajectory that would send it close to the Martian satellite Phobos.

From the U.S.S.R. Konstantin Feoktistov stated that his nation had no plans for a manned landing on Mars: "I don't think that a flight to Mars will be carried out earlier than 10–15 years from now. There is no sign of the kind of goals that would make a flight to Mars necessary." However, he added that if unmanned probes to Mars found evidence of life, the U.S.S.R. would consider a manned mission to that planet.

The Sun as a target for probes. A signal detected from late 1983 through early 1984 from Voyager 1, which passed Saturn in 1980 and then headed out of the Solar System, provided scientists with what may be the first remote observation of an area where the interstellar and solar cosmic winds collide. William Kurth of the University of Iowa made the analogy: "If you think about a guitar and you pluck the strings, you don't really hear the strings vibrating. You hear sound waves that propagate through the air to your ear." He continued, "In this analogy, the electron plasma oscillations are the vibrating string, and the radio waves we detect are the sound waves." Studies by Kurth and others concluded that the source of the signal was some 6.4 billion km (4 billion mi) from Earth.

In other news concerning solar probes the ESA International Solar Polar Mission (ISPM) was placed into storage in a specially built canister filled with the inert gas nitrogen. The action was taken because of changes in the U.S. space shuttle, the launch vehicle for the probe. Scientific instruments were removed from the probe before it was placed in storage. Removal of the probe from its container was planned for March 1985, with the replacement of its instruments and subsequent requalification for the mission set for May 1986.

If all goes as planned, the ISPM will begin a 14-month trajectory toward Jupiter, where that planet's gravitational field will deflect it into a high-inclination trajectory out of the ecliptic plane (the plane of the Earth's orbit extended to meet the celestial sphere) toward the Sun. Some two and a half years later the probe will pass over one of the poles of the Sun. About eight months later it will recross the

Tass/Sovfoto

Vega 1 probe, shown in the drawing above, and Vega 2 were launched in December by the Soviet Union for a flyby mission to Venus. The probes will then rendezvous with Halley's Comet in June 1986.

plane of the ecliptic on its way over the other pole.

ESA planned to provide the spacecraft and half its scientific instrumentation, while NASA was to supply the other half of the instrumentation, the radioisotope thermonuclear generator for electrical power, the space shuttle launch vehicle and Centaur injection rocket, and probe-tracking by the deep space tracking network. The launch delay of the ISPM led to a launch conflict with Galileo. NASA promised ESA that the ISPM would have priority and that Galileo would be scheduled for a later launch.

In the autumn of 1984 ESA and NASA announced that the name ISPM had been changed to Ulysses. The name was chosen not only with reference to Homer's mythological hero but also to Dante's description of Ulysses' urge to explore "an uninhabited world behind the Sun."

Cometary probes. In April 1984, a meeting of the International Technical Commission took place in Prague, Czech. Attended by six Soviet-bloc nations as well as Austria, West Germany, and France, it was a review of the state of technological progress, largely in the area of instrumentation developed for the Vega probe. Scientists from each country presented papers on the status of that nation's progress in completing specific instrumentation. The work of all nations was coordinated by the Soviet Academy of Science's Institute of Space Research.

Early in 1984 a working group of scientists from the U.S. and Europe recommended to NASA that it launch a probe to Comet Kopff in July 1990, to rendezvous with it in May 1994. The comet orbits the Sun every six and a half years. Unlike other cometary probes that are designed to pass close to the comet or pass through its tail, this satellite would be placed into an orbit that would match that of Comet Kopff. It would study the comet from a distance of 10 km (6.2 mi) and would travel with it for several years. The probe could also be placed on a

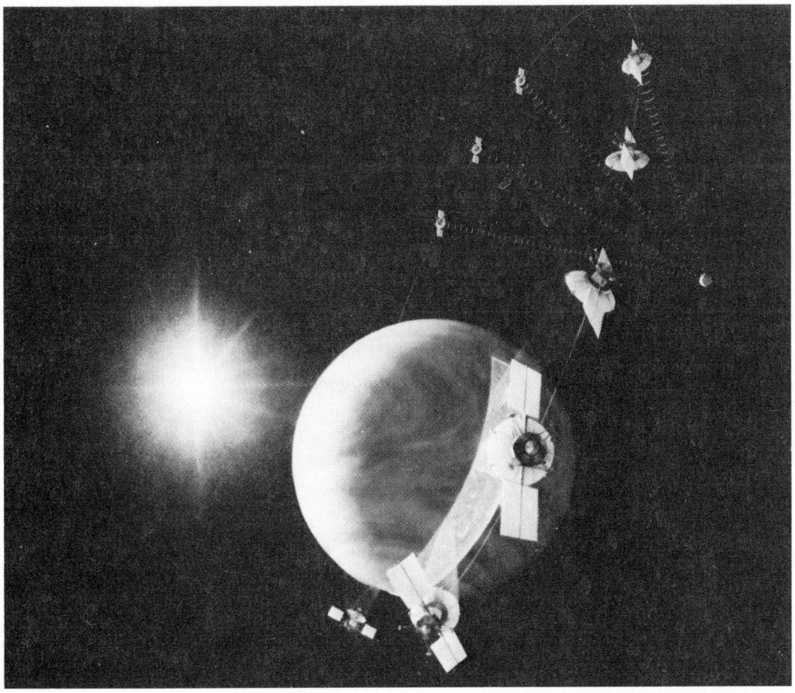

NASA

The Venus Radar Mapper, scheduled for launch in 1988, is being designed to penetrate the thick cloud cover of Venus and map more than 90% of the planet's surface. The probe will be placed in an elliptical orbit and will achieve a radar resolution of about 200 meters (656 feet) at its closest approach to Venus of 300 kilometers (185 miles). The probe's antenna was designed for dual use as a radar transceiver to collect mission data and as a communications link with Earth.

trajectory that would permit a flyby of the asteroids Lucia and Namaqua.

While the U.S. had no probe for a mission to Halley's Comet, it called into service for limited studies of the comet the Pioneer Venus Orbiter, launched in 1978 and still providing data on the Venusian atmosphere. All observations of the comet would be made from the orbit of Venus. Pioneer Venus was in a favorable position to make such observations. Both Venus and Halley's Comet will be on the far side of the Sun from Earth when the comet comes closest to the Sun. Thus, direct observation from the Earth at that time will be impossible.

In preparation for the observations of Halley's Comet, which will take place from six to eight weeks in February and March 1986, Pioneer Venus performed a dress rehearsal using Comet Encke. In April mission controllers at NASA's Ames Research Center tilted the probe 37° so that its ultraviolet spectrometer could observe the comet. The instrument was also able to make observations of the star Spica. The sensor observed the two bodies for a day and a half.

Somewhat to the surprise of scientists at Ames, it appeared that the comet was losing water at a rate three times higher than expected based on earlier measurements. Analysis of data from the satellite International Ultraviolet Explorer, which observed Encke from the end of April to mid-May, confirmed the findings of Pioneer Venus. It was theorized that the loss could be explained by the way that dust and ice are distributed in the head of the comet. Since comets are not smooth, round objects, they have hills and valleys. As they rotate in passing the Sun, various features are exposed. Those with more ice than dust boil off more water.

Comet Encke is believed to be about 2 km (1.2 mi) in diameter. Observations by Pioneer Venus were made while the comet was approximately 85,300,000 km (53,000,000 mi) from the Sun and Pioneer Venus was 109,450,000 km (68,000,000 mi) distant.

The exercise proved invaluable in perfecting the techniques for observing Halley's Comet and in calibrating instrumentation on the Pioneer Venus orbiter. Once it is tilted as it was for viewing Comet Encke, the Venus probe will be able to observe the gases and dust emanating from Halley's Comet. Data from the orbiter's ultraviolet spectrometer will characterize the comet's gas composition, its water vaporization rate, and its dust-to-gas ratio.

In Europe on July 12, ESA signed an agreement with Arianespace for the launch of ESA's Giotto probe to Halley's Comet in July 1985. The probe was scheduled to arrive at the comet on March 13, 1986. It will then pass within 500 km (310.5 mi) of the coma of the comet on the sunward side. The encounter phase will last for only four hours.

Scientific objectives of the Giotto mission include determination of components in the coma; study of the chemical reactions and physical processes that take place in Halley's ionosphere and atmosphere; determination of the composition of dust particles accompanying the comet; measurement of rate of gas production, dust flux, and size/mass distribution in order to calculate the dust-to-gas ratio in the comet; study of the system of plasma flows that results from the interaction of plasma from the comet and the solar wind; and the taking of as many photographs as possible in order to determine the size of the nucleus as well as its mass and rate of rotation.

Elsewhere in the world Japan continued work on its two probes for Halley's Comet. MS-T5 was scheduled for launch from the nation's Uchinoura Space Center in January 1985. A companion probe with different instrumentation was scheduled for launch from the same site in August 1985.

—Mitchell R. Sharpe

See also Feature Articles: THE COSMOS ACCORDING TO IRAS; THE INFLATIONARY UNIVERSE.

Transportation

Wind power as auxiliary propulsion for merchant ships was extended during the past year to transoceanic vessels. And in the United States the deregulated environment for transportation stimulated both operational and technological innovations by carriers of all modes. The adequate supplies of petroleum fuels that were available at stable prices slowed somewhat the fuel-conservation technology program, thus freeing research and development funds for other areas.

Air transport. Manufacturers during recent months were divided on the issue of future aircraft sizes. Some believed that planes must be larger to handle the expected growth in traffic under the conditions of more constricted airways traffic control, while others maintained that smaller aircraft would be preferable because they could be operated with higher load factors, on more frequent schedules, and with greater flexibility. The current trend appeared to be toward smaller transports, in the range of 150–130 seats or less.

To illustrate, several airlines canceled orders for the new-generation B-767, a twin-engine, widebody, 211–255-seat transport made by the Boeing Co., and ordered instead Boeing's B-737-300, a smaller, twin-engine, 128–140-seat aircraft. Airbus Industrie, concerned about lagging sales of its A300-310, widebody, 200-seat-plus jets, won approval from its European backers—Great Britain, West Germany, France, and Spain—to produce the A320, a twin-engine, 150-seat transport equipped with ad-

Avtek Corporation

The all-composite Avtek 400 turboprop airplane is made mostly of Nomex sandwiched between layers of Du Pont Kevlar aramid fiber. A canard for achieving stability is mounted on top of the fuselage.

vanced engines built especially for aircraft of that size. Plans called for its initial flight in early 1987, with deliveries by the spring of 1988.

The new-generation B-737-300 entered passenger service during the year. Its major technological gains resulted from its CFM56-3 engine, built by CFM International, a joint venture of General Electric Co. and SNECMA of France. Boeing claimed that the aircraft's noise is actually a whisper and is "pretty much confined to the airport boundaries." The transport consumes less fuel than the 737-200, even though it carries 21 more people.

Boeing began assisting airlines in modifying B-767s in anticipation of final approval by the Federal Aviation Administration (FAA) of a proposal that would allow twin-engine transports to operate over long-distance overwater flights, *i.e.*, transatlantic routes. The FAA, in recognition of the safety and economy of new-generation, twin-engine aircraft, would allow them to operate over such routes if they could reach a suitable airport within two hours at single-engine speed; the present rule requires one hour. The Boeing equipment provided Trans World Airlines for this purpose was designed to strengthen the landing gear and lower fuselage; add another automatic direction finder, hydraulic power generator, and fire-suppression system; and install a high-frequency communications system and door-mounted evacuation slides and rafts.

The highly publicized test crash of a B-720 jet airliner that had been designed primarily to prove that an additive—antimisting kerosene—could create a fire-free fuel unexpectedly ended in a huge fireball that destroyed the airliner's cabin. The additive had previously proved successful when used in fuel which was powering smaller aircraft that crashed on a horizontal track. The B-720 test program originated in 1978 and cost the sponsoring FAA and U.S. National Aeronautics and Space Administration (NASA) $11.8 million. It included many other tests that should provide valuable information for better aircraft safety in the future.

As examples, special fabrics and plastics were used in the test craft's seats and interior to evaluate their "fire-blocking" ability—thus giving passengers more time to evacuate before the materials deteriorate under fire into deadly carbon monoxide and cyanide gas. Special fire-resistant windows were used to protect passengers from outside flames. The 72 dummies seated in the plane, some in rearward-facing seats and others fitted with electronic impact sensors, were screened by onboard high-speed cameras. The cameras were in protected cases that in most instances allowed the film to be preserved.

A few months before the test crash the FAA adopted a new cabin safety rule giving airlines three years to install a layer of fire-resistant material over the foam core of passenger seats. The material must delay by about 40 seconds the "flashover" point, the time at which flammable vapors trapped in the cabin ignite and consume the oxygen needed for passenger survival. The FAA earlier proposed such a rule only to find that the most highly recommended fabric released very toxic fumes when exposed to fire.

According to the McDonnell Douglas Corp. the next-generation transport may have to be powered by advanced-technology propfans that in order to compete with redesigned existing jetliners must be at least 20% more fuel efficient. A spokesman claimed that an MD-80 (formerly DC-9 Super 80) would burn about 40% less fuel if powered by propfans. Emphasis was being placed on minimizing noise and vibration, two liabilities of past propeller-driven planes.

General Electric Co. planned full-scale ground tests in 1985 of an unducted fan propulsion system that would use a core engine to drive two turbines rotating in opposite directions (contrarotate) on the same shaft, with each driving a set of contrarotating fans. General Electric claimed that the arrangement would remove airflow swirl caused by the forward set of rotating blades so as to boost efficiency as much as 8%. Each blade would be reversible for use in "breaking" when landing. Plans called for flight tests in late 1986.

British Aerospace began production of what it called the first Advanced Turboprop (ATP) transport, which was scheduled to be placed in commercial service in late 1987. ATP was a stretched version of that company's HS. 748 with 64 seats in comparison with the latter's 48. As of early 1985 it was the only transport being developed with that seating capacity.

AP/Wide World

B-720 airliner flown by remote control bursts into flames when it is deliberately crashed at Edwards Air Force Base in December 1984. The crash was a test primarily of an antimisting kerosene, which, it had been hoped, would create a fire-free fuel.

The ATP was to use two turboprop engines built by Pratt & Whitney of Canada, each driving six-bladed propellers developed jointly by British Aerospace Dynamics and United Technologies.

Highway transport. With the U.S. Congress approving two sharp increases in federal highway diesel fuel taxes—from 4 cents per gallon to 9 cents in 1982 and 15 cents in 1984—the trucking industry again began seeking ways to cut consumption; fuel costs, according to industry spokesmen, represent from 35 to 50% of total operating costs of a major trucking company. Most truck-trailer combination vehicles consume diesel fuel at a rate of approximately 8 km (5 mi) per gallon, with newer models averaging 7 mpg. A potential savings of $10,000 a year per unit was predicted if the present overall average of 5.5 mpg could be increased to 10 mpg. This would amount to a total savings of $12.5 billion a year for the nation's trucking industry.

Gelco Truck Leasing, in a cooperative project with International Harvester Co. (tractors), Cummins Engine Co. (engines), and TRW, Inc. (computerized data), completed a 4,481-km (2,783-mi) Los Angeles-to-New York City fuel "Economy Run" with four tractor-trailer combinations of varying power and configurations. The purpose was to demonstrate the potential savings in fuel consumption and costs that could be achieved through technological innovations. One truck did not contain any special fuel-stretching technology and was used as a control vehicle. It averaged 6.02 mpg. The other three units, which averaged 8.46, 9.59, and 11.06 mpg, respectively, all contained different fuel-economy devices. These included cab extenders to reduce aerodynamic drag from the normally open space between the cab and trailer; side skirts to eliminate the drag under the trailer; the use of eight shorter and wider tires to replace the standard 16 dual tires (excluding the two steering tires); and the addition of a computerized information system that included an electronic speed/cruise control for 89 km/h (55 mph).

General Motors Corp. announced during the year that it would discontinue production of diesel engines for its mid-size, full-size, and luxury automobiles starting with the 1986 models. It said that this action resulted from rising costs of diesel fuel compared with gasoline, the plentiful supply of unleaded gasoline at stable prices, and the tougher new U.S. government standards on particulate emissions from diesel engines.

Deregulation of the trucking industry in most areas of customer services stimulated many innovations to help attract traffic. Some examples included an expandable and skeletonized (to cut weight) truck container chassis, produced by Vulcan Trailer Manufacturing Co., to handle units ranging from 6 m (20 ft) in length to the 14.5 m (48 ft) now permitted on most major highways; a convertible hopper trailer, built by the Wilson Trailer Co., with four hinged floor sections that can be used to reinforce the floor for heavy loads or that can be slid to form two sections with 70° sloping sides for self-unloading bulk cargo; and a computerized, audio-response freight-rating system by Ryder/P-I-E Nationwide, to permit customers to obtain freight rates in 90 seconds by furnishing the carrier four facts: origin and destination, ZIP codes, weight, and the freight's classification by the National Motor Freight Traffic Association.

Testing programs on electric cars continued on a modest scale, although major changes in conditions that had previously encouraged their development slowed research and development significantly. A spokesman for General Motors Research Laboratories said that General Motors had withdrawn its prediction that it would have an electric car on the consumer market in the late 1980s. He said that "the technology is not there" and a resurgence in interest would require a significant increase in the cost of gasoline or a reduction in its availability.

The major technological roadblock remained the lack of a battery capable of providing the power needed over a reasonable length of driving distance. Most batteries suited for this purpose must be replaced after 16,000 to 24,000 km (10,000 to 15,000 mi) at a cost of several thousand dollars. Programs to develop a long-lasting battery experimented with such components as zinc-nickel oxide, zinc-chlorine, zinc-bromine, nickel-iron, and hot molten salt.

Battery makers acknowledged that there continued to be difficulty in linking effectively the main battery pack and the smaller 12-volt battery that powers accessories such as headlights, windshield wipers, radios, and air conditioners. The smaller battery drew off so much power from the main battery that it prevented the use of air conditioners in electric cars. (*See* Feature Article: BATTERIES: A TECHNOLOGY RECHARGED.)

Pipelines. Operators of oil pipelines, the most automated and computerized of the transport systems, continued seeking ways to improve flow and increase throughput at the lowest possible cost. One method was the use of additives that reduce drag through the pipeline. In 1979 the CDR 101 flow improver of Conoco Inc. was successfully tested as a way to boost throughput in the huge trans-Alaska crude oil pipeline. Conoco later improved this chemical additive by cutting its concentration, making it appropriate for use in products pipelines.

An improved CDR 102 flow improver was put into commercial use in 1982. It proved effective for increasing output in lines from offshore oil platforms that often have insufficient capacity for short-term production demand and was used successfully in offshore platforms in the North Sea, off the Gulf Coast of the United States, and in the Bass Strait off the Australian coast.

Aircraft developed at the Massachusetts Institute of Technology won the Kremer World Speed Competition Prize in 1984 for making the first human-powered flight around a 1,500-meter (5,000-foot) course in under three minutes.

Steve Finberg

The drag-reducing additive is stored in a container and pressurized to permit flow into the injection pumps. These pumps raise the pressure to that of the pipeline—often in the range of 600 to 2,000 psi—through use of variable-speed drives. The additive is actually injected beyond the mainline pumps and downstream of constrictions such as control valves. The injection rate is monitored by a flowmeter, and the additive passes through check valves before entry into the pipeline.

Pipelines are cleaned with "pigs," so called because the first models caused a squealing sound when moving through the lines. These devices are inserted into the pipeline and move along with the flow of the product being transported. S.U.N. Engineering Inc. developed an all-plastic cleaner that it claimed was more efficient than previous models and that lasts three to ten times as long. This new "Super-Pig" is made of a special polyester pre-polymer produced by Mobay Chemical Corp. Manufactured in 37 sizes to permit cleaning of pipelines ranging in diameter from 5 cm (2 in) to 150 cm (60 in), it consists of two or four cups made of urethane mounted on a flexible but strong plastic shaft made of urethane elastomer, with steel wire brushes interspersed in some configurations. The edges of the cups and, when used, the carbon steel or polyurethane brushes scrape and clean the inside of the pipeline. Approximately every 50 mi (80 km) the pig is removed for cleaning and removal of debris.

The proponents of coal-slurry pipelines suffered a setback when the ETSI Pipeline Project—a proposed $3 billion, 2,900-km (1,800-mi) slurry line from coal mines in Wyoming to utilities in Oklahoma and Texas—ceased operations because of the withdrawal of financial backing and opposition from railroads. ETSI lost about $145 million over an 11-year period without laying a foot of pipeline. In 1984 railroads, environmentalists, and other groups again blocked legislation that would have given slurry lines federal eminent domain rights to cross railroad and other properties. The Slurry Transport Association as well as the National Coal Association said that the setbacks would not stop them from trying again, claiming that the technological and economic feasibility of such lines had been clearly established. They contended that removal of the interstate construction barrier would open the door to the building of a half dozen or more major slurry lines.

MRAZ Technologies Inc., which had been developing its "Pneumatic Extrusion Process" for 12 years, announced its availability for short-haul movements of bulk commodities such as coal, cereals, cocoa, and plastics. The firm claimed that short-haul coal movements could be made at a small fraction of the cost of moving them by present methods and that this process would be more economical than coal-slurry lines at long distances because of the costly watering and dewatering processes that are necessary when using the latter.

The MRAZ innovation is based on the principle of controlling the flow of air through a tube so that it can be used for "floating" bulk solids economically and with much less drag friction and abrasion than in a slurry line. It has been used for short-haul movements of phosphate rock. In 1985 MRAZ was seeking to build a 9.5-km (6-mi) pneumatic line from mines in West Virginia to finishing stations, claiming that it would cut the cost of current conveyor belt-truck movements by more than one-third.

Railroad transport. The Association of American Railroads, expressing concern about the ability of railroads to compete in the future in the new deregulated environment, launched a program to stimulate interest by rail equipment manufacturers to develop jointly a High-Productivity Integral Train. According to the association the train would be "designed as a total system to perform some specialized task, whether in the haulage of bulk commodities or intermodal traffic"—with potential savings in line-haul operating costs of 35% for coal and other bulk commodities and up to 50% for intermodal service (in which more than one type of transport equipment is used on a single journey). The train would ideally fit in with the current trend to create huge rail systems that are able to operate over extensive territory as through trains and thus avoid delays in classification and assembly yards. The association said that 13 suppliers had indicated plans to design a prototype train.

A ruling by a federal labor arbitrator allowing removal of cabooses in as many as 25% of through trains in the U.S. resulted in substantial cost savings on affected trains. The Union Pacific System maintained that safety would not be diminished by the action, because the functions of the caboose were being handled by a portable electronic device that fitted over the coupler of the last car for monitoring rear-end air brake pressure, status of train movement, and operation of the rear marker light. This information was then automatically communicated to the engineer. The Grand Trunk Western Railroad installed similar electronic rear-of-train monitors in its new cabooseless two-man piggyback train between Chicago and Detroit.

The Illinois Central Gulf Railroad became the first railroad to train locomotive engineers on a locomotive simulator, located at the Illinois Institute of Technology Research Institute in Chicago. A joint project of IIT and the U.S. Federal Railroad Administration, the simulator consists of a fully equipped locomotive cab and a separate control room. It can give trainees the realistic feeling of controlling trains having a wide range of configurations—up to ten locomotive units in front and as many as 150 freight cars of different sizes and weights.

The Burlington Northern Railroad and Rockwell International Corp. began a joint research and testing project to apply avionics technology for the purpose of developing faster and safer trains with lower fuel consumption and smoother handling. This would be done by advanced on-train computers and

High Speed Surface Transport undergoes testing in Tsukuba, Japan, a city developed for scientific research. The HSST is magnetically levitated about one centimeter (0.4 inch) above the track and rides on a cushion of air. The virtually noise-free vehicle is designed to achieve a speed of 301 km/h (187 mph).

AP/Wide World

a sophisticated communications system. An important feature was the capability of train dispatchers at central locations to determine, through satellite communications, specific train speeds and locations—reportedly within 1.6 km/h (1 mph) and within 45.5 m (150 ft). This information can then be instantly transmitted to an onboard computer to enable the train engineer to respond quickly.

The fastest growing area of rail freight transport continued to be container and trailer service on flatcars (COFC/TOFC), which was completly deregulated in the U.S. from the standpoint of rates and services. This stimulated such technological innovations as skeletonized flatcars to reduce weight, double-decker cars to increase capacity, and articulated cars that provide greater operating flexibility.

The trend toward use of double-stack COFC flatcars continued, with American President Lines, a major steamship company, starting service with them on special trains between Los Angeles and Chicago. Each of the four train sets already acquired is made up of 20 deep-well cars; each of these cars, in turn, consists of five articulated platforms able to handle two 12-m (40-ft) containers for a total of 200 containers per train. With such cars, operating productivity is almost doubled according to American President Lines, since no extra trainmen are needed and locomotive power is only slightly increased.

The use of such double-stack cars in Canada met with a mixed reception. Operating runs by both Canadian National Railways and CP Rail indicated potential usefulness, although problems such as the inability to load a full container on top of an empty one and the need for special loading/unloading equipment were cited. A partial solution may be the new flatcar designed by FMC Corp. Called TWIN-STACK, it is a five-unit, well-type version able to handle containers ranging from 6 to 13.5 m (20 to 45 ft) in length. The articulated car contains bulkheads at the ends of each unit to provide restraint for the upper container.

Another example of added flexibility in COFC/TOFC service was the introduction into service by the Canadian National Railways of new articulated flatcars able to handle trailers ranging in length from 12 to 14.5 m (40 to 48 ft). According to Canadian National it was the first North American carrier to buy the longer 14.5-m trailers. Canadian National also announced another first, the ability of the new cars to accommodate tri-axle trailers. The cars were to operate between Montreal and Toronto.

A special train transported 100 head of cattle from Jacksonville, Fla., to Amarillo, Texas, in two stacked containers on a flatcar. The containers, each measuring 12 m by 3.5 m by 2.4 m (40 ft by 12 ft by 8 ft), were equipped with automatic feeding and watering systems and had doors at each end for quick self-loading and unloading. Television monitoring provided 24-hour surveillance of the cattle. All arrived in good condition after the 2,900-km (1,800-mi), 52-hour trip, with less weight loss than 74 additional animals that made the same trip by highway, covering 2,495 km (1,550 mi) in 32 hours without facilities for food or water. With approximately 500,000 head of cattle moving each year from Florida to Texas, railroads called the prospects for recapturing traffic lost to the trucking industry "very promising."

Despite successes reported by government-backed high-speed rail passenger trains in Japan, Great Britain, and France, the outlook for such service in the U.S. was not favorable, according to a study by the Congressional Office of Technology Assessment. The OTA noted that such service, in addition to government subsidies, required very high population density—a characteristic of all those three nations' systems—that exists in the U.S. only in the Washington–New York–Boston corridor already served by relatively fast rail passenger service. The highly publicized and promoted plans for 200-km/h (125-mph) rail passenger service in California between Los Angeles and San Diego failed to obtain adequate funding, and the American High Speed Rail Corp. went out of business. Feasibility studies continued for a number of such projects, however, including one between New York City and Montreal; another between Las Vegas, Nev., and Los Angeles; and a third, in Florida, between Tampa and Miami by way of Orlando.

The more technologically advanced maglev train, which moves a fraction of an inch above the surface of track in a magnetic field providing both lift and propulsion, entered another stage of development. British Rail opened what it called the world's first maglev for revenue service. It connected the railroad's Birmingham station and the city's airport, a distance of 0.6 km (0.4 mi). While the speed was to be only 54.1 km/h (33.6 mph) for a one-way trip of 90 seconds, the operations should provide vital information for future large-scale projects. Test operations of a maglev train near Emsland, West Germany, backed by the government of West Germany and a consortium of West German companies, reported a world record speed of 302.13 km/h (187.66 mph).

Adequate supplies of diesel fuel and stable prices dampened the enthusiasm for development of a new coal-burning locomotive, but the General Electric Co., in cooperation with the Burlington Northern and Norfolk Southern railroads, continued to pursue a modest research and development program. Originally concentrated on developing a diesel-burning fuel laced with micropulverized coal, it was broadened to include fluidized-bed combustion and steam and gas turbines. A study by the California Institute of Technology's Jet Propulsion Laboratory, funded by the U.S. Department of Energy, predicted that direct burning of coal in railroad locomotives could produce billion-dollar energy savings.

Water transport. The first deep-sea sail-assisted freighter, "Aquacity," crossed the Pacific Ocean from Tokyo to Vancouver, B.C. The $11.5 million, 31,000-ton deadweight (dw) Japanese ship, built by Nippon Kokan, encountered headwinds for most of its crossing, which allowed use of its two computer-controlled sails for only 72 hours. Under normal North Pacific wind conditions the sails were expected to reduce average diesel fuel consumption by at least 10% as well as increase the ship's speed.

In France Pechiney S.A., a nationalized aluminum and metals group, and the Fondation Cousteau, a nonprofit marine ecology association, signed a pact to develop and promote "turbovoiles," or turbosails, as auxiliary power for vessels up to 40,000 tons dw. The two organizations claimed that their system would provide fuel savings three to four times that of Nippon Kokan's rigid-sail system. One prototype was tested successfully, and another was planned for a transatlantic crossing in 1985.

The French system employed the principle of the "magnus effect," used as long ago as the 1920s when it provided propulsion by the rotation of a cylinder in a moving substance such as air or water. The turbovoiles, however, use large stationary cylinders mounted on the top deck that contain openings on one side for air intake, supplemented by an intake fan; the air is then expelled over the top of the cylinder.

In the containership field the trend toward bigness continued. Barber Blue Sea started service with its SuperCarriers, which it described as "the world's most advanced cargo ships." The three second-generation SuperCarriers are 33.8 m (111 ft) longer than Barber Blue Sea's six original roll-on/off ships. Each can carry 2,400 6-m (20-ft) containers, 1,500 automobiles, and 72,480 cu m (2,560,000 cu ft) of baled cargo. A huge, 35-m (115-ft)-wide stern ramp permits rapid loading/unloading, and a crane on the weather deck can lift up to 40 tons of cargo at a time.

United States Lines put new containerships into service that could carry 4,400 6-m (20-ft) containers, and Sea-Land Service Inc. planned to order, in 1985, six new containerships able to carry 1,700 to 1,800 12-m (40-ft) containers. Sea-Land also announced that it would "jumboize" (slice in half and enlarge with new mid-bodies) 12 of its older containerships in order to increase carrying capacity by 37% and to reduce each vessel's cost per container-mile by 20%.

The U.S. Navy's long-term resistance to containerships—because they need sophisticated shoreside loading/unloading facilities and thus are not suited for service to most ports and in times of emergency—may be overcome by the use of special crane ships. Under a joint program with the U.S. Maritime Administration, a fleet of 11 such ships was planned. The first of the new ships had six cranes grouped in pairs on its starboard side. Each crane could lift 33 tons and handle a 12-m container. The cranes were designed to work singly or jointly. For example, a pair could lift an M-60 tank, and four could handle a 90-ton load. The remotely controlled cranes permit the loading and unloading of 300 containers a day.

The Port of Houston Authority put into service a new $4.7 million ship-loading system, which it claimed would reduce turnaround time for ships and allow handling of a far greater range of cargoes. It featured a $3.4 million ship loader able to handle up to 3,000 tons per hour. The loader can travel the entire length of the wharf, and the loading chute mounted on its boom can extend 20 m (66 ft) from the wharf. A new dust-suppression system being installed will enable the port to handle new cargo such as petroleum coke, soda ash, alfalfa pellets, and soybean meal. The new system pulls the dust-filled air into a scrubber system that sprays it with a mist, turning the dust into a sludge for easy discharging.

—Frank A. Smith

U.S. science policy

"I think it's positive, terrific, unique, the envy of the world," said Frank Press, president of the prestigious National Academy of Sciences. "It's the American giant awakening from its slumber, you might say." He was talking about the explosive growth of U.S. industrial investment in research and development (R and D). As science advisor to Pres. Jimmy Carter, he had tried to encourage greater collaboration between industry and government in the funding of research. Now it seemed to be happening without direct government intervention, and he liked what he saw.

His successor in the White House, George A. Keyworth II, saw further cause for optimism. "What's more important to me," he said, "is not just the magnitude of the investment but the fact that it's been more innovative. The best thing I've seen in my years here is a renewed vigor, vitality, and imagination in a great number, even a majority, of American industries."

Filling the funding gap. The Battelle Memorial Institute, a private research organization, forecast that total U.S. expenditures on R and D in 1985 would reach a new high of $107.3 billion, 10.6% more than the estimated total for 1984. According to its estimate the nation's industries would fund 51.3% of the research and development conducted in the U.S. in 1985, compared with 45.4% by the govern-

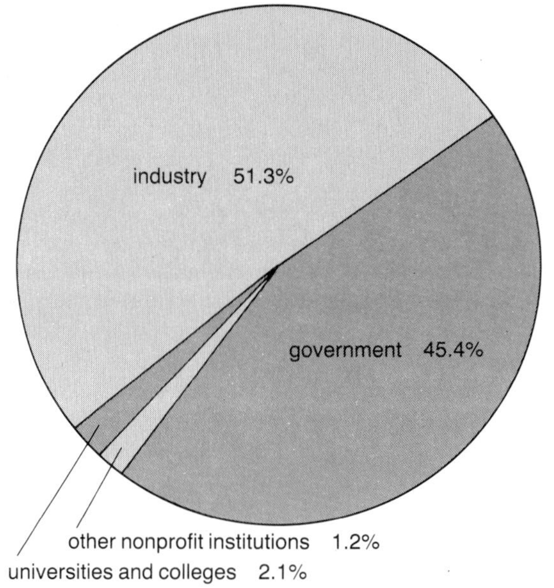

Of the $107.3 billion to be spent in 1985 on research and development in the United States, 51.3% will be funded by private industry, 45.4% by government, 2.1% by universities, and 1.2% by other nonprofit institutions.

ment, 2.1% by universities and colleges, and 1.2% by other nonprofit institutions. In addition, industry would perform nearly three-quarters of all the R and D that year.

Continuing the trend of recent years, much of the growth in research funding by the federal government could be attributed to national defense. Of the $52.7 billion included in Pres. Ronald Reagan's 1985 R and D budget, defense programs accounted for 70%, the highest proportion since 1962. The overall R and D budget increased a healthy 19% over 1984, but defense-related R and D jumped 26%. Funding of space activities kept pace with the overall growth rate, but R and D in energy, natural resources, and the environment fell behind.

Although scientific leaders who considered themselves political realists were thankful that scientific research overall was not doing too badly in a time of severe federal budget trimming, there were strong signals that some of the long-term costs would be high. Directors of national laboratories and university researchers expressed their anxieties to government agencies in Washington in terms of the ultimate costs to actual research productivity. For example, because new construction was virtually frozen in place by the Department of Energy, a new linear collider would start up in 1987 at the Stanford (Calif.) Linear Accelerator Center (SLAC) with a detector built a decade earlier and at lower energy than was currently available. As SLAC's Burton Richter put it: "It does not seem wise to spend so much money building new machines without at the same time making provision for the detectors required to use them properly."

The response in Washington, especially on Capitol Hill, was that physicists were being unrealistic. For example, a panel of the National Academy of Sciences had identified $1 billion of big new materials sciences facilities that "needed" to be built, and the Nuclear Science Advisory Committee had recommended nearly $1 billion of electron, heavy-ion, and kaon machines. All of these plus the Superconducting Super Collider, which was expected to carry a price tag of $6 billion by the time it was built in 1993 or 1994, upset some government officials.

The medical sciences, which had been granted only a tiny increase in the fiscal year 1985 budget—less than the rate of inflation—suffered a second blow after the budget emerged from the legislative process. The administration informed the National Institutes of Health that it would be permitted to award only 5,000 competitive research grants, rather than the 6,500 that had been authorized by Congress. The cutbacks affected medical studies on cancer, heart disease, immunology, arthritis, genetics, neurological disorders, and stroke. Thus was there reason to be even more grateful for increased R and D expenditures by industry, since a substantial fraction of it was being invested by pharmaceutical firms in fundamental research at major universities.

Another sign of industry's greater role in national R and D was the presence of some of its leaders in positions formerly open only to the academic community. For the first time in its 35-year history the National Science Foundation (NSF) was headed by an industrial scientist, Erich Bloch, an IBM Corp. vice-president. At the same time NSF's policy-making National Science Board was chaired by Roland Schmitt of General Electric, and the White House Science Council was headed by Solomon J. Buchsbaum, executive vice-president of Bell Laboratories.

Commercialization of science. Yet not everyone was enthusiastic about the boom in industrial R and D. William D. Carey, executive officer of the American Association for the Advancement of Science (AAAS), worried about all the enthusiasm for basic research as a tonic for productivity and the economy. He told the *New York Times:* "I worry about this because the rhetoric tends to pull the resources after it. For the time being, it's a honeymoon, and industry is responding by putting more money into the universities. But if this were to go on for six or eight more years, the mix and spread and versatility of basic research would narrow very dangerously. You could get an overemphasis on areas of basic research that are likely to produce near-term economic benefits."

Although most of the impetus for increased R and D funding by industry derived from its own need for innovation and productivity, there was also a substantial push from academic science itself. Donald Kennedy, president of Stanford University, saw a growing dependence on industrial support as a product of academic involvement in so-called Big Science. Heavy initial investments by universities in major research facilities, he said, created a 16-year capital crunch just as federal funding of academic science began to dry up. "We now find ourselves," he said, "caught in a mismatch between the needs and expectations of scientific research, on the one hand, and the willingness and capacity of the public to support it, on the other."

The mismatch had led to the very problems that many science-policy observers had come to fear. Academic science had always benefited from industrial support but, in the gentle words of a Twentieth Century Fund Task Force on the Commercialization of Research, "the new relationships differ in magnitude and extent from past relationships, introducing the possibility of conflicts of interest."

The conclusions of the task force were based, in part, on an exhaustive review conducted for the Twentieth Century Fund by Nicholas Wade, a member of the *New York Times* editorial board. Wade saw much that was potentially troubling. He found, for instance, that "there has been an abrupt shift in the attitude of [academic] researchers from open disdain for commercial involvement to implicit acceptance of it." And, Wade noted, commercial involvement eroded some of the basic self-regulatory processes that serve to protect science from fraud and error. He went on to tick off other dangers—pressure on universities to emphasize applied over pure science, a threat to the easy circulation of ideas and the broad availability of new techniques, and a new skepticism concerning the objectivity of scientific advice.

One further worry that may have bulked larger to practicing journalists was that, with increasing frequency, scientific advances were not announced through publication in peer-reviewed journals but at well-advertised press conferences. (Even journal publication was no guarantee of objectivity: *The New England Journal of Medicine* announced a policy that required disclosure by its authors of any financial interests they may have had in the research being reported, as well as the source of their funding.)

Erich Bloch, an IBM Corp. vice-president, was the first industrial scientist to be named head of the National Science Foundation.

National Science Foundation, Washington, D.C.

Science and the pork barrel. Pressures on university administrators to pick off a larger share of the federal research budget continued, in 1984, to tempt some away from the scientific tradition of peer review and toward the political tradition of the pork barrel. According to a "pork barrel scorecard" published by the AAAS in November 1984, universities electing to fund campus projects through congressional appropriations rather than the customary peer-review procedures had benefited by almost $100 million over the past two years.

Boston University demonstrated how the system could be made to work. Working with a Washington, D.C., consulting firm with excellent political connections, the university had nevertheless failed the year before to secure funding from the Department of Education for an engineering center. In 1984 it stayed with its consultants but changed its target agency. In August Sen. Edward M. Kennedy (Dem., Mass.) successfully proposed an amendment to build the center as part of a Department of Commerce program to "revitalize economically depressed areas."

Once more the mandarins of science spoke out against the practice. The American Physical Society issued this statement in January 1985:

> We deplore the inclusion in the federal budget of major scientific projects that have not been subjected to the normal process of proposal submission and expert review. The consequences go beyond the diversion of scarce resources from projects of higher priority. Institutions are increasingly persuaded that political maneuvering is more important than scientific justification in securing federal funds. Yet each year many worthy scientific projects must be abandoned or deferred for lack of funds. The developing pattern of funding major projects without subjecting them to impartial expert scrutiny diminishes science.
>
> Quite clearly Congress is responsible for considering a variety of factors—such as geographic balance, economic need, and affirmative action—in allocating funds for major projects of any sort. We urge, however, that review by impartial scientific experts be an important consideration in any allocation of funds for research facilities.

The physicists' plea was joined not only by their historic allies in the scientific community but also, this time, by the White House science adviser and the science chiefs of the NSF, the Department of Defense, and the Department of Energy. On August 10 they sent letters to 30 leading members of Congress on both sides of the aisle.

By early December only two replies had been received. One was from Don Fuqua (Dem., Fla.), chairman of the House Committee on Science and

Technology. Questions on the funding of education and training of scientists, regional and national economic development, or the construction of a major scientific facility, he said, must "be considered in a broader decision-making context." A second reply, in action rather than text, was offered only a few days later on the Senate floor when Richard G. Lugar (Rep., Ind.) and Jake Garn (Rep., Utah) successfully engineered legislation that required the Department of Education to provide $6 million for a Center of Excellence in Science and Mathematics Education at Indiana University and $4 million to the University of Utah for construction of a research center to study the health effects of nuclear energy and other energy technologies.

In the words of Robert M. Rosenzweig, president of the Association of American Universities: "It's a mess. Even as we express our outrage, the problem grows more serious."

Security restrictions. In 1982 the academic research community raised formal complaints about the inhibiting effect of government restrictions on the free communication of scientific findings. Initial responses from federal officials led to early optimism that a working accommodation between academia and government would meet the needs of both science and national security. By mid-1984 much of that hope had faded. A government review of the problem, begun in December 1982 and supposedly to be completed in two months, lay dead in the water. More to the point, a number of major new federal policy initiatives indicated that government concerns about the unwanted transfer of technological advances to potential enemies continued to have priority over any concern about freedom of scientific communication.

In the spring a proposed new Department of Defense policy that would allow some military control over the right to publish the results of unclassified military research brought a sharp reaction from the presidents of Stanford University, the California Institute of Technology, and the Massachusetts Institute of Technology. In their response they notified the White House and the Department of Defense (DoD) that they would refuse to conduct unclassified research for the federal government if the governing contracts in any way restricted their right to publish openly.

The issue loomed much larger to the academic community than it did to the Defense Department. Academic research absorbs only a small part of the department's annual research budget, but—at a figure approaching $1 billion—it is not inconsiderable to academia. For this and other reasons the battle between the campus and the Defense Department resembled a coal-mine fire. Only the smoke was visible from time to time.

In fact, the proposed restriction had never been published, nor was the letter from the three university presidents ever turned over to the news media. In May Edith W. Martin, deputy undersecretary of defense for research and advanced technology, said that it was the responsibility of the government to classify research in advance. "No restrictions," she declared, "may be placed upon the conduct or reporting of fundamental research that has not received national security classification." But near the end of 1984 Secretary of Defense Caspar Weinberger noted an exception: "No restriction will be placed on the conduct or reporting of unclassified contracted research . . . but for those rare exceptions where there is a high likelihood of disclosing performance characteristics of military systems, or of manufacturing technologies that are unique and critical to defense."

Weinberger's exceptions did not appear that rare to the organizers of U.S. scientific conferences, especially those with international attendance. There was a growing trend to restrict certain sessions to U.S. citizens. According to conference organizers the impetus for closing sessions normally came from potential authors, most of whom were from Defense Department laboratories or defense contractors, who informed the organizers that under the provisions of their contract they could not present the work at an open meeting.

In other cases the impetus came from potential authors whose research contracts had restrictive clauses forbidding "export" of the information obtained. (These restrictions were of particular concern to Europeans when they applied to projects being supported by the 16 member states of NATO. European governments worried that rigid controls on technological exchange would result in waste of precious funds on needlessly duplicative research projects, while European high-tech firms feared that they would be closed out of competitive bidding on U.S. weapons contracts.) Weinberger, however, disavowed responsibility for the exclusionary rules. "Some professional societies," he said, "have attempted to address this issue by independently adopting 'U.S. only' conference policies. DoD does not support this approach because it fails to recognize our national obligation to work with our friends and allies toward mutual attainment of national security."

Defense agencies were still seeking a modus vivendi with academic scientists when Harvard University fired a mortar round. In a telling review of the principal federal restrictions on the free flow of academic information and ideas, John Shattuck, vice-president for government and public affairs, saw an effort to insert prepublication review requirements even into contracts for research clearly unrelated to matters of national defense, presumably for purposes of quality control. Still more worrisome to academic scientists was the potential for abuse by suppression of unwelcome findings.

In addition, new clauses were introduced in defense contracts for unclassified research prohibiting the hiring of foreign nationals to work on the contract. Several major research universities, including Cornell, Harvard, and the University of Illinois, refused to sign contracts with such restrictions. As the year ended, the federal government once more announced the creation of an advisory panel to consider scientists' concerns about national security controls over scientific communication.

International science. As in past years, U.S. scientists had good reason to be concerned over treatment of their colleagues in other countries. The condition of the noted Soviet physicist Andrey Sakharov became more troubling to his colleagues in the West when friends reported that his health and that of his wife were deteriorating as a result of mistreatment.

On May 24 the Royal Society of London, the French Academy of Sciences, the Royal Swedish Academy, and the National Academy of Sciences sent a cable to the Soviet Academy of Sciences in which they called upon their colleagues "to act as effectively as possible to help Academician Sakharov and his wife in getting the health care they require and request, and subsequent permission to devote themselves to their scientific work." The move had special significance for the U.S. Academy since its president had told members only a month before that, after a period of "consensus-building," the Academy leadership was ready to enter into new negotiations with the Soviet Academy following a suspension of several years.

By the end of the year, however, it appeared that the U.S. Academy, at least, was prepared to accept a stalemate on the matter of the Sakharovs. In January 1985 the U.S. Academy signed a draft protocol with the Soviet Academy outlining a two-year bilateral agreement for cooperation in certain advanced fields of science. (Meanwhile, there were other grounds for concern. The American Mathematical Society sent inquiries to the Chilean government concerning the fate of three Chilean mathematicians who had disappeared during government roundups in November and December. The American Physical Society also expressed its concern about the case of two physicists at the University of Santiago who were reportedly pressured into resigning their posts.)

Cooperative programs in space research seemed to be exempt from political considerations. University of Chicago astronomers announced at the end of the year that they had worked closely with Soviet counterparts in the installation of instruments aboard two Soviet spacecraft heading for a rendezvous with Halley's Comet in 1986. Data from the instruments would help a European team and the designers of two Japanese spacecraft to determine how close an approach to the comet would be feasible.

Also on the horizon was formal acceptance by research ministers of major European nations of an agreement to participate in the design and construction of the proposed U.S. space station. Although a number of scientists questioned whether the potential scientific output of the facility warranted the resources it would require, the concept had other attractions. European aerospace industries anticipated lucrative spinoffs in terms of space-based manufacturing opportunities, and their political leaders felt that the technological relationship would serve to strengthen political alliances as well.

Another strong signal that mutual cooperation in science held political advantages came from a September meeting in Paris of 21 European research ministers, which produced a pledge to increase scientific collaboration across political borders. Among the proposals launched at the meeting were the issuance of common trade and travel privileges to all European research workers, a fellowship scheme that would encourage the exchange of postgraduate research assistants, and the creation of a European science academy.

Not all European nations were prepared to undertake new financial commitments. Great Britain, especially, was experiencing economic difficulties, and its scientific leadership warned that without substantial growth in research funds, Britain would have to withdraw from some major existing cooperative activities. In November John Kingman, chairman of the Science and Engineering Research Council, announced the start of a major reexamination of the SERC program and said the scientific community would have to prepare itself for "major surgery." French scientists, on the other hand, were surprised to find themselves suddenly favored. The Mitterrand government embraced science and technology as a way to get a handle on a balky economy and ordered the research budget increased by a hefty 6.3% over inflation. Some 600 new positions were to be created in government research laboratories.

At the end of the year the withdrawal of the United States from the United Nations Educational, Scientific and Cultural Organization (UNESCO) became an accomplished fact. When it was announced, the administration gave its assurance that the funds that had been going to UNESCO would be redirected to other international projects. Later, however, Ambassador Roger Kirk acknowledged that no more than $2,750,000 was actually earmarked for projects in lieu of UNESCO support. It was clear, moreover, that no additional funds for international science and education could be expected.

—Howard J. Lewis

Scientists of the Year

Honors and awards

The following article discusses recent awards and prizes in science and technology. In the first section the Nobel Prizes for 1984 are described in detail, while the second is a selective list of other honors.

Nobel Prize for Chemistry

R. Bruce Merrifield of Rockefeller University in New York City was awarded the 1984 Nobel Prize for Chemistry for developing an efficient method for preparing the chainlike molecules of polypeptides and proteins from their basic constituents, the amino acids. All of the enzymes and many of the hormones that are present in the tissues of humans and other animals are members of these families of nitrogenous organic compounds. Merrifield devised his procedure—called solid-phase peptide synthesis, or SPPS—so that it could be automated, and commercial versions of his apparatus are now in service throughout the world. The principles of SPPS also have been extended to the preparation of substances not based on amino acids, such as polynucleotides, which play a vital role in studies of the structure of the nucleic acids.

The distinction between polypeptides and proteins is not sharp; their chemical structures are the same, but the molecules of proteins ordinarily are larger and commonly are folded and looped into complex three-dimensional shapes that are essential for their biological activities. The German chemist Emil Fischer (who won the Nobel Prize for Chemistry in 1902 for studies of sugars and purines) introduced the term polypeptide about 1899 for products he obtained by combining small numbers of amino acids. During the 20th century it was shown that all polypeptides and proteins are made up of various combinations of about 20 amino acids, strung together like the letters of different words.

In 1955 the U.S. biochemist Vincent du Vigneaud won the Nobel Prize for Chemistry for his research on the pituitary hormone oxytocin. In this project, which occupied several years, he showed that the hormone is made up of nine amino acids; he found the order in which these components are arranged in the molecule; and, starting with the invividual amino acids, he synthesized the compound. In 1958 the British biochemist Frederick Sanger won the first of his two Nobel Prizes for Chemistry for determining the pattern in which the 51 amino acids of the hormone insulin are joined together. By 1964 chemists had prepared synthetic insulin.

The method used by Merrifield's predecessors to synthesize a polypeptide from, say, 20 amino acids required 19 steps, each involving a chemical reaction between the amino group of one amino acid and the carboxyl (acid) group of another. Each of these reactions ordinarily entailed several distinct procedures—measuring the proper amounts of the starting materials, dissolving them in a solvent, mixing the solutions, waiting for the reaction to proceed, separating the product from the solvent and any by-products and unconsumed starting materials, purifying the new compound, and verifying its identity. Many of these operations could be carried out easily and without loss of material, but others—notably, the purification of previously unknown substances—had to be painstakingly developed for each step. In most cases recovery of the desired pure product was not complete, and even if only 3% was wasted at each stage, the 19-step synthesis would yield only 56% of the amount of the polypeptide that would have been obtained if no losses had occurred.

To streamline this cumbersome method, Merrifield used a solid substance with special properties. This material—just as a sponge is permeated by water without being dissolved—is penetrated but not dissolved by organic liquids that dissolve amino acids. The solid also combines with the carboxyl group of an amino acid, leaving the amino group available for the next stage of the synthesis. To begin the preparation of a polypeptide Merrifield treated the solid with a solution of the first amino acid of the chain. The ensuing reaction made the amino acid part of the solid, which was isolated and purified by simply filtering it and washing it with fresh solvent. The solid, which remained in the same vessel during the whole procedure, was immediately ready for treatment with a solution of the next amino acid. The same steps were repeated for each successive amino acid until the entire polypeptide chain was assembled. A final reaction detached the completed chain from its solid support.

Because every step could be carried out under the same conditions, varying only the amino acid to be used, the whole process could be performed

Ingbert Grüttner

R. Bruce Merrifield

by a machine programmed to dispense the proper volumes of solutions and solvents, mix them for the necessary time, filter, wash, and start again. The apparatus could be allowed to operate day and night without close attention.

Robert Bruce Merrifield was born July 15, 1921, in Fort Worth, Texas. He graduated from the University of California at Los Angeles in 1943 and received a Ph.D. in chemistry from the same institution in 1949. He then joined the Rockefeller Institute for Medical Research (now Rockefeller University), becoming professor of biochemistry in 1966 and John D. Rockefeller, Jr., Professor in 1983.

Merrifield proposed the idea of solid-phase peptide synthesis to his laboratory chief in 1959; within three years he had achieved substantial results. In 1964 he announced that his apparatus had produced a sample of the polypeptide bradykinin, composed of nine amino acids. A greater triumph was his synthesis of the enzyme ribonuclease, which consists of a chain of 124 amino acids.

Nobel Prize for Physics

The Nobel Prize for Physics was divided equally in 1984 between Carlo Rubbia and Simon van der Meer, two of the principal members of the staff of CERN, the European Organization for Nuclear Research, whose laboratories are located near Geneva. (The initials CERN have been retained from the original French title—Conseil Européen pour la Recherche Nucléaire—of the establishment, which was founded in 1952 and is funded by 12 Western European nations.) The Nobel citation of Rubbia and van der Meer praised "their decisive contributions to the large project which led to the discovery of the field particles W and Z, communicators of the weak interaction." The prizewinning achievement, which was reported in 1983, had been immediately recognized as both the end of one major chapter in the history of physics and the opening of another. In Europe the achievement was regarded as satisfying evidence that the home team could compete on a par with U.S. physicists, who had occupied center stage at the Nobel ceremonies with great regularity for many years.

The forces that govern the interactions of subatomic particles have engaged the attention of the world's physicists for more than five decades. Until about 1930 the only forces known to act on material bodies were gravitation, described by Isaac Newton in 1687, and electricity and magnetism, unified by James Clerk Maxwell about 200 years later. Gravitation causes any two particles to attract each other with a force directly proportional to the product of the masses and inversely proportional to the square of the distance between them. Maxwell's electromagnetic theory subsumed the discovery—made by Charles-Augustin de Coulomb in 1785—that electric charges either attract or repel each other; the magnitude of the force, like that of gravitation, conforms to an inverse-square law involving the product of the charges instead of the masses. The steady motion of the planets around the Sun demonstrates that gravity acts between bodies even if they are millions of miles apart, and there is no reason to doubt that it acts at any distance whatever. Maxwell showed that his electromagnetic theory accounts for the rapid propagation of light through space, and the visibility of the planets and stars suggests that the range of electromagnetic forces, like that of gravity, is infinite.

Around the beginning of the 20th century a succession of major discoveries, including the electron, X-rays, radioactivity, quantized radiation, the atomic nucleus, and isotopes, launched a new academic specialty—atomic physics. All these new phenomena were signs that atoms are not simply small, structureless bits of matter but are complex systems composed of still smaller parts that influence one another and interact with the rest of the world. Studies of how individual atoms absorb and emit light, for example, revealed that their inner workings do not conform to the laws of Newtonian mechanics and Maxwellian electromagnetism. No one seriously doubted that atoms are the true building blocks of ordinary matter, and the goal of the atomic physicists was first to find the principles underlying the structure and properties of atoms and their constituents and then to show how, for great numbers of atoms, those principles form the basis of the laws that apply to everyday objects.

Ernest Rutherford's discovery that the atom must be mostly empty space, with a tiny, dense, positively charged nucleus at the center and one or more electrons somehow arranged around it, attracted little attention when he announced it in 1911. The Danish physicist Niels Bohr, however, quickly refined Rutherford's idea. In 1913 he proposed that in the simplest atom, that of hydrogen, the electron could move in any of a finite number of orbits about the nucleus. As long as it remained in a single orbit, it would not emit any radiation, although Maxwell's theory demanded that the electron continuously radiate away its energy and quickly spiral into the nucleus. Bohr's suggestion was that the atom would absorb or emit radiation only when the electron jumped from one orbit to another. The frequency of the radiation should be directly proportional to the energy of the electron's motion in the two orbits involved, and because only a finite number of orbits are permitted, only a finite number of radiation frequencies should be observed. Bohr's theory provided the first satisfactory explanation of how the hydrogen atom absorbs and emits radiation of particular frequencies. Bohr won a Nobel Prize for Physics for

Courtesy, CERN

(Left) Carlo Rubbia and Simon van der Meer (right).

his achievement in 1922; by then his quantum mechanical theory was being refined and generalized into a form that, although mathematically complicated, was fully recognizable as the submicroscopic basis of Newton's macroscopic mechanics.

Quantum mechanics was further developed by Arthur Compton, Werner Heisenberg, Paul Dirac, Erwin Schrödinger, Wolfgang Pauli (all winners of the Nobel Prize), and others into a deeper understanding of the electromagnetic force. Their results supported the view that this force is transmitted by the exchange of photons, discrete packets (quanta) of energy. According to their new theory, called quantum electrodynamics, the infinite range of the electromagnetic force is a consequence of the fact that photons have no mass. Within a few years of its conception quantum electrodynamics was rising toward the status it now has reached; of all the physical theories that deal with subatomic phenomena, it is the only one that provides results that agree with experimental observations as closely as such observations can be made.

This unparalleled accuracy arises from the explicit allowances made for various effects, including the apparent spin of subatomic particles about their own axes, the increase in a particle's mass when it moves at speeds close to that of light, and the details of the interaction between a charged particle and an electromagnetic field. As early as 1928 Dirac's work with the new theory indicated that there should exist a particle just like the electron, but with opposite electric charge. The discovery of this antiparticle, the positron, in 1932 was one of the early triumphs of quantum electrodynamics. Since 1932 ample experimental evidence has shown that every subatomic particle is a member of such a pair and that when a particle encounters its corresponding antiparticle, the masses disappear and an electromagnetic photon appears. The energy of this photon can be calculated from the lost mass by Einstein's equation $E = mc^2$.

By the 1930s it was evident that gravitation and electromagnetism were not sufficient to explain either the stability or the instability of atomic nuclei. It became necessary to propose two entirely different kinds of forces, one (called the strong force) to account for the tight binding of nucleons (positively charged protons and uncharged neutrons) in nuclei and another (the weak force) to rationalize such processes as the decay of the neutron to a proton, an electron, and a neutrino. Hideki Yukawa of Japan was awarded the Nobel Prize for Physics in 1949 for formulating one of the first fruitful theories concerning these forces. In 1935, when he first made his proposals, quantum electrodynamics had already become the most useful theory for dealing with the interactions of charged particles. Yukawa adopted that theory as a model, introducing modifications to accommodate the fact that the range of the strong and weak forces is restricted to distances comparable to the diameter of an atomic nucleus. He suggested that such short-range forces arise through the exchange of quanta that possess substantial mass and that have very short lifetimes. These proposed particles became known as mesons—from the Greek word meaning "middle"—because Yukawa's calculations indicated that their masses should be intermediate between those of the electron and the nucleons.

While Yukawa was working out his meson theory, Enrico Fermi of Italy (and later the United States) had in 1934 published a different analysis of the weak force and how it controlled the decay of the neutron. Fermi, who won the Nobel Prize for Physics in 1938 for other research on neutrons, also based his theory on quantum electrodynamics but did not duplicate Yukawa's assumption that the weak force is transmitted by an exchange of particles that had not been experimentally observed. Such an exchange necessitates that one of the interacting particles emit an entity that is absorbed by the other. Fermi proposed instead that when a neutron decays, it simultaneously emits an electron and a neutrino; no subsequent absorption process was involved.

Fermi also suggested an alternative picture of the weak decay, in which two analogues of electric currents interact. In this version of the process the emission of the neutrino can be replaced by its equivalent, the absorption of an antineutrino. Now the weak decay involves the disappearance of two particles (neutron and antineutrino) and the appearance of two new particles (proton and electron). One of Fermi's currents was the motion—in time, rather than space—of the heavy particles (neutron and proton); the other current was the corresponding motion of the light particles (electron and antineutrino). The weak interaction has the effect of transferring a unit of electric charge from one current to the other;

the uncharged neutron becomes the positive proton, and the uncharged antineutrino becomes the negative electron. The electromagnetic counterpart of the interacting currents is the attraction or repulsion that exists between wires through which electricity is flowing. The electromagnetic interaction does not transfer charge from one current to the other, but in the case of neutron decay the weak interaction does.

Neither Yukawa's theory nor Fermi's had the sustained value of quantum electrodynamics, but they were far from wasted efforts. They provided partial answers to a difficult question and stimulated a great amount of further investigation.

Since the 1930s more than a dozen further Nobel Prizes have been awarded to experimentalists and theoreticians who devised new techniques for investigating subatomic particles and for interpreting the findings. The primary tools of the particle physicists were more and more powerful accelerators, in which beams of electrons, protons, and other charged species could be raised to very high energies and directed onto targets composed of any desired material. Each successive generation of accelerator made it possible to study interactions involving finer details of the structure of nuclei and to detect new particles created during the conversion of energy into matter. The mesons predicted by Yukawa—and many more, besides—were identified, as were increasing numbers of short-lived species. Most of these new particles have been classified as hadrons, a term adopted for all particles that are susceptible to the strong force. The hadrons are subdivided into baryons, which include the proton and all other particles whose decay products include a proton, and mesons, whose decay products do not include a proton. (This modern definition of mesons differs from the older one, based on mass, used for the quanta of the Yukawa theory.) The few particles that are unaffected by the strong force are called leptons; this group includes the electron, the muon, the tau particle, and—one associated with each of these—three species of neutrinos.

The list of hadrons grew so long that it literally offended the physicists' sense of reasonableness and provoked an intensive search for some still deeper layer of structure. The goal of this quest was the identification of a small number of more fundamental building blocks that combine according to some orderly pattern that can account for the profusion of hadrons. In 1947 a few particles were observed that were produced by strong interactions but decayed by weak interactions. In the context of the much more common sequence of creation and destruction by the same kind of interaction, the new phenomenon was strange; the particles involved, both baryons and mesons, have been called strange ever since. The origin of this particular strangeness was proposed in 1953 by Murray Gell-Mann and Abraham Pais of the United States and, independently, by Kazuhiko Nishijima of Japan. These scientists identified strangeness as a basic property of hadrons; it could be absent, as electric charge is absent from neutrons, or it could be present in integral multiples of a fundamental amount. Interactions in which the starting particles differ in total strangeness from the final particles can be observed, but they are much slower than interactions in which the total strangeness does not change. That is, strangeness is said to be both quantized and strongly conserved, although not absolutely conserved, as electric charge appears to be.

To any conserved property of a physical system there always corresponds some form of symmetry in the laws that describe the system. The discovery of the conservation of strangeness therefore provided a valuable clue in the search for a proper mathematical description of the hadrons. In 1961 Gell-Mann and Yuval Ne'eman of Israel independently published a theory, which Gell-Mann called the Eightfold Way, that organized all the known hadrons into families based on recognizable relationships. Some of these families were not complete, but the theory indicated the properties of the missing members; the discovery of one of these in 1964 confirmed the validity of the Eightfold Way. In the same year Gell-Mann and George Zweig proposed three hypothetical particles, the quarks, and their antiparticles; they showed that every meson could be regarded as a combination of a quark and an antiquark, and every baryon as a combination of three quarks. The Eightfold Way and the quark theory brought order to the hadrons in the same way that the periodic law a century earlier had brought order to the chemical elements.

The price that had to be paid for a profound revision of the picture of the nature of subatomic particles was a reexamination of the ideas concerning the nature of the forces that act between particles. What force is involved in binding three quarks into a proton or a neutron? Does the strong force, which holds atomic nuclei together, act between protons and neutrons or between the quarks? Does the weak decay of a neutron into a proton involve only a single quark? In brief, if an apparent simplification in the description of particles translates into a complication in the description of forces, is any real progress being made?

While quantum electrodynamics was becoming a more and more accurate description of the electromagnetic force, it also served as an increasingly useful guide in the formulation of theories of the other fundamental interactions. For example, any such theory must satisfy mathematical conditions ensuring that the calculated values of observable properties—such as the charge or mass of the electron—are not infinite. The necessary restrictions on the equations for quantum electrodynamics, giving them the quality called renormalizability, were worked out during the 1940s by Richard Feynman and Julian Schwinger in the United States and Shin'ichiro Tomonaga in Japan; these three scientists received the Nobel Prize in 1965.

During the 1960s Sheldon Glashow and Steven Weinberg of the United States and Abdus Salam, working in Trieste, Italy, explored the idea of a family of four particles that would account for both electromagnetic and weak forces. One member of this family was the familiar massless photon of quantum electrodynamics; the others were all massive particles—one positive (called W^+), one negative (W^-), and one neutral (Z^0). The charged W particles would be involved in the well-known weak interactions that change neutrons to protons and vice versa. The Z particle should manifest itself in weak interactions in which electric charge is not exchanged, although no such interaction had ever been reported.

The new electroweak theory was not taken seriously until Peter Higgs of the University of Edinburgh gave a mathematical justification for the large masses of the W and Z particles and Gerard 't Hooft of the State University of Utrecht, Neth., showed in 1971 that the equations, like those of quantum electrodynamics, could be renormalized. These features made the theory too persuasive to be ignored, and the search for experimental corroboration began in earnest.

In 1973 two independent research groups reported evidence for the kind of weak interaction that, according to the electroweak theory, required the existence of the Z particle. Rubbia was a leader of one of these groups, working at the Fermi National Accelerator Laboratory (Fermilab) in Batavia, Ill. It was clear to him and his collaborators that the next goal was the production of the W and Z particles themselves, but no accelerator in the world was capable of bringing about processes that would release enough energy for the necessary mass to materialize. They saw that the quickest way to their objective would be to modify the proton accelerators at Fermilab and CERN so that a proton beam moving in one direction could be made to collide with an antiproton beam moving in the opposite direction. In the mutual annihilation of a proton and an antiproton much more energy would be released than in the alternative procedure of directing a beam of either kind of particles onto a stationary target. Rubbia formally proposed the changes to both laboratories in 1976. Fermilab was already committed to another expensive project, but CERN accepted the suggestion and began to implement it.

The major difficulties to be overcome were the design and construction of two novel pieces of equipment. One of these was a detector system that could provide unequivocal evidence that the W and Z particles had indeed been formed; the other was a means of generating an intense beam of antiprotons. Antiprotons can be produced by allowing a beam of accelerated protons to strike a fixed target. The process is inefficient: about one million protons must hit the target for each antiproton formed. A more serious drawback is that the antiprotons fly out of the target in all directions, with a broad range of speeds. To make Rubbia's experiment work, this chaotic group of particles must be organized into a tightly focused stream moving in one direction at a uniform speed. In 1972 van der Meer had described a procedure, called stochastic cooling, for accomplishing this result. At that time there had been no pressing need to verify his theory, but Rubbia's proposal put the situation in a different light. Van der Meer promptly assembled equipment to collect the antiprotons into a diffuse beam that could be steered into a circular accelerator with a wide aperture. He provided this device with detectors that measured the nonuniformity in the speeds of the particles while successive groups of them passed and, for each group, relayed an appropriate correcting signal across the ring in time to nudge each group toward the desired path. As the antiprotons continued to circulate, they gradually became focused into a beam that finally could be transferred to the large proton synchrotron; there the magnets that controlled the paths of protons in one direction also controlled the paths of the antiprotons in the other direction.

While van der Meer was perfecting his technique for manipulating antiproton beams, Rubbia supervised the assembly of the huge detection system. Testing of the whole array was begun in 1981, and in late 1982 data were recorded for about 1,000,000,000 individual collisions. Stringent analysis of these records winnowed out a handful of events that satisfied all the criteria for the production of W particles. In another experiment that was performed in 1983 the colliding beams were made even more intense, and the data provided evidence for the formation of Z particles in more than five events. The properties of both kinds of particles were found to agree very well with those calculated from the electroweak theory.

Rubbia was born on March 31, 1934, in Gorizia, Italy. He was educated at the Normal School of Pisa and the University of Pisa, receiving a doctorate from the latter in 1957. He taught there for two years before moving to Columbia University in New York City as a research fellow. Returning to Italy, he joined the faculty of the University of Rome in 1960 and was appointed senior physicist at CERN in 1962. In 1970 he became professor of physics at Harvard University and thereafter divided his time between Harvard and CERN.

Van der Meer was born on Nov. 24, 1925, in The Hague, Neth. He graduated from the Higher Technical School in Delft and then worked as an engineer for the Philips Co., a manufacturer of electrical equipment. After receiving a degree in physical engineering from the Delft University of Technology, he joined the staff of CERN in 1956.

Nobel Prize for Physiology or Medicine

The Nobel Prize for Physiology or Medicine was shared equally in 1984 by three immunologists: Niels Kai Jerne, who claimed both British and Danish citizenship and was professor emeritus of the Basel Institute of Immunology in Switzerland; Georges J. F. Köhler, a West German who, at the time the prize announcement was made, had already decided to transfer from the Basel Institute to the Max Planck Institute of Immune Biology at Freiburg, West Germany, where he was named a co-director; and César Milstein, a citizen of both Argentina and the U.K. who is affiliated with the Medical Research Council in Cambridge, England.

Jerne, who observed his 73rd birthday two weeks after accepting his prize, retired to France after a long career. He was recognized long ago as one of the principal architects of modern immunology and one of those rare scientists with the knack of recognizing the need for a simple but crucial experiment that will guide the paths of other researchers for a long time. He is well remembered for formulating three productive theories concerning the functioning of the immune system. Milstein and Köhler were cited in the prize announcement for their discovery, described in 1975, of a method for producing antibodies of unparalleled homogeneity. Their technique revolutionized many diagnostic and medical screening procedures and was being widely used in the development of therapeutic agents for use in the treatment of many intractable disorders.

The immune system, like the nervous system, is not only very complex but is dispersed among the other organs and tissues of the body. Its distinctive function is to safeguard the integrity of the body by identifying and initiating the expulsion of potentially harmful substances. The system, regarded as a diffuse organ, consists of cells called lymphocytes and protein molecules called antibodies. There are millions of species of lymphocytes, each of which can secrete a single kind of antibody in response to the appropriate stimulus. The lymphocytes originate in the bone marrow and proliferate by cell division in the thymus, the spleen, and the lymph nodes. In their usual state of inactivity they are spherical cells, about a hundredth of a millimeter in diameter, secreting no antibody. They are transported continuously throughout the body, being carried to the tissues in the bloodstream and back to the lymph nodes in a separate set of vessels called the lymphatic ducts. From the lymph nodes they return to the blood vessels.

The function of the antibodies is to attach themselves to cells or large molecules that originate outside the body of the individual in whom the antibodies have formed. Such foreign substances, which may range from the components of snake venom to a transplanted heart or kidney, are called antigens. An antibody combines with an antigen when a distinctively shaped site on the antibody molecule encounters a complementary site, called an epitope or an antigenic determinant, on the antigen, matching as do adjacent pieces of a jigsaw puzzle. To protect the body from the enormous diversity of possible antigens, the immune system maintains a corresponding variety of antibodies and automatically increases the supply of those that are called into action.

Jerne advanced the first of his major theories concerning the operation of the immune system in 1955. It offered an explanation of the mechanism by which the system supplies antibodies appropriate to neutralize any kind of foreign matter, regardless of its structure. The preceding version of this process held that any incoming antigen induced the formation of antibodies by serving as a template upon which the

(Left) Niels Kai Jerne, (center) Georges J. F. Köhler, (right) César Milstein

Photos, The Nobel Foundation, Stockholm

antibody molecules were synthesized. Jerne's theory was based on the alternative proposition that the genes controlling the origin of the lymphocytes in any individual contain all the information needed to ensure the formation of an adequate variety of lymphocytes and antibodies. McFarlane Burnet and Peter Medawar carefully investigated the implications of Jerne's theory; they showed that the ability of antibodies to distinguish between components of their host and those of other organisms is not genetically fixed but, instead, can be modified during the embryonic stage of prenatal development. Burnet and Medawar won the Nobel Prize for Physiology or Medicine in 1960 for their research.

A second theory, proposed by Jerne in 1971, dealt with the steps that occur during the development of the immune system in the course of the life of each individual. He pointed out that all lymphocytes arise from precursor cells formed in the bone marrow but that only about half of them pass through the thymus on their way into the circulation. These, called T lymphocytes, do not secrete antibodies; the others, the B lymphocytes, do. The T cells are not left without a role, however; in different circumstances they can either suppress the response of B cells to particular antigens or enhance that response.

In 1974 Jerne proposed a third concept, now called the network theory. In it he emphasized that the self-regulation of the immune system depends on a duality in the way lymphocytes and antibodies interact with antigens and with each other. Not only do these entities recognize and dispose of foreign matter, but they also possess features by which they themselves can be recognized and left unscathed.

The research of Milstein and Köhler solved a problem that had plagued immunologists for decades. The only technique that had been available for the preparation of substantial quantities of an antibody was to inject an antigen into an animal, wait for antibodies to form, draw blood from the animal, and isolate the antibodies. The proteins obtained by this procedure were almost never pure, because a typical antigen possesses several epitopes, each of which stimulates a different lymphocyte. The result was the simultaneous secretion of different antibodies that were practically impossible to separate. Milstein and Köhler saw that if a way could be found to clone lymphocytes—to cause them to subdivide indefinitely in a culture medium—then the antibody molecules secreted by the resulting population would all be identical. Lymphocytes, however, are short-lived and cannot be cultured satisfactorily. Milstein and Köhler took advantage of a discovery made in 1962 by Michael Potter of the National Cancer Institute in the United States. Potter had found a way to induce the formation in mice of myelomas, which are tumors that attack the cells of the blood plasma. Other scientists had found ways to foster the growth of these tumors in cultures.

Milstein and Köhler found that lymphocytes would fuse with the myeloma cells, forming a hybrid cell that retained the two desired properties: like the lymphocytes, these hybrids secreted a single species of antibody molecule and, like the myeloma cells, they perpetuated themselves in culture, providing potentially unlimited amounts of any desired antibody. Among many examples of the application of the so-called monoclonal antibodies produced by this technique, the antibodies produced by these cells have been found superior to human blood sera for use in identifying blood types A, B, and O.

Jerne was born of Danish parents on Dec. 23, 1911, in London. He grew up in Denmark but attended the State University of Leiden in The Netherlands, receiving a bachelor's degree at the age of 16. He has written that it then took him 12 years to decide on a profession. He was engaged as a researcher at the Danish State Serum Institute from 1943 to 1955 while studying medicine at the University of Copenhagen, which granted him a degree in 1951. He served as a chief medical officer for the World Health Organization from 1956 to 1962, then accepted research and teaching appointments in Switzerland, the United States, and West Germany. After helping organize the Basel Institute of Immunology, he became its director from 1969 until 1980. He served a final year at the Pasteur Institute in Paris before retiring.

Köhler was born in Munich, West Germany, on April 17, 1946. He attended the University of Freiburg, receiving a doctorate in biology in 1974. He then joined Milstein at the Laboratory of Molecular Biology in Cambridge, where they discovered the method of preparing monoclonal antibodies. In 1976 Köhler joined the staff of the Basel Institute, and in 1984 he was appointed one of three directors of the Max Planck Institute.

Milstein, who was born Oct. 8, 1927, in Bahía Blanca, Arg., was educated at the universities of Buenos Aires and Cambridge, receiving a doctorate from the latter in 1960. He served as a member of the staff of Argentina's National Institute of Microbiology from 1957 until 1963, when he returned to England to join the Laboratory of Molecular Biology of the Medical Research Council in Cambridge. He was initially interested in the chemistry of enzymes but shifted his attention to antibodies when their structures were deciphered. He was a senior scientist at the laboratory when Köhler arrived as a postdoctoral fellow in 1974. Milstein then became director of the division of protein and nucleic acid chemistry, having shared that position with Frederick Sanger until Sanger's retirement.

—John V. Killheffer

AWARD	WINNER	AFFILIATION
ARCHAEOLOGY		
MacArthur Prize Fellow Award	Peter Mathews	Peabody Museum of Archaeology and Ethnology, Cambridge, Mass.
ARCHITECTURE		
Honor Award of the American Institute of Architects	Bohlin Powell Larkin Cywinski	Wilkes-Barre, Pa.
Honor Award of the American Institute of Architects	Theodore M. Ceraldi	Nyack, N.Y.
Honor Award of the American Institute of Architects	Cooper-Lecky Partnership	Washington, D.C.
Honor Award of the American Institute of Architects	Croxton Collaborative/Hammill-Walter Associated Architects	New York, N.Y.
Honor Award of the American Institute of Architects	CRS/Caudill Rowlett Scott	Houston, Texas
Honor Award of the American Institute of Architects	Gwathmey Siegel & Associates	New York, N.Y.
Honor Award of the American Institute of Architects	Hammond Beeby and Babka, Inc.	Chicago, Ill.
Honor Award of the American Institute of Architects	Kohn Pederson Fox/ Perkins & Will	New York, N.Y.
Honor Award of the American Institute of Architects	Richard Meier	Richard Meier & Partners, New York, N.Y.
Honor Award of the American Institute of Architects	Moore Ruble Yudell	Marina del Rey, Calif.
Honor Award of the American Institute of Architects	I. M. Pei & Partners	New York, N.Y.
Honor Award of the American Institute of Architects	Tigerman Fugman McCurry	Chicago, Ill.
Honor Award of the American Institute of Architects	Venturi, Rauch and Scott Brown	Philadelphia, Pa.
International Pritzker Architecture Prize	Richard Meier	Richard Meier & Partners, New York, N.Y.
ASTRONOMY		
Annie Jump Cannon Award	Harriet Dinerstein	University of Texas, Austin
Balzan Prize	Jan H. Oort	State University of Leiden, Neth.
Catherine Wolfe Bruce Medal	Olin C. Wilson (Emeritus)	Mount Wilson Observatory, Calif.; Las Campanas Observatory, Chile
Dannie Heineman Prize	Martin J. Rees	University of Cambridge, England
Dirk Brouwer Award	Walter Fricke	Astronomisches Rechen-Institut, Heidelberg, West Germany
George Ellery Hale Prize	John W. Evans (Retired)	Sacramento Peak Observatory, Sunspot, N.M.
Helen B. Warner Prize	Scott D. Tremaine	Massachusetts Institute of Technology, Cambridge
Henri Chretien Award	Almas Chalabaev	Meudon Observatory, France; European Southern Observatory, Cerro La Silla, Chile
Henri Chretien Award	Barry F. Madore	David Dunlap Observatory, University of Toronto, Ont.

AWARD	WINNER	AFFILIATION
Henry Medal	George B. Field	Smithsonian Astrophysical Observatory, Cambridge, Mass.
Kuiper Prize	Eugene M. Shoemaker	U.S. Geological Survey
MacArthur Prize Fellow Award	Alar Toomre	Massachusetts Institute of Technology, Cambridge
Muhlmann Prize	François and Monique Spite	Mauna Kea Observatory, Hawaii; Paris Observatory
National Medal of Science	E. Margaret Burbidge	University of California, San Diego
Newton Lacy Pierce Prize	Alan Dressler	Mount Wilson Observatory, Calif.; Las Campanas Observatory, Chile
Simon Newcomb Award	Dennis D. McCarthy	U.S. Naval Observatory, Washington, D.C.
Steacie Prize	William G. Unruh	University of British Columbia, Vancouver
Urey Prize	David Stevenson	California Institute of Technology, Pasadena
CHEMISTRY		
A. Cressy Morrison Award	George W. Flynn	Columbia University, New York, N.Y.
Albert Lasker Clinical Medical Research Award	Paul C. Lauterbur	State University of New York, Stony Brook
American Chemical Society Award for Nuclear Chemistry	Gregory R. Choppin	Florida State University, Tallahassee
American Chemical Society Award in Inorganic Chemistry	F. G. A. Stone	University of Bristol, England
Charles Frederick Chandler Medal	Rudolph A. Marcus	California Institute of Technology, Pasadena
Charles Goodyear Medal Award	Herman E. Schroeder (Retired)	E. I. du Pont de Nemours
Charles Herty Medal	Eugene C. Ashby	Georgia Institute of Technology, Atlanta
Creative Invention Award	Ralph Milkovich	PPG Industries, Inc.
Davy Medal	Sir Sam Edwards	University of Cambridge, England
Earle K. Plyler Prize	Jon T. Hougen	National Bureau of Standards, Washington, D.C.
George Hevesy Medal	Garman Harbottle	Brookhaven National Laboratory, Upton, N.Y.
George Hevesy Medal	Edward Sayre	Brookhaven National Laboratory, Upton, N.Y.
Harold Lamport Award	Richard A. Mathies	University of California, Berkeley
Irving Langmuir Award in Chemical Physics	Robert W. Zwanzig	University of Maryland, College Park
James Flack Norris Award in Physical Organic Chemistry	Paul G. Gassman	University of Minnesota, Minneapolis
National Medal of Science	Roald Hoffmann	Cornell University, Ithaca, N.Y.
National Medal of Science	George C. Pimentel	University of California, Berkeley
National Medal of Science	Richard N. Zare	Stanford University, Calif.
Oesper Award	Fred Basolo	Northwestern University, Evanston, Ill.
Perkin Medal	John H. Sinfelt	Exxon Research & Engineering Co.

AWARD	WINNER	AFFILIATION
Peter Debye Award in Physical Chemistry	Stuart A. Rice	University of Chicago, Ill.
Priestley Medal	Henry Taube	Stanford University, Calif.
Roger Adams Award in Organic Chemistry	Donald J. Cram	University of California, Los Angeles
Rothschild Prize	Meir Wilchek	Weizmann Institute of Science, Rehovot, Israel
Royal Medal	Alan R. Battersby	University of Cambridge, England
Welch Award in Chemistry	Kenneth S. Pitzer	University of California, Berkeley
Willard Gibbs Medal	Elias J. Corey	Harvard University, Cambridge, Mass.
EARTH SCIENCES		
Appleton Prize	Keith D. Cole	La Trobe University, Bundoora, Australia
Arthur L. Day Medal	Wallace S. Broecker	Lamont-Doherty Geological Observatory of Columbia University, Palisades, N.Y.
Carl-Gustaf Rossby Research Medal	Tiruvalam N. Krishnamurti	Florida State University, Tallahassee
Charles Franklin Brooks Award	Robert G. Fleagle	University of Washington, Seattle
Clarence Leroy Meisinger Award	Louis W. Uccellini	Goddard Space Flight Center, Greenbelt, Md.
Cleveland Abbe Award for Distinguished Service to the Atmospheric Sciences	Sir (David) Arthur Davies (Emeritus)	World Meteorological Organization, Geneva, Switz.
Japanese Meteorological Society Award	Tetsuji Yamada	Los Alamos National Laboratory, N.M.
Jule G. Charney Award	Keith A. Browning	Meteorological Office, Bracknell, England
Jule G. Charney Award	Richard S. Lindzen	Massachusetts Institute of Technology, Cambridge
National Medal of Science	Helmut E. Landsberg (Emeritus)	University of Maryland, College Park
National Medal of Science	Walter H. Munk	University of California, San Diego
Outstanding Contribution to the Advance of Applied Meteorology	E. Philip Krider	University of Arizona, Tucson
Penrose Medal	Donald E. White (Retired)	U.S. Geological Survey
Sverdrup Gold Medal	S. George Philander	National Oceanic and Atmospheric Administration, Princeton, N.J.
William Bowie Medal	Akimoto Syuniti	University of Tokyo, Japan
ELECTRONICS AND INFORMATION SCIENCES		
A. M. Turing Award	Niklaus Wirth	Institut für Informatik, Zürich, Switz.
Arthur Von Hippel Award	Walter L. Brown	AT&T Bell Laboratories
Capt. James M. Gilliss Award	George H. Kaplan	U.S. Naval Observatory, Washington, D.C.
Fredkin Foundation Prize	Joe Condon	AT&T Bell Laboratories
Fredkin Foundation Prize	Ken Thompson	AT&T Bell Laboratories

AWARD	WINNER	AFFILIATION
George E. Pake Prize	Arthur G. Anderson	IBM Corp.
Harry Goode Memorial Award	Ralph E. Gomory	IBM Corp.
Joseph Fraunhofer Award	Peter K. Runge	AT&T Bell Laboratories
Merit Award of the Nuclear and Plasma Sciences Society	Frederick Goulding	University of California, Berkeley
Merit Award of the Nuclear and Plasma Sciences Society	Veljko Radeka	Brookhaven National Laboratory, Upton, N.Y.
National Academy of Engineering Founders Award	John Bardeen (Emeritus)	University of Illinois, Urbana
National Medal of Science	Herman H. Goldstine (Retired)	IBM Corp.
National Medal of Science	William R. Hewlett	Hewlett-Packard Co.
National Medal of Science	John G. Trump (Deceased)	Massachusetts Institute of Technology, Cambridge
ENERGY		
Charles Porter Award	Allen Laskin	Exxon Research & Engineering Co.
Enrico Fermi Award	Georges Vendryès	French Atomic Energy Commission
Leverhulme Medal	John F. Davidson	University of Cambridge, England
ENVIRONMENT		
Conservationist of the Year	National Academy of Sciences	Washington, D.C.
Craaford Prize	Daniel H. Janzen	University of Pennsylvania, Philadelphia
George M. Darrow Award	Walter J. Kender	Citrus Research and Education Center, Lake Alfred, Fla.
George M. Darrow Award	Robert C. Musselman	Air Pollution Research Center, Riverside, Calif.
George M. Darrow Award	Lewis Weinstock	TRC Environmental Consultants
Gulf Oil Conservation Award	Jack B. Woody	U.S. Fish and Wildlife Service
Hal Borland Award	Archie Carr	University of Florida, Gainesville
J. E. Bussart Memorial Award	John Borden	Simon Fraser University, Burnaby, B.C.
J. Paul Getty Wildlife Conservation Prize	Jane Goodall	Gombe Stream Research Center, Tanzania
New York Academy of Sciences Award	Arthur H. Westing	Stockholm International Peace Research Institute, Sweden
Sarah L. Poiley Memorial Award	Arthur C. Upton	Institute of Environmental Medicine, New York University, New York
Seth Gordon Award	Charles D. Kelley	Conservation and Natural Resources Department, State of Alabama
Tyler Prize for Ecology	Roger R. Revelle (Emeritus)	Scripps Institution of Oceanography, University of California, San Diego
Tyler Prize for Ecology	Edward O. Wilson	Harvard University, Cambridge, Mass.
FOOD AND AGRICULTURE		
Alexander von Humboldt Award	Vernon Ruttan	University of Minnesota, Minneapolis

AWARD	WINNER	AFFILIATION
Bio-Serv Award in Experimental Animal Nutrition	Carl L. Keen	University of California, Davis
Borden Award in Nutrition	Carlos L. Krumdieck	University of Alabama, Birmingham
Bristol-Myers Award for Distinguished Achievement in Nutrition Research	Hector DeLuca	University of Wisconsin, Madison
Conrad A. Elvehjem Award for Public Service in Nutrition	George K. Davis (Emeritus)	University of Florida, Gainesville
Cyrus Hall McCormick Gold Medal	Carl W. Hall	National Science Foundation, Washington, D.C.
J. Roger Porter Award	Herman J. Phaff	University of California, Davis
John J. Carty Award	Robert H. Burris	University of Wisconsin, Madison
Lederle Award in Human Nutrition	Lucille S. Hurley	University of California, Davis
Mead Johnson Award for Research in Nutrition	Allen S. Levine	University of Minnesota; Veterans Administration Medical Center, Minneapolis
Mead Johnson Award for Research in Nutrition	John E. Morley	University of Minnesota; Veterans Administration Medical Center, Minneapolis
Osborne and Mendel Award of the Nutrition Foundation	Willard J. Visek	University of Illinois, Urbana
LIFE SCIENCES		
Adolph Lomb Medal	Edward H. Adelson	David Sarnoff Research Center, RCA Laboratories, Princeton, N.J.
Albert Lasker Basic Medical Research Award	Georges J. F. Köhler	Basel Institute of Immunology, Switz.
Albert Lasker Basic Medical Research Award	César Milstein	Medical Research Council, Cambridge, England
Albert Lasker Basic Medical Research Award	Michael Potter	National Institutes of Health, Bethesda, Md.
Charles Reid Barnes Award	Martin Gibbs	Brandeis University, Waltham, Mass.
Darbaker Prize	Mark M. Littler	Smithsonian Institution, Washington, D.C.
Darwin Medal	Ernst Mayr (Emeritus)	Harvard University, Cambridge, Mass.
Edward D. Tillyer Award	Mathew Alpern	University of Michigan, Ann Arbor
Eli Lilly Award	Linda L. Randall	University of Uppsala, Sweden
Ernest Orlando Lawrence Memorial Award	John J. Dunn	Brookhaven National Laboratory, Upton, N.Y.
Finsen Prize	Kendric C. Smith	School of Medicine, Stanford University, Calif.
Gairdner Foundation International Award	Alfred G. Gilman	University of Texas, Dallas
Gairdner Foundation International Award	Krešimir Krnjević	McGill University, Montreal, Que.
Gairdner Foundation International Award	Martin Rodbell	National Institutes of Health, Bethesda, Md.
George R. Cooley Award	Douglas E. Soltis	Washington State University, Pullman

AWARD	WINNER	AFFILIATION
Gold Medal of the Entomological Society of Canada	Kenneth G. Davey	York University, Downsview, Ont.
Gregory Aminoff Medal	David Harker (Emeritus)	Medical Foundation of Buffalo, N.Y.
Harold Lamport Award	Richard A. Mathies	University of California, Berkeley
Henry Allen Gleason Award	Carl S. Keener	Pennsylvania State University, University Park
Henry Allen Gleason Award	Rogers McVaugh	University of Michigan, Ann Arbor
Isabel C. Cookson Paleobotanical Award	Jon J. Hamer	Ohio University, Athens
Kettering Award in Photosynthesis	Daniel I. Arnon	University of California, Berkeley
Louis and Bert Freedman Foundation Award	Susan E. Leeman	University of Massachusetts Medical School, Worcester
Louisa Gross Horwitz Prize	Michael S. Brown	Health Science Center, University of Texas, Dallas
Louisa Gross Horwitz Prize	Joseph L. Goldstein	Health Science Center, University of Texas, Dallas
MacArthur Prize Fellow Award	Shelly Bernstein	Harvard University Medical School, Boston, Mass.
MacArthur Prize Fellow Award	Matthew Meselson	Harvard University, Cambridge, Mass.
MacArthur Prize Fellow Award	Roger Payne	World Wildlife Fund
MacArthur Prize Fellow Award	Arthur T. Winfree	Purdue University, West Lafayette, Ind.
Merit Award of the Botanical Society of America	Donald R. Kaplan	University of California, Berkeley
Merit Award of the Botanical Society of America	Theodore T. Kozlowski	University of Wisconsin, Madison
Merit Award of the Botanical Society of America	Rolla M. Tryon	Harvard University, Cambridge, Mass.
National Medal of Science	Howard L. Bachrach (Retired)	U.S. Department of Agriculture
National Medal of Science	Paul Berg	Stanford University, Calif.
National Medal of Science	Wendell L. Roelofs	Cornell University, Ithaca, N.Y.
Ralph E. Alston Award	Jonathan Gershenzon	University of Texas, Austin
Robert W. Kleemeier Award	Caleb E. Finch	University of Southern California, Los Angeles
Senior U.S. Scientist Award	Winslow Briggs	Carnegie Institution of Washington, D.C.
Stephen Hales Prize	Bernard O. Phinney	University of California, Los Angeles
Texas Instruments Foundation Founders' Prize	Patricia Pearce Jones	Stanford University, Calif.
MATERIALS SCIENCES		
A. L. Patterson Memorial Award	Herbert A. Hauptman	Medical Foundation of Buffalo, N.Y.
A. L. Patterson Memorial Award	Jerome Karle	Naval Research Laboratory, Washington, D.C.
American Physical Society International Prize for New Materials	Joseph P. Remeika	AT&T Bell Laboratories

AWARD	WINNER	AFFILIATION
Ernest Orlando Lawrence Memorial Award	Siegfried S. Hecker	Los Alamos National Laboratory, N.M.
MATHEMATICS		
Leroy P. Steele Prize	Lennart Carleson	Mittag-Leffler Institute, Stockholm, Sweden
Leroy P. Steele Prize	Joseph L. Doob (Emeritus)	University of Illinois, Urbana
Leroy P. Steele Prize	Elias M. Stein	Princeton University, N.J.
MacArthur Prize Fellow Award	Michael Freedman	University of California, San Diego
MacArthur Prize Fellow Award	Fritz John (Retired)	New York University, N.Y.
National Medal of Science	Isadore M. Singer	Massachusetts Institute of Technology, Cambridge
Waterman Prize	Harvey Friedman	Ohio State University, Columbus
MEDICAL SCIENCES		
Achievement Award in the Life Sciences	Maclyn McCarty	Rockefeller University, New York, N.Y.
Albert Lasker Public Service Award	Henry J. Heimlich	Xavier University, Cincinnati, Ohio
American Chemical Society Award in Medicinal Chemistry	Paul A. J. Janssen	Janssen Pharmaceutica, Beerse, Belgium
Bristol-Myers Award for Distinguished Achievement in Cancer Research	William S. Hayward	Memorial Sloan-Kettering Cancer Center, New York, N.Y.
Bristol-Myers Award for Distinguished Achievement in Cancer Research	Philip Leder	Harvard University Medical School, Boston
Charles Thom Award	Richard P. Elander	Bristol-Myers Co.
Distinguished Service Award of the American Medical Association	William D. Holden	Case Western Reserve University, Cleveland, Ohio
Ernest Orlando Lawrence Memorial Award	Kenneth N. Raymond	University of California, Berkeley; Lawrence Berkeley Laboratory, Calif.
Ernst W. Bertner Memorial Award	Leroy E. Hood	California Institute of Technology, Pasadena
Gairdner Foundation International Award	J. Michael Bishop	University of California, San Francisco
Gairdner Foundation International Award	Yuet Wai Kan	University of California, San Francisco
Gairdner Foundation International Award	Robert L. Noble	University of British Columbia, Vancouver
Gairdner Foundation International Award	Harold E. Varmus	University of California, San Francisco
Gairdner Foundation Wightman Award	Douglas G. Cameron (Retired)	McGill University, Montreal; Montreal General Hospital
Garvan Medal	Catherine C. Fenselau	Johns Hopkins University, Baltimore, Md.
General Motors Cancer Research Foundation Prize	J. Michael Bishop	University of California, San Francisco
General Motors Cancer Research Foundation Prize	Robert C. Gallo	National Institutes of Health, Bethesda, Md.

AWARD	WINNER	AFFILIATION
General Motors Cancer Research Foundation Prize	Barnett Rosenberg	Michigan State University, East Lansing
General Motors Cancer Research Foundation Prize	Harold E. Varmus	University of California, San Francisco
Heath Memorial Award	Raymond Yesner (Emeritus)	School of Medicine, Yale University, New Haven, Conn.; currently Veterans Administration Hospital, New Haven
Hoechst-Roussel Award	Hiroshi Nikaido	University of California, Berkeley
Jeffrey A. Gottlieb Memorial Award	Daniel E. Bergsagel	Ontario Cancer Institute, Princess Margaret Hospital, Toronto, Ont.
Joanne Vandenberge Hill Award	Leopold G. Koss	Albert Einstein College of Medicine, Yeshiva University, New York, N.Y.
Joseph B. Goldberger Award in Clinical Nutrition	Lewis A. Barness	University of South Florida, Tampa
Lita Annenberg Hazen Award	Yuet Wai Kan	University of California, San Francisco
MacArthur Prize Fellow Award	Shelly Bernstein	Harvard University, Cambridge, Mass.
MacArthur Prize Fellow Award	Curtis Hames, Sr.	Medical College of Georgia, Claxton
MacArthur Prize Fellow Award	Edward Roberts	World Institute on Disability
National Medal of Science	Berta Scharrer (Emeritus)	Albert Einstein College of Medicine, Yeshiva University, New York, N.Y.
National Medal of Science	John G. Trump (Deceased)	Massachusetts Institute of Technology, Cambridge
Public Service Award of the Harvard School of Public Health	Vincent L. Gregory, Jr.	Rohm & Haas Co.
Public Service Award of the Harvard School of Public Health	Lewis Thomas (President emeritus)	Memorial Sloan-Kettering Cancer Center, New York, N.Y.
Royal Medal	Mary F. Lyon	Medical Research Council, London
Scientific Achievement Award of the American Medical Association	Maurice J. Jurkiewicz	Emory University, Atlanta, Ga.
Selman A. Waksman Award in Microbiology	Purnell W. Choppin	Rockefeller University, New York, N.Y.
U.S. Steel Foundation Award in Molecular Biology	Geoffrey M. Cooper	Dana-Farber Cancer Institute, Boston, Mass.; Harvard University Medical School, Boston, Mass.
U.S. Steel Foundation Award in Molecular Biology	Robert A. Weinberg	Massachusetts Institute of Technology, Cambridge
Wellcome Foundation Prize	Edward R. Andrew	University of Florida, Gainesville
Wellcome Foundation Prize	J. M. S. Hutchison	University of Aberdeen, Scotland
Wellcome Foundation Prize	J. R. Mallard	University of Aberdeen, Scotland
Wellcome Foundation Prize	P. Mansfield	University of Nottingham, England
William D. Coolidge Award	Robley D. Evans (Emeritus)	Massachusetts Institute of Technology, Cambridge
Wilson S. Stone Memorial Award	Jeffrey A. Drebin	Harvard University Medical School, Boston, Mass.
Wolf Prize	Donald F. Steiner	University of Chicago Medical Center, Ill.

AWARD	WINNER	AFFILIATION
OPTICAL ENGINEERING		
C. E. K. Mees Medal	William H. Steel	Commonwealth Scientific and Industrial Research Organization, Lindfield, Australia
Charles Hard Townes Award	S. E. Harris	Stanford University, Calif.
Cledo Brunetti Award	Abe Offner	Perkin-Elmer Corp.
David Richardson Medal	Norman J. Brown	Lawrence Livermore National Laboratory, Calif.
Frederic Ives Medal	Emmett N. Leith	University of Michigan, Ann Arbor
R. W. Wood Prize	David H. Auston	AT&T Bell Laboratories
Rumford Medal	Harold H. Hopkins	University of Reading, England
Simrad Prize in Electro-Optics	Tycho Jaeger	Norwegian Defense Research Establishment
PHYSICS		
American Physical Society Prize in Fluid Dynamics	Frank E. Karasz	University of Massachusetts, Amherst
American Physical Society Prize in Fluid Dynamics	William J. MacKnight	University of Massachusetts, Amherst
Beams Award	Rufus H. Ritchie	Oak Ridge National Laboratory, Tenn.; University of Tennessee, Knoxville
Bingham Medal	Bernard D. Coleman	Carnegie-Mellon University, Pittsburgh, Pa.
Boris Pregel Award	Gordon Danby	Brookhaven National Laboratory, Upton, N.Y.
Copley Medal	Subrahmanyan Chandrasekhar	University of Chicago, Ill.
Dannie Heineman Prize for Mathematical Physics	David P. Ruelle	Institut des Hautes Études Scientifiques, Bures-sur-Yvette, France
Distinguished Service Award of the Optical Society of America	Karl G. Kessler	National Bureau of Standards
Enrico Fermi Award	Robert R. Wilson (Emeritus)	Cornell University, Ithaca, N.Y.
Ernest Orlando Lawrence Memorial Award	Robert W. Conn	University of California, Los Angeles
Ernest Orlando Lawrence Memorial Award	Peter L. Hagelstein	Lawrence Livermore National Laboratory, Calif.
Ernest Orlando Lawrence Memorial Award	Robert B. Laughlin	Lawrence Livermore National Laboratory, Calif.
Frank Isakson Prize for Optical Effects in Solids	Manuel Cardona	Max Planck Institute for Solid-State Research, Stuttgart, West Germany
Hewlett-Packard Europhysics Prize	Gerd Binnig	IBM Research Laboratory, Zürich, Switz.
Hewlett-Packard Europhysics Prize	Heinrich Rohrer	IBM Research Laboratory, Zürich, Switz.
Hughes Medal	R. P. Kerr	University of Canterbury, Christchurch, N.Z.
I. E. Tamm Prize	Efim S. Fradkin	P. N. Lebedev Physical Institute, Moscow, U.S.S.R.
International Centre for Theoretical Physics Medal and Prize	Ganapathy Baskaran	University of Madras, India

AWARD	WINNER	AFFILIATION
International Crystal Growth Award	Bruce Chalmers (Emeritus)	Harvard University, Cambridge, Mass.
J. Robert Oppenheimer Memorial Prize	John A. Wheeler	University of Texas, Austin
Karl Scheel Prize	Manfred Rosenzweig	Technische Universität Berlin, West Germany
King Faisal International Prize in Science	Gerd Binnig	IBM Research Laboratory, Zürich, Switz.
King Faisal International Prize in Science	Heinrich Rohrer	IBM Research Laboratory, Zürich, Switz.
Kistemaker Prize	W. Werner	Institute of Applied Physics, Delft, Neth.
MacArthur Prize Fellow Award	Sidney Drell	Stanford University, Calif.
MacArthur Prize Fellow Award	Mitchell Feigenbaum	Cornell University, Ithaca, N.Y.
MacArthur Prize Fellow Award	David Nelson	Harvard University, Cambridge, Mass.
Marian Smoluchowski Medal	Adriano Gozzini	National Research Council, Italy
Marian Smoluchowski Medal	Wladyslaw Opechowski	University of British Columbia, Vancouver
Max Born Award of the Optical Society of America	Roy J. Glauber	Harvard University, Cambridge, Mass.
Max Born Prize	Amand Faessler	Institute for Theoretical Physics, University of Tübingen, West Germany
Max Born Prize	Andrew Keller	University of Bristol, England
Max Planck Medal	Nicholas Kemmer (Emeritus)	University of Edinburgh, Scotland
Medal of Honor of the Institute of Electrical and Electronics Engineers	Norman F. Ramsey	Harvard University, Cambridge, Mass.
Morris E. Leeds Award	Erich P. Ippen	Massachusetts Institute of Technology, Cambridge
Morris E. Leeds Award	Charles V. Shank	AT&T Bell Laboratories
National Medal of Science	Maurice Goldhaber	Brookhaven National Laboratory, Upton, N.Y.
National Medal of Science	Frederick Reines	University of California, Irvine
National Medal of Science	Bruno B. Rossi	Massachusetts Institute of Technology, Cambridge
National Medal of Science	J. Robert Schrieffer	University of California, Santa Barbara
Nishina Memorial Prize	Taiji Yamanouchi	Fermi National Accelerator Laboratory, Batavia, Ill.
Norsk Data Award	Jon Magne Leinaas	University of Oslo, Norway
Peter Mark Medal	Henry Kolm	Massachusetts Institute of Technology, Cambridge
Peter Mark Medal	Richard Marshall	Australian National University, Canberra
Rutherford Medal and Prize	Peter Higgs	University of Edinburgh, Scotland
Rutherford Medal and Prize	Tom Kibble	Imperial College of Science and Technology, London
State Prize of the Soviet Union	Efim S. Fradkin	P. N. Lebedev Physical Institute, Moscow, U.S.S.R.

AWARD	WINNER	AFFILIATION
Tomalla Foundation Prize	Subrahmanyan Chandrasekhar	University of Chicago, Ill.
Tomalla Foundation Prize	Andrey Sakharov	Soviet Academy of Sciences
Walter Schottky Prize	Klaus Sattler	Universität Konstanz, West Germany
Wigner Medal	Louis Michel	Institut des Hautes Études Scientifiques, Bures-sur-Yvette, France
William F. Meggers Award	Theodor W. Hänsch	Stanford University, Calif.
Wright A. Gardener Award	Robert P. Bauman	University of Alabama, Tuscaloosa
PSYCHOLOGY		
American Association for the Advancement of Science Socio-Psychological Prize	David P. Phillips	University of California, San Diego
MacArthur Prize Fellow Award	Amos Tversky	Stanford University, Calif.
Presidents' Award of the British Psychological Society	Peter E. Bryant	University of Oxford, England
SPACE EXPLORATION		
Distinguished Public Service Medal of the National Aeronautics and Space Administration	Charles A. Barth	Laboratory for Atmospheric and Space Physics, University of Colorado, Boulder
National Medal of Science	George M. Low (Deceased)	Rensselaer Polytechnic Institute, Troy, N.Y.
TRANSPORTATION		
Winthrop-Sears Award	George Gregory	Products Research & Chemical Corp.
SCIENCE JOURNALISM		
American Association for the Advancement of Science-Newcomb Cleveland Prize	Gerald M. Rubin	University of California, Berkeley
American Association for the Advancement of Science-Newcomb Cleveland Prize	Allan C. Spradling	Carnegie Institution, Washington, D.C.; Johns Hopkins University, Baltimore, Md.
American Association for the Advancement of Science-Westinghouse Science Journalism Award	Byron Spice	*Journal,* Albuquerque, N.M.
American Association for the Advancement of Science-Westinghouse Science Journalism Award	James S. Trefil	*Smithsonian*
American Association for the Advancement of Science-Westinghouse Science Journalism Award	Hill Williams	*Times,* Seattle, Wash.
Dorothea Klumpke-Roberts Award	Helen Sawyer Hogg (Emerita)	University of Toronto, Ont.
James T. Grady Award for Interpreting Chemistry for the Public	Joseph W. Alper	Chicago, Ill.

AWARD	WINNER	AFFILIATION
MISCELLANEOUS		
Arthur M. Bueche Award	Edward E. David, Jr.	Exxon Research & Engineering Co.
Daniel Giraud Elliot Medal	G. Evelyn Hutchinson	Yale University, New Haven, Conn.
Distinguished Public Service Award of the National Science Foundation	Philip H. Abelson (Retired)	*Science*
Distinguished Public Service Award of the National Science Foundation	Klaus D. Timmerhaus	University of Colorado, Boulder
Leonardo da Vinci Award	H. Grady Rylander	University of Texas, Austin
Olympia Prize	Francesco Nicosia	Centro di Restauro Della Soprintendenza Archeologica Dell'Etruria, Italy
Olympia Prize	S. Dillon Ripley	Smithsonian Institution, Washington, D.C.
Presidential Award of the New York Academy of Sciences	Matthew Meselson	Harvard University, Cambridge, Mass.
Vannevar Bush Award	Roger R. Revelle (Emeritus)	Harvard University, Cambridge, Mass.; University of California, San Diego
Westinghouse Science Talent Search	1. Alan J. Hu	La Jolla High School, Calif.
	2. Anna A. Penn	University of Chicago Laboratory Schools, Ill.
	3. Michael Friedman	Stuyvesant High School, New York, N.Y.
	4. Michael S. Graziano	City Honors High School, Buffalo, N.Y.
	5. John S. Kuo	Bronx High School of Science, New York, N.Y.
	6. Anthony M. Ciabarra	Cheltenham High School, Wyncote, Pa.
	7. Mark Kantrowitz	Maimonides School, Brookline, Mass.
	8. Michael W. Gesner	Cardinal Spellman High School, Brockton, Mass.
	9. Audrey Zelicof	Stuyvesant High School, New York, N.Y.
	10. Allan M. Goldstein	Cheltenham High School, Wyncote, Pa.

Obituaries

Adams, Sir John Bertram (May 24, 1920—March 3, 1984), British engineer, was twice director general of the European Organization for Nuclear Research (CERN) at Geneva and was a leading figure in the assemblage of CERN's particle accelerators. He worked during World War II on radar development before moving to the U.K. Atomic Energy Research Establishment at Harwell, where in 1949 he made a major contribution to the creation of the 180-MeV cyclotron. Adams joined CERN in 1953, became director of its Proton Synchrotron Division the following year, and headed the team that built the high-energy accelerator inaugurated in 1959. In 1960 he was appointed CERN's director general but left soon afterward to return to Britain as director of the newly formed U.K. Atomic Energy Authority's (UKAEA's) Culham Laboratory. When it opened in 1965, Culham, as the site of the Joint European Torus (an experimental nuclear fusion device), became an internationally renowned center for atomic research, and Adams, through his association with the Ministry of Technology and the UKAEA, became one of the British government's chief advisers on nuclear physics. He returned to CERN in 1969 to build the 450-GeV Super Proton Synchrotron and served as CERN's director general again from 1976 to 1980. A fellow of the Royal Society from 1963, Adams received the Faraday Medal of the Institution of Electrical Engineers in 1977 and was knighted in 1981.

Antonov, Oleg Konstantinovich (Feb. 7, [new style; Jan. 25, old style], 1906—April 4, 1984), Soviet aircraft designer, after World War II designed a range of highly successful transport aircraft remarkable for their versatility. His inventions included the An-2 biplane, which with a minimum speed of 64 km/h (40 mph) could be used for civil or military passenger transport, communications, crop spraying, forestry patrol, and other purposes. Other Antonov designs were the An-10 turboprop passenger transport, the An-12 medium cargo transport, and the huge An-22 long-range transport. A graduate of the Leningrad M. I. Kalinin Polytechnic Institute, Antonov began his career designing gliders. In 1946 he was put in charge of experimental aircraft design, and in 1962 he was appointed designer general for the Soviet aircraft industry. Antonov joined the Communist Party of the Soviet Union in 1945 and was a deputy to the Supreme Soviet (parliament) during 1958–66. Among the many awards he received were the State Prize (1952) and the Lenin Prize (1962).

Cori, Carl F(erdinand) (Dec. 5, 1896—Oct. 20, 1984), U.S. biochemist, with his wife, Gerty, determined how glycogen (animal starch) is catalytically converted, a discovery that led to a greater understanding of diabetes mellitus and earned them the

EB Inc.

Carl Cori, with his wife, Gerty.

1947 Nobel Prize for Physiology or Medicine (a prize they shared with the Argentine physiologist Bernardo Houssay). Cori obtained his scientific training at the German University in Prague, where he met Gerty and where the two earned M.D. degrees in 1920, the same year that they married. After immigrating to the U.S. in 1922, the couple joined the Institute for the Study of Malignant Disease, Buffalo, N.Y., where he was a biochemist and she served as an assistant pathologist and as an assistant biochemist from 1922 to 1931. In 1931 they moved to the Washington University Medical School in St. Louis, Mo., where they conducted their pioneering research. Cori served as professor of pharmacology and biochemistry and as chairman of the biochemistry department (from 1947). There in 1936 they isolated from muscle tissue a previously unknown compound of the simple sugar glucose, glucose 1-phosphate (phosphate bound to a specific carbon atom on the glucose molecule). They found that this compound, also referred to as the Cori ester, represents the first step in the conversion of glycogen into glucose and also, because the reaction is reversible, in some cases the last step in the conversion of blood glucose to glycogen. In 1942 the Coris isolated and purified polysaccharide phosphorylase, the enzyme responsible for catalyzing the glycogen–Cori ester reaction. They used this enzyme to synthesize glycogen in a test tube. With this achievement they were able to formulate the "Cori cycle," in which they postulated that liver glycogen is converted to blood glucose that is then reconverted to glycogen in muscle tissue. The subsequent breakdown of the glycogen to lactic acid provides the energy used in muscle contraction. The lactic acid is used to reform glycogen in the liver to complete the cycle. The Coris then studied the ways in which hormones affect carbohydrate metabolism in animals. They determined that the hormone adrenaline (epinephrine) induces the formation of a type of phosphorylase enzyme that encourages the conversion of glycogen to glucose and also that the hormone insulin causes the removal of sugar from the blood by promoting the addition of phosphate to glucose. The latter discovery was essential to a better understanding of the disease diabetes mellitus. After the death of his wife in 1957 Cori undertook research on the physicochemical mode of action of enzymes involved in the breakdown of glycogen to lactic acid.

Deacon, Sir George Edward Raven (March 21, 1906—Nov. 16, 1984), British scientist, served as director of the National Institute of Oceanography from 1949 to 1971. Educated at the City Boys' School, Leicester, and at King's College, University of London, he was a member of the scientific staff of the Discovery Investigation Committee (1927–39). He conducted research on whales both in laboratories at home and in the royal research ships "William Scoresby" and "Discovery II." During his four voyages Deacon studied the varying salinity and temperatures of the Antarctic Ocean; he received a doctorate for his contributions to the *Discovery Reports* and was elected a fellow of the Royal Society (1944). He then supervised a study on waves for the Admiralty Research Laboratory, assessing readings of wave and swell movements by means of spectral analysis. Deacon was a founder of the National Institute of Oceanography, and while serving as its director he attracted colleagues from his previous projects; he also extended the scope of the institute's studies to include marine geology and geophysics and oceanic circulation. He served as president (1960–63) of the International Association for the Physical Sciences of the Ocean and was knighted in 1971.

Dirac, Paul Adrien Maurice (Aug. 8, 1902—Oct. 20, 1984), British physicist, was Lucasian professor of mathematics at the University of Cambridge from 1932 to 1969 and the creator of the complete theoretical formulation of quantum mechanics. He developed the work of Louis de Broglie, Werner Heisenberg, and Erwin Schrödinger by linking it to classical mechanical theory and advancing the theory of positrons, the basis of antimatter. Dirac published *The Principles of Quantum Mechanics* in 1930 and was awarded the Nobel Prize for Physics three years later. The son of a Swiss father and an English mother, he attended the University of Bristol, where he studied electrical engineering before conducting research in mathematics at St. John's College, Cambridge. A paper in 1925 was his first outstanding contribution to quantum theory. He was made a fellow of the Royal Society in 1930 and awarded the Royal Society's Royal Medal in 1939 and its Copley Medal in 1952. Dirac continued to refine quantum theory, but it was left largely to others to develop his later work on quantum electrodynamics. He became professor of physics at Florida State University in 1971 and was appointed to the Order of Merit in 1973.

Paul Adrien Maurice Dirac

Ramsey & Muspratt Ltd., Cambridge

Ehricke, Krafft A. (March 24, 1917—Dec. 11, 1984), German-born aeronautical engineer and physicist, was a leading rocket specialist who during World War II served as a propulsion engineer in Germany's V-2 rocket program. Ehricke graduated from the Technical University of Berlin with a degree in aeronautical engineering; he also studied celestial mechanics and nuclear physics at the University of Berlin. In 1942 he began working at the secret Peenemünde rocket center, where the V-2s that devastated London in 1944 were developed and tested. In the U.S. after the war Ehricke worked (1947–52) for the Army's missile program before joining the Convair division of the General Dynamics Corp. in San Diego, Calif. There he played a vital role in the design and development of such rockets as the Atlas (the nation's first intercontinental ballistic missile) and the Centaur (the first rocket to use liquid hydrogen as a propellant). A space enthusiast who envisioned flights to the Moon and planets and well-equipped space stations, Ehricke was also committed to what he termed the "extraterrestrial imperative" (man's obligation to expand the environmental and resource base beyond Earth in order to preserve life). He was chief scientific adviser to the space division of the Rockwell International Corp. and later headed his own consulting concern, Space Global. In 1984 he was honored with the Goddard Astronautics Award, given by the American Institute of Aeronautics and Astronautics.

Geschwind, Norman (Jan. 8, 1926—Nov. 4, 1984), U.S. neuroscientist, conducted pioneering research on the structure and function of the brain that led to a greater understanding of behavioral patterns, body activity, and learning disorders. After earning his M.D. from Harvard University, Geschwind was associated with various hospitals in Boston before being named James Jackson Putnam Professor of Neurology at Harvard Medical School in 1969. His work established that the left and right hemispheres of the brain have specific functions that influence language, left-handedness, and physical development. He concluded that left-handed people were more susceptible to learning disabilities, allergies, stuttering, migraine headaches, and autoimmune diseases, but also suggested that more research was needed to discover the advantages of left-handedness. He also found that dyslexia (a reading disorder) was related to brain function, and that mental illness as diagnosed in some 40% of the 300,000 mental patients in the U.S. was probably due to physical disorders of the brain rather than emotional problems. From 1975 until his death Geschwind served as director of neurology at Beth Israel Hospital in Boston.

Gibberd, Sir Frederick (Jan. 7, 1908—Jan. 9, 1984), British architect, was the architect-planner of Harlow, Essex, one of eight new towns built near London after World War II to accommodate overflows of population and industry from the metropolis. He also designed such well-known public buildings in London as the Central Mosque (1977) in Regent's Park, the London (Heathrow) Airport passenger terminal buildings and control tower (1950–69), Coutts Bank in the Strand (1978; England's first modern "atrium" building, constructed behind the existing John Nash Regency facade), and the Roman Catholic Metropolitan Cathedral (1967) in Liverpool. Trained at the Birmingham School of Architecture, Gibberd entered private practice in London in 1930. His first buildings, such as the Pullman Court apartment block in Streatham, London, were in the modern reinforced-concrete style pioneered in Britain by his contemporary, F. R. S. Yorke. Gibberd's special talents, however, were for planning and landscape design, as exemplified in the Harlow project, which he headed from 1947 to 1980. He was a member of the Royal Fine Art Commission (1950–70) and of the Royal Institute of British Architects council (1959–76) and president of the Building Centre (1959–76). He was knighted in 1967 and became a royal academician in 1969. Gibberd's publications included *The Architecture of England* (1938), *Town Design* (1953), *Modern Flats* (with F. R. S. Yorke, 1958), and *Harlow: The Story of a New Town* (1980).

Giglioli, Marco (April 21, 1927—March 3, 1984), Guyanese-born entomologist, served as director of the Cayman Islands Mosquito Research and Control Unit since its founding in 1965 and was an international authority on the control of insect pests. He also helped to attract research students to the Cayman Islands, which became a center for biological study. Giglioli, the son of a noted malariologist, studied at McGill University, Montreal, and earned a Ph.D. at the London School of Hygiene and Tropical Medicine. From 1954 to 1957 he worked for the World Health Organization in Liberia and then for the Medical Research Council in The Gambia, studying malaria. Though the Cayman Islands mosquito is not a malaria carrier, its presence posed a threat to the development of tourism and other industries, and Giglioli's work eliminated the threat. At the same time his encouragement of exchanges with British universities and his work with the Cayman Islands Natural Resources Study made a substantial contribution to international studies in marine biology. Giglioli served as a consultant to international organizations and to governments in the West Indies, Central America, and the Middle East. He served as regional director (1976–78) of the American Mosquito Control Association and received the organization's award for meritorious service in 1979. He was made an Officer of the Order of the British Empire in 1971.

Habachi, Labib (1906—Feb. 18, 1984), Egyptian archaeologist, undertook notable excavations at Aswan and Karnak and used his expertise to aid the Egyptian government in the takeover of the foreign-led Egyptian Antiquities Service and to administer it successfully after the Free Officers' coup of 1952. Habachi studied at Cairo University and became an inspector in the Antiquities Service, visiting many major and less-known sites and acquiring an encyclopedic knowledge of antiquities, especially in the Nile Delta. His major excavations, the Heqaib complex and the Kamose stela, were completed in 1946 and 1954. An outstanding scholar, greatly respected by his Egyptian and foreign colleagues, he retired in 1960 and devoted his time to writing and advising visiting archaeological expeditions.

Hathaway, Starke Rosencrans (Aug. 22, 1903—July 4, 1984), U.S. psychologist, codeveloped with J. C. McKinley the Minnesota Multiphasic Personality Inventory (MMPI) in 1937, a psychological test yielding scores for deviate personality trends based on scales derived from the performance of persons with diagnosed mental illness. Hathaway earned (1932) a Ph.D. from the University of Minnesota, where he served as lecturer in psychology (1930–37), assistant professor (1937–40), associate professor (1940–47), professor (1947–71), and director of the division of clinical psychology from 1951 until his retirement in 1971. He introduced his MMPI test in the U.S. in 1943, and it was first used by the military during World War II to screen people with mental disorders. The test, consisting of more than 500 true-false questions, became the most widely used psychological examination in the English-speaking world. Some of the statements used in the MMPI include: "I believe I am being plotted against," "I am a special agent of God," "I am not afraid of picking up a disease or germs from doorknobs," and "Evil spirits possess me at times." At the time of Hathaway's death a four-year project to update the language of the test and expand the number of items in order to include other clinical areas was nearly half completed by James N. Butcher.

Hooker, Sir Stanley George (Sept. 30, 1907—May 24, 1984), British aeronautical engineer, joined Rolls-Royce Ltd. in 1938 and was involved in many crucial developments in aircraft-engine design. He studied mathematics at Imperial College, University of London, and aeronautics as a research fellow at Brasenose College, Oxford. He worked in the Admiralty research department before joining Rolls-Royce, where he headed its research into superchargers. Hooker made a vital improvement in the Merlin engine used in the Spitfire fighter aircraft, increasing its speed significantly. After World War II he moved to production and design of jet engines but resigned in 1948 after the failure of the Avon engine. After he became chief engineer at the Bristol Aeroplane Co. Ltd. in 1951, Hooker developed the Olympus engine for the Vulcan bomber (predecessor of the Concorde engine) and the Pegasus engine, capable of horizontal and vertical thrust. The latter was used in the Kestrel aircraft and eventually led to the engine of the Harrier. He returned to Rolls-Royce when the company was merged with Bristol but retired in 1967. When Rolls-Royce collapsed in 1971, he was recalled and made group technical director and a director of the company to oversee the development and production of the RB-211 engine. He continued as technical adviser and consultant after his retirement in 1976. Hooker became a fellow of the Royal Society in 1962, was made a Commander of the Order of the British Empire in 1964, and was knighted in 1974. His autobiography, *Not Much of an Engineer,* was published in 1984.

Hunsaker, Jerome Clarke (Aug. 26, 1886—Sept. 10, 1984), U.S. aeronautical engineer, was an aviation pioneer who revolutionized the designs of lighter-than-air ships, seaplanes, and carrier-based aircraft, and founded in 1914 the first college course in aeronautical engineering, at the Massachusetts Institute of Technology (MIT), Cambridge. Hunsaker, who graduated in 1908 from the U.S. Naval Academy at Annapolis, Md., served in the naval construction corps from 1909 to 1926. The Navy sent him to MIT in 1909 to study naval architecture, and Hunsaker spent two years in Europe studying the dynamics of aeronautical engineering. When he returned to the U.S., he established a course at MIT and in 1916 was awarded the institute's first Ph.D. in aeronautical engineering for developing the first modern wind tunnel in the U.S. During the same year he was recalled by the Navy to head the construction department of its new aircraft division. Hunsaker worked on zeppelins, notably the "Shenandoah" (which he designed to use helium instead of hydrogen), and he oversaw the construction of some 1,000 seaplanes during World War I. In 1921 he became chief of design in the Bureau of Aeronautics, where he supervised the development of launching catapults and arresting gear for aircraft, the development of the torpedo plane, and the improvement of radial air-cooled aircraft engines. After resigning from the Navy in 1926 he developed a communications system for aircraft for Bell Telephone Laboratories and served as vice-president of the newly formed Goodyear-Zeppelin Corp. In 1933 he returned to MIT where he became head in 1939 of the department of aeronautical engineering. Hunsaker retired in 1951.

Kapitsa, Pyotr Leonidovich (July 8 [June 26, old style], 1894—April 8, 1984), Soviet physicist, shared the Nobel Prize for Physics in 1978 in recognition of his contributions to low-temperature physics, notably his research on the liquefaction of helium, during a career spanning some 60 years. Kapitsa studied at the Petrograd Polytechnic Institute and after graduation served as a lecturer there until 1921. In that year his professor, Abram Joffe, an eminent physicist, was instrumental in arranging Kapitsa's admittance to the Cavendish Laboratory in Cambridge, England, where he was befriended by Sir Ernest Rutherford and remained for almost 13 years. As director of magnetic research there from 1924, he designed apparatus that produced a magnetic field of 500,000 gauss—a strength not exceeded until 1956. Kapitsa was made a fellow of Trinity College, University of Oxford, in 1925 and four years later was elected to fellowship of the Royal Society—the first foreign scientist in 200 years to be so honored. At Rutherford's instigation the Royal Society set up its Mond Laboratory for low-temperature and magnetic research, which Kapitsa directed (1930–34). There he designed apparatus that produced liquid helium more efficiently than had previously been possible. In 1934, however, his research at Cambridge ended; after spending a summer vacation in the U.S.S.R., he

Pyotr Leonidovich Kapitsa

Tass/Sovfoto

was refused permission to return to England. A year later he accepted the post of director of the Soviet Academy of Science's newly established S. I. Vavilov Institute of Physical Problems in Moscow. Rutherford agreed to the transfer of Kapitsa's apparatus from the Mond Laboratory to the Vavilov Institute, and Kapitsa continued his research into the properties of liquid helium, establishing in particular its "superfluidity," or virtual lack of resistance to flow. During World War II he designed equipment for the mass liquefaction of oxygen for use in the steel industry. In 1946 his moral convictions led him to refuse to work on nuclear weapons research, and as a result he was dismissed from his post. He remained under house arrest until after Stalin's death but was reinstated as head of the Vavilov Institute in 1955. He then worked on the application of low-temperature physics to the production of liquid rocket fuels for the Sputnik satellite program and more recently on the problems of controlled nuclear fusion.

Kaplan, Henry Seymour (April 24, 1918—Feb. 4, 1984), U.S. radiologist, revolutionized the treatment of Hodgkin's disease as the coinventor in 1955 (with Edward Ginzton) of the medical linear accelerator, which became the foundation of modern radiation therapy. Kaplan obtained his medical degree (1940) at Rush Medical College in Chicago and then received additional training at the University of Minnesota, Yale University, and the U.S. National Cancer Institute. In 1948 he joined the staff of Stanford University, where he designed the first medical linear accelerator in the Western Hemisphere; the six-million volt device produced high-dosage radiation that could be beamed at various cancers with far greater accuracy and effectiveness than older, lower-powered X-ray machines. The medical linear accelerator was especially effective in treating patients suffering from Hodgkin's disease, which was once fatal. Kaplan was also responsible for developing a technique known as total lymphoid irradiation, an experimental treatment used for kidney transplants and rheumatoid arthritis. Kaplan devoted himself to cancer research primarily because his father died of lung cancer at age 45. He also was credited with discovering that leukemias and cancer of the lymphatic system in mice could be induced by X-radiation or chemicals, which suppressed the function of the animal's immune system. Kaplan, who spent most of his professional career at Stanford, was the author of volumes on heart disease caused by birth defects, Hodgkin's disease, and other cancers of the lymphatic system. His many laurels included admission to the National Academy of Sciences in 1972 (the first radiologist so honored) and the Charles F. Kettering Prize in 1979, given by the General Motors Cancer Research Foundation. Kaplan died of lung cancer, though he was a nonsmoker.

Kastler, Alfred (May 3, 1902—Jan. 7, 1984), French physicist, won the Nobel Prize for Physics in 1966 for his work on "optical pumping," a method of stimulating matter so that it attains higher energy states. This allowed him to study the structure of the atoms in the matter and also, since the light energy used to stimulate the atoms was reemitted, marked an important step toward the creation of the maser and the laser. Kastler himself denied being the "father of the laser," for the development of which Charles Townes was awarded the Nobel Prize for Physics in 1964. Kastler's achievement was the discovery of a new method for studying atomic structures and in pointing the way toward various areas of further research. Perhaps his most lasting contribution to physics in France was as a teacher, from 1941 at the University of Paris and at the École Normale Supérieure and from 1968 as director of research at the Centre National de la Recherche Scientifique, where he trained a generation of physicists. He was also notable for his uncompromising stands on moral and political issues. Opposed to the Vietnam war, he was active from the 1950s in the antinuclear movement and played a significant part in the formulation of treaties to limit the spread of nuclear weapons, especially through the Pugwash Conferences. He condemned the attitudes of both the United States and the Soviet Union and was opposed to the French nuclear deterrent. Kastler was elected a member of the Académie des Sciences in 1964.

Lea, Sir Frederick Measham (Feb. 10, 1900—July 7, 1984), British chemist, achieved international recognition for his research in the chemistry of cement and as a pioneer in modern building techniques. Director of building research at the Department of Scientific and Industrial Research (1946–65), he also held appointments as president of the International Council for Building Research, Studies, and Documentation (1955–57) and of the International Union of Testing and Research Laboratories for Materials and Structures (1957–58). He served as chairman (1953–59) of the Concrete Committee of the International Commission on Large Dams. Lea was the first person outside North America to receive (1964) the Walter C. Voss award of the American Society for Testing and Materials. His book *The Chemistry of Cements and Concrete* (1956) became a standard text. He studied chemistry at the University of Birmingham and in 1925 joined the Department of Scientific and Industrial Research. After a year's research (1928) in the U.S., he worked on the uses of industrial waste products such as slag. Later he turned to the field of buildings, considering not only materials but aspects such as lighting, heating, and soundproofing. His *Science and Building* (1971) outlined current thinking and practice.

Low, George Michael (June 10, 1926—July 17, 1984), Austrian-born aeronautical engineer, was a visionary spacecraft manager who used his ingenuity to nurture and propel the NASA manned space program from 1949 to 1976. After Low earned (1950) an M.S. in aeronautical engineering from Rensselaer Polytechnic Institute, Troy, N.Y., he joined the National Advisory Committee for Aeronautics, where he specialized in aerodynamic heating research. He served successively at NASA as chief of the special projects branch (1949–58), assistant director for manned spaceflight programs (1958–63), deputy associate administrator for manned spaceflight (1963–64), program manager of the Apollo spacecraft project (1967–69), and deputy administrator of NASA (1969–76). In his capacity as an engineer he conducted experiments in boundary layer theory and transition, and in internal flow in supersonic and hypersonic aircraft. As manager of the manned spaceflight programs he oversaw Mercury (1961–63; the first series of manned U.S. spacecraft); Gemini (1965–66; a series of ten two-man spacecraft sent into orbit); and, until his promotion in 1969, Apollo (the first spacecraft—Apollo 11—to land men on the Moon). It was Low who in 1961 suggested and then helped to prepare a memorandum to Pres. John F. Kennedy advising that a man could be put on the Moon and returned to Earth by the end of the decade. This initiative led to the Apollo project, and Low was placed in charge of redesigning the Apollo spacecraft after three astronauts were killed on a launching pad in 1967. Low worked 90 hours each week to ensure the success of the project and personally oversaw each detail. In 1976 he became president of his alma mater, a position he held at his death. He was named to receive posthumously the Presidential Medal of Freedom for his contributions to education and the nation's space program.

Merrill, John Putnam (March 10, 1917—April 4, 1984), U.S. physician, made medical history in 1954 as the supervisor of a surgical team that accomplished the first successful kidney transplant. As a specialist in internal medicine at the Peter Bent Brigham Hospital (now the Brigham and Women's Hospital) in Boston, Merrill became an integral part of the transplant team with his knowledge of the vital role that the body's immune system plays in transplant surgery. Merrill, who earned an M.D. from Harvard Medical School in 1942, was also the flight surgeon for the B-29 that dropped the atomic bomb on Hiroshima. He then spent his entire professional career at Peter Bent Brigham, where the pioneering kidney transplant surgery was performed on Dec. 23, 1954. The operation marked the first time that a kidney of a healthy person was transplanted into a person suffering from advanced kidney disease. The transplant surgery, which was between identical twin brothers, was a milestone because it demonstrated that the main obstacle to such an operation was immunological rejection of the organ rather than the difficulty of the surgery itself. In 1959 a kidney transplant was performed between fraternal twins, and two years later drug therapy, to counteract rejection reactions, paved the way for performing transplant surgery among persons less closely related. Merrill also helped develop artificial kidneys and for his work was generally regarded as being the founder of nephrology, the scientific study of the kidney and its diseases. He was drowned in a sailing accident while on vacation in The Bahamas.

Owings, Nathaniel Alexander (Feb. 5, 1903—June 13, 1984), U.S. architect and engineer, as the cofounder in 1936 of the prestigious architectural firm of Skidmore and Owings (from 1939, Skidmore, Owings & Merrill) used his organizational expertise to preside over more than $3 billion in construction during a prolific 40-year career. Owings earned a B.S. degree from Cornell University, Ithaca, N.Y., and, together with Louis Skidmore, an outstanding designer, worked on the Chicago Century of Progress Exposition in 1933. The two entered into partnership in 1936; three years later they hired engineer John Merrill as a partner. By using the group approach, the firm was able to tackle large-scale commissions. Its first big break came during World War II when the firm was hired by the U.S. government to build a secret town for 75,000 residents—Atom City, Oak Ridge, Tenn., headquarters of the U.S. atomic energy project. The firm's specialty, however, was skyscrapers. With the completion in 1952 of Lever House in New York City, which helped usher in an era of tall glass curtain-wall business buildings, Skidmore, Owings & Merrill became one of the foremost architectural firms in the nation. Owings, who made no claims to being a first-rate designer, nonetheless guided the firm through decades of prosperity. Some of the firm's other innovative commissions included the John Hancock Center, Chicago; the Crown Zellerbach Building, San Francisco; the Chase Manhattan Bank, New York City; and the Sears Tower, Chicago (the world's tallest building). Owings also wrote two books, *The American Aesthetic* (1969) and *The Spaces in Between: An Architect's Journey* (1973), and was awarded the American Institute of Architects' Gold Medal in 1983.

Randall, Sir John Turton (March 23, 1905—June 16, 1984), British biophysicist, made a vital contribution to the development of radar that made possible its use for the detection of submarines during World War II. With H. A. H. Boot he developed the cavity magnetron; this device enabled the use of radar transmitters in aircraft, and it played a decisive role in the battle of the Atlantic. Randall studied at the University of Manchester, specializing in work on

X-rays and working on fluorescent lighting before becoming a fellow at the University of Birmingham. After the war he became Wheatstone professor of physics (1946–61) and then professor of biophysics (1961–70) at King's College, University of London, where he applied his discoveries in physics to research biology. The work conducted in his laboratory was crucial to the discovery in 1953 of the structure of DNA (deoxyribonucleic acid), as well as other advances in biology. His imaginative breaching of traditional barriers between disciplines led to the creation of a department of biophysics and a School of Biological Sciences at King's College. After his retirement he moved to Edinburgh, where he set up a research group in the zoology department of the university. A fellow of the Royal Society from 1946, Randall was knighted in 1962.

Richards, Audrey Isabel (July 8, 1899—June 29, 1984), British anthropologist, was director of studies in anthropology at Newnham College at the University of Cambridge and founder and director (1956–67) of the Centre of African Studies at the university. From 1930 she established her reputation as a pioneer in African anthropology with her work on East Africa, which began with her study of the Bemba tribe. She graduated in natural sciences from Cambridge and then became a research student at the London School of Economics and Political Science, where she was a lecturer from 1931 to 1937. After a lectureship (1939–41) at the University of the Witwatersrand in South Africa and a period in the British Colonial Office, she returned to the University of London and was a reader there from 1946. From 1950 to 1956 she served as director of the East African Institute of Social Research at Makerere College in Kampala, Uganda. Richards was the first woman president (1959–61) of the Royal Anthropological Institute, president (1964–65) of the African Studies Association, and a recipient of the Wellcome Medal and the Rivers Memorial Medal. On her retirement in 1967 she was elected a Fellow of the British Academy. Her many books include *Land, Labour and Diet in Northern Rhodesia* (1939), *The Changing Structure of a Ganda Village* (1966), and *The Multicultural States of East Africa* (1969). With Jean Robin she published a study of Essex village life in *Some Elmdon Families* (1974). She was appointed a Commander of the Order of the British Empire in 1955.

Richards, Owain Westmacott (Dec. 31, 1901—Nov. 10, 1984), British scientist, served as head of the department of zoology and applied entomology at Imperial College, London (1953–67), and was a distinguished entomologist who made important contributions in evolutionary theory, taxonomy, and ecology. Educated at Hereford Cathedral School and at Brasenose College, University of Oxford, he held a research scholarship at Oxford (1924–27) before entering Imperial College as a research assistant (1927) and becoming lecturer (1930) and reader (1937). With G. C. Robson he criticized natural selection theories in *The Variation of Animals in Nature* (1936). Richards conducted extensive studies on the population dynamics of British insects and also traveled widely, especially in California, Brazil, and Guyana. He was a special authority on Sphaerocerid flies, but perhaps his most important book was *The Social Wasps of the Americas* (1978). He was co-editor of the tenth edition (1977) of *Imms's General Textbook of Entomology*. A past president of both the Royal Entomological Society of London and of the British Ecological Society, Richards was elected a fellow of the Royal Society in 1959.

Rock, John (March 24, 1890—Dec. 4, 1984), U.S. obstetrician-gynecologist, as a pioneer in the study of human reproduction was an expert on fertility and captured world attention as the developer with Gregory Pincus and M. C. Chang of the first effective oral contraceptive in the 1950s. After several ill-fated business ventures Rock entered Harvard University and then Harvard Medical School. His first research addressed the problem of infertility, and in 1944 with Mirian F. Menkin he achieved the first successful fertilization of a human ovum in a test tube. He used his knowledge of infertility to aid him in the development of a birth-control pill, which revolutionized sexual mores, population control, and the status of women after it was approved by the Food and Drug Administration in 1960. A devout Roman Catholic yet a leading supporter of population control, Rock found himself the center of controversy within the church for his work on the pill. In his 1963 book, *The Time Has Come: A Catholic Doctor's Proposals to End the Battle over Birth Control*, Rock called for a change in the church's stand against birth control and also for new research into improved natural contraceptive methods that would be acceptable to all religious faiths and inexpensive enough for the poor. He steadfastly maintained that the pill was a "natural" form of contraception because its ingredients, including estrogen and progesterone, were substances found in the human body. After Pope Paul VI issued the 1968 encyclical that banned all forms of artificial contraception, Rock accused the pope of abdicating "responsibility for the ultimate welfare of all." During his longtime career Rock served as founding director (1926–56) of the Fertility and Endocrine Clinic at the Free Hospital for Women in Brookline, Mass., professor of gynecology for 30 years at Harvard Medical School, and director from 1956 of the Rock Reproductive Clinic, also in Brookline.

Russell, Sir Frederick Stratten (Nov. 3, 1897—June 5, 1984), British marine biologist, was for more than 40 years with the Plymouth Laboratory of the Marine Biological Association and from 1945 to 1965 served as its director. His studies of variations in productivity in the English Channel (which, in the long term, follow what has since been termed the Russell Cycle) led to his election as fellow of the Royal Society in 1938. His studies of planktonic migration as a whole were an important contribution to the understanding of marine fauna. Russell studied at the University of Cambridge before serving with distinction as an observer in the Royal Naval Air Service and Royal Air Force during World War I. After the war he spent two years as assistant director of fisheries research in Egypt before joining the Plymouth Laboratory in 1924. Apart from the Great Barrier Reef Expedition of 1928 and service during World War II with Air Staff Intelligence in London, he remained in Plymouth for the rest of his career pursuing the work on marine biology recorded in such volumes as *The Medusae of the British Isles* (two volumes, 1953 and 1970) and *The Eggs and Planktonic Stages of British Marine Fishes* (1976). Russell was editor of *Advances in Marine Biology* and coauthor with C. M. Yonge of *The Seas* (1928). He was made a Commander of the Order of the British Empire in 1955 and knighted in 1965.

Ryle, Sir Martin (Sept. 27, 1918—Oct. 14, 1984), British radio astronomer, was astronomer royal from 1972 to 1982 and in 1974 was awarded (jointly with Antony Hewish) the Nobel Prize for Physics for his pioneering work in radio astronomy. After earning a degree in physics at the University of Oxford in 1939 he was engaged during World War II in the development of radar. As a lecturer in physics at the University of Cambridge he began to search for methods of studying radio waves from distant astronomical bodies, developing the technique of aperture synthesis by using a combination of parabolic reflectors and then analyzing the results by computer. At the Mullard Radio Astronomy Observatory, Cambridge, where he served as director from 1957 to 1982, he was able to give a clear picture of the shape of radio galaxies and to continue his work in physics, using his observations in astronomy. His name was also associated with the model of the "evolving universe" that was gradually accepted by a majority of scientists, in opposition to the theory of "continuous creation." He was professor of radio astronomy at Cambridge from 1959 to 1982 and at the Mullard Observatory led an outstanding team of astronomers. Ryle became a supporter of nuclear disarmament—he was the author of *Towards the Nuclear Holocaust* (1981)—and a campaigner for the development of alternative sources of energy to replace both traditional and nuclear fuels. A foreign member of the Soviet Academy of Sciences and of the U.S. National Academy of Sciences, he was made a fellow of Trinity College, Cambridge, in 1949 and a fellow of the Royal Society in 1952 and was knighted in 1966.

Sir Martin Ryle

The Nobel Foundation, Stockholm

Sharon, Arieh (May 28, 1902—July 24, 1984), Israeli architect and town planner, was director and chief architect of the National Planning Agency of Israel (1948–53) and president of the Israeli Association of Architects and Engineers (1965–71). Among his notable public buildings were the regional hospital at Beersheba and the Ichilov hospital and the Jewish Agency's headquarters at Tel Aviv. He also designed buildings for the Hebrew University of Jerusalem and the Weizmann Institute of Science at Rehovot. In 1973 Sharon contributed to a study for planned development of the Old City of Jerusalem (his recommendations were later ignored). Sharon won international recognition with his appointment as a member (1963–67) of the executive committee of the International Union of Architects and by his election as a fellow of the American Institute of Architects and as a member of the Royal Institute of British Architects. An early Jewish immigrant (1920) to Palestine, he studied architecture in Germany at Dessau and Berlin before returning to practice in Tel Aviv in 1932.

Simpson, George Gaylord (June 16, 1902—Oct. 6, 1984), U.S. paleontologist, established himself as an expert on mammalian paleontology with his studies of continental migration of South American animals but was perhaps better known for his work on the modern biological theory of evolution and the philosophical implications of the acceptance of evolutionary theory as discussed in his most famous work, *The Meaning of Evolution* (1949). Simpson earned a Ph.D. from Yale University in 1926 and the following year began a 35-year association with the American Museum of Natural History in New York City, first as field assistant, then curator, and finally chairman of the museum's department of geology and paleontology. During his tenure he conducted numerous expeditions in search of fossils, includ-

Harvard University News Office

George Gaylord Simpson

ing journeys to Patagonia (1930–31 and 1933–34), Venezuela (1938–39), and an especially fruitful mission to Colorado (1953). There he discovered eight skulls of 38-cm (15-in)-high creatures named dawn horses, a species that had flourished some 50 million years ago. Simpson also served as professor of vertebrate paleontology at Columbia University until 1959 when he became professor of vertebrate paleontology at Harvard University, a post he held until 1970. He then served as professor at the University of Arizona until his retirement in 1982. Simpson published hundreds of scientific articles and such books as *Tempo and Mode in Evolution* (1944), *Major Features of Evolution* (1953), *The Principles of Animal Taxonomy* (1961), *Fossils and the History of Life* (1983), and *Discoverers of the Lost World* (1984).

Spedding, Frank Harold (Oct. 22, 1902—Dec. 15, 1984), Canadian-born chemist, was credited with creating the technology to produce metallic uranium on a large scale, a crucial stage in the development of the atomic bomb. Spedding, who earned a Ph.D. in chemistry (1929) from the University of California at Berkeley, joined Iowa State University as a faculty member in 1937. In 1947 he was named the first director of the Iowa State University Institute for Atomic Research (now Ames Laboratory), a position he held until he retired in 1968. Spedding both organized and directed the chemistry division of the plutonium project for the atomic bomb research that was conducted in Chicago in 1942 and 1943; his contributions led to the development of plutonium, an essential ingredient of the atomic bomb. The recipient of numerous awards, Spedding was nominated several times for the Nobel Prize for Chemistry.

Sugden, Sir (Theodore) Morris (Dec. 31, 1919—Jan. 3, 1984), British physical chemist, was physical secretary of the Royal Society from 1979 and master of Trinity Hall, University of Cambridge, from 1976. His work on the chemistry of combustion derived originally from studies of gun flash and rockets that he conducted during World War II for the Ministry of Supply after graduating (1940) from Jesus College, Cambridge. He continued this work when he returned to Cambridge as a research student at Pembroke College. Sugden eventually became reader (1960–63) in the department of physical chemistry, fellow of Queen's College, and a fellow of the Royal Society from 1963. In 1963 he left academic life to work for Shell Research Ltd. at the Thornton Research Centre, where he assembled a team of scientists in the specialized field of combustion research. His achievements were recognized in 1967, when he was made director of the center, and seven years later when he became chief executive of Shell Research Ltd. Returning to Cambridge in 1976, he pursued a variety of activities there including chairmanship of the governing body of the Cambridge University Press. In 1983 he was made deputy vice-chancellor of the university. Sugden was awarded the Royal Society's Davy Medal in 1975, appointed a Commander of the Order of the British Empire in the same year, and knighted in 1983.

Ulam, Stanislaw Marcin (April 3, 1909—May 13, 1984), Polish-born mathematician, was a mathematical wizard who was instrumental in helping Edward Teller ("father of the hydrogen bomb") solve the problem of how to initiate the fusion of hydrogen atoms in order to release the devastating energy of the bomb. After Ulam earned a Ph.D. in mathematics at the Polytechnic Institute in Lvov, Poland (now in the Soviet Union), he lectured in Europe. In 1936 he came to the United States as a visiting member of the Institute for Advanced Study at Princeton, N.J. He then taught at Harvard University and the University of Wisconsin before becoming a member of the team that developed the design for the hydrogen bomb at the Los Alamos National Laboratory in New Mexico. While there he also devised what is known as the "Monte-Carlo method," a technique of calculating approximate solutions of physical or mathematical problems by means of statistical sampling procedures that are based on the use of random numbers. His method was found to be readily adaptable for use with electronic computers and by the 1960s was used in weapons design, mathematical economy, and operations research. With J. C. Everett, Jr., Ulam also proposed the "Orion" plan, a technological innovation for nuclear propulsion of space vehicles. In 1965 he joined the faculty of the University of Colorado, where he served as professor of mathematics and chairman of the department. At the time of his death Ulam was professor of biomathematics at the University of Colorado medical school. His writings include *A Collection of Mathematical Problems* (1960), *Sets, Numbers and Universes: Selected Works of Stanislaw Ulam* (1974), and *Adventures of a Mathematician* (1976).

Contributors to the Science Year in Review

Peter J. Andrews *Chemistry: Physical chemistry.* Science writer, Rockville, Md. Winner in 1984 of the National Award of Excellence of the Society for Technical Communications.

D. James Baker *Earth sciences: Oceanography.* President, Joint Oceanographic Institutions Inc., Washington, D.C.

Fred Basolo *Chemistry: Inorganic chemistry.* Morrison Professor of Chemistry, Northwestern University, Evanston, Ill.

Keith Beven *Earth sciences: Hydrology.* Hydrologist, Institute of Hydrology, Wallingford, Oxon, United Kingdom.

Eric Block *Chemistry: Organic chemistry.* Professor of Chemistry, State University of New York at Albany.

David Boore *Earth sciences: Geophysics.* Consulting Professor, Department of Geophysics, Stanford University, Stanford, Calif.

Harold Borko *Electronics and information sciences: Information systems and services.* Professor, Graduate School of Library and Information Science, University of California, Los Angeles.

Robert Boylestad *Electronics and information sciences: Electronics.* Professor of Electrical Engineering Technology, City University of New York, New York, N.Y.

D. Allan Bromley *Physics: Nuclear physics.* Henry Ford II Professor of Physics, Yale University, New Haven, Conn.

James W. Canan *Defense research.* Senior Editor, *Air Force Magazine,* Washington, D.C.

Douglas E. Comer *Electronics and information sciences: Computers and computer science.* Professor of Computer Science, Purdue University, West Lafayette, Ind.

John Davis *Architecture and civil engineering.* Editor, U.S. Army Intelligence and Threat Analysis Center, U.S. Army Intelligence and Security Command, Arlington, Va.

Warren D. Dolphin *Life sciences: Zoology.* Professor of Zoology and Executive Officer, Biology, Iowa State University, Ames.

F. C. Durant III *Electronics and information sciences: Satellite systems.* Aerospace Historian and Consultant, Chevy Chase, Md.

Robert G. Eagon *Life sciences: Microbiology.* Professor of Microbiology, University of Georgia, Athens.

Robert L. Forward *Physics: General developments.* Senior Scientist, Hughes Research Laboratories, Malibu, Calif.

David R. Gaskell *Materials sciences: Metallurgy.* Professor of Metallurgical Engineering, Purdue University, West Lafayette, Ind.

Richard L. Gordon *Energy.* Professor of Mineral Economics, Pennsylvania State University, University Park.

Stig B. Hagstrom *Physics: Solid-state physics.* Laboratory Manager, Xerox Palo Alto Research Center, Palo Alto, Calif.

Robert Haselkorn *Life sciences: Molecular biology.* F. L. Pritzker Distinguished Service Professor, Department of Molecular Genetics and Cell Biology, University of Chicago, Ill.

Lawrence W. Jones *Physics: High-energy physics.* Chairman and Professor, Department of Physics, University of Michigan, Ann Arbor.

John Patrick Jordan *Food and agriculture: Agriculture.* Administrator, Cooperative State Research Service, U.S. Department of Agriculture, Washington, D.C.

Lou Joseph *Medical sciences: Dentistry.* Senior Science Writer, Hill and Knowlton, Inc., Chicago.

George B. Kauffman *Chemistry: Applied chemistry.* Professor of Chemistry, California State University, Fresno.

John V. Killheffer *Scientists of the Year: Nobel prizes.* Associate Editor, *Encyclopædia Britannica.*

David B. Kitts *Earth sciences: Geology and geochemistry.* Professor of the History of Science, University of Oklahoma, Norman.

Mina W. Lamb *Food and agriculture: Nutrition.* Professor emeritus, Department of Food and Nutrition, Texas Tech. University, Lubbock.

Susan V. Lawrence *Medical sciences: General medicine.* Free-lance medicine and science writer; Contributing Editor, *Medical World News;* Chevy Chase, Md.

Howard J. Lewis *U.S. science policy.* Managing Editor, National Association of Science Writers, Bethesda, Md.

Melvin H. Marx *Psychology* (in part). Professor Emeritus of Psychology, University of Missouri, Columbia.

Charles W. McNett, Jr. *Anthropology.* Chairman and Professor, Department of Anthropology, American University, Washington, D.C.

Louis Nashelsky *Electronics and information sciences: Electronics.* Professor of Electrical Technology, City University of New York, New York, N.Y.

Colin V. Newman *Psychology* (in part). Scientific and Professional Secretary, British Psychological Society, Leicester, England.

Roger A. Pielke *Earth sciences: Atmospheric sciences.* Associate Professor of Atmospheric Science, Colorado State University, Fort Collins.

W. M. Protheroe *Astronomy.* Professor of Astronomy, Ohio State University, Columbus.

Robert R. Shannon *Optical engineering.* Professor and Director, Optical Sciences Center, University of Arizona, Tucson.

Mitchell R. Sharpe *Space exploration: Space probes.* Historian, Alabama Space and Rocket Center, Huntsville.

Albert J. Smith *Life sciences: Botany.* Professor of Biology, Wheaton College, Wheaton, Ill.

Frank A. Smith *Transportation.* Senior Associate, Transportation Policy Associates, Washington, D.C.

J. F. Smithcors *Medical sciences: Veterinary medicine.* Editor, American Veterinary Publications, Santa Barbara, Calif.

Lynn Arthur Steen *Mathematics.* Professor of Mathematics, St. Olaf College, Northfield, Minn.

Robert E. Stoffels *Electronics and information sciences: Communications systems.* Editor, *Telephone Engineer & Management* magazine, Geneva, Ill.

Norman M. Tallan *Materials sciences: Ceramics.* Chief, Metals and Ceramics Division, Materials Laboratory, Wright-Patterson Air Force Base, Ohio.

Kenneth E. F. Watt *Environment.* Professor of Zoology, University of California, Davis.

Terry White *Space exploration: Manned flight.* Public Information Specialist, NASA Johnson Space Center, Houston, Texas.

James D. Wilde *Archaeology.* Assistant Manager, Cultural Resources Management Services, Brigham Young University, Provo, Utah.

Index

This is a three-year cumulative index. Index entries to feature and review articles in this and previous editions of the *Yearbook of Science and the Future* are set in boldface type, *e.g.,* **Astronomy.** Entries to other subjects are set in lightface type, *e.g.,* Radiation. Additional information on any of these subjects is identified with a subheading and indented under the entry heading. The numbers following headings and subheadings indicate the year (boldface) of the edition and the page number (lightface) on which the information appears. The abbreviation "il." indicates an illustration.

Astronomy 86–261; **85**–256; **84**–256
  cosmic voids **85**–48
  honors **86**–402; **85**–403; **84**–404
  optical engineering **86**–361; **85**–362; **84**–361
  quasars **84**–160

All entry headings are alphabetized word by word. Hyphenated words and words separated by dashes or slashes are treated as two words. When one word differs from another only by the presence of additional characters at the end, the shorter precedes the longer. In inverted names, the words following the comma are considered only after the preceding part of the name has been alphabetized. Names beginning with "Mc" and "Mac" are alphabetized as "Mac"; "St." is alphabetized as "Saint." Examples:

Lake
Lake, Simon
Lake Charles
Lakeland

A

C

E

G

M

N

O

T

U

V

W

X

Y

Z

Acknowledgments

8–9	Harald Sund; (inset) Harry Wadsworth
12–13	(Far left, bottom left, right) SCALA/Art Resource; (top left) Brown Brothers; (far right, top and bottom) The National Maritime Museum, Haifa, Israel
31	(Top) Adapted from information obtained from the Jet Propulsion Laboratory; (center and bottom) JPL
106–107	Art background reproduced by courtesy of LeMessurier/SCI
119	(Right) Illustration by Charlotte D'Almaine
126–127	© Erich Hartmann—Magnum
206	Adapted from "Some Physical and Chemical Factors in the Metabolism of Lakes," D. S. Rawson, *Problems of Lake Biology,* F. R. Moulton (ed.), pp. 9–26, 1939, American Association for the Advancement of Science, Publication 10; illustration by Charlotte D'Almaine
220	(Clockwise from top left) The Granger Collection; The Museum of Modern Art Film Still Archives, New York; Edison National Historical Society, West Orange, New Jersey; Harvard University News Office; Brown Brothers; EB Inc.

IN THE BRITANNICA TRADITION OF QUALITY...

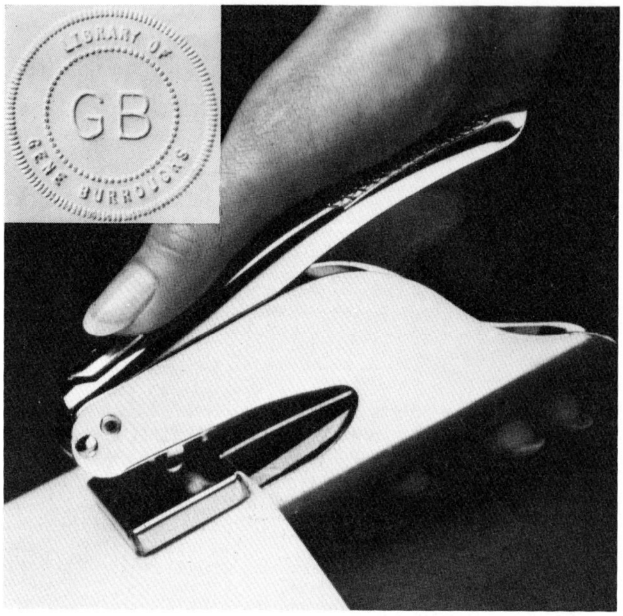

PERSONAL EMBOSSER

A mark of distinction for your fine books. A book embosser just like the ones used in libraries. The 1½″ seal imprints "Library of ________" (with the name of your choice) and up to three centered initials. Please type or print clearly BOTH full name (up to 26 letters including spaces between names) and up to three initials.
Please allow six weeks for delivery.

Just **$20.00**

plus $2.00 shipping and handling

This offer available only in the United States.
Illinois residents please add sales tax

Britannica Home Library Service, Inc.
